NON-EUCLIDEAN GEOMETRIES

János Bolyai Memorial Volume

Mathematics and Its Applications

Volume 581

NON-EUCLIDEAN GEOMETRIES

János Bolyai Memorial Volume

Edited by

ANDRÁS PRÉKOPA
Rutgers Center for Operations Research,
Piscataway, New Jersey, USA

EMIL MOLNÁR
Budapest University of Technology and Economics,
Hungary

Springer

ISBN-10: 1-4614-9771-X ISBN: 0-387-29555-0 (eBook)

ISBN-13: 978-1-4614-9771-4

Printed on acid-free paper.

Softcover re-print of the Hardcover 1st edition 2006

9 8 7 6 5 4 3 2 1

springeronline.com

This plaquette depicting János Bolyai was made by Kinga Széchenyi in commemoration of the 200th anniversary of Bolyai's birth.

Contents

Part VI Physics

PREFACE

János Bolyai is the greatest figure in the history of Hungarian mathematics. He solved the more than two thousand year old problem in connection with Euclid's fifth postulate and discovered non-Euclidean geometry. The glory of the discovery is shared by the Russian Nicolai Ivanovich Lobachevskii but it brought more pain and bitterness than joy to the innovators.

Non-Euclidean geometry fundamentally changed our views about geometry and mathematics, in general. Some historians state that since the time of the ancient Greeks there has never been such a great revolution in mathematics than the one originating in the works of Bolyai and Lobachevskii. It became clear that geometry and reality may be different and geometry does not belong to natural sciences. The same is true for the other branches of mathematics. By the time the famous Greek mathematicians Thales and Pythagoras introduced deductive reasoning into mathematics and their followers, primarily Euclid, systematized mathematical knowledge and clarified which are the assertions that we accept without proof and which are the ones we need to prove, it was only a matter of time to learn that our mathematical way of thinking is based on abstract structures. Mathematics does not address reality in a direct manner but substitutes real life objects by abstract ones, determines their relations to each other and then solves the problems within their structures. These structures or axiomatic systems may or may not adequately describe reality. In successful cases they do and provide us with powerful tools for theoretical and practical problem solutions. For example, non-Euclidean geometrical structures, those created by Bolyai and Lobachevskii as well as more general ones, allowed for the development of modern physical theories in the twentieth century.

By the end of the nineteenth century almost all branches of mathematics became collections of axiomatic systems and the deductive consequences of the statements within. The advent of computers made mathematics more powerful and contributed tremendously to its applicability. Interestingly, applications also began to use axiomatic systems.

In fact, when we start to solve a practical problem first we create a mathematical model, a collection of notions, that represent real life objects, and their relations. Then we elaborate on it, derive its mathematical properties and solve the computational problems.

To commemorate the 200th anniversary of the birth of the great scientist, the Hungarian Academy of Sciences, together with other institutions in Hungary and abroad, organized an international conference on hyperbolic geometry on July 6-12, 2002, in Budapest, at the headquarters of the Academy. Besides the Conference this volume is a tribute to the great scientist and his world-famous scientific achievements.

The 100th, 150th and 175th anniversaries were commemorated in Kolozsvár, Budapest and Budapest, respectively, but only lectures in the Hungarian language were presented. The 200th anniversary is special not only because of the round number of years that have elapsed since 1802 but because the Bolyai research reached a significant stage.

When János Bolyai died he left behind 14,000 pages of manuscript, out of which 3,000 contain his mathematical notes and the rest his utopian ideas about science and society. Some of these pages have been scrutinized earlier, mostly about 100 years ago, and important letters (e.g., ...from nothing I have created a new, different world ...) and theories (e.g., foundations of the theory of complex numbers) have been discovered. However, most of the 14,000 pages remained unread until 1952. In that year Samu Benkő, professor of history in Kolozsvár (Cluj) began to arrange the manuscripts (which were put into chests after János Bolyai's death on order of the commanding officer of the Marosvásárhely garrison). His work lasted sixteen years. During this time he also scrutinized the nonmathematical texts that had remained unread. Similarly, Elemér Kiss, professor of mathematics in Marosvásárhely (Târgu Mureş) studied the mathematical texts during the 1990s. Both scholars have found interesting and important ideas in the manuscripts and presented them to the world.

There are two places, where Bolyai manuscripts, documents and memorabilia are collected: the Teleki Library in Marosvásárhely and the Library of the Hungarian Academy of Sciences in Budapest. Simultaneously with the conference an exhibition of the most important pieces of the latter was organized in the Gallery of the Academy.

At the Conference there were 300 participants from 25 countries. We all were honoured that the most famous geometer of the time, the 95 years old Canadian professor H.S.M. Coxeter, came to Budapest to participate at the Conference and delivered the first plenary talk (the next year we learned the sad news that he had passed away).

On the occasion of the anniversary Kinga Széchenyi made a plaquette of János Bolyai (accepting the relief of the mathematician on the façade of the Palace of Culture of Marosvásárhely to be authentic). Copies of it were given to the main speakers and those who have done outstanding research in connection with János Bolyai. The picture of the plaquette can be seen on page 20. Fig.1 of this volume. A special edition of the Appendix, sponsored by the Hungarian Academy of Sciences, printed in Latin, English and Hungarian, was published and given to all participants. The Hungarian National Bank issued a 3000 HUF face value silver coin designed by György Kiss and the Hungarian Post issued special stamps for the anniversary. The latter could be purchased on site during the conference.

We express our special thanks to the Hungarian Academy of Sciences for allowing the use of the main building of the Academy, together with its equipment. We are also grateful to the Manuscript Section of the Library of the H.A.S., especially to its head Marianne Rozsondai and researchers Béla Mázi and Károly Horányi, for the organization of the exhibition of the "Bolyai Collection" of the Academy. For the excellent organizational work in connection with the Conference our thanks should go to the Conference Organizing Group of the Computing and Automation Institute of the H.A.S., in particular to its head Gusztáv Hencsey and his associate Viktor Richter, who were mainly in charge. Many thanks should go to the members of the Program and Organizing Committees as well as those who contributed to this volume, including Dr Attila Bölcskei and Ms Ildikó Szabó, who made the collection of papers ready for print. Last but not least we express our thanks to Springer Publishers for the publication of this memorial volume.

András Prékopa
member of the Hungarian Academy of Sciences
chairman of the János Bolyai Conference
editor

Emil Molnár
editor

I

HISTORY

THE REVOLUTION OF JÁNOS BOLYAI

András Prékopa
Member of the Hungarian Academy of Sciences
University Professor
RUTCOR, Rutgers Center for Operations Research
640 Bartholomew Road, Piscataway NJ, 08854-8003
prekopa@rutcor.rutgers.edu

Department of Operations Research, Loránd Eötvös University of Budapest
1117, Budapest, Pázmány Péter Sétány 1/c
prekopa@cs.elte.hu

1. Summary

János Bolyai is the greatest figure of Hungarian science; many think he is the Copernicus of geometry. In his 26-page work published in 1831 and generally referred to as the Appendix, (which was published as an appendix to Vol. 1 of Tentamen, the two-volume monumental monograph of his father, Farkas Bolyai) he made a revolutionary achievement by the creation of the so-called non-Euclidean geometry. With this work János Bolyai broke the monopoly of Euclidean geometry and paved the way for humanity to think about space in a different fashion. Through his findings in axiomatic thinking Bolyai considerably formed the history of mathematics as a whole. The development of modern mathematics in the 19th and 20th centuries can, to a large extent, be attributed to János Bolyai's discovery. However, the importance of his results was recognized only after his death but even then not without resistance. In his lifetime almost no one understood his brilliant ideas, which matured in him by the time he was 21. He presented them with the revolutionary bravery of youth, having no fears for the criticism of the scientific community. Naturally, he exhibited a great degree of naiveté, because he thought that great discoveries, including his, would lead to recognition and fame. But the only individual who understood Bolyai's ideas, Gauss, 'the prince of mathematicians', was unfair to János Bolyai when he formed his opinion of the Appendix in 1832. He wrote in his letter to Farkas Bolyai that he was unable to praise János's work because in so doing he would be praising himself. Gauss reasoned that János Bolyai's way of thinking and results coincided almost entirely with the ideas he had been developing for the last thirty-five years. After Gauss's death in 1855, no written proof of the aforementioned statement was found. Gauss behaved reprehensibly on yet another occasion. When he learned that the Russian Lobachevskii, whose election to be a foreign corresponding member of the Royal Society of Göttingen (arranged in 1842), made the same discovery as János Bolyai, he failed to inform Lobachevskii that there was another person who had achieved almost the same results.

For many years scientists thought that although after his retirement in 1833 János Bolyai produced some work including an important theory on the foundation of complex numbers, the lack of recognition pushed him into a state of depression and he renounced creative mathematical research. It was Elemér Kiss, Professor at Marosvásárhely (now Targu Mureş) who refuted this misconception. Having consulted Bolyai's manuscripts he found significant mathematical 'gems' in them that were new at their birth.

The scientists discovered János Bolyai's greatness first abroad and it was recognized in Hungary later. His work became widely known on the European Continent by the turn of the 19th and 20th centuries. Also, in the Anglo-Saxon countries there were some who knew his work and were enthusiastic about it but they were fewer than those on the Continent. After World War II the world became bipolar. The Russians did not mention János Bolyai much but emphasized the merits of Lobachevskii. In the USA – as has been mentioned above – our scholar was less known.

The year 1977 when the 200th anniversary of Gauss's birth was celebrated all over the world became a turning-point. Although Gauss had always been regarded as the primary discoverer of non-Euclidean geometry by the authors of numerous studies, this tendency became stronger, pushing even Lobachevskii into the background. Russian authors have managed to contest these opinions in the interest of Lobachevskii. We, Hungarians, have the duty to show the rest of the world where János Bolyai's place is in the history of mathematics and universal culture. Therefore, the relevant documents and research results should be presented to the world.

2. Introductory notes

The Hungarian territories, held by the Turks for 150 years were reoccupied by the end of the 17th century. The Treaty of Karlóca, (now

Karlovac) 1699, sealed the new world order. The subsequent Hapsburg attempt to colonize the country was averted by the Rákóczi war of independence but the country's sovereignty remained rather limited in several respects. Transylvania, which had been an autonomous principality during the Turkish rule, did not become again part of Hungary. It continued to be a principality but was treated by the Hapsburg monarch as a province. Maria Teresa promoted it a Grand Duchy in 1765. The Chancellery of Transylvania was in Vienna and, at home there was the Gubernium headed by the governor, a separate General Headquarters (G.H.Q.), the autonomy of the counties and a national assembly. While Croatia belonged to Vienna only through Hungary, Transylvania was subject to the Hapsburgs directly.

However, the existence of a G.H.Q. in Transylvania did not mean that Transylvania was independent in a military sense. In Hungary, too, military affairs came under the authority of the central power, the monarch. János Bolyai was granted the title 'Imperial and Royal Military Engineer' because there was only 'Imperial and Royal Army'.

In the second half of the 18th century both in Transylvania and Hungary middle-class mentality commenced to develop. An excellent book by Kosáry (2001) describes a true picture of the period between the Treaty of Szatmár of 1711 and the Compromise of 1867 and, within this, that of the Bolyais as well. However, in order to do justice to our subject matter, some other details need to be mentioned.

Both Transylvania and Hungary were considerably influenced, from political and cultural points of view, by the movements of the German speaking territories. Most of the German-speaking world belonged to the Holy Roman Empire and only a smaller part of it to the monarch of Vienna. However, from the time of the coronation of Leopold I in 1658 to the fall of the empire, the Austrian duchies, the Hungarian and the Czech kingdoms, the principality of Transylvania, etc. The Holy Roman Empire was a loose political formation. After the Peace Treaty of Westphalia (1648), which marked the end of the Thirty Years' War, the Reichstag, a parliamentary body, without much legislative power over its 143 years of existence, was set up. From the aspect of legislation the individual German states – Baden, Bavaria, Saxony, Prussia etc. – were more important. There were a lot of people among the 17th century Germans who considered the ties to the Empire important. For them the Reich was equal to Germany and they considered themselves not Saxons or Prussians, etc. but, primarily, Germans.

The curricula and teaching methods of German schools and universities were very different from one another. There was no uniform standard. It was not specified how the various levels of education should be built on each other. Few people attended schools for long and even fewer went to universities. The number of students enrolled in universities decreased during the 18th century. Universities were small but their number was not. In 1780 there were six universities in Austria and twenty-four in the other German states. At that time there were 360 students registered in Leipzig, 400 in Halle and 810 in Göttingen (in 1787, fifty years after its foundation) (see Sheehan, 1944). The latter was the most dynamically developing university and, at the same time, it was one of the elite institutions. Noble families sent their sons willingly to this university, which they also supported financially to a great extent. However, most of the students were the sons of lawyers, professors and officials.

In his paper 'Göttingen, Gauss and Erdély' (1979, pp. 294-313) Samu Benkő provides an insight into the relationship between the University of Göttingen and the Transylvanians. Among others this university was famous for the freedom of thought. This is why Transylvanians, who had become used to the freedom of religion at home, were inclined to attend this university. Among the numerous Transylvanian students who studied at Göttingen, Farkas Bolyai was one of the most renowned, spending three years there from 1796 to 1799. Between 1795 and 1798 Gauss, too, studied there and the two became lifelong friends.

Generally, university education was expensive, especially in Göttingen. If a family wanted to have their son educated, they had to be either wealthy, or find a protector who covered the expenses. Historians have also mentioned that drinking, duels, and irresponsible behavior were in fashion at universities, therefore many families sent their sons to universities with reluctance. However, debauchery may have been more moderate in Göttingen where the magistracy kept watch over morals. By the end of the 18th century most states regulated the lives of the universities in some way but they were not allowed to violate academic autonomy.

With the partitioning of Poland in 1772, Prussia became a connected territory. The ruling Hohenzollern dynasty had many talented members. From our point of view, Frederick the Great and his reign (1740-1786) are of special importance. It was at the end of his reign and in the subsequent years that the main works of Immanuel Kant, one of the greatest philosophers of all times, had been published. Kant (1724-1804) was born in Königsberg; he lived and died there. Königsberg was a great city of German culture; many great scholars and artists were born and lived there. The first edition of Kant's greatest work, the 'Critique of Pure Reason' was published in 1781 and the second edition in 1787. He published forty books altogether. In the series entitled 'Critique' there are two more books: the 'Critique of Practical Reason' (1788) and the 'Critique of Judgment' (1790). Kant had an enormous cultural influence. His works were read and studied at all universities of Central Europe; his ideas were discussed in social circles and at dinner tables. Visitors to Königsberg were interested in Kant's newest habits and thoughts or, at least, they wanted to see the famous philosopher.

Kant is a representative of the German Enlightenment, the Auferklärung; Rousseau heavily influenced him in this respect. But having read Hume, he diverged partly from the world of the rationalist philosophy. (As a German representative of this field Kant mentioned C. Wolff.)

With regard to the ideas of the Enlightenment, Kant's critique was related to the tolerant attidude of Frederick the Great. In the preface of the Critique of Pure Reason he explained that his age was the age of critique in which institutions, such as the State and the Church could be criticized freely and ideas could be presented to the public without any restraint. After the death of Frederick the Great in 1786 Frederick William II acceded to the throne. Under his rule the situation changed.

In Kant's philosophy his ideas concerning space are the most important for our purposes. In order to explain Kant's theories we must clarify some of his definitions in advance.

Kant differentiates analytic and synthetic judgments; according to him this difference can be traced in their contents. An analytic judgment only reveals the object of the judgment, while a synthetic one contributes something to it. According to Kant, mathematical judgments are all synthetic. Frege (1884) criticized this distinction claiming that not the content of propositions but their proofs are important. From our point of view, however, this is not significant. Kant formulated his thoughts in the above-mentioned way, primarily, in order to criticize the philosophy of C. Wolff (1679-1754).

Another concept is intuition. Aristotle distinguishes intuitive knowledge from demonstrative knowledge. In his opinion the prime force of scientific knowledge is intuition; it is a direct and immediate cognition, contrary to scientific knowledge, which is acquired through the process of proofs and demonstrations. Yet, the question remains unanswered whether intuition means knowledge inherent in our mind or acquired through experience. Kant combines these two, an example of which is his concept of space expounded in the Chapter 'Transcendental Aesthetics' in the Critique of Pure Reason.

First, Kant states that geometry is a field of science that determines the characteristics of space synthetically and *a priori*. In his opinion the origin of the concept of space should be found in intuition that is *a priori*, i.e., inherent in us before any experience is acquired about an object and it has a coordinating role in the development of the form of our external perception. Space is neither the property of objects, nor is it a determinant of their relations to each other. To put it in a different way, space is in no way attached to objects, i.e., space exists only for man. If we disregard the state of the subject that makes external intuition possible, the representation of space becomes empty. Still, is the geometry of space Euclidean or may it be something else? In his thesis Kant (1770) thinks it possible that the structure of space may be different from Euclidean geometry. Later, however, he abandoned that view and in the Critique of Pure Reason, he maintains that the structure of space is Euclidean. We may observe this in the argumentations in which he discusses the difference between the methods of philosophy and those of mathematics.

Now the question should be posed: What was the mathematics of the 18th century, especially in the second half of it, like? Was it similar, in style, to today's mathematics or that of the 19th and 20th centuries? Was Herman Hankel right when he said (Hankel, 1884): 'In most sciences newer generations destroy what the former built and discard their theories. It only occurs in mathematics that a new generation erects a new story on the old structure'. Hankel's statement is only partially true. The most appropriate contrast to this, which will be treated more thoroughly later, is provided by geometry. A further example can be found when the level of exactness of mathematics in the 18th century and that of the subsequent centuries are studied. The mathematicians of the 18th century: Leibniz, the members of the Bernoulli family, Euler, Taylor and Lagrange were not particulary concerned about the exactness of their results. For them, results were more important than precise proofs. That had been typical not only of the mathematics of the 18th century but also of earlier times, starting from the Renaissance (Grabiner, 1974).

This result-oriented period that lasted till the end of the 18th century, more or less, came after the discovery of the formula of the roots of the cubic equation in 1545. The discovery of differential and integral calculus as well as probability theory took place during that period. Although new results in their first forms often lack a high degree of exactness, the founders of the above sciences did not even think to strive for it.

What made mathematicians turn their attention to exactness from the turn of the 18th and 19th centuries? Two causes may be mentioned. One of them is that by the end of the 18th century the pace of new mathematical discoveries slowed down considerably. By that time, a great number of results had been accumulated. They had to be systematized, unified and this could not be achieved at the former level of exactness. The other reason is that since the French Revolution the financial state of the rulers and patrons were shaken, and mathematicians were obligated to teach. In order to earn a living, however, only logically constructed and attractive material can be taught with success. By the end of the 18th century one of the manifestations of the increasing demand for exactness was that in 1784 the Academy of Berlin, where Lagrange was the director of the Mathematical Section, offered a prize for the clear and precise foundation of mathematical infinity (Grattan-Guinness, ed., 1980). The prize was awarded to Simon L'Huilier in 1786.

Cauchy, Bolzano, Peacock, Babbage and others worked hard to make the theory of functions and analysis exact. In Cambridge, 1813, the latter two with Herschel founded the Analytical Society that, in addition to making analysis exact, set itself the goal to modernize the Newtonian notational system widespread in England.

It was also high time to clarify the axiom of parallels. The scientific world awaited an elegant proof of the statement, i.e., they expected somebody to deduce it from the other axioms. However, something quite different happened. The most brilliant chapter in the history of mathematics in modern times began, and it was shown that what the overwhelming majority of people had expected was impossible. The Hungarian János Bolyai and the Russian Nikolai Ivanovich Lobachevskii achieved the breakthrough. Their works will be analyzed later. Now, in connection with Lobachevskii, we discuss, at some length, the state of Russian universities in the early 19th century.

Alexander I became the czar of Russia in 1801. He made great efforts to develop, among other things, universities. He reopened the university of Dorpat (= Tartu, Estonia) and founded new ones in four cities: Vilna (1802), Kazan (1804), Kharkov (1804) and St. Petersburg (1819). He laid great emphasis on teaching sciences and mathematics. Universities followed the German system; the professors were obliged to do scientific research, and they had to acquire the knowledge of new scientific results and had to include them in their lectures. The number of students was not high: 40 students were enrolled in 1809 at the University of Kazan and 135 students at the University of Moscow. At the new universities as well as at the old ones there were many foreign professors. Their number declined considerably after 1815 because a decree issued in that year stipulated that the language of academic instruction had to be Russian. Simultaneously, it was forbidden for the would-be scholars to study in Germany and, later, it was also forbidden for universities to employ professors who had studied in Germany. The justification for this was that in Germany universities were atheistic and professors made the youth skeptical and hate authorities. As a consequence, foreign professors left the universities voluntarily (Boyer, 1991).

Johann Martin Bartels (1769-1836), a compatriot of Gauss was born in Braunschweig, and was Gauss' instructor at school. Bartels was appointed professor at Kazan, and Lobachevskii became his disciple. Bartels made close friendship with Gauss and may have been aware of the importance of the problem of parallels. It is also possible that Lobachevskii might have learnt about it from Bartels. However, the latter as a foreign professor had to leave Kazan and, thus, could not have been there when Lobachevskii started to work seriously on the solution of the problem in the 1820's. It is also known (see Szénássy, 1977/1980) that Lobachevskii was primarily influenced by Legendre's work (1794). Farkas Bolyai could have learnt about the problem from Kaestner, professor of Göttingen since the latter was an expert of the topic, as his book shows (see Kaestner, 1790).

3. The lives of the Bolyais

We know a lot about the lives of the two Bolyais. It was Ferenc Schmidt, an architect of Temesvár (now Timişoara), and later Budapest, who was the first and the most thorough and devoted researcher of the topic. His father, Antal Schmidt, also an architect of Temesvár, met János Bolyai, as a military engineer working in Temesvár between 1823 and 1826. Thus, Ferenc Schmidt also heard many interesting things about János from his father and he spared no effort to gather all information on the Bolyais. His favorite pastime was to study mathematics and natural sciences. He corresponded with scientists of several Western countries, requesting them to inform him on the newly published scientific books in their countries.

Guillaume Jules Hoüel, a young professor of the history of mathematics in Bordeaux, who became one of the first discoverers of János Bolyai's work, was his French connection. Hoüel translated the Appendix into French and enclosed to it János Bolyai's biography, written by Ferenc Schmidt. The biography also was published in German (see Schmidt, 1868). Ferenc Schmidt provided the bulk of information to Paul Stäckel's two-volume book (1913). According to Barna Szénássy, these pieces of information make up two-thirds of the first volume. Among the first references on the Bolyais, János Bedőházi's book (1897) and articles of Péter Szabó (1910) and Lajos Schlesinger's (1903) should be mentioned. From later literature the books by Lajos Dávid (1922, 1979), Samu Benkő (1968, 1971, 1978, 1979, 2002), Barna Szénássy (1970), Ferenc Kárteszi (1973, 1977), Tibor Weszely (1981, 2002), Tibor Ács

(1997, 2002, 2004) and Elemér Kiss (1999) as well as the papers by Ernő Sarlóska (1965, 1973) and Barna Szénássy (1977, 1980, 1983) may be considered the most important sources.

In this article, it is not our primary goal to elaborate on the lives of the two Bolyais. However, to give a full picture about János Bolyai's life and work, we need to deal with it to some extent.

Farkas Bolyai was born on February 9, 1775 in Bolya near Nagyszeben. He came from a Hungarian family of ancient lineage. The fortified castle of Bolya was given to the family in the early 14th century. Members of the family were gallant soldiers but in the first half of the 17th century another János Bolyai lost the castle while in captivity in Turkey. They became more and more impoverished and Gáspár Bolyai, Farkas' father, inherited only a small estate near Bolya, which belonged to the County of Nagy-Küküllő in those days. A small estate close to Domáld, a village near Marosvásárhely, which comprised the heritage of Krisztina Pávai Vajna, wife of Gáspár Bolyai, was added to their wealth. Between the ages of 6 to 13 Farkas Bolyai was a pupil in the Lutheran and Calvinist College of Nagyenyed. Then Baron Simon Kemény, Sr. hired Farkas as a fellow-student to his son Baron Simon Kemény, Jr. Farkas Bolyai made a lifelong friendship with the young baron. Starting from 1790, they had studied together for five years in the Calvinist School of Kolozsvár. Meanwhile, Farkas' mathematical talent became more and more obvious but he was interested in music, drawing, and acting as well. In the fall of 1795, he set off with Simon Kemény to continue his studies in Göttingen, Germany. However, because of illness, he had to return and was able to join Simon Kemény again only in the spring of 1796. First, they spent some months in Jena, and later enrolled at the University of Göttingen in October. The position as 'a fellow-student' assured the costs of living and the possibility of learning for Farkas. In Göttingen, he made a lifelong friendship with Gauss. After Gauss' death their correspondence became a collection of documents for the history of mathematics. A selection was also published in Hungarian (Bolyai Letters, 1975, The Correspondence of Bolyai and Gauss, 2001). After the years in Göttingen, Farkas went to Kolozsvár in 1799, where he was a family tutor for a short time. He had married Zsuzsanna Árkosi Benkő, the daughter of a chirurgus of Kolozsvár (called 'barber' that time). The newlywed couple moved to Domáld and returned to Kolozsvár only in the fall of 1802, expecting the great event: the birth of János.

Gáspár Bolyai died in 1804. Just before his father's death, Farkas accepted the position of a professor at the Calvinist College of Marosvásárhely where he taught mathematics, physics and chemistry. He held the position until his retirement in 1851. Samu Benkő writes about the process of selecting the best candidate for the job (1979, pp. 155-182). According to him, there were ten applicants, who were supported by a total of twenty-three recommendations. The following three applicants: Pál Sipos, Farkas Bolyai and Mihály Marussi had most supporters. On January 22, 1804, at the consistorial meeting, Farkas Bolyai received eight out of twelve votes and Mihály Marussi received four. Thus, Farkas Bolyai was elected, by a majority vote, and the letter of appointment was prepared on that very day.

The original College of Marosvásárhely was established in the mid-16th century (Koncz, 1887). Its building was erected partly on the ruins of the Holy Church named after St. Nicholas. The church was destroyed by Basta's looting around 1600 (Balázs Orbán, 1868, Vol. 4, p. 134.). The reconstruction of the original school occurred in the early 18th century, when the Calvinist students expelled from Sárospatak were given shelter there.

Two children were born from Farkas Bolyai's first marriage: János and a daughter who died in early childhood. The marriage was unhappy. According to Farkas, his mother-in-law was a troublemaker and wanted to take away Zsuzsanna from him. On the other hand, Zsuzsanna was neurotic; there were signs of the problem already in the first years of the marriage and they grew worse after 1817. She died in 1821 after long suffering.

When Farkas Bolyai was appointed professor, he was paid partly in wheat, wine, salt, pig, lamb, honey, and wood as well as a large house with a garden – and partly in money: his annual salary was 400 Hungarian forints. In four years' time a more spacious and solid house was built for him. That was destroyed in c.100 years, in 1909.

Vásárhely – as Marosvásárhely was called at that time – was a settlement with a local government. It did not belong to the landed aristocracy or the county. It was the largest city of the Székely people, the Hungarians of Eastern Transylvania. Its Gothic castle church dates back to the 15th century and the castle built around it was a reconstructed former monastery of the Black Friars (the Dominicans). In 1571, in the very same church the freedom of religion for Unitarians was proclaimed, strengthening the law on religious tolerance that had been passed under the rule of Prince János Zsigmond in 1560.

Farkas became married for the second time in 1824, to Teréz Somorjai Nagy, who was twenty-two years younger than he was and the daughter of a merchant in Marosvásárhely. Two children, Gergely and Berta were born to them. The latter died in her childhood and the former lived at Bolya as an adult. The second wife, too, was of weak health; she died young in 1833. However, this marriage was more relaxed than the first one.

Farkas Bolyai was a very talented man. As a mathematician he was famous and one of the forerunners of the discovery of non-Euclidean geometry. He devoted his life to proving Euclid's 5th postulate, which, as it is known, is impossible. His main work is the two-volume 'Tentamen' published in 1832/33. It was an outstanding summary on mathematics of the age. Gauss, too, spoke of this work highly, pointing out the author's precise way of discussion. Tentamen served as a textbook for the students of the higher classes in the College of Marosvásárhely but it contained much more than the obligatory curriculum. Farkas Bolyai was elected a corresponding member of the Learned Society (former name of the Hungarian Academy of Sciences) on March 9, 1832 but not in the Department of Mathematics, as has been mentioned in several of his biographies, but in the Department of Natural Sciences. The basis for the election was his book entitled the 'Elements of Arithmetics' published in Hungarian in 1830, and not his Tentamen written in Latin. This is clear from the letter of Secretary Gábor Döbrentei of August 29, 1833 to Farkas Bolyai. The letter states that the Secretary was not able to recommend either Farkas as a full member or János as any kind of member because their works were written in Latin (see Vekerdi, 2001). The Learned Society was founded for the purpose of promoting and developing the Hungarian language. This explains the attitude of the Secretary in this matter.

Farkas Bolyai was not only a very talented mathematician but a many-sided genius. Due to his plays, he acquired a place in the history of Hungarian literature. Another favorite pastime of his was designing stoves and ovens. Having heard that the construction of economical stoves was on the agenda in Vienna, he set out to solve the problem. He invented stoves of different types and had them made or he himself built them. So the Bolyai stoves came into fashion in Transylvania. He had many other inventions, too: a seat on wheels that had to be driven by feet and a stick; 'a coach home' placed on wheels and covered with wooden tiles, which was the forerunner of mobile homes. He gave private music lessons and delivered lectures on the theory of music as well. In addition to Hungarian, he spoke German, Latin and Romanian fluently and knew several other languages. When an opening for the Inspector General of the Forests of the Chamber in Transylvania was advertised, he applied for it in order to lessen his financial troubles. He was not appointed to the job but in an attempt to obtain it he pursued studies in forestry and wrote one of the first books on the topic in Hungarian.

As a witty and good conversationalist, he was a favorite guest of the local high society.

Farkas Bolyai died on November 20, 1856. Pursuant to his will, at his funeral there was no other ceremony but 'the ringing of the school's bell'. Also, his grave was unmarked: only a 'pojnik' apple tree, he introduced in his homeland, was planted on it (Orbán Balázs, 1868, Vol. 4, p. 133), because the famous professor was an expert in horticulture, too. After the funeral, János wrote an essay about his father. There certainly were disputes between father and son but that was not the main characteristic of their relationship. János said his father was remarkably athletic and a man of universal genius. In several of his works, too, he ranks his father among the greatest (see Elemér Kiss, 1999). When discussing János' life, we will see in what a high esteem the father held his son as well.

János Bolyai was born on December 15, 1802 in Kolozsvár where his parents moved from Domáld, to obtain better medical care during the delivery. János was born in a house that had belonged to his mother's family. It is still in existence and there is a memorial tablet on it. In two years' time the family moved to Marosvásárhely when Farkas was appointed professor to the local College.

The genius of János already manifested itself in his childhood. When he was six, he learned to read nearly alone. A year later he learnt German and to play the violin. He was nine years old when his father began to teach him mathematics; at 14, he was well versed in higher mathematics and worked with differential and integral calculus easily and skillfully. This is documented in his father's letter of April 16, 1816 to Gauss. At the same time János made remarkable progress in playing the violin; he already played difficult concert pieces. At 12 he became a regular student of the College. He skipped the first three grades and was enrolled in the fourth grade. This corresponds to the eighth grade of today's elementary school. He passed his final exam in June 1817.

We have already touched on the problem János faced regarding higher education. In Transylvania at that time there were no universities and at the University of Pest and the University of Vienna there were no mathematics professors whose level of instruction would have benefited the young genius. It was clear to Farkas that he had to send János to Gauss in Göttingen. We do not know whether like his father before him, János was offered a contract as a fellow-student to the son of a well-to-do family, which would have provided him with money for living and tuition. At that time, many of the students at German universities led loose lives. Farkas was just aware of that and, perhaps this is why he wanted János to stay in Gauss' house. Note that János was only 15 years old in 1817 and Farkas was 21 when he had gone to Göttingen in 1796. Expecting that his son's higher education was to start in 1817 in Göttingen, Farkas wrote a letter to Gauss on April 10, 1816 in which he asked Gauss to let his son stay in his house for three years, and offered to reimburse him for his expenses. But after this request, he destroyed everything when he asked Gauss to answer the following questions sincerely: '1. Have you not a daughter who may turn (reciproce) to be dangerous...? 2. Are you healthy and not poor? Are you satisfied and not grumbling? And, primarily, is your wife exceptional among women? Is she not more changeable than a weathervane? Is she not unpredictable just like the change of a barometer?' Gauss did not answer this letter.

After this, the possibility that János would study at the Vienna Academy of Military Engineers came up. During his journey to Göttingen, Farkas visited the Academy and fell in love with it so much that he almost stayed there. Thus, he could wholeheartedly recommend it to his son. However, he was unable to raise the necessary money immediately, therefore, he had János enrolled at the Faculty of Arts in the College of Marosvásárhely in 1817. Later, Count Miklós Kemény (1791-1829), the president of the College, together with other benefactors, provided the necessary money for János' education in Vienna. After he passed the admission examination in 1818, János was allowed to begin his studies in the eight-year program of the Imperial and Royal Academy of Military Engineers. One could start either in the fourth year or lower. János was registered in the fourth and he was expected to complete his studies in four years study. Although Count Miklós Kemény and others covered his tuition, there were additional expenses (e.g., for horseback riding); therefore, the father's financial contribution was needed, too. That was not an easy task as the economic situation in Transylvania also began to deteriorate due to the French wars that had been going on since 1792. In 1817, banknotes were devalued to two-fifths. We know that Farkas' annual salary was 200 silver Rhenish Ft around 1820; however, he did not always receive his salary in full or on time. The annual costs of János' education was about 900 Ft. Of this amount, 130 Ft had to be paid in silver. The items János needed to begin his student life cost nearly 220 Ft.

It was Ernő Sarlóska (1965, 1973) who first wrote about János Bolyai's military career. Sarlóska's articles discuss Bolyai's years spent at the Academy of Military Engineers, and his next ten years of his military service. In the books of Tibor Ács (1997, 2002, 2004) there are detailed reports on Bolyai's years at the Academy of Military Engineers. Here, it may be sufficient to mention that János was an excellent student who was ranked first among the students by his professors although his classmates ranked him second. He stayed at that place in the overall rankings. The main reason why he was not ranked first was his performance in drawing: János was just bored by drawing. During the years at the Academy of Military Engineers, from 1820 on, he had been concerned intensively with the research of parallels. He wanted to prove the fifth postulate, which his father had long sought to prove. Farkas warned his son against doing that in his letter of April 4, 1820:

'You must not attempt this approach to parallels: I know this way to its very end. I have traversed this bottomless night, which extinguished all light and joy of my life. For God's sake! I entreat you leave parallels alone, abhor them like indecent talk, they may deprive you [just like me] from your time, health, tranquility and the happiness of your life. That bottomless darkness may devour a thousand of tall towers of Newton and it will never brighten up in the earth.... I thought I would sacrifice myself for the sake of the truth. I was ready to become a martyr who would remove the flaw from geometry and return it purified to mankind. I accomplished monstrous, enormous labors; my creations are far better than those of others and yet I have not achieved complete satisfaction.... I turned back when I saw that no man can reach the bottom of this night. I turned back unconsoled, pitying myself and all mankind.... I have traveled past all reefs of this infernal Dead Sea and have always come back with broken mast and torn sail. The ruin of my disposition and my fall date back to this time. I thoughtlessly risked my life and happiness – *aut Caesar aut nihil.*'

János Bolyai finished his studies at the Academy of Military Engineers in 1822, but he was permitted to stay there to pursue further studies for one more year, as he was one of the two best students.

In early September of 1823 he was nominated sub-lieutenant, and was assigned to the Directorate of Fortification of Temesvár. From here he wrote his father his letter of November 3, 1823, which became widely known all over the world: 'My dear Father! I have so much to write you about my new findings that, for the time being, I cannot help but avoid their discussion here in depth and I am going to write you on a quarto.... I am determined to publish a work on parallels as soon as having arranged and prepared it and, there is an opportunity to do so; for the time being, it is not found out yet but the way I have gone has promised to achieve my goal if possible at all; it is not ready yet but I revealed such superb things that I myself was astonished, and it would mean everlasting shame to let them lost for ever; if you, my dear Father, see them, you will acknowledge them; now I cannot say anything else: from nothing I have created a new different world; all other things that I have sent to you are just a house of cards compared to a tower.'

As it is already known today, this 'new different world' is the magical world of absolute and hyperbolic geometry. At the beginning of 1825, János visited his family at Marosvásárhely. He had great success there. The aristocratic society was fascinated by the personality and violin playing of the elegant officer. His father took delight in his son, primarily because of his mathematical genius. In his letter of February 27, 1825, to Pál Bodor he writes that János is a handsome young man of great and tough nature. In addition, János was an excellent fencer; he had become famous for that in his student years. Once, during his stay in Arad, 13 cavalry officers challenged him to a duel. He accepted the challenge under the condition that after every two duels he might play the violin. He vanquished all thirteen duelers. If this story is true, and if the men used heavy cavalry swords, – the challengers were cavalry officers –, we may conclude that János was a young man of great physical strength.

Fate willed that when he was transferred to Arad in 1826, his superior became Johann Wolter Eckwehr, who had been his professor of mathematics at the Academy of Military Engineers. János had corresponded with Eckwehr before. That year, he handed over his manuscript in German to his former professor, in which he summarized his investigations in non-Euclidean geometry. Regrettably, this manuscript has been lost.

In 1831, János was transferred to Lemberg, and in 1832, Olmütz became the last station of his military career. On his way to Lemberg, he visited his father in Marosvásárhely.

In Arad, János had recurrent fever. Presumably, he caught malaria because there was a marshland around the town. Later, he suffered from cholera, too; his health had deteriorated significantly. This was aggravated by the fact that on his way from Lemberg to Olmütz, his coach turned over and he suffered a serious head injury. He had already neglected his job because he was uninterested in the routine drafting he was required to do. Instead, he tried to use all his spare time to work on solution of mathematical problems. He tried to apply for a 3-year leave from service to pursue his mathematical research. In 1832, his application was forwarded to Archduke János, who rejected it. Finally, in 1833, he was discharged with a pension as a second-class captain. A further reason for his discharge was that on his way from Lemberg to Olmütz he had an argument with customs officers at the border because he refused to open his trunk. The officers then reported him to the authorities.

Coming back to the year of 1831, the most important event was the publication of the Appendix, as a preprint, in Latin. Volume 1 of Tentamen, written also in Latin and bound together with János' Appendix, was published in 1832. Volume 2 was published in 1833. It is an important fact that the date of the imprimatur of Tentamen is October 12, 1829.

Farkas Bolyai sent Gauss a copy of the Appendix, almost immediately after its first publication in April, 1831, asking him for his opinion. That copy had been lost, but the accompanying letter, dated June 20, 1831, exists (see Bolyai Letters, 1975, 168-173). Farkas sent another copy on January 16, 1832. Gauss' reply of March 6, 1832, is widely known. In one of the most devastating parts of the letter he writes:

'Regarding your son's work: If I began by saying that I am unable to praise this work, you would certainly be surprised for a moment. To praise it would be to praise myself. Indeed, the whole contents of the work, the path taken by your son, the results to which he is led, coincide almost exactly with my own ideas I have been developing for thirty to thirty-five years'.

Then he continues: he, too, intended to write all down eventually, so that at least it would not finally perish with him. In other letters, Gauss denotes an earlier date when he had already been concerned with non-Euclidean geometry. But in his letter to Gerling, he recognizes that his ideas in 1798 were far from the maturity found in the work of János Bolyai. Gauss also praised János Bolyai and his work but the praises could not alleviate the pain the first letter had caused the young Titan.

János Bolyai moved to his father's house in Marosvásárhely in 1833, but a year later he went to Domáld, where he lived until 1846. From 1834, he cohabited with Rozália Kibédi Orbán. Legal marriage was out of the question because they were unable to raise the money for a deposit called 'caution money' that was required as János was an army officer. They had two children: Dénes (1837-1913) and Amália (1840-1893). Amália had no children but Dénes had several from his three marriages. Among several other descendants, János Bolyai the great-grandchild of our János, lives in Edelény, Hungary.

The year of 1837 brought a significant event in the lives of both Bolyais. The Jablonowski Scientific Society of Leipzig announced a competition for the foundation of the theory of imaginary numbers (the original text of the competition is rather long and it appears strange by modern standards). The Bolyais learned about it not long before the deadline in November 1837, but both of them submitted competition papers. Beside them Ferenc Kerekes, professor of the College of Debrecen (1784-1850) took part in the competition. The Bolyais did not win but Kerekes was granted half the prize. This work of János Bolyai, known as Responsio, is based on principles similar to those of Hamilton's, who founded the theory of complex numbers. Although János Bolyai submitted his paper in 1837, his theory was complete in 1831, which is earlier than Hamilton's submission of his own paper to the Academy of Dublin.

János Bolyai had several other new mathematical results which have been discussed in a recently published book by Elemér Kiss (1999).

In 1846, János Bolyai moved to Marosvásárhely with his family because his father was discontent with János's management of the estate of Domáld and leased the estate to a tenant.

The year of 1848 gave a surprise to János. He read Lobachevskii's work, published in 1840 in German, whose content coincided with that of the Appendix to a large extent. First, he was suspicious that he had been a victim of theft but later he commented on the work. Paul Stäckel and József Kürschák published the comments in 1902.

During the war of independence in 1849, when caution money was not required, he married Rozália Orbán. But after the war the Army did not recognize the legality of the marriage.

In 1852, János Bolyai moved away from his family, leaving the house to his wife and giving a considerable amount of money for the support of the children. However, he continued looking after them. He was in bad health and taken care of by a servant, Júlia Szőts.

In 1857, with his half brother Gergely, who ran the estate at Bolya, he sold Domáld for 1600 Rhenish Ft.

On January 27, 1860, Júlia Szőts wrote a letter to Gergely, asking him to come urgently because János was unwell. Having signed the letter, she looked at her master and continued: 'While I was writing this letter, he died, thus, there is nothing to be said: the Captain is gone'.

In addition to the obligatory military escort, there were three civilian persons present at the funeral. Below the formal records in the registry of the Calvinist Church the following notes were added: 'He was a famous mathematician of great mind. He was first even among the first. It is a pity that his talent was buried unused' (Kiss, 1999).

However, as we will show later, people much more competent than the aforesaid registrar, were unable to assess the greatness of János Bolyai's personality and work at the time.

No pictures of János Bolyai exist. There was a picture that showed him in uniform, however, at one occasion, Bolyai cut it to pieces with his sword in rage. Recently, the opinion that one of the reliefs on the top of the façade of the Palace of Culture at Marosvásárhely portrays him has gained acceptance. Out of the six reliefs in total in five have been proved to represent the individuals, whose names are inscribed on them. Below the sixth one the name of János Bolyai is inscribed and it is just next to that of Farkas Bolyai's relief. There is further proof, namely, the testimonies of those who had known János Bolyai personally at the time the Palace was built. Moreover, there is a striking similarity between the relief and the portrait of György Klapka. It is known that János Bolyai resembled György Klapka, a general of the Hungarian revolutionary army of 1848-49. Kinga Széchenyi made a plaquette of János Bolyai for the anniversary of 2002 based on this relief (see Figure 1).

Figure 1. János Bolyai. The plaquette made by Kinga Széchenyi.

4. Forerunners of the non-Euclidean geometry

The word 'geometry' derives from Old Greek 'geometrein' that meant land surveying. Originally, geometry was the collection of simple rules that had been obtained through experimentation, data-collection, and intuition. The Egyptians, Babylonians, and Chinese were also aware of such rules but the Greeks were the first who deduced geometric propositions from known or evident ones. Among them Thales of Miletos should primarily be mentioned as the main inventor of the deductive proof in geometry and mathematics. He established the first logical geometry.

Pythagoras, the founder of a mystic religious sect and the discoverer of the theorem named after him, followed Thales in this respect.

Around 400 B.C., Hippocrates, the mathematician (not the physician), summarized the Phythagoreans' knowledge on the geometry of plane. The work was entitled 'Elements'. Although this work is no longer in existence: it is probable that the first four books of Euclid are based on it.

In 387 B.C., Plato established his famous school one mile from the town gate of Athens. Since the estate where the school was located used to be owned by Academos and had been called 'Academy' even before the foundation of the school, the name of the school became the same. In Plato's Academy instruction of mathematics was intensive. You may obtain a comprehensive insight in Fowler's book (1999). Euclid, who published his 13 volume work entitled 'Elements' about 300 B.C., was a student of that school. Science historians have since pointed out that Elements was based, to a large extent, on earlier works, namely, on those of Architas, Eudoxos and Theaetetos.

The Elements of Euclid is such a masterpiece that it has been the source of geometric knowledge for over 2000 years. The greatest merit of the work is its uniform logical system, based on a few given propositions, which through deductive reasoning, lead to new statements, all built on one another. Some of the relatively few initial propositions are postulates, while the others are simple common notions. Some commentators claim that a postulate is nothing but a premise, i.e., an arbitrary starting-point, whose truth or falsity we do not question, while a common notion or axiom is a proposition that is obvious for everyone. Yet, others state that the postulates in 'Elements' are mainly geometric in nature, and the axioms are universal. In today's discussion no distinction is made between them and all initial propositions are called axioms. This approach has been followed for a long time.

Euclid's fifth postulate, the starting-point of various geometrical systems, is János Bolyai's 11th axiom. Over the years Elements has had several editions, some of which of have been supplemented. In this article, the terms 'fifth postulate' and 'eleventh axiom', will be used interchangeably.

In addition to the axioms in Elements, there are also definitions and theorems. Euclid defines the concepts of 'point' and 'straight line' as follows: A 'point' is that which has no parts and a 'straight line' is a length without width. These concepts are not defined in today's axiomatic system of geometry. We only define the relations of the objects with each other rather than the objects themselves. In Euclid's time most of our mathematical notations were unknown. Euclid proves that the sum of the angles of the triangle is 180°, i.e., $\alpha+\beta+\gamma = 180°$. However, he does not use an equation but states that: the sum of the angles is two right angles. Neither the '+' sign nor the degree sign was used in his time. He proves the Pythagorean theorem, i.e., $a^2 + b^2 = c^2$ but he expresses this with the areas of the squares laid on the sides of a right triangle. While he proves a great number of theorems, his reasoning is sometimes objectionable by standards of today. Part of the reason for this is the incompleteness of the axiomatic system of geometry in Euclid's Elements. David Hilbert (1862-1945) formulated the axiomatic system that meets today's standard of exactness, in 1899. Bolyai, Lobachevskii, and Gauss still thought within the axiomatic system of Elements.

However, this does not mean that their theorems are untrue. It only means that the old theorems and their proofs should be put into the new framework and they should be taught today in a different way. E.g., in Bolyai's time one was satisfied with the visually evident meaning of the assertion that point C on a line is between the points A and B on the same line. In the axiomatic system of Hilbert the concept of order, too, has an axiomatic foundation.

Hilbert's axioms are grouped into five categories. They are the axioms of (1) coincidence, (2) ordering, (3) congruence, (4) parallelism and (5) continuity. The last but one group contains only one axiom, which, in that form, was first formulated by John Playfair (1748-1819).

Playfair's axiom of parallelism is the following: In the plane there is one and only one line that goes through a given point P and is parallel to a given line l where P is not located on l.

The above axioms jointly determine Euclidean geometry. The original form of the axiom of parallelism – Euclid's 5th postulate – is different. First, Euclid interprets the concept of parallel lines: they are in one plane and if they are lengthened to infinity into both directions, they will not intersect. Then he proves that if the angles between the lines and a transversal line are equal, then they are parallel to each other. To prove a reverse proposition he formulates and uses the postulate as it is presented below:

Euclid's 5th postulate: If two lines in the plane are intersected by a third one and the sum of the interior angles on the same side is smaller than two right angles, they will intersect when extended to infinity.

One of the best-known models of Hilbert's axiomatic system is plane analytic geometry of Descartes where the geometrical objects are created by numbers: a point is an ordered pair of numbers (x, y) and a line is a set of ordered pairs of numbers (x, y) which satisfy the equation $y = ax + b$ or $x = a$ where a and b are fixed real numbers. Similarly, we may obtain the space as the collection of all ordered number triples (x, y, z), and the

n-dimensional Euclidean space as the collection of all ordered number n-tuples.

Euclid's 5th postulate differs from the first four to the extent that we cannot check its validity empirically. It is only possible to draw sections of straight lines; infinite lines cannot be drawn. Until Bolyai, Lobachevskii, and Gauss' time that postulate was considered a defect of the theory and there was a universal belief that it could be deduced from the other axioms. Bonola (1911) and Stäckel (1914) summarize the various equivalent reformulations of the axiom of parallelism, which have appeared over history, as well as the attempts of their proofs. Here we are concerned with this only briefly.

Proclos (410-485) commented on Euclid's first book and he mentioned the early attempts to prove the 5th postulate. Among others he mentions Ptolemy (2nd century, A.D.), and criticizes one of Ptolemy's attempts. However, he himself makes a mistake as well. The Arabs, who continued the Greeks' work in the field of mathematical discoveries, were also concerned with the 5th postulate. Al-Nirizi (9th century, A.D.) wrote a commentary on the first book of Euclid. Nasireddin (1201-1279) gave the following equivalent formulation to the 5th postulate: If all points of a curve in a plane are at an equal distance from the points of a given straight line in the same plane, then the curve is also a straight line.

In Western Europe Elements became well known through Arab textbooks which were translated into Latin in the 12th and 13th centuries. Later, in the 15th and 16th centuries, translations based on the original text were published. From the 16th century to the early 18th century F. Commandino (1509-1575), C. S. Clavio (1537-1612), P. A. Cataldi (?-1626), G. A. Borelli (1608-1679), G. Vitale (1633-1711) and J. Wallis (1616-1703) were the most important mathematicians who dealt with the 5th postulate. They reformulated it and attempted to prove it. They paved the way for the forerunners of the discovery of non-Euclidean geometry.

Among the above-mentioned West-Europeans' contributions, Wallis' result is the most interesting. Assuming that there are similar triangles to a given triangle (their respective angles are the same and the lengths of their respective sides are proportional) he deduced the 5th postulate. Here an interesting fact of the Bolyai–Lobachevskii geometry appears: within this geometry the size of a triangle cannot be diminished or enlarged keeping the angles.

The opinions of those who are concerned with the history of hyperbolic geometry differ regarding whom they consider founders and whom forerunners of the field. We are not biased towards Bolyai and Lobachevskii but led by the norms applied in the history of mathematics when we claim: There are two groups of the forerunners in modern times while there is only one group of the discoverers. To the first group of the forerunners belong: Saccheri, Lambert, Legendre and Farkas Bolyai; to the second one: Schweikart, Taurinus and Gauss. The discoverers are Bolyai and Lobachevskii.

This differs from the classic grouping by Stäckel and Engel (1895) because we ranked Gauss amongst the forerunners and not the discoverers.

Also, there are other opinions, e.g., those of Bonola (1911), and Kline (1990). Bonola differentiates two groups of discoverers. In the first group there are Schweikart, Taurinus and Gauss; in the second one Bolyai and Lobachevskii. Kline's otherwise very informative book is superficial in connection with Bolyai and Lobachevskii.

As far as Bonola's opinion is concerned: If Schweikart, Taurinus and Gauss were ranked among the discoverers, the description of the history of the other great mathematical discoveries should be adjusted to this norm as well. Thus, e.g., not only Newton and Leibniz but, at least, Kepler, Galilei, Cavalieri, Saint-Vincent, Roberval, Fermat, Toricelli, Descartes, Wallis, Barrow, Child et al. should be considered as the discoverers of the differential and integral calculus in modern times. In this respect the names of many other mathematicians of similar importance may be mentioned going back to Archimedes.

Below we briefly discuss the results of the forerunners and provide the grounds for our statement.

Gerolamo Saccheri (1667-1733), an Italian mathematician and member of the Jesuit Order, published two studies. In one of them (1733) he was concerned with the proof of the 5th postulate, in the other one (1697) he actually established a non-Euclidean geometry, according to which, in the plane, there is no parallel to any given line, that would go through a point outside the line. Also, he wanted to prove the 5th postulate in an indirect way. However, instead of arriving at a contradiction, he laid the foundations of elliptic geometry (as we call it today). Saccheri's starting point is a quadrangle, in which two sides of equal length, extending from the base, enclose angles with equal size and those with the opposite to the base may be right, obtuse, or acute angles. Considering these as right, obtuse, or acute angle hypotheses, Sacchieri proves: If any of the three hypotheses is true for a quadrangle, it is true for all quadrangles. It follows from this that if in a triangle the sum of the angles equals two right angles or is greater or smaller than that, then it is true for all triangles. This result is an important and beautiful theorem of absolute geometry (where we do not assume or negate the validity of the 5th postulate). But at the end of his book, published in 1733, Saccheri provides a simple but false proof for the 5th postulate, as a conclusion of his studies. Imre Tóth (2000) is of the opinion that Sacchieri did it for fear of the Inquisition.

Johann Heinrich Lambert (1728-1777), a mathematician and philosopher of Swiss origin, spent most of his life in Berlin. His work on parallels was published in 1786, after his death. Similarly to Saccheri, Lambert, too, took a quadrangle as a basis of his investigation but his quadrangle had three right angles while the fourth one might be any of the three possibilities. Depending on the assumption on the type of the fourth angle he also considered the acute angle, the obtuse angle, and the right angle hypotheses.

One of the most important findings of Lambert is connected with the measure of geometric figures. In Euclidean geometry length, area, volume are additive measures. By this we mean that the length (area, volume) of a section (planar or spatial geometrical object) that is the union of sections (planar, spatial geometrical objects) is the sum of the individual lengths (areas, volumes). Preserving additivity we can extend the measure to more complicated sets. The measure obtained from length, area and volume, by extension, is called the Jordan-Lebesgue measure. Now, in Euclidean geometry there exist measures with the property of additivity different from the Jordan-Lebesgue measure. Lambert discovered that in the case of the acute angle hypothesis – that can be equivalently formulated as: the sum of the angles of a triangle is smaller than two right angles – disregarding a universal positive constant factor, there is a unique additive measure. Lambert also gave a vague proof for that.

Further on, he discovered that in the case of the acute angle hypothesis the area of the triangle can be obtained through formal manipulation from the formula of the area of the spherical triangle. If the angles of a spherical triangle are α, β, γ, then its area is $r^2(\alpha+\beta+\gamma-\pi)$ where r is the radius of the sphere. Substituting ir for r (where i is the imaginary unit), the former formula is transformed into: $r^2(\pi-\alpha-\beta-\gamma)$. If we subdivide a triangle into smaller triangles, then the area of the large triangle will be equal to the sum of the areas of the smaller triangles, i.e., this measure has the additivity property. In the same way one can show that the measure is unique, disregarding a universal positive multiplier. Thus, the above-mentioned formula provides us with the area of the triangle in hyperbolic geometry (as we call it today), up to a universal constant factor. The measure may be extended to unions of triangles and more complicated sets. The above argument is a clarified version of the one in Lambert's book. Actually, Lambert did not even write down the formula $r^2(\pi-\alpha-\beta-\gamma)$. On the other hand, Lambert regarded the above-mentioned result as absurd because, as he wrote, a circle or a sphere of imaginary radius did not exist in geometry.

One of Legendre's (1752-1833) major contributions is his excellent geometry textbook that became the first serious rival to Elements since Euclid's time. It is important to note that Legendre, too, discovered the theorem that had been mentioned in connection with Saccheri, hence the name Saccheri-Legendre theorem.

Farkas Bolyai intensely tried to prove the 5th postulate. In his main work, entitled Tentamen, he presented its equivalent reformulations. Among them the two below are the most interesting:

- Four points that are not on a plane are on a sphere.
- Three points that are not on a line are on a circle.

In the second group of the forerunners, first the work of Schweikart (1780-1859), professor of law at Marburg, can be mentioned. In 1819, through Gerling, a mathematician in Marburg, Schweikart sent Gauss a brief manuscript in which he described what he called 'astral geometry'. It was different from Euclidean geometry. Gauss' answer, written in the same year, was not favorable in this case either. In his reply Gauss expounds, without any proof, that in astral geometry (when in the plane to a given line through a given point outside of the line more than one parallel can be drawn), the area of the triangle is proportional to the defect of the angle sum $(\pi - \alpha - \beta - \gamma)$ and the upper bound of the area of the triangle is

$$\pi C^2 / \left[\log\left(1 + \sqrt{2}\right)\right]^2 ,$$

where C is Schweikart's universal constant: the limiting value of the height of an isosceles right triangle when the length of the sides goes to infinity. As we know it today this upper bound can be reached when all sides are asymptotic to each other. It is not clear why in Gauss's formula log.hyp stands instead of simply log.

Taurinus (1794-1874) found the relationship between C and the constant k in the area of the triangle (see the next section):

$$k = \frac{C}{\log\left(1 + \sqrt{2}\right)} .$$

He published it in his book of 1826. Taurinus got quite far with his results. For instance, he obtained trigonometric formulas for the hypothesis of acute angles. Nevertheless, he did not accept the new geometry as a possible geometry of space in the physical sense.

Now we take a look at Gauss' relevant activities in more details. Gauss published nothing about non-Euclidean geometry. His results may be reconstructed from his legacy and letters addressed to others. In what follows we primarily rely on the very informative article by Barna Szénássy (1977/1980).

First of all, it is important to mention that in the second half of the 19th century Bolyai was pushed into the background unduly, in comparison with Lobachevskii. By the early 20th century the picture began to clear up, at least on the European Continent. However, recently, under the influence of the conferences held at the 200th anniversary of Gauss' birth and the books and papers written for this occasion, both Bolyai and Lobachevskii became overshadowed by Gauss. Even today many consider Gauss the real discoverer of non-Euclidean geometry.

Gauss, who is called the prince of mathematicians, is really a giant amongst the giants. Not only was he a mathematician but also an astronomer, physicist, and geodesist. His contribution was fundamental both in pure and applied science. His collected works fill twelve thick volumes. But Sartorius von Walterhausen, Gauss' first biographer, stated that Gauss was not interested in developing a geometry that differed from Euclidean (see Gauss, Werke). According to another opinion he was interested, primarily, in three sciences starting with the letter a: arithmetics, analysis and algebra.

In Gauss' legacy there are about 25 instances where he deals with the foundations of geometry and/or non-Euclidean geometry but his notes on these topics would make up altogether 12 pages. In addition, in his diary written from March 30, 1796, to July 9, 1814, we can find a note that consists of one sentence: 'We have made excellent progress in the foundations of geometry. Braunschweig, Sept. 1799.'

Gauss' notes attest that he knew Lambert's book (1786) thoroughly. Today we also know that in 1795, in the first year of his studies in Göttingen and also in 1797 he borrowed it from the library of the university (Dunnington, 1955). Gauss formulated the concept of parallels, which, undoubtedly, is similar to the parallelism of Bolyai and Lobachevskii. As has already been mentioned, in his reply of 1819 to Gerling (actually, to Schweikart) Gauss gave the upper bound of the area of the hyperbolic geometrical triangle without a proof. In his letter of July 12, 1831, to Schumacher he presented the circumference of the hyperbolic geometrical circle without proof. This can be obtained by formal manipulation from a corresponding formula of the spherical circle with radius r by substituting ir for r. In his letter of March 6, 1832, to Farkas Bolyai, in which he reflects on the Appendix, he gives a sketch of a proof for the formula of the area of the hyperbolic triangle. This is

Gauss' most comprehensive study on non-Euclidean geometry. Though his proof contains ingenious ideas, it is incomplete because he uses the statement, without proof, that the area of the hyperbolic triangle has a finite upper bound. Finally, between 1840, and 1846, he made comments on Lobachevskii's work (1840). His comments are valuable but they rely on some ideas already included in the Appendix.

Thus, from the time before the Appendix was published, as a reprint in 1831, few documented results can be attributed to Gauss.

It is worth noting that Gauss knew Lambert's book. However, we know from János Bolyai's writings (see Stäckel, 1914, pp. 221-223) what were those earlier books on the theory of parallels which were known to him but Lambert's book is not on the list. Accordingly, either Gauss did not mention to Farkas Bolyai the existence of Lambert's book or the father became aware of the book but did not tell his son about it. I think the latter case may be excluded. It seems more likely that the discussion between Gauss and Farkas Bolyai were not deep enough and Gauss did not tell Farkas Bolyai all that he knew about the question of parallels. It is important to note this because several authors claim that János Bolyai published what his father had heard from Gauss. There are a large number of arguments too to refute this allegation. (For further details see Szénássy's above cited paper and Császár, 1978.)

5. The revolution of János Bolyai

Twenty years ago an article entitled 'Hyperbolic geometry: the first 150 years' was published by John Milnor the renowned mathematician of Princeton. He stated in that article that non-Euclidean geometry was in an uncertain state over its first 40 years. Later, it was integrated into the more established branches of mathematics through Gauss' theory on curved surfaces, and Riemann's theory on curved manifolds of higher dimension. What Milnor wrote is only a brief description of the story. A few more remarks can be made.

The theory of the curvature and geometry of surfaces and manifolds of higher dimension do not lead out of classical geometry and mathematics, at least, not in a great degree. The interpretation of the curvature of surfaces and the study of their features could be placed into the existing system of mathematics. However, the case with Riemann's curved manifolds or geometry is different. In this theory a general approach to geometry appeared but only one quarter of a century later than the discoveries of Bolyai and Lobachevskii. Riemann discussed his theory in his habilitation paper in 1854. By that time it began to be obvious what had not been at the time of the publications of Bolyai and Lobachevskii on non-Euclidean geometry that geometry and reality may be different. Geometry may be conceived as a class of abstract theories, not giving up the intention of application, because its structures can arbitrarily be interpreted and studied just like functions or other mathematical objects. Riemann's paper was published in 1868, after his death.

Before the discovery of non-Euclidean geometry, the science of geometry described the real world around us, was inseparable from it. Geometry was a kind of natural science. The point, the right line and the plane were what our imagination forcefully imposed on us. It should be remembered that Euclidean axioms were born only for the sake of order in our thinking. They have been formulated so that we should find our way in the chaos of concepts and statements and clarify what is evident and what is to be proved. Obvious statements and axioms should be the least numerous as possible. Statements that can be deduced from the others should not be regarded as axioms.

Before the discovery of János Bolyai, mathematicians expected a genius to come and give an ingenious proof for the 5th postulate, relying on the other axioms. Even the immediate predecessors, Saccheri and Lambert, supposed this postulate to be untrue just in order that they should arrive at a contradiction through indirect proof, for the world is Euclidean. They did not put it this way but thought like that. From the greatest philosopher of the age Immanuel Kant to the man of the street that was the conviction. Today, it is known that the teaching of the theory of relativity is different and there are experimental proofs of it but this fact, too, is known only by the more erudite. Our everyday lives and activities are based on Euclidean geometry. When a child draws lines in his copybook, when someone surveys our land, he does not need to consider whether more parallels may be drawn to a given straight line through a point lying outside of it.

Bolyai sent geometry to the world of abstract theories. He showed that logically more than one geometries are possible. As he said in his letter of November 3, 1823, sent from Temesvár to his father: 'I have created a new different world from nothing'. An abstract world, of course.

But if the world follows Euclidean geometry, what is the use of the new invention at all? Gauss dared not publish his results in connection with non-Euclidean geometry, which were partial results compared to those of Bolyai's. However, Bolyai was a revolutionary, he had the courage of his convictions. But for the sake of objectivity it also should be mentioned that Bolyai was convinced he would be understood and would receive the deserved recognition, based on his work.

After the letter of November 3, 1823, Bolyai wrote his results in German and gave his paper to Johann Wolter Eckwehr, his one-time professor in Vienna and supervisor in Arad in the year of 1826. His father urged him to write his paper in Latin, too, and publish it because other people may have had the same ideas by that time. He wrote to his son: 'When the time is ripe for certain things, this things appear in different places in the manner of violets coming to light in early spring.' So we can credit the father for János' work appearing in due time, even if the place where it appeared was not the best one to call the attention of the international mathematical community. It was published as an appendix to Tentamen, the two-volume monumental work of Farkas Bolyai. Its full title is as follows: 'Appendix, Scientiam Spatii absolute veram exhibens: a veritate aut falsitate Axiomatis XI. Euclidei (a priori haud unquam decidenda) independentem; adjecta ad casum falsitatis, quadratura circuli geometrica'. In English: 'Appendix, The absolute true Science of Space exhibited: independently of the 11th Euclidean axiom (that can never be decided a priori) being true or false: for the case of being false the geometric squaring of the circle is supplemented.

János Bolyai did not attempt to publish his work in any of the leading mathematical periodicals of his age. Though his father would have had connections with the help of Gauss; the Bolyais did not pursue that avenue, perhaps luckily for János, because Gauss wrote his already mentioned letter to Farkas about Appendix, forwarded to him in 1832, which disappointed János. Although Gauss had a good opinion about the results in the Appendix, he also said he had already discovered them. We have quoted some parts of this letter.

Later the Appendix was published in Hungarian and other languages several times. George Bruce Halsted of Texas translated it into English in 1891. This translation was included, together with the translator's preface in the English translation of Bonola's book (1911) that had been written in Italian originally. Excluding the title page, etc. The Appendix is a work of 24 pages. Professor Halsted says in his preface: 'These are the most extraordinary two dozen pages in the history of thinking'. In addition to our laudation of János Bolyai, let us get acquainted with some of the characteristic results of his work.

As has been mentioned, Bolyai was still thinking within the system of the Euclidean axioms. Hilbert's more complete axiomatic system was published only in 1899. However, as far as Bolyai's deductions in the Appendix and his methodology in general are concerned, he had utilized the great inventions of the former centuries, first and foremost, the analytical geometry of Descartes as well as the differential and integral calculus of Newton and Leibniz. In a sense, the former meant to be a new and higher level of exactness, not only a solution of geometrical problems by means of algebra as Bos (2001) convincingly explains.

Bolyai first, rejecting the 5th postulate (which bears the name axiom XI in his work), defines parallelism. Consider a line l and a point P outside the line. If, starting from P, we draw a half line which intersects l in one direction, moveing the intersection into infinity, there will be a limiting position, where the half line does not intersect l any more (see Figure 2). We can do the same in the other direction. If we continue both half lines into the other directions as well, then we obtain two lines both of which are parallel to l. If they are distinct, then there are infinitely many lines between the two that do not intersect l. All of them have a common perpendicular with l. The geometry corresponding to this case is called hyperbolic.

At this point we remark that Bolyai's lines are not lines in the everyday sense even though we visualize them as such in Figure 2. Lines in the Bolyai–Lobachevskii geometry may be any geometrical objects satisfying the axioms.

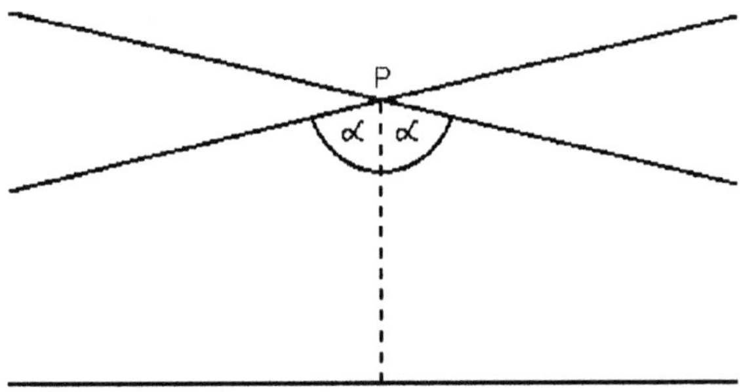

Figure 2. Parallel lines.

Bolyai developed the absolute geometry that is independent of the 5th postulate. The theorem stated below belongs to the absolute plane geometry. If a point P has distance d from line l and α is the angle between the line incident to P and orthogonal to l and the limiting position line parallel to l, then Bolyai's formula is:

$$\cot \frac{\alpha}{2} = e^{\frac{d}{k}} \ .$$

Here k is a universal constant, independent of the choices of l and P and may also be equal to infinity in which case we obtain the Euclidean geometry. In a small neighborhood of a point d is also small, hence the geometry is approximately Euclidean.

Bolyai developed hyperbolic trigonometry and applied it to calculate length and area. For example, the circumference of the circle of radius

r in hyperbolic geometry is given by

$$\pi k \left(e^{\frac{r}{k}} - e^{-\frac{r}{k}} \right) = 2\pi k \sinh \frac{r}{k} ,$$

where k is the already mentioned universal constant. In later mathematical works this is identified with the reciprocal value of the curvature of the space. If $k \to \infty$, then as limiting value we obtain $2r\pi$ which is the well-known formula for the circumference of the circle in Euclidean geometry.

One of Bolyai's most beautiful absolute geometry theorems is the following: The sines of the angles of a triangle are in the same proportion to each other as the circumferences of the circles whose radii are equal to the sides opposite to the angles. If A, B, C designate the angles, a, b, c the sides and $\mathrm{O}r$ the circumference of the circle of radius r, then Bolyai's theorem can be stated as:

$$\mathrm{O}a : \mathrm{O}b : \mathrm{O}c = \sin A : \sin B : \sin C .$$

In Euclidean geometry

$$\mathrm{O}r = 2\pi r ,$$

thus, the above formula takes the form:

$$a : b : c = \sin A : \sin B : \sin C .$$

On the other hand, in hyperbolic geometry we have

$$\mathrm{O}r = 2\pi k \sinh \frac{r}{k} ,$$

from which it follows that

$$\sinh \frac{a}{k} : \sinh \frac{b}{k} : \sinh \frac{c}{k} = \sin A : \sin B : \sin C .$$

Now, consider two parallel lines: a, b and take one point on each of them: A, B. Earlier we mentioned that the lines have directions too, let us designate them as M, N (see Figure 3). Assume that the angle MAB is equal to the angle NBA. Then the points A, B are called isogonal corresponding, or briefly corresponding points (Gauss' terminology) and the fact is expressed by the relation $A \simeq B$ (notation of János Bolyai). This relation is independent of the 5th postulate, belongs to the realm of absolute geometry, and has the reflexivity, symmetry, and transitivity properties: $A \simeq A$; if $A \simeq B$, then $B \simeq A$; if $A \simeq B$ and $B \simeq C$, then $A \simeq C$. If a relation has the above properties, then it is called an equivalence relation. It is well known that any equivalence relation in a

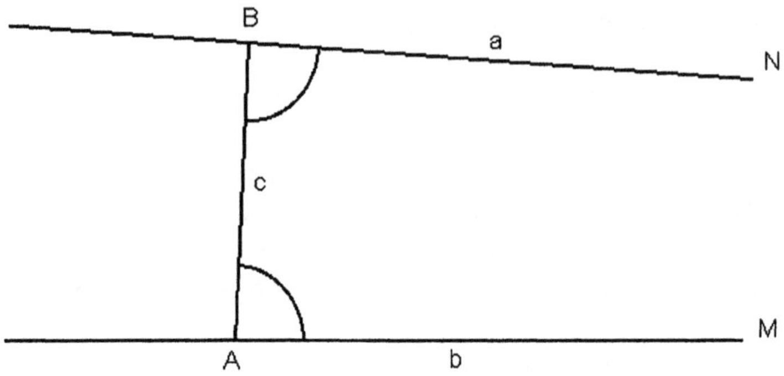

Figure 3. Corresponding points

set gives rise to a subdivision of the set into disjoint subsets. These are called equivalence classes.

Each equivalence class, produced by the correspondence relation, is a planar set, which is called a horocycle.

Similar is the definition of the horosphere in a space. In a horosphere, as Bolyai demonstrated it, the Euclidean plane geometry is valid and the horocycles are the lines. Today we say that any horosphere, in the absolute space, models the Euclidean plane geometry. The horocycle and horosphere can be regarded as circle and sphere. respectively, of infinite radius.

If the angles of a triangle are α, β, γ, then in Euclidean geometry $\alpha + \beta + \gamma = \pi$, in hyperbolic geometry, however, $\alpha + \beta + \gamma < \pi$. The difference $\pi - (\alpha + \beta + \gamma)$ is called the defect of the triangle. Bolyai proved that the area Δ of the triangle is equal to the quantity:

$$\Delta = k^2(\pi - (\alpha + \beta + \gamma)) ,$$

where k is the universal constant. This formula was known to Lambert but Bolyai gave an exact proof to it.

One further interesting theorem of Bolyai is the following: for the legs a, b and hypotenuse c of a right triangle (angle between two lines means the one at the intersection of the lines) we have the formula:

$$\cosh \frac{c}{k} = \cosh \frac{a}{k} \cosh \frac{b}{k} .$$

If $k \to \infty$, then, as limiting case, we obtain the formula $c^2 = a^2 + b^2$ which is Pythagoras' theorem.

Farkas Bolyai, on a few pages in the Tentamen, added some remarks to the Appendix. Among them he gave a more detailed proof for the above limiting relation. (Ungar, 1999 published an interesting Pythagoras' theorem in the framework of the Poincaré disk model of hyperbolic geometry which, however, differs from the above-mentioned formula essentially.)

Finally, it should be mentioned that in the Appendix, Bolyai is concerned also with constructions within hyperbolic geometry.

Other mathematical studies of János Bolyai are discussed in depth by the following authors: Paul Stäckel (1913), Lajos Dávid (1922, 1979), B. Kerékjártó (1937, 1944, 1955, 1966), Pál Szász (1973), Ferenc Kárteszi (ed., 1952, 1973, 1977), Tibor Weszely (1981, 2002) and Elemér Kiss (1999). Bolyai's impact on the development of geometry, and mathematics, in general, is discussed in the papers written by Ottó Varga (1953) and András Rapcsák (1953). Milnor's above-mentioned paper (1982) is the latest survey on the results in hyperbolic geometry.

The other great discoverer of non-Euclidean geometry is the Russian Lobachevskii (1793-1856). The difference between the works of Bolyai and Lobachevskii may be summarized the following way. While Bolyai developed a system of absolute geometry, in addition to hyperbolic geometry, in a rigorous and concise manner, Lobachevskii provided more details about hyperbolic trigonometry. In later years Lobachevskii elaborated an elegant calculus for the volumes of hyperbolic tetrahedra, based on his famous function. Bolyai also found the formula for the volume of the hyperbolic tetrahedron, for the case of an ortho-scheme. There is no reason for the debate over priority. However, in order to see the difference, let us consider the following:

Lobachevskii's first publication on non-Euclidean geometry appeared in Russian in Kazan Messenger in 1829-1830. Bolyai's Appendix was published as a reprint in 1831, but the year of the imprimatur of the whole Tentamen is 1829. It is known that Bolyai outlined his geometry in 1823, and the full German text was ready in 1826. Since the latter has been lost and the former is only a report of the discovery in a letter, there are no earlier documents on Bolyai's discovery than the Appendix. On the other hand, it is also true that Lobachevskii held a lecture on the relevant topic at the University of Kazan in 1826. But if the title of this is scrutinized, we may see that the lecturer intended to prove the 5th postulate at that time (Kiss, 1999).

According to some authors, the geometry of Bolyai and Lobachevskii is the criticism of the Kantian concept of space; according to others, it is a refutation of the concept. Their argument is as follows: If in our mind there is room for both Euclidean geometry and hyperbolic geometry, it is not possible that our concept of space is *a priori* in our mind, independently of our experiences about objects.

Undoubtedly, the absolutization of Euclidean geometry in Kantian philosophy proved to be a dead-end not so much because of the Bolyai and Lobachevskii geometry but because of the twentieth century results in physics. However, Kant's view that space is Euclidean, can be separated from his other views concerning space, which are subtler and different from the one in the aforesaid contrary opinion. Kant did not deny that more than one abstract mathematical theory on space might also be formulated.

Still, in the age of Gauss, Bolyai and Lobachevskii, Kantian philosophy was considered a definite supporter of Euclidean geometry. Unlike Gauss, who was afraid to publish his results because of the attacks of the Boeotians, who were regarded as pleasure-seeking and stupid people by the Athenians, Bolyai and Lobachevskii did not fear to do so: both of them were revolutionaries. They revealed their scientific convictions to the world with courage.

6. Reception and afterlife of Bolyai – Lobachevskii geometry

In his book published in 1840, Lobachevskii laments the little interest shown in his writings. With Bolyai the situation was different – he was even shocked by Gauss' reaction. We have already cited from Gauss' letter: 'I am unable to praise this work.... To praise it would be to praise myself.'

However, others may have had a positive opinion on Bolyai's work. How did the most authentic body, the Mathematical Department of the Hungarian Academy of Sciences, express its recognition? Antal Vállas, mathematician member of the Academy wrote a comprehensive article on the results of Hungarian mathematics in 1836. He was complaining that Hungarians had suffered by the Tatar invasion and the long Turkish rule, etc.; therefore Hungarian science had not been able to develop properly. In connection with Bolyai, he quotes the works: Tentamen, Elements of Arithmetics, and the Elements of Arithmetics, Geometry and Physics. In the text he mentions that 'Bolyai stands out with his eccentricity'. From the enumeration of the works, we can see that it is Farkas and not János that is written about. But in the title of Tentamen the wording 'Cum Appendice triplici' is included, so we may raise the question: Is János' work mentioned somehow in the enumeration? Regrettably, this is not the case. The three appendices mentioned belong to Volume 2, i.e., they can be found at the end of the complete book. The Hungarian prospectus attached to Tentamen informed the customers where the three appendices indicated could be found. Thus, the work of János Bolyai is not mentioned in the survey paper of Antal Vállas and it is not mentioned in the bibliography, either. The fact that the author of the aforesaid paper did not understand the essence and significance of János Bolyai's discovery may somehow be accepted, but it is unacceptable that Vállas had not even mentioned it, even though Farkas Bolyai sent Tentamen to the Academy in 1832. Apparently, the work lay on the shelves in the library, untouched.

Alexits (1977) mentions that the Mathematical Department of the Academy offered a prize for the solution of the problem of parallels in 1844. But this is not true. The text inviting applications to win a prize, which was accepted in the General Assembly of December 26, 1844, went like this: 'What are the properties of imaginary quantities and their analytical and geometrical meanings?' Although both János and Farkas Bolyai, had written papers on this topic, neither of them had published it, just sent it to the Jablonowski Society of Leipzig with the aim to win a prize. Thus, the mathematicians of the Academy cannot be blamed in this respect. In his article, Tibor Ács (1998) discusses the history of the foundation of the prize in detail. In those days military science was included in the Mathematical Department and Captain Károly Kiss, member of the Department concerned with military science, would have been pleased if a prize had been offered in his field, too. He proposed the development of a defense project for the case of a probable attack from the North. His goal was to obtain more appreciation for military science within the Mathematical Department. But his efforts failed and, finally, the above-mentioned prize was offered.

Thus, neither the scientific community nor others were very much interested in János Bolyai's work. The following is an excerpt from Farkas Bolyai's letter of October 3, 1836, to Gauss:

'Here nobody needs mathematics; out of my students there are only a few who have a sense for it; I use my work as waste paper for wrapping up things and the like; it has been of great use for me recently during the time of raging cholera from which I, too, suffered for a month but without vomiting and spasm; I was only very depressed. I was off my food and wine; I was craving for drink, for fresh water and I had a heavy diarrhea and still have it'. (Bolyai Letters, 1975, p. 188.)

The story of the discovery of János Bolyai started abroad. The sorting of Gauss' papers after his death, in 1855, played a great role in the discovery. Among his papers, the works of Bolyai and Lobachevskii as well as Gauss' letters to others and letters to Gauss were found. Gauss' work was sorted by Sartorius von Waltershausen, professor of mathematics in Göttingen to whom Farkas Bolyai, too, sent Gauss' letters addressed to him, so they were included in the papers as well. The picture slowly started to unfold.

The first words of appreciation published came from Baltzer, a professor of mathematics in Dresden. They were included in his influential work, well known at that time, entitled 'Die Elemente der Mathematik' (1867). In the same year Hoüel, professor in Bordeaux published a translation of the Appendix in French in full with the enclosure of Ferenc Schmidt's article on the Bolyai's biography.

The year 1868 aroused the interest of others in the Bolyais as well. That year the Italian Beltrami's article in which he gave a model for the Bolyai-Lobachevskii geometry was published. The two pioneers left open the question of the absence of contradictions in their system of axioms. This question may be answered by providing the concepts not interpreted in axioms (point, line, plane) with concrete contents, i.e., we frame a model for the new geometry. If there is at least one model, the axioms are exempt from contradiction; they make sense.

Riemann laid the foundations of his geometry in 1854. As a onetime Gauss-disciple in Göttingen, he based his geometry on the surface theory worked out by Gauss earlier. The comprehension and recognition of Riemann's geometry did not cause any difficulty in the German and international mathematical communities. Beltrami, who placed the Bolyai-Lobachevskii geometry into the framework of the Riemannian geometry, contributed to the acceptance of hyperbolic geometry through this step considerably. Since that time, there has been more interest in the topic. The great mathematicians of the age Poincaré, Klein, Fuchs and others published new models and further results. Consequently, new mathematical studies also emerged. However, regrettably enough, they made references only to Lobachevskii most of the time. Bolyai's recognition was still delayed.

Here we pose the question: Did the representatives of science – primarily, the mathematicians, at best, after several decades of the discovery – comprehend the real significance of Bolyai and Lobachevskii's work? Cayley (1821-1895), who was President of the British Association at the time, said in his presidential address held in September 1883: 'It is well known that Euclid's twelfth axiom, even in Playfair's form of it, has been considered as needing demonstration; and that Lobachevsky constructed a perfectly consistent theory, wherein this axiom was assumed not to hold good, or say a system of non-Euclidean plane geometry. There is a like system of non-Euclidean solid geometry. My own view is that Euclid's twelfth axiom in Playfair's form of it does not need demonstration, but is part of our notion of space, of the physical space of our experience – the space, that is, which we become acquainted with by experience but which is the representation lying at the foundation of all external experience.'

Cayley knew an edition of the Elements in which the fifth postulate was referred to as the twelfth axiom. As has been mentioned, Playfair's axiom is equivalent to the fifth postulate. In Cayley's citation Kant's influence can be traced, but the essence is that in his opinion non-Euclidean geometry has no sense at all since the world is Euclidean.

Until Gauss' death in 1855, the scientific world had not paid any attention to Lobachevskii's work, although he published his results in German in 1840, and Gauss did not begrudge him praise, albeit he praised Lobachevskii only in letters and not in publications. Nevertheless, as a token of his esteem, Gauss had Lobachevskii elected a corresponding member of the Royal Society of Göttingen in 1842. However, at home, in Russia, Lobachevskii was pushed to the side because of his revolutionary theory. His career started promisingly, as he became a professor of Kazan University very young, at the age of 23. In 11 years' time, in 1827, he became a rector there. He did his utmost to develop his university both as a scholar and an administrator, and he did it successfully. During the cholera epidemic in 1830, he managed to protect the professors of his university, their families, and the students from the disease. The death rate among them was only 2.5 %. Despite all this, at the age of 54, without any reason, he was dismissed from the positions as rector and professor. Might Bolyai have had the same fate if he had succeeded in getting a professorship due to his talent and scientific results in his homeland? On the life and work of Lobachevskii see Rosenfeld (1988), Kárteszi (ed., 1953).

Having cited Cayley's quotation above, that reflected the Kantian spirit, let us recall what Riemann said about the geometrical concept of space. In his lecture held in 1854 in Göttingen and published in 1868, he explains the following view:

'I have set the aim to create the theory of manifolds with the help of their general concept. From this it will follow that a manifold is able to realize various metrical relations. Consequently, space is only a special case of the three dimensions but from this it follows inevitably that the postulates of geometry cannot be deduced from the general concept of dimension and the properties which differentiate space from the other three-dimensional objects can be justified only by experience.'

The role of experience in the relationship between geometry and reality was even more accentuated by Helmholz. The citation below has been taken of his lecture held at the anniversary of the foundation of the Frederick William University (the forerunner of today's Humboldt University):

'If we wish to base geometry on empirical facts where their value is only physical correspondence, merely the statements of the field of science which I have called physical geometry may be applied. For everybody who deduces axioms from experience, up till now, our geometry has been a physical geometry, indeed.'

A bit further on, criticizing Kant, Helmholz writes that the propositions included in the axioms of geometry may be applied to the relations of the real world only if their validity has been justified and stated.

Lobachevskii sought to acquire proof of the validity of the new geometry through astronomical measurements, but he failed. However, the validity of a geometry cannot be proven experimentally; at best, it can be refuted. It is a widespread notion (see, e.g., Born, 1965) that while measuring the angles of the triangle of the three peaks in the neighborhood of Hannover, i.e., those of Brocken, Hohen Hagen and Inselberg, Gauss wanted to know whether their sum was 180° or less, and the result (within the error limit) was 180°. These peaks are at 107 km, 69 km and 85 km distances from each other, so the above-mentioned result would not be surprising. However, as Szénássy (1977/1980) mentions, Gauss made measurements not only with a single triangle but carried out a series of measurements, whose goal was different from deciding whether space is Euclidean or not. In 1816, Gauss was commissioned by the government of Hannover to make a map of the country and it was completed by 1841. Setting out from the afore-mentioned triangle, he constructed subsequent triangles, whose vertices served as triangulation points. Measurements were smoothed by the method of least squares and the order of magnitude of the accuracy was a fraction of a second. Gauss could not have thought that measurements in the case of such small triangles might indicate equality to or deviation from the 180° sum of the angles. On the other hand, the method of least squares – although it may be considered as a numerical procedure independent of any geometry – has a close relationship with Euclidean geometry. Gauss must have known this, and it is improbable that he would have applied this method if his objective had been the experimental justification of geometries.

Gauss considered the method of least squares as one of his most significant scientific results. Because of that he had a priority debate with Legendre, another great mathematician of his age.

It may be added to what is written above that the last German banknote of 10 DM was devoted to Gauss' memory. On one side of the banknote is the picture of Gauss and beside it are the so-called Gauss curve and the landscape of Göttingen and, on the other side is the sextant for measuring angles as well as the sketch of the map made by Gauss.

Although Gauss made his measurements to create maps and not to justify geometries, somehow he may have stated that he did not experience deviation from Euclidean geometry. Namely, the latter statement was criticized by philosophers claiming that if a deviation had been experienced it would not have justified the invalidity of Euclidean geometry (as agreement within error limit would not prove its validity either) because the result can be the consequence of an unknown light deflection.

The critics touched on the essence of the theory. Einstein's special theory of relativity came to light in 1905, and his general theory of relativity was published in 1916. The latter relies to a great extent on non-Euclidean geometry and its Riemannian extension. However, Einstein's theories are not based on different kinds of geometries of space; rather they reject the separation of space and time and start from the new concept of 'space-time'. Although the theory of relativity, too, induced a revolution in science, it brought recognition and world fame instead of neglect to the discoverer. Einstein was able to achieve this because his theory was brilliantly supported by experiments.

The general theory of relativity identifies gravitation with the curvature of space and one of the manifestations of this is that light close to a great mass is deflected, compared to the Euclidean line. The question was raised: How can this be justified experimentally in a way that good agreement would be shown between computed values and experimental data?

Einstein pointed out that if light, coming from a distant star, passes by the Sun, its path, compared to the Euclidean line, will deflect slightly due to the gravitational effect of the Sun. Therefore we see the star in a position different from its actual one. The position where we see it may be obtained if the line of the shaft of light is extended in the reverse direction, as shown in Figure 4. The phenomenon can be justified through observation only when there is a solar eclipse because at other times we are unable to observe a star in the neighborhood of the Sun due to the strength of its light.

In 1919 Eddington, English astronomer, organized two expeditions to study the phenomenon: one of them made observations on the western coast of Africa and the other on the northern part of Brazil on May 29. That day there was a solar eclipse on both sites. Based on former observations and calculations it was known that certain stars were behind the Sun at that time but in the photos taken, they were seen next to the Sun. The results proved the hypothesis of the theory of relativity according to which mass produces a curvature of space and the shortest path between two points, on which light travels, is not an Euclidean line.

After the end of World War I, society was craving for culture and scientific knowledge, which represented human values of higher standards. Einstein's success was overwhelming; his name was on the front pages of newspapers. He became world famous at once. By the end of the 20th century, in 2000, he was selected the Man of the Century by Time Magazine.

Since Einstein's theory of relativity, several other theories of our physical space have been born. The latest sensation is that although space has a curvature locally, in cosmic sizes it is shown to be Euclidean within error limit!? Szalay and Gray (2001) report that on the Internet there will be a 'virtual observatory', which renders the observation of planets possible while sitting in front of a computer. Their effort helped to derive the above conclusion from existing data.

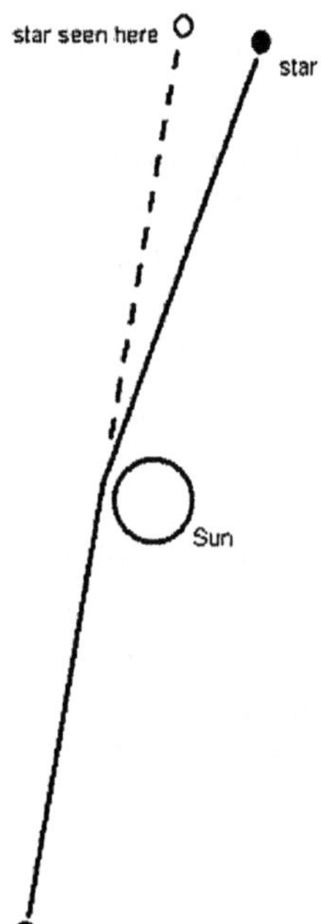

Figure 4. Deflection of the light from a star by the Sun

When did the official Hungarian science community obtain knowledge of János Bolyai and his work? Nine years after János Bolyai's death in 1869, Baron József Eötvös, who was President of the Hungarian Academy of Sciences and also Minister of Culture at that time, received a letter from Italy, written by Prince Baldassare Boncompagni (1821-1894), a historian of mathematics. He informed Eötvös that the biographies of János and Farkas Bolyai along with Appendix were translated into Italian and would be sent to Eötvös under a special cover. He added that the mathematicians in Rome consider the new discovery described in the Appendix the greatest mathematical achievement in the 19th century. Eötvös did not know whether to be pleased or blush – he wrote in his letter to his son.

In the background of the writing of the letter there was Hoüel, who tried to get a reply to his letter, in which he made inquiries about the Bolyais, from Marosvásárhely. Hoüel was astonished at the negligence of the Hungarians. In one of his letters he wrote: 'I am grieved to see how little Hungary appreciates her own scientific results... '

József Eötvös felt the importance of the Bolyai issue and began to patronize it. He helped transport the chests containing the János Bolyai papers to Pest. The chests were deposited in the Archives of the Academy, where they stayed for a quarter of a century. A committee tried to identify the scientific results dormant in the papers without significant success. Nevertheless, during that time, Ferenc Schmidt found János Bolyai's letter of November 3, 1823, to his father ('I have created a new different world from nothing') among the documents. Later, in Marosvásárhely, a part of the works with mathematical content, written after the Appendix, were found among the manuscripts. These are the following: Responsio, the material sent to the competition of the Jablonowski Society in Lepizig; Supplements (to Appendix); Cubature (volume) of the Tetrahedron (in hyperbolic geometry); Studies in the Absence of Contradictions; Comments (on Lobachevskii's work); Raumlehre.

It was Samu Benkő, a professor of history in Kolozsvár, who arranged the Bolyai manuscripts with the help of Gusztáv Abafáy. It took Benkő 16 years of work to sort the 14000 page manuscript. He was assisted by János Bolyai's 'watch-words', the last words on the bottom of several pages copied onto the top of the following pages. Samu Benkő read through the manuscripts and published a book on the non-mathematical texts entitled 'Confessions of János Bolyai' in 1968. Elemér Kiss, a professor in Marosvásárhely read the 3000 page mathematical manuscript in the 1990's, and he reported his findings in a book and an article in 1999. In the manuscript, Kiss discovered significant new mathematical results, such as the number theoretical theorem, published by Jeans 38 years after Bolyai's death, which is today textbook material. Both professors did invaluable work. Bolyai wrote part of his manuscripts on papers already filled with other things, e.g., playbills, back sides of envelopes, notebooks of his son, so not only their interpretation but their mere reading was tremendous work. For the time being, the publication of all János Bolyai's manuscripts is in progress under the editorship of Samu Benkő.

Besides the Bolyai collection of Marosvásárhely, there is another one at the Library of the Hungarian Academy of Sciences. This has emerged mainly through the gift of Péter Szabó. His father, Sámuel Szabó, was a professor in Marosvásárhely, who studied the Bolyai manuscripts and took home part of them. They remained with him, and after his death they were found in his bequest. Péter Szabó handed over the material to the Academy but he himself also collected many manuscripts and letters in connection with the Bolyais. The collection was arranged by Mrs. Fráter (1968) and entered in a catalogue. Recently, Vekerdi (2000) has written a very interesting paper about the collection.

Szénássy (1977/1980) wrote a comprehensive paper about the earlier history of the discovery of the Bolyais.

Respect towards the Bolyais, especially towards János and his work increased more and more. In 1897, the Appendix was published in Hungarian, too, (translated by Rados). In the same year the book entitled 'The Two Bolyais' was published by Bedőházy, the successor of Farkas Bolyai in the College of Marosvásárhely. In 1897, the first volume of Tentamen edited by Gyula Kőnig and Mór Réthy was republished and, its second volume edited by József Kürschák, Mór Réthy and Béla Tőtössy was published again in 1904. János Bolyai's Appendix was attached to the second volume. For the 100th anniversary of János Bolyai's birth, a memorial session was held in Kolozsvár. In Mathematikai és Physikai Lapok (Mathematical and Physical Papers) there were several articles on János Bolyai and his geometry by Manó Beke, Mór Réthy, Péter Szabó and Lajos Schlesinger. (The latter was the main speaker of the Kolozsvár celebration.) In 1902, the Academy established the Bolyai Prize that was given to Poincaré in 1905, and to Hilbert in 1910. Both of them achieved important results in the foundations of geometry and in hyperbolic geometry. (It was at that time when Poincaré discovered that János Bolyai was Hungarian and not Bulgarian as he wrote it in his book entitled 'La valeur de la science' in 1905.)

On the occasion of the 150th anniversary of János Bolyai's birth, a memorial meeting was held at the Hungarian Academy of Sciences, whose material was published in the Publications of the Mathematical and Physical Department of the Hungarian Academy of Sciences. The authors of the articles are Kárteszi, Varga, Szász, Alexits, Kalmár and Hadamard. At the time of the 175th anniversary, there was a smaller meeting at the Academy. The speakers were Császár, Bretter, Sarlóska, Gazda, Lambrecht and J. Molnár and their papers are published in Természet Világa (The World of Nature). A 190th anniversary session was also organized by the Academy. In 2002, the 200th anniversary was celebrated in Budapest, by a large international conference, organized by the Hungarian Academy of Sciences, János Bolyai Mathematical Society, Loránd Eötvös Society of Physics, Loránd Eötvös University of Budapest, University of Debrecen, University of Szeged, Babes-Bolyai University of Kolozsvár (Cluj-Napoca), Sapientia University of Transylvania, the Mathematical Institute of the H.A.S. and the Computing and Automation Institute of the H.A.S.

Bolyai and Lobachevskii discovered and systematically developed hyperbolic geometry, the first non-Euclidean geometry (if we disregard earlier unfinished attempts.) However, these two great scholars have not proved that the new geometry is free of contradiction.

The absence of contradiction can be proved by presenting a model, i.e., a collection of mathematical objects, together with their relations to each other, in such a way that they satisfy the axioms, in the present case, those of hyperbolic geometry. If there is at least one realization of the system of axioms, it cannot be contradictory.

First, the Italian Beltrami (1868) created a model of hyperbolic geometry. In doing so, he used Minding's work (1838) and the apparatus of the Riemannian geometry. The model is the so-called pseudosphere, which is the surface of revolution of a tractrix. The tractrix curve is characterized in such a way that the section between the point of contact and the vertical axis of the tangent drawn to any point of the curve is of the same length. (See Figure 5)

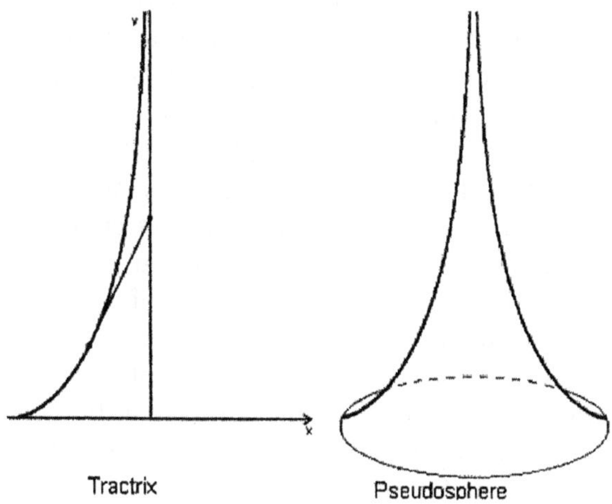

Figure 5.

Huygens called the tractrix a dog's curve because if we pull a reluctant dog on a leash along the axis y, the dog will move along the tractrix. The 'line segment' on the surface of the pseudosphere is characterized as the shortest path between any two points of it.

Poincaré's models are much simpler. In one of them, called Poincaré's disk model, the points are those of the (open) circle without boundary and the lines are the circular arcs within the (open) circle such that they meet the boundary perpendicularly as well as the Euclidean lines going through the center of the circle. In the other model, called Poincaré's half plane model, the points are those of the open upper half plane, the part above axis x. In this model a line is a semi-circle such that it meets axis x at two points perpendicularly or an Euclidean line which is perpendicular to axis x. (See Figure 6)

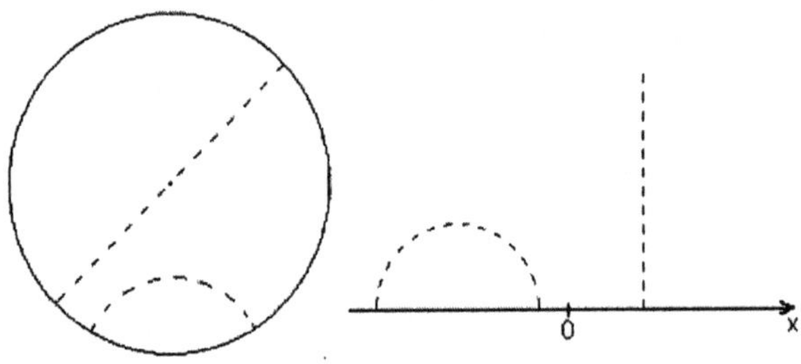

Figure 6. The disk and half-plane models of Poincaré

Hyperbolic geometry has many applications both in mathematics and other sciences. In mathematics it is an indispensable tool in the theory of complex functions when the so-called Riemann surfaces are discussed. Variants of hyperbolic geometry appear in discrete geometry, topology, group theory etc. Among the physical applications of hyperbolic geometry the special theory of relativity should be mentioned first. Next, its indirect application is also very important: its creation opened the way for the Riemannian and other geometries that provide us with the mathematical foundations to the general theory of relativity. It is also used in statistical physics. A nice connection between hyperbolic geometry and the special theory of relativity is presented in Ungar (1997). Recently, the arrangement of the Internet objects in branching trees on the computer screen has been proposed, by the use of Poincaré's disk model of hyperbolic geometry (see Gunn, 1993).

The world of hyperbolic geometry exerts a great influence not only on scientists but artists as well. Among them, Escher should be mentioned first. Several drawings and engravings of Escher's are concerned with configurations that can be found within a circle. In some of them smaller and smaller circles are approaching the periphery. In the immediate neighborhood of the periphery, Escher placed tiny circlets, however, they cannot fill in the circle. Having studied the figures of hyperbolic geometry of H.S. MacDonald Coxeter, the famous Canadian geometer, and having made his acquaintance, Escher accepted the former's advice concerning his pictures, and then created his marvelous 'circle limit' engravings. On this the 'lines' of Poincaré's disk model of hyperbolic geometry (arc in an Euclidean sense) converge to the periphery while shrinking (see Escher 1989, 1992, 1995). In Figure 7 we reproduce Coxeter's figure (1998) and Figure 8 shows Escher's picture Circle Limit III. Coxeter (1979) returned to this engraving in a paper and revealed its relation to non-Euclidean geometry.

Figure 7. Coxeter's hyperbolic tessalation

The work of Konrad Polthier of Berlin is another work of art on hyperbolic geometry. It is based on Reynold's article (1993) in which the author describes a model of hyperbolic geometry realized on the two-sheeted hyperboloid. The picture may be seen on the cover of Stillwell's book (1996).

Finally, it should be mentioned that, recently, the Mathematical Research Institute (MSRI) of the University of Berkeley, California in the USA has had a mathematical sculpture made. This has been erected in the yard of MSRI. The statue represents the so-called Felix Klein quartic curve and its title is: 'The Eightfold Way'. On the pedestal there is a colored tile mosaic, calling the disk model of hyperbolic geometry to mind; it was made by Lajos Biczó, a Hungarian, remembering that one of the discoverers of hyperbolic geometry was a Hungarian. About the sculpture, an entire book (ed. by Lévy, 2001) was written.

7. On the axiomatic method

In the history of mathematics non-Euclidean geometry became the first closed contradiction free logical system that has no direct relation to reality. Its creation started from physical reality but it became an independent mathematical theory. The same is true about Euclidean geometry and the other geometries as well. This does not mean that geometry lost its applications; on the contrary, it became applicable to a great many problems in mathematics, physics, computer science and almost all fields of science and engineering.

Non-Euclidean geometry is not the only independent mathematical theory that was created in Bolyai's time. In the mid-19th century, abstract algebra, also called symbolic algebra, emerged in England, Ireland and France. Already at the beginning of the 19th century, mathematicians did operations with real and complex numbers, although none of them, especially complex numbers, had exact mathematical foundations in today's sense. Therefore, it was necessary to incorporate the rules of operations (associativity, commutativity, etc.) in a clear and well-arranged system. Simultaneously, there was a need to extend operations in some ways, e.g., to make it possible to raise a positive real number to a real power. Algebraists carried out operations with symbols, while keeping the rules of game. Also, the problem emerged that those rules should be regarded as abstract and applied not only to numbers but also to other mathematical objects.

Kant had already discussed in his works the role of mathematical symbols. Lambert intended to deduce the Euclidean fifth postulate from the other postulates and axioms merely symbolically (1786, §11).

George Peacock (1791-1858), a mathematician at Cambridge and, later, the dean of Ely Cathedral was the first in England who raised the necessity to give foundations for mathematical manipulations. He

Figure 8. Escher's Circle Limit III, woodcut

did not restrict the problem to algebra but also extended it to differential calculus. Remaining with algebra, Duncan Gregory (1813-1844) is the next important individual in the establishment of symbolic algebra. In one of his works written in 1840, he said that he regarded symbolic algebra as a science of the combinations of operations. He determined it not by the facts of what they were and what they resulted in but by the rules which they obeyed. Thus, algebra departed from what it had been used for before, i.e., from the operations with positive real numbers; its sphere of operation was extended.

De Morgan (1806-1871) made the next important step. He was the first who proposed the application of the new algebra in logic. The symbolic logical operations, called De Morgan rules, originate from him.

Hamilton (1805-1865), an Irishman by birth, gave exact foundation to the theory of complex numbers, created the notion of quaternion, and developed the relevant theory. (Hamilton sent Gauss his work on quaternions and got a reply similar to what János Bolyai received concerning Appendix.)

George Boole's work (1815-1864) and, primarily, his book entitled 'The Laws of Thought' are also important. The algebra named after him is almost universally known; its application is well known in mathematics, logic, the theory of circuits etc.

Arthur Cayley (1821-1895) and James Joseph Sylvester (1814-1897), important members of the algebraic school of Great Britain, participated in the elaboration of matrix theory and group theory. The creation and development of the latter is also credited to the French mathematician Evariste Galois (1811-1832).

Now that we have created abstract geometry, abstract algebra etc., the question may be posed if mathematics is not concerned any more with concrete things but only with abstract structures? Is this what mathematics has become after the solution of so many serious problems? Is there only abstract mathematics and has it not anything to do with reality? No, this is incorrect to say. However, in mathematics we substitute real objects with mathematical objects, formulate models in which we solve the problems. However, we must return to reality. Our results should be applied all the more so because reality is the major source of the emergence of new ideas, models and structures. Of course, we also have wrong tracks, over which there are great debates. Thom's (1971) paper can be mentioned as an interesting polemical essay in which he criticizes the mathematical activities deviating from reality and the far too abstract educational methods. The reader is also advised to look at the book by Morris Kline: 'Mathematics, The Loss of Certainty', where detailed discussion can be found about the state of the art.

In the middle of the 19th century, the application of the axiomatic method became quite frequent and not just in mathematics. The 'fever of axiomatization' became widespread in science and technology. Naturally, axiomatization in physics was not a novelty but now, surprisingly enough, some engineers, too, used axiomatization. The German Ferdinand Redtenbacher (1809-1863) and the Swiss Franz Reuleaux (1829-1905) were the two outstanding representatives of this trend. They systematized the variables and possible movements in mechanisms, determined their relations to each other, and proved theorems. Although the practical use of this theory was questioned at that time, the theory helped its creators in the construction of new mechanisms (see Brentjes, 1982).

However, axiomatic thinking has its barriers. Gödel (1931) proved that in every axiomatic system that meets some general conditions, there exists a problem that cannot be answered positively or negatively. The problem of Euclid's fifth postulate is the most classical one in this respect.

Another barrier is that mathematicians are generally inclined to deal with problems emerging up in the frameworks of axiomatic systems and frequently neglect to look at the relations of theories to reality. About this topic we may read very interesting thoughts in the book of von Mises (1957).

Axiomatic thinking and problem solving have become everyday tasks for those too who apply science. When starting to solve a problem, we collect our variables and constants and identify their relations to one another. Essentially, we set up an axiomatic system. In this we solve the mathematical problems needed for the application. We also develop new, or use already existing algorithms. Problem solutions go via man-machine systems. Nowadays, a machine means an electronic computer. The simplest "machines" that have been used from ancient times are the compass and the ruler, which are analogue machines. When a geometer draws a geometrical object, he uses the man-machine system in which the compass and the ruler constitute the machine. The abacus and, later, the computers of Pascal, Leibniz, Babbage and Zuse (See Korte, 1981, Ceruzzi, 2003) made the solution of more and more complicated problems possible with increasing efficiency. Nowadays, we use the high-performance successors of the first electronic computer ENIAC (Electronic Numerical Integrator and Computer) and its successors developed by Eckert, Mauchly, John von Neumann and others.

The technique of modern problem solution works efficiently. In this – as it has been mentioned – axiomatic thinking is a fundamental and indispensable tool because each problem is described by a special axiomatic system. The general use of axiomatic thinking is one of the great achievements of János Bolyai's revolution.

Thus, the work of János Bolyai created 'novum' not only in geometry, other fields of mathematics, physics and the history of culture, but also opened the way to a new approach of practical problem solution technique.

Bibliography

[1] Alexits Gy. (1952). János Bolyai. *Matematikai Lapok (Mathematical Papers)* 3, 107-110 (in Hungarian).

[2] Alexits Gy. (1953). *The Life and Works of János Bolyai.* MTA Mat. Fiz. Tud. Oszt. Közl. (Publ. of the Math. Phys. Dept. of the H.A.S.) 3, 131-150 (in Hungarian).

[3] Alexits Gy. (1977). *The World of János Bolyai.* Akadémiai Kiadó (Publ. House of the H.A.S.), Budapest (in Hungarian).

[4] Alexits Gy. (1977/1980). *The World of János Bolyai.* Matematikai Lapok (Mathematical Papers), 28, 1-9 (in Hungarian).

[5] Anderson, J.W. (1999), *Hyperbolic Geometry.* Springer Verlag, London.

[6] Ács T. (1997, 2002). *János Bolyai at the Imperial and Royal Academy of Military Engineers,* Vienna. Bolyai János Katonai Műszaki Főiskola (János Bolyai Military Technical College) Budapest (in Hungarian).

[7] Ács T. (1998). *The Fate of An Unknown Competition of the Academy.* Hadtörténeti Közlemények (Proceedings of Military History) 111, 284-291 (in Hungarian).

[8] Ács T. (2004). *The new face of János Bolyai - the military engineer.* Akadémiai Kiadó (Publ. House of the H.A.S.), Budapest (in Hungarian).

[9] Baltzer, R. (1867). *Die Elemente der Mathematik.* Zweite Anflage. Hirzel, Leipzig.

[10] Beke M. (1903). *The Trigonometry of Bolyai.* Mathematikai és Physikai Lapok (Math. Phys. Papers) 12, 30-49 (in Hungarian).

[11] Bell, E. T. (1972). *The Development of Mathematics.* Dover Publications Inc., New York.

[12] Bell, E. T. (1986). *Men of Mathematics.* Simon and Schuster, New York.

[13] Beltrami, E. (1868). *Saggio di interpretazione della geometria non-euclidea.* Giornale di Matematiche VI., 284-312.

[14] Beltrami, E. (1868). *Teoria fondamentale degli spazii di curvatura costante.* Ann. Mat. Pura Appl. Ser. II, 232-255.

[15] Benkő S. (1968). *The Confessions of János Bolyai.* Irodalmi Könyvkiadó (Literary Publisher), Bucharest, 2nd ed. Kritérion Publisher, Bucharest, 1972, 3rd ed. Mundus Publisher, Budapest, 2002 (in Hungarian).

[16] Benkő S. (1971). *Fate-forming Mind.* Papers in Cultural History. Kritérion Publisher, Bucharest (in Hungarian).

[17] Benkő S. (1978). *Father and Son. (Bolyai Studies).* Magvető Publisher, Budapest (in Hungarian).

[18] Benkő S. (1979). *Progress and Survival. Studies in Cultural History.* Szépirodalmi Publisher, Budapest (in Hungarian).

[19] Bolyai F. (1830). *Elements of Arithmetics.* Marosvásárhely (in Hungarian).

[20] Bolyai Farkas (1832-1833). *Tentamen juventutem studiosam in elementa matheseos purae, elementaris ac sublimioris, methodo intuitiva, evidentiaque huic propria, introducendi.* Cum Appendice triplici. I., II. Maros Vásárhely.

[21] Bolyai Farkas (1897, 1904). *Tentamen.* (2nd ed. Kőnig Gy. and Réthy M. eds. Vol. 1; Kürschák J., Réthy M. and Tőttössy B. eds. Vol. 2). Hungarian Academy of Sciences, Budapest.

[22] *Correspondence of Bolyai and Gauss.* (Selected and ed. by Nagy F.) Better, Püski Publisher, Budapest (in Hungarian).

[23] Bolyai Joannes (1831). *Appendix, Scientiam Spatii absolute veram exhibens; a veritate aut falsitate axiomatis XI. Euclidei (a priori haud unquam decidenda) independentem; adjecta ad casum falsitatis quadratura circuli geometrica.* Maros Vásárhely.

[24] Bolyai Joannes (1832). *Appendix, Scientiam Spatii absolute veram exhibens: a veritate aut falsitate Axiomatis XI. Euclidei (a priori haud unquam decidenda) independentem: adjecta ad casum falsitatis quadratura circuli geometrica.* In: Bolyai F., Tentamen, Tom. I. Maros Vásárhely.

[25] Bolyai János (1868). *La science absolue de l'espace...* (Transl. from Latin by Hoüel.) Bordeaux Mem. 5, 207-248.

[26] Bolyai János (1868). *Sulla scienza della spazio assolutamente vera...* (Transl. from Latin by Battaglini). Giorn. Mat. 6, 97-116.

[27] Bolyai János (1891). *The Science of Absolute Space...* (Transl. by G. B. Halsted). Austin, Texas. Further eds. 1892, 1894, 1896.

[28] *The Bequest of the Manuscripts of János Bolyai.* Bolyai-Teleki Library, Marosvásárhely.

[29] *Bolyai Letters (1975).* Selected, the introduction written and notes compiled by Samu Benkő. Kritérion Publisher, Bucharest (in Hungarian).

[30] Bolyai János (1977, 1987). *Appendix, the Science of Space.* (Kárteszi Ferenc ed.) Publ. House of the H.A.S., Budapest (in Hungarian, English Translation).

[31] *Remembering János Bolyai on the 175th Anniversary of His Birth.* (Staar Gyula ed.). Budapest Branch of the Society for the Popularization of Scientific Knowledge, Budapest, 1978 (in Hungarian).

[32] *Bolyai (2001). Biography, Library and Bibliography.* (Nagy F. ed.) Better, Püski Publisher, Budapest (in Hungarian).

[33] Bonola, R. (1911). *Non-Euclidean Geometry.* Dover, New York.

[34] Boole, G. (1854). *An Investigation of the Laws of Thought on which are Founded the Mathematical Theories of Logic and Probabilities.* Macmillan, Cambridge.

[35] Born, M. (1965). *Einstein's Theory of Relativity.* Dover, New York.

[36] Bos, H. J. M. (2001). *Redefining Geometrical Exactness. Descartes Transformation of the Early Modern Concept of Construction.* Springer, New York.

[37] Boyer, C. B. (1991). *A History of Mathematics.* (2nd ed.). Wiley, New York.

[38] Brentjes, S. (1982). *Relations between Mathematics and Engineering in the 19th Century in Germany* (manuscript.)

[39] Bretter Gy. (1978). *The Logic of Wearing Out.* Természet Világa (The World of Nature), 109, 92-93 (in Hungarian).

[40] *Briefwechsel zwischen Carl Friedrich Gauss und Christian Ludwig Gerling (1927).* Clemens Shaeffer, Berlin.

[41] *Briefwechsel zwischen C. F. Gauss – Friedrich Wilhelm Bessel (1975).* Georg Olms Verlag, Hildesheim, New York.

[42] *Briefwechsel zwischen C. F. Gauss und H. C. Schumacher, l-3 (1975).* Georg Olms Verlag, Hildesheim, New York.

[43] Burton, D. M. (1985). *The History of Mathematics. An Introduction.* Allyn and Bacon Inc. Boston, London.

[44] Cajori, F. (1991). *A History of Mathematics.* Chelsea Publ. Co., New York.

[45] Caygill, H. (2000). *A Kant Dictionary.* Blackwell, New York.

[46] Cayley, A. (1854). *On the Theory of Groups.* Philosophical Magazine 7, 40-47.

[47] Ceruzzi, P. E. (2003). *A History of Modern Computing* (2nd edition). The MIT Press, Cambridge, MA.

[48] Coxeter, H.S.M. (1977). *Gauss as a Geometer.* Historia Mathematica 4, 379-396.

[49] Coxeter, H.S.M. (1979). *The non-Euclidean symmetry of Escher's picture 'Circle Limit III'.* Leonardo 12, 19-25.

[50] Coxeter, H.S.M. (1998). *Non-Euclidean Geometry.* (6th ed.). Mathematical Association of America, Washington, D.C.

[51] Császár Á. (1978). *János Bolyai and Gauss.* Természet Világa (The World of Nature), 109, 89-92 (in Hungarian).

[52] Dávid L. (1959). *In memoriam Wolfgang Bolyai.* MTA Mat. Fiz. Tud. Oszt. Közl. (Publ. of the Math. Phys. Dept. of the H.A.S.) 9, 215-236 (in Hungarian).

[53] Dávid L. (1979). *The lives and works of the two Bolyais.* (2nd enlarged ed.). Gondolat Publisher, Budapest (in Hungarian).

[54] Deé Nagy Anikó (2000). *The library of the two Bolyais.* Erdélyi Múzeum (Museum of Transylvania) Nos 1-2, 19-45 (in Hungarian).

[55] Dunnington, G. W. (1955). *Gauss: Titan of Science.* Hafner, New York.

[56] Escher, M. C. (1989). *Escher on Escher: exploring the infinite.* Harry N. Abrams, New York. With contribution by J. W. Vermuelen; transl. by Karin Ford.

[57] Escher, M. C. (1992). *The Graphic Work: Introduced and Explained by the Artist.* Benedikt Tashen Verlag.

[58] Escher, M. C. (1995). *The M. C. Escher sticker book.* Harry N. Abrams, New York.

[59] Ewald, V. (1999). *From Kant to Hilbert.* A Source Book in the Foundations of Mathematics Vols, I, II. Clarendon Press, Oxford.

[60] Euclid (1956). *Elements.* Vols 1-3. Dover, New York.

[61] Fowler, D. (1999). *The Mathematics of Plato's Academy. A new reconstruction.* (2nd ed.) Clarendon Press, Oxford.

[62] Fráter Jánosné (1968). *The Bolyai Collection.* The Library of the H.A.S. (K22-K30) (in Hungarian).

[63] Frege, G. (1884). *Die Grundlagen der Arithmetik, eine logisch-mathematische Untersuchung über den Begriff der Zahl.* Koebner, Breslau. (English transl.: The Foundations of Arithmetic. A Logico-Mathematical Enquiry into the Concept of Number. Basil Blackwell, Oxford, 1950, revised 2nd ed. 1953; translator: John Langshaw).

[64] Gauss, C. F. (1870-1929). *Werke.* Vols. 1-12, (2nd reprint). Göttingen, Gotha, Leipzig, Berlin.

[65] Gödel, K. (1931). *Über formal unentscheidbare Sätze der Principia Mathematica und verwandte Systeme.* Monat. Math. Phys. 38, 173-198.

[66] Grabiner, J. V. (1974). *Is Mathematical Truth Time-Dependent?* American Math. Monthly 81, 354-365.

[67] Grattan-Guinness, I. (1998). *The Norton History of the Mathematical Sciences.* Norton and Co. New York.

[68] Gray, J. (1979). *Ideas of Space.* Clarendon Press, Oxford.

[69] Gray, J. (1979). *Non-Euclidean Geometry – A Re-Interpretation.* Historia Mathematica 6, 236-258.

[70] Greenberg, M. J. (1993). *Euclidean and Non-Euclidean Geometries.* (3rd ed.). Springer, New York.

[71] Gregory, D. F. (1840). *On the nature of symbolical algebra.* Transactions of the Royal Society of Edinburgh 14, 208-216.

[72] Gunn, C. (1993). *Discrete groups and visualization of three-dimensional manifolds.* In: ACM SIGGRAPH Computer Graphics Proceedings, Annual Conference Series, 255-262.

[73] Hadamard, J. (1953). *Non-Euclidean Geometry and Axiomatic Definitions.* MTA Mat. Fiz. Tud. Oszt. Közl. (Publ. of the Math. Phys. Dept. of the H.A.S.) 3, 199-208. (in Hungarian).

[74] Hadamard, J. (1999). *Non-Euclidean Geometry in the Theory of Automorphic Functions.* (J. J. Gray and A. Shenitzer eds.). History of Math. Vol. 17, Amer. Math. Soc., London Math. Soc.

[75] Hamilton, W. R. (1837). *Theory of conjugate functions, or algebraic couples; with a preliminary and elementary essay on algebra as the science of pure time.* Transactions of the Royal Irish Academy 17, 293-422.

[76] Hamilton, W. R. (1853). *Lectures on Quaternions.* Hodges and Smith, Dublin.

[77] Hamilton, W. R. (1931-1967). *The Mathematical Papers of Sir William Hamilton.* (H. Halberstam and R. E. Ingram eds.). Cambridge University Press.

[78] Hankel, H. (1884). *Die Entwicklung der Mathematik in den letzten Jahrhunderten.* (2nd ed.) Verlag und Druck von Franz Vues, Tübingen.

[79] Hartshorne, R. (2000). *Geometry: Euclid and Beyond.* Springer, New York.

[80] Helmholz, H. (1878). *Die Tatsachen in der Wahrnehmung.* Berlin.

[81] Helmholz, H. (1884). *Vorträge und Reden I, II.* Vieweg, Braunschweig.

[82] Hollingsdale, S. (1994). *Makers of Mathematics.* Penguin Group, New York.

[83] Iversen, B. (1992). *Hyperbolic geometry.* Cambridge University Press, Cambridge, UK.

[84] Jelitai J. (1939). *János Bolyai's Draft Report of 13 May 1849.* Matematikai és Természettudományi Értesítő (Math. Sci. Bulletin) 58, 708-715 (in Hungarian).

[85] Kaestner, A. G. (1790). *Was heisst in Euclids Geometrie möglich?* Philosophisches Magazin herausgegeben von J. A. Eberhardt. Halle a. S.

[86] Kalmár L. (1953). *The Influence of the Bolyai-Lobachevskii Geometry on the Development of Axiomatic Method.* MTA Mat. Fiz. Tud. Oszt. Közl. (Publ. of the Math. Phys. Dept. of the H.A.S.), 3, 235-242 (in Hungarian).

[87] Kant, I. (1781, 1787). *Kritik der reinen Vernunft.* (1st, 2nd eds.) Johann Friedrich Hartknoch, Riga.

[88] Kant, I. (1788). *Kritik der praktischen Vernunft.* Johann Friedrich Hartknoch, Riga.

[89] Kant, I. (1790). *Erste Einleitung in die Kritik der Urteilskraft.* I. Kant's Works Vol. 5 Publ. by E. Cassirer, 1922.

[90] Kárteszi F. (1953). *The Life and Works of Lobachevskii.* MTA Mat. Fiz. Tud. Oszt. Közl. (Publ. of the Math. Phys. Dept. of the H.A.S.) 3, 189-197 (in Hungarian).

[91] Kárteszi F. (ed., 1951). *Geometrical Studies in the Field of the Theory of Parallels.* (With Introduction, Comments and Supplement by V. F. Kagan.) Publ. House of the H.A.S., Budapest (in Hungarian).

[92] Kárteszi F. (1976). *Bolyai's Lifework.* Matematikai Lapok (Mathematical Papers) 27, 59-63 (in Hungarian).

[93] Kárteszi F. (1980/1983). *The precedents and scientific results of Appendix.* Matematikai Lapok (Mathematical Papers) 31, 15-22 (in Hungarian).

[94] Kerékjártó B. (1937). *On fundamentals of geometry I, The elementary foundation of Euclidean geometry,* Publ. House of the H.A.S., Szeged (in Hungarian)

[95] Kerékjártó B. (1944). *On fundamentals of geometry II, Projective geometry,* Publ. House of the H.A.S., Budapest (in Hungarian).

[96] Kerékjártó B. (1955). *Les fondements de la géométrie I, La construction élémentaire de la géométrie Euclidienne,* Maison d'édition de l'Académie des Sciences de Hongrie.

[97] Kerékjártó B. (1966). *Les fondements de la géométrie II, Géométrie projective,* Maison d'édition de l'Académie des Sciences de Hongrie.

[98] Kiss E. (1994). *The Latest Secrets of the 'Bolyai Chests'.* Természet Világa (The World of Nature) 125, 405-408 (in Hungarian).

[99] Kiss E. (1999). *Mathematical gems from the Bolyai chests.* Academic Publ. and Typotex Publ. Budapest.

[100] Kiss E. (1999). *Notes on János Bolyai's Researches in Number Theory.* Historia Mathematica 26, 68-76.

[101] Klein, F. (1872). *Vergleichende Betrachtungen über neuere geometrische Forschungen (Erlanger Programm).* Akademische Verlagsgesellshaft, Leipzig, 1974.

[102] Klein, F. (1873). *Über die sogenannte Nicht-Euklidische Geometrie.* Math. Ann. 6, 112-145.

[103] Klein, F. (1928). *Vorlesungen über Nichteuklidische Geometrie.* Springer, Berlin.

[104] Kline, M. (1990). *Mathematical Thought from Ancient to Modern Times I., II., III.* Oxford University Press, Oxford UK.

[105] Koncz J. (1887). *The history of one hundred years of the Printing House of the Lutheran and Calvinist College of Marosvásárhely.* Marosvásárhely (in Hungarian).

[106] Koncz J. (1896). *The history of the Lutheran and Calvinist College of Marosvásárhely.* Marosvásárhely (in Hungarian).

[107] Kosáry D. (2001). *The Reconstruction and the Development of Bourgeois Civilization 1711-1867.* História/Holnap Publisher, Budapest (in Hungarian).

[108] Korte, B. (1981). *Zur Geschichte des maschinelles Rechnens.* Bouvier Verlag, Herbert Grundmann, Bonn.

[109] Lambert, J. H. (1786). *Theorie der Parallellinien.* In: Engel und Stäckel (1895).

[110] Lambrecht M. (1973). *János Bolyai as a Romantic Hero.* Természet Világa (The World of Nature) 104, 504-505 (in Hungarian).

[111] Lánczos K. (1976). *The Development of the Concept of Geometric Space.* Gondolat Publisher, Budapest (in Hungarian).

[112] Legendre, A. M. (1794). *Eléments de géométrie.* Paris.

[113] Lévy, S. ed. (2001). *The eightfold Way. The beauty of Klein's quartic curve.* Cambridge University Press, Cambridge, UK.

[114] Lobachevskii, N. I. (1837). *Géometrie imaginaire.* J. reine und angewandte Math. 17, 295-320.

[115] Lobachevskii, N. I. (1840). *Geometrische Untersuchungen zur Theorie der Parallellinien.* Berlin.

[116] Lobachevskii, N. I. (1951). *Geometrical Studies in the Field of the Theory of Parallels.* (With the Introduction, Comments and Supplement by V.F. Kagan.) Publ. House of the H.A.S., Budapest (in Hungarian).

[117] Milnor, J. (1982). *Hyperbolic geometry: The first 150 years.* Bull. (New Series) Amer. Math. Soc. 6, 9-24.

[118] Minding F. (1838). *Über die Biegung krummer Flächen.* J. reine und angewandte Math. 18, 365-368.

[119] von Mises, R. (1981). *Probability, statistics and truth.* Dover, New York.

[120] Molnár E. (1975). *János Bolyai and the 'Science of Space'.* Természet Világa (The World of Nature) 106, 469-470 (in Hungarian).

[121] Nagy I. (1858). *Families of Hungary with coats of arms and genealogical trees.* Vol. 2 (in Hungarian).

[122] Orbán B. (1868). *The description of Székely land in eastern Transylvania from historical, archeological and ethnological aspects.* Ráth Mór's Consignation. Pest (in Hungarian).

[123] Pálffy I. and Pálffy M. (1962). *Bibliographia Bolyaiana 1831-1960.* The list of publications on the Bolyai geometry. National Széchényi Library, Central Library of the Technical University of Budapest (in Hungarian).

[124] Peacock, G. (1830). *A Treatise on Algebra.* Deighton, Cambridge.

[125] Poincaré, H. (1960). *Mathematical Creation.* In: The World of Mathematics, Vol. 4 (J. R. Newman ed.). George Allen and Unwin Ltd., London, 2041-2050.

[126] Poincaré, H. (1985). *Papers on Fuchsian Functions.* Springer Verlag.

[127] Poincaré, H. (1905, 2001). *La valeur de la science.* English transl.: The value of science. The Modern Library, New York.

[128] Redtenbacher, F. (1852). *Prinzipien der Mechanik und des Maschinenbaus.* Mannheim.

[129] Rényi A. (1952). *János Bolyai the great revolutionary of science.* Matematikai Lapok (Mathematical Papers) 3, 174-178 (in Hungarian).

[130] Réthy M. (1903). *Discussion of 'the new other universe' of János Bolyai.* Mathematikai és Physikai Lapok (Math. Phys. Papers) 12, 1-29 (in Hungarian).

[131] Reuleaux, F. (1875). *Theoretische Kinematik.* Braunschweig.

[132] Reynolds, W. F. (1993). *Hyperbolic Geometry on a Hyperboloid.* American Mathematical Monthly 100, 442-455.

[133] Riemann, B. (1868). *Über die Hypothesen, welche der Geometrie zu Grunde liegen.* Abhandlungen der Königlichen Gesellschaft der Wissenschaften zu Göttingen 13, 133-152.

[134] Riemann, B. (1876). *Gesammelte Mathematische Werke und Wissenschaftlicher Nachlass.* (H. Weber and R. Dedekind eds.) Teubner, Leipzig.

[135] Riemann, B. (1921). *Über die Hypothesen, welche der Geometrie zu Grunde liegen.* (New ed. with H. Weyl's notes.) Julius Springer, Berlin.

[136] Rosenfeld, B. A. (1988). *A history of non-Euclidean geometry.* Springer, New York.

[137] Saccheri, G. (1920). *Euclides vindicatus.* (Transl. by G. B. Halsted). Open Court, Chicago.

[138] Sarlóska E. (1965). *János Bolyai – the Soldier.* MTA Mat. Fiz. Tud. Oszt. Közl. (Publ. Math. Phys. Dept. of the H.A.S.) 15, 341-387 (in Hungarian).

[139] Sarlóska E. (1973). *The 150-year old 'Bolyai'.* Természet Világa (The World of Nature) 104, 482-484, 507 (in Hungarian).

[140] Sarlóska E. (1973). *Recollection on Samu Benkő's book.* Természet Világa (The World of Nature) 104, 504 (in Hungarian).

[141] Schlesinger L. (1903). *János Bolyai.* Mathematikai és Physikai Lapok (Math. Phys. Papers) 12, 57-88 (in Hungarian).

[142] Schlesinger L. (1903). *On the birthplace of János Bolyai.* Mathematikai és Physikai Lapok (Math.Phys. Papers) 12, 53-56 (in Hungarian).

[143] Schmidt, F. (1868). *Aus dem Leben zweier ungarischer Mathematiker, Johann und Wolfgang Bolyai von Bolya.* Grünerts Archiv der Math. und Phys. Greifswald 48, 217-228.

[144] Schmidt, F. (1899). *Briefwechsel zwischen Carl Friedrich Gauss und Wolfgang Bolyai.* Lepzig.

[145] Sheehan, J. J. (1994). *German History 1770-1866.* The Oxford History of Modern Europe. Clarendon Press, Oxford.

[146] Staar Gy. (1990). *The mathematics lived, pp. 59-80 : Man with light – Tibor Weszely.* Gondolat Publisher, Budapest (in Hungarian).

[147] Stäckel, P., Engel, F. (1895). *Die Theorie der Parallallinien von Euklid bis auf Gauss.* Teubner, Leipzig.

[148] Stäckel, P. (1902). *Studies in the field of absolute geometry in the Bolyai manuscripts.* Matematikai és Természettudományi Értesítő (Math. Sci. Bulletin) 20, 160-186 (in Hungarian).

[149] Stäckel, P. (1902). *János Bolyai's comments on Nicholas Lobachevskii's investigations concerning parallels.* Matematikai és Természettudományi Értesítő (Math. Sci. Bulletin) 20, 40-67 (in Hungarian).

[150] Stäckel, P. and Kürschák J. (1902). *János Bolyai's comments on Nicholas Lobachevskii's investigations concerning parallels.* Matematikai és Természettudományi Értesítő (Math. Sci. Bulletin) 20, 41-67. (in Hungarian).

[151] Stäckel, P. (1903). *János Bolyai's theory of space.* Matematikai és Természettudományi Értesítő (Math. Sci. Bulletin) 21, 135-145 (in Hungarian).

[152] Stäckel, P. (1913). *Wolfgang und Johann Bolyai.* Geometrische Untersuchungen, Vols. 1,2. Teubner, Leipzig.

[153] Stillwell, J. (1996). *Sources of Hyperbolic Geometry.* History of Math. Vol. 10, Amer. Math. Soc., London Math. Soc.

[154] Strommer Gy. (1977/1980). *On the theory of construction of Bolyai geometry.* Matematikai Lapok (Mathematical Papers) 28, 65-67 (in Hungarian).

[155] Szabó P. (1903). *On a basic theorem of absolute geometry.* Mathematikai és Physikai Lapok (Math. Phys. Papers) 12, 321-326 (in Hungarian).

[156] Szalay, A. S., Gray, J. (2001). *The World Wide Telescope.* Science, 203, 2037-2038.

[157] Szász P. (1973). *An introduction to the Bolyai-Lobachevskii geometry.* Publ. House of the H.A.S., Budapest (in Hungarian).

[158] Szász P. (1953). *The various elementary presentations of hyperbolic trigonometry.* MTA Mat. Fiz. Tud. Oszt. Közl. (Publ. Math. Phys. Dept. of the H.A.S.) 3, 209-225 (in Hungarian).

[159] Szénássy B. (1960). *János Bolyai.* A Matematika Tanítása (Teaching Mathematics) 7, 34-39 (in Hungarian).

[160] Szénássy B. (1970). *The history of mathematics in Hungary.* (From ancient times to the early 20th century). Publ. House of the H.A.S., Budapest (in Hungarian).

[161] Szénássy B. (1975). *Farkas Bolyai.* Publ. House of the H.A.S., Budapest (in Hungarian).

[162] Szénássy B. (1978). *Questions and answers.* In: Bolyai Jánosra emlékezünk! TIT Budapesti Szervezete Matematikai Szakosztályának Kiadványa (Remembering János Bolyai. Publication of the Mathematical Dept. of the Budapest Branch of the Society for the Popularization of Scientific Knowledge), 29-40 (in Hungarian).

[163] Szénássy B. (1978). *János Bolyai.* Publ. House of the H.A.S., Budapest (in Hungarian).

[164] Szénássy B. (1977/1980). *Comments on Gauss' results on non-Euclidean geometry.* Matematikai Lapok (Mathematical Papers) 28, 133-140 (in Hungarian).

[165] Szénássy B. (1977/1980). *Contributions to the history of the discovery of the two Bolyais.* Matematikai Lapok (Mathematical Papers) 29, 71-95 (in Hungarian).

[166] Szénássy B. (1980/1983). *The careers of the Bolyais and the scientific significance of Tentamen.* Matematikai Lapok (Mathematical Papers) 31, 3-14 (in Hungarian).

[167] Szőkefalvi-Nagy Gy. (1953). *János Bolyai's trisection of angles.* Matematikai Lapok (Mathematical Papers) 4, 84-86 (in Hungarian).

[168] Taurinus, F. A. (1826). *Geometriae prima elementa.* Köln.

[169] Thom, R. (1971). *'Modern' Mathematics: An Educational and Philosophic Error?* American Scientist 59, 695-699.

[170] Thurston, W. P. (1982). *Three-dimensional manifolds, Kleinian groups and hyperbolic geometry.* Bull. Amer. Math. Soc. (New Series) 6, 357-381.

[171] Thurston, W. P. (1997). *Three-Dimensional Geometry and Topology.* Vol. I. (Silvio Levy ed.). Princeton University Press, Princeton, N.J.

[172] Toró T. (1973). *Non-Euclidean geometries in modern physics and relativistic cosmology.* Korunk Annual, Kolozsvár, 245-256 (in Hungarian).

[173] Torretti, R. (1996). *Relativity and Geometry.* Dover, New York.

[174] Tóth Imre (1965). *From the prehistory of non-Euclidean geometry.* Matematikai Lapok (Mathematical Papers) 16, 300-315 (in Hungarian).

[175] Tóth Imre (2000). *God and geometry.* Osiris, Budapest (in Hungarian).

[176] Tymoczko, T. (1998). *New Directions in the Philosophy of Mathematics (revised and expanded paperback edition).* Princeton University Press, Princeton N.J.

[177] Ungar, A. (1997). *Thomas Precession: Its Underlying Gyrogroup Axioms and their Use in Hyperbolic Geometry and Relativistic Physics.* Foundations of Physics 27, 881-951.

[178] Ungar, A. (1999). *The Hyperbolic Pythagorean Theorem in the Poincaré Disk Model of Hyperbolic Geometry.* The American Mathematical Monthly 106, 759-763.

[179] Varga O. (1953). *Impact of the Bolyai-Lobachevskii Geometry on the Development of Geometry.* MTA Mat. Fiz. Tud. Oszt. Közl. (Publ. of Math. Phys. Dept. of the H.A.S.) 3, 151-171 (in Hungarian).

[180] Vállas A. (1836). *The History of Literature.* Tudománytár (Archive of Science), 143-17 (in Hungarian).

[181] Vekerdi L. (1969). *Wandering about the history of sciences.* Magvető Publisher, Budapest (in Hungarian).

[182] Vekerdi L. (1981). *Changes in the Bolyai research.* Természet Világa (The World of Nature) 112, 56-58 (in Hungarian).

[183] Vekerdi L. (2001). *The Bolyai Collection in the Bolyai Research.* Our Heritage, Our Living Past. Collections in the Library of the Hungarian Academy of Sciences. Library of the H.A.S., Budapest 85-114 (in Hungarian).

[184] Weszely T. (1974). *Farkas Bolyai the mathematician.* Tudományos Könyvkiadó (Scientific Publisher), Bucharest (in Hungarian).

[185] Weszely T. (1981). *Mathematical works of János Bolyai.* Kritérion Publisher, Bucharest (in Hungarian).

[186] Weszely T. (1980/1983). *On János Bolyai's mathematical works left behind in manuscripts.* Matematikai Lapok (Mathematical Papers) 31, 29-37 (in Hungarian).

[187] Weszely T. (1985). *In Remembrance of János Bolyai.* In: Bolyai Jánosra emlékezünk! TIT Budapesti Szervezete Matematikai Szakosztályának Kiadványa (Remembering János Bolyai. Publication of the Mathematical Dept. of the Budapest Branch of the Society for the Popularization of Scientific Knowledge).

[188] Weszely T. (2002). *János Bolyai.* The first 200 years. Vince Publisher, Budapest (in Hungarian).

GAUSS AND NON-EUCLIDEAN GEOMETRY

Jeremy Gray
Centre for the History of the Mathematical Sciences
Open University
Milton Keynes, MK7 6AA
U.K.

Introduction

The claim, made on Gauss's behalf, that he was a, or even the, discoverer of non-Euclidean geometry is very hard to decide because the evidence is so slight. It is nonetheless implicit in the excellent commentaries of Stäckel [23] and Dombrowski [7], as it is in Reichardt's book [21] and the broader but slighter survey by Coxeter [6]. Proponents of this view, with Dunnington, are happy to tie documents written in the late 1820s and 1830s to cryptic claims made by Gauss for early achievements, and to equally elusive passages from the 1810s. In fact, the evidence points in another direction. It suggests that Gauss was aware that much needed to be done to Euclid's *Elements* to make them rigorous, and that the geometrical nature of physical Space was regarded by Gauss as more and more likely to be an empirical matter, but in this his instincts and insights on this occasion were those of a scientist, not a mathematician.

I shall also argue that the whole question of what it is to discover a new geometry of space requires much careful thought, and if definitive agreement on the matter cannot be reached – and perhaps it cannot – then at least positions can profitably be made explicit. This has not usually been done when discussing the discovery of non-Euclidean geometry.

1. The evidence

Gauss was 22 when he confided to Wolfgang Bolyai that he was doubtful of the truth of geometry. He had already found too many mistakes in other people's arguments in defence of the parallel postulate to be so confident any longer in their conclusion. He had begun to consider the fundamental assumptions of geometry at least two years earlier, in July

27, 1797, when he wrote in his Mathematical Diary only too cryptically that he had 'demonstrated the possibility of a plane'. It is tempting to connect this with fragments of arguments dating from 1828 to 1832 in which Gauss investigated whether the locus of a line perpendicular to a fixed line and rotating about that fixed line has all the properties of a plane, because, in a famous letter to Bessel of January 1829, where Gauss claims to have harboured these thoughts for almost 40 years, he wrote that

"apart from the well-known gap in Euclid's geometry, there is another that, to my knowledge no-one has noticed and which is in no way easy to alleviate (although possible). This is the definition of a plane as a surface that contains the line joining any two of its points. This definition contains more than is necessary for the determination of the surface, and tacitly involves a theorem which must first be proved..." [8, Gauss to Bessel, VIII, p. 200].

One knows from the later history of geometry, most clearly from the remarks of Pasch ([20]) that trying to spell out what exactly elementary Euclidean geometry is about is extremely difficult.

By 1808 Gauss was aware that in the hypothetical non-Euclidean geometry similar triangles are congruent, and therefore there is an absolute measure of length. But at this stage, according to Schumacher, he found this conclusion absurd, and therefore held that the matter was still unclear. As he put it in 1813: "In the theory of parallels we are no further than Euclid was. This is the shameful part of mathematics, that sooner or later must be put in quite another form". Evidently he did not then feel confident in a non-Euclidean geometry. By April 1816 he had shifted his opinion

"It seems to be something of a paradox that a constant line can at the same time be given a priori, but I find nothing self-contradictory in this. It would be remarkable if Euclid's geometry were not true, because then we would have a general a priori measure [of length], for example one could take as the spatial unit the length of the side of an equilateral triangle with angle $59°; 59'59.99999''$ [8, Gauss to Gerling, VIII, pp. 168-169].

As Dunnington correctly observed, being remarkable is consistent with being attractive. But still there is no evidence that Gauss deduced anything specific about the new geometry.

In 1816 we do get a glimpse of what Gauss knew as reported by his former student Wachter. On a certain (unspecified) hypothesis, Wachter wrote to Gauss, the opposite of Euclidean geometry would apparently be true, which would involve us with an undetermined constant, a sphere of infinite radius which nonetheless lacks some properties of the plane, and the use of a transcendent trigonometry that probably generalises or underpins spherical trigonometry. Gauss now, as he wrote to Olbers in April 1817, was coming

"ever more to the opinion that the necessity of our geometry cannot be proved, at least not with human understanding. Perhaps in another life. ... but for now geometry must stand, not with arithmetic which is pure a priori, but with mechanics." [8, Gauss to Olbers, VIII, pp. 177].

The 'transcendent trigonometry' is usually taken to be the hyperbolic trigonometry appropriate to non-Euclidean geometry, but there is very little evidence to support any interpretation. Accordingly, when Gauss replied to Schweikart in March 1819 that he could "do all of astral geometry once the constant is given" we cannot be sure what precisely, Gauss had formulae for. The only one dated to this period is the one in his reply to Schweikart for the maximum area of a triangle in terms of Schweikart's Constant (the maximum altitude of an isosceles right-angled triangle).

All that his correspondence with Taurinus reveals is that, by 1824, Gauss was more comfortable than ever with the idea of a new geometry. In November 1824, in the course of explaining his views about non-Euclidean geometry, and his reluctance to be drawn in public, Gauss wrote to Taurinus that

"... the assumption that the angle sum is less than 180° leads to a geometry quite different from Euclid's, logically coherent, and one that I am entirely satisfied with. It depends on a constant, which is not given a priori. The larger the constant, the closer the geometry to Euclid's and when the constant is infinite they agree. The theorems are paradoxical but not self-contradictory or illogical. [...] All my efforts to find a contradiction have failed, the only thing that our understanding finds contradictory is that, if the geometry were to be true, there would be an absolute (if unknown to us) measure of length [...] As a joke I've even wished Euclidean geometry was not true, for then we would have an absolute measure of length a priori." [8, Gauss to Taurinus, VIII, pp. 187].

But, Gauss went on, to reject the geometry on that ground would be to confuse the unnatural with the absolutely impossible.

There is very little evidence of Gaussian contributions to trigonometry in non-Euclidean geometry before the letter to Schumacher of 12 July 1831, where he says that the circumference of a semi-circle is $\frac{1}{2}\pi k \left(e^{r/k} - e^{-r/k}\right)$ where k is a very large constant that is infinite in Euclidean geometry. In particular, there is no evidence that Gauss derived the relevant trigonometric formulae from the profound study of differential geometry that occupied him in the 1820s. What he did say in the *Disquisitiones generales circa superficies curvas* [1828] is summed up in what he regarded as one of the most elegant theorems in the theory of curved surfaces:

"The excess over 180° of the sum of the angles of a triangle formed by shortest lines on a concavo-concave surface, or the deficit from 180° of the sum of the angles of a triangle formed by shortest lines on a concavo-convex surface, is measured by the area of the part of the sphere which corresponds, through the direction of the normals, to that triangle, if the whole surface of the sphere is set equal to 720 degrees." [8, IV, p. 246]

What weight can two trigonometric formulae be made to carry? They are not difficult to obtain and manipulate if, as for example Taurinus did, one assumes that non-Euclidean geometry is described by the formulae of hyperbolic trigonometry – a natural enough assumption. To introduce hyperbolic trigonometry into the study of non-Euclidean geometry properly is, as Bolyai and Lobachevskii found, a considerable labour of which no trace remains in Gauss's work. It is more plausible to imagine that he made the assumption, but did not derive it from basic principles. So, perhaps by 1816, or, at the latest, 1824, Gauss was convinced of ideas like these:

- there could be a non-Euclidean geometry, in which the angle sum of triangles is less than π,
- the area of triangles is proportional to their angular defect and is bounded by a finite amount,
- the trigonometric formulae for this geometry are those of hyperbolic trigonometry, and the analogy with spherical geometry and trigonometry extends to formulae for the circumference and area of circles.

2. The question of the empirical test

Did Gauss, however, as a scientist, make an empirical test of the matter? This is one of the most discussed questions in the whole subject of Gauss and non-Euclidean geometry. Those who believe that he did quote Sartorius von Waltershausen's reminiscence, where on p. 81, he states that Gauss did check the truth of Euclidean geometry on measurements of the triangle formed by the mountains Brocken, Hohenhagen, and Inselsberg (BHI), and found it to be approximately true. This claim was most recently advanced by Scholz [22], on the basis of a number, quoted by von Waltershausen elsewhere in his reminiscence, relating to the very close agreement between the measurements of this triangle and the predictions of Euclidean geometry (once the mountain tops are treated as three points on a sphere). Scholz concludes that "there is no longer any reason to doubt that Gauss *himself* conducted such a test of the angle sum theorem." (Scholz [22, p. 644]).

Those who dispute that Gauss made such a test argue that the problem that occupied Gauss, and figures so prominently at the end of the *Disquisitiones generales circa superficies curvas*, is the question of the spheroidal or spherical shape of the Earth, and that von Waltershausen was simply confused about the hypothesis that Gauss found to be approximately confirmed. This is the opinion of Miller [17].

The most thorough analysis of the question is Breitenberger [5]. He confronts the question: 'if von Waltershausen was not simply confused in some way, what was he saying?' and he gives it an elegant answer. Surveying Hanover threw up many triangles and many numbers (a figure of a million is sometimes mentioned). Conclusions were drawn (and maps made) on numbers which are the result of many calculations, and at every stage discrepancies between real and expected results lay within expected error bounds (Gauss analysed the errors quire carefully). Not only was Euclidean geometry never called into question, because the errors were only what was to be expected, each calculation amounted to a tacit defence of Euclidean geometry. But the measurements of the BHI triangle were not fed into such a mill. They show that, within experimental error, Space is described by Euclidean geometry. To be sure "as a single instance it proves very little, but it has been designed so as to be transparent, and hence it will drive a point home" (Breitenberger [5, p. 288]). Newton dropped an apple in conversation to similar purpose and effect. The myth, Breitenberger concludes, is that the BHI triangle was surveyed as part of a *deliberate* test of Euclidean geometry. But it did incidentally show that Euclidean geometry is true to within the limits of the best observational error of the time. Put that way, the gap between Scholz and Breitenberger may be quite small.

3. Interpretations

What then is it to discover – or, if you prefer, invent – non-Euclidean geometry? One way of thinking about this question must be ruled out straight away. Had it just been a question of exhibiting an axiom system for something fairly geometrical, then spherical geometry would have done. One needs, of course, to strike out two of Euclid's axioms: the parallel postulate and the indefinite extendibility of the straight line. That this was not done suggests that the question in 1800 was not one about 'axioms for geometry'. It makes it clear that what was to be investigated was the geometry of physical space. The ongoing question was not 'Is the parallel postulate independent of the other axioms of geometry'?, but 'Is the parallel postulate independent of the other axioms of geometry when giving an account of space?'. This is a different enterprise from the much more overtly logical one in fashion around 1900. But in 1900 axioms were very fashionable, Hilbert's *Grundlagen der Geometrie* was not the only book to exemplify the merits of thinking axiomatically, and from first to last, Bonola's account of the origin and development of non-Euclidean geometry is rooted in an analysis of axioms; their equivalence and their independence. It was published in Italian in 1906 and has become the standard account of the subject in English (into which it was translated in 1912), and it has many merits. Indeed, an Italian geometer, and a pupil of Enriques, writing between 1900 and 1911, would naturally see geometry as a matter of axioms, and so see history as a history of axioms.

It will bring the question into sharper focus if we ask a seemingly absurd other question: Why do we not simply say that non-Euclidean geometry was discovered by Ferdinand Karl Schweikart, the professor of Jurisprudence at Marburg, and communicated to Gauss in 1818? If the answer is simply that Gauss already knew what Schweikart told him, then what Gauss knew in 1818 counts as a discovery, and Bolyai and Lobachevskii are condemned to come second. They have only the honour of being the first to publish – and much good did it do them. A better reply would be that what Schweikart did is too flimsy, that it had no chance of convincing anyone, and does not amount to the sort of activity that counts as a discovery. It is validated only after the event, by the subsequent discovery of a system of geometry that is, in all its glory and detail, what Schweikart had merely glimpsed. Nothing in his account should have swayed the prudent sceptic, unwilling to abandon Euclid for a wish list of new figures.

Any one seriously interested in the question of non-Euclidean geometry by the 1810s could have found out quite a lot just by asking around and reading in a decent library. Certainly, people were interested. 'Not a year goes by', Gauss himself wrote in a book review, 'without a book being published on the parallel postulate' [8, VIII, p. 170]. The problem had a striking degree of visibility. Even if one assumes, as is likely, that German authors only read Germans, and the French only French, that leaves the many attempts of Legendre and the open-ended speculations of Lambert. Both of these were major authors, and one supposes that Legendre's repeated attempts hinted clearly enough that he was not convinced of any but his most recent version. Lambert, even more clearly, knew that his attempts had not led to the defence of the parallel postulate that he sought, and that the matter remained unresolved. If one concluded that all this failure was in the nature of the problem, that indeed a specific non-Euclidean geometry was possible, one could copy out quite a number of oddities and simply proclaim them, not as steps on the way to the final refutation, but as fine new theorems. One could be courageous in so doing, or simply giving up too soon from a failure to see deeply enough into the problem.

The fairest answer is surely that to discover non-Euclidean geometry is to describe a system of geometry that is convincing and persuasive (even if it did not, in its day, persuade, and even if by the much higher standards of a later time it was in fact imperfect). Applied to Schweikart and his memorandum, this leads to what is indeed the charitable consensus, which is to grant Schweikart the courage, and the freshness of mind to see what too many had not been willing to see (and what, indeed, his nephew Taurinus was to avert his gaze from some six years later). But one cannot find in his work that persuasive character required in a true discoverer.

But if a mere memorandum does not clinch it, did Gauss know so much more that he could claim the prize? Does his ability to draw together Schweikart's ideas better than Schweikart had done rest on sufficiently more than the lawyer already possessed? If we recall that Gauss's reply to Schweikart is only too close to the famous reply he was to send the Bolyais in its assertion that he knew the ideas already, then we might conclude there was a chance that sending good ideas to Gauss was, at the very least, likely to stimulate him to add to what he already knew, and quite innocently assume that he was simply restating what he had known for some time. The point is worth noting, because as we have seen, there is no other evidence to corroborate his assertions that at these dates he had a coherent theory drawing together the consequences of assuming some standard definition of an alternative geometry to Euclid's.

We have to ask: What kind of knowledge, at what time, is Gauss to have before he can be regarded as a true discoverer of non-Euclidean geometry? Is it a worked-out theory that looks consistent and has some chance of being true of Space? Is it a deep empirical dissatisfaction with Euclidean geometry as an account of Space? What degree of conviction should it impart to all but the most logically, even legalistically, minded? In what ways should it surpass Schweikart's level of knowledge or belief? How much should it resemble what Bolyai and Lobachevskii were to do? Or was it, perhaps, significantly different?

When, in 1808, Gauss was aware that in the hypothetical non-Euclidean geometry similar triangles are congruent, and therefore there is an absolute measure of length, he found this conclusion absurd, and therefore held that the matter was still unclear. This is a long way from believing that there is indeed a meaningful non-Euclidean geometry.

The historical sources then go quiet until 1816, and there is no evidence at all. In 1816 Gauss wrote to Gerling that the idea of an absolute measure of length is somewhat paradoxical but not self-contradictory, and that it would be remarkable if Euclid's geometry was not true, because then we would have an a priori measure of length. As Dunnington correctly observed, being remarkable is consistent with being attractive.

Also in 1816 there is a glimpse of what Gauss knew as seen through Wachter's eyes. But it is less convincing than is often thought. It is hard to know precisely what Gauss's letter to Olbers (quoted above) actually means. 'Our geometry' is surely Euclidean geometry. Gauss was a devout man, not given to presuming to understand the mind of the Divinity. Human understanding would inevitably fall short of God's. But what would a proof – of a kind that surpasses human understanding – be that establishes the necessity of our (Euclidean) geometry? How would it differ from a proof that does not surpass human understanding? Would it be some argument, compelling even to God, that made Euclidean geometry the right geometry for Space? Would Gauss understand it in the afterlife, or is merely a further century of this mortal life going to be enough for somebody to resolve the matter? Arithmetic, it seems, has an apodictic status. It is pure a priori. Whatever that might mean, whether Kantian or Friesian terminology should be imputed here or some equation of arithmetic with logic, the truth of arithmetic is being said to be of a higher kind than the truth of geometry, which is down with mechanics. The problem is the twist of thought conveyed in the two words 'for now'. The sense of the passage is that the status of the truth claims of Euclidean geometry is unclear, and might remain so either forever, or only for a while. The passage does not say that there *are* two geometries at some logical level and some experiment must choose between them. That would be the meaning of a remark like this: "... the necessity of our geometry cannot be proved. Geometry stands, not with arithmetic which is pure a priori, but with mechanics." The passage is agnostic, not heretical. Knowledge is lacking, says Gauss; he does not claim to possess new knowledge, of a new geometry. In the context, that surely means that even the ideas he was discussing with Wachter he considered to be hypothetical, and capable of proving to be false.

The 'transcendent trigonometry' is usually taken to be the hyperbolic trigonometry appropriate to non-Euclidean geometry, but there is very little evidence to support any interpretation. The geometry apparently depends on an undetermined constant, which indeed non-Euclidean geometry does, but the implication that this constant was unproblematic is unwarranted. Why should it not yield to two incompatible determinations, and thus the contradiction most previous writers on the subject had hoped to find?

On 25 January 1819 Gerling passed Schweikart's note on to Gauss, who replied in March that he was "uncommonly pleased with the note, and can do all of astral geometry once the constant is given." The angle sum of a triangle is proportional to its angular defect, and there is a maximum area for triangles which is attained by trebly asymptotic triangles. This does not make clear if Gauss now possessed formulae from which all the elementary mensuration of a non-Euclidean geometry can be derived, or if he knew only that part which resembles Euclidean geometry. What, precisely, was covered by the phrase 'all of astral geometry'? And whatever Gauss was claiming, it leaves the time of his discoveries vague. Should it be part of the transcendent trigonometry that he discussed in 1816 with Wachter? Could it be something that only came to him in the year between receiving Schweikart's note and replying to it? The first alternative is the more likely, but even if it is true the trigonometry per se does not seem to have clarified everything in Gauss's mind.

In November 1824, writing to Taurinus, Gauss said that "...the assumption that the angle sum is less than 180° leads to a geometry quite different from Euclid's, logically coherent, and one that I am entirely satisfied with." Taurinus, of course, was looking for reasons to deny the possibility of non-Euclidean geometry, and Gauss did not want to lend his name in any way to that enterprise. But even if we allow that by 1824 Gauss was comfortable with a novel geometry for space, we learn nothing from this letter about what theorems the new geometry might contain, and what its implications might be for science, geodesy, or astronomy.

There is an unfortunate disparity throughout this part of Gauss's work between the material that can be precisely dated, and his claims in letters to his friends that he has known this or that fact for 30, even 40 years. These claims may well be true, in the sense that the teen-age Gauss could well have found some arguments in elementary geometry inconclusive and begun to harbour suspicions about them. They are harder to square with interpretations that impute specific knowledge to the young man, especially when they contradict later, datable, evidence. For example, there is a proof which Gauss himself noted he found on 18 November 1828 that the angle sum of a triangle cannot exceed 180°. This may mark the occasion on which Gauss discovered what, for him, was a new and particularly perspicacious proof of a result he already knew. Equally well, it may mark the discovery of a proof of something he had hitherto only suspected.

The same is true of the passages on the theory of the line and the plane, which Stäckel very plausibly dated to the years 1828-1832. Here Gauss noted that many of the statements mathematicians make about the plane conceal theorems which, if not difficult, are not entirely trivial either. The assumption that Gauss had thought his way through this tangle in the 1790s is unwarranted; what then was he doing writing it down in the late 1820s? The same is true of the notes on the theory of parallel lines, which Stäckel dated to 1831. Stäckel connected them to letters that Gauss wrote to Schumacher in May of that year, where Gauss explained that he had had these ideas for 40 years but never written them down. It is hard to believe that Gauss could write down, after such a long period of time, exactly what he had had in mind when he was 14. If one takes 40 as a round number, even if one replaces it with 32 (because in 1799 Gauss was involved with the elder Bolyai on the question of the foundations of geometry), it is still implausible that Gauss was just writing down ideas from memory. It is not what good mathematicians do. They polish the ideas even as they recall them, and writing for themselves alone, as they are, they are under no obligation to prevent subsequent historians from becoming a little confused. We can grant that Gauss, as he said, was already concerned in the 1790s about the foundations of geometry, without supposing that the notes of 1831 are a true and faithful record of what he had believed so many years before.

On the other hand, it is worth noting that these pages are firmly in a style which may be called classical, with the implication that adherence to a classical formulation denies trigonometric methods a fundamental role. A 'classical style' or 'classical formulation' is an approach to geometry that regards terms such as point, line, plane, distance, and angle as undefinable, primitive terms. The relationship between them, the properties of these objects and of figures composed of them, may be obscure and in need of elucidation, an author in this style may feel that basic questions may have been begged by all previous writers, but the fundamental terms are not to be reduced to others (for example, numbers). The disentangling of the ideas will be done by patient combing, not by radically new definitions. Gauss's surviving early notes are very much in this classical style.

Of course, the problem of parallels is prominent among these difficulties. But the famous letter to Bessel of January 1829, where Gauss claims to have harboured these thoughts for almost 40 years, does not give parallels pride of place. There Gauss wrote that he hadn't written up his extensive researches, and perhaps he never would, because he feared the howl of the Boetians, however "my opinion that we cannot ground geometry completely a priori has become, if anything, even stronger."

Would it be fair to point out that, to someone of Gauss's high standards, a strong opinion becoming even stronger is still an opinion that falls short of certainty? And while the well-known gap is that staple of the literature, the problem of parallels, but it is firmly situated here, as it is throughout the years 1828 to 1832, in the context of a number of other problems in the foundations of geometry. Bessel's reply encouraged Gauss to state that geometry has a reality outside our minds whose laws we cannot completely prescribe a priori. This is entirely consistent with the classical formulation. The concepts of point, line, plane and so forth are formed in whatever way concepts are generated (by abstraction, through experience, introspection, it doesn't matter) and whatever problems there might be in saying how this is done belong to philosophy, not mathematics. The tacit implication is that forming these concepts is among the simpler pieces of concept formation people do, and can be treated, for mathematical purposes, as entirely unproblematic. The task of the mathematician is to get them truly clear in the mind, which, Gauss suggested, is liable to involve one in two kinds of activity. One is teasing out tacit assumptions and fitting them up with proofs (Gauss did not suggest there were likely to be any erroneous beliefs). The other, which concerns parallels, is the elaboration of new ideas about a hitherto unsuspected species of geometry, which might nonetheless turn out to be (for some value of an unknown constant) the true geometry of Space.

4. Evidence in geodesy

Can one then connect transcendent trigonometry with Gauss's work on differential geometry and geodesy? The famous Hanover survey kept Gauss occupied for most of the three years from 1822 to 1825 (too much of it, he complained to Olbers in October 1825) and in the last three months of that year he wrote the *Disquisitiones*. The book is famous for the discovery that Gaussian curvature is intrinsic, which, in the *Disquisitiones*, has a very computational proof, giving the (Gaussian) curvature in terms of the coefficients of the first fundamental form and their derivatives with respect to the coordinates. He then went on to deduce the elegant theorem already quoted:

"The excess over 180° of the sum of the angles of a triangle formed by shortest lines on a concavo-concave surface, or the deficit from 180° of the sum of the angles of a triangle formed by shortest lines on a concavo-convex surface, is measured by the area of the part of the sphere which corresponds, through the direction of the normals, to that triangle, if the whole surface of the sphere is set equal to 720 degrees."

Now Gauss knew very well that the area of a non-Euclidean triangle is proportional to the deficit from 180° of the sum of the angles of the triangle. It would therefore be tempting to suppose that Gauss would connect this elegant theorem with the study of non-Euclidean geometry, by considering a concavo-convex surface of constant negative Gaussian curvature. But such a conclusion is highly speculative; Minding in 1839, produced just such a surface without making that connection, and Codazzi in 1859 even showed that on such a surface the appropriate trigonometric formulae are the hyperbolic analogues of the formulae in spherical trigonometry. Neither man observed the connection with non-Euclidean geometry.

If we grant that Gauss might have suspected that non-Euclidean geometry was the geometry on a surface of constant negative curvature, we must none the less note that Gauss did not develop the trigonometry of triangles on surfaces of constant (positive or negative) curvature until after 1840, when he had read Lobachevskii's *Geometrische Untersuchungen.* Moreover, he did not have such a surface to hand; there is every reason to suppose that Minding was the first to discover it. Minding's example, moreover, is a surface of revolution, and therefore has a number of topological properties that rule it out as a model of space, notably self-intersecting geodesics and pairs of geodesics that meet in more than one point. Minding's example also has singular points, at which the surface comes to a halt, which marks a significant difference from the sphere.

It is not at all obvious that these properties of Minding's surface do not suggest that there can, after all, be no surface obeying the rules of non-Euclidean geometry. Why should there not be local models of parts of a 'non-Euclidean' space, but no global model? But even if a surface had been known to Gauss, and even if (in contradiction to Hilbert's later theorem about surfaces of constant negative curvature) it had not had any unfortunate properties, what would it establish? Only that there is a surface in Space whose intrinsic geometry is non-Euclidean. It would not establish that Space could be non-Euclidean, because Space is three-dimensional. There is no sign that Gauss had any of the concepts needed to formulate a theory of differential geometry in three-dimensions.

While it is always dangerous to speculate on what Gauss could not do, it is worth noticing that adherence to the classical formulation makes thinking of novel three-dimensional geometries very difficult indeed, if not impossible. According to the classical formulation, Space is our source of knowledge of all primitive geometric terms. It is difficult enough to follow out the implications of this for surfaces in Space, to find formulae for geodesics on surfaces and so forth. At the end of such an analysis, Gauss discovered that although all geometric concepts are impressed on the surface from the ambient Space, some, such as curvature, are after all intrinsic. To imagine that there is another type of Space altogether with any chance of proving theorems about it, it would be best to take a big step backwards and consider how to do geometry at all. One cannot (easily, at least) think one's way from Euclidean three-dimensional Space to another three-dimensional Space. What, in the end, Riemann did, was to step back and create new ways of thinking about geometry that made it possible to think of many, many kinds of Space.

Riemann also produced the first pieces of essential mathematical machinery for doing differential geometry in three or more dimensions. There is no sign that Gauss did that, but without it almost nothing useful can be said.

To conclude: Gauss's work on differential geometry in the 1820s, remarkable as it is, does not connect with any kind of transcendent trigonometry. Nowhere in the work on differential geometry did Gauss even hypothesise, much less study, a surface of constant negative curvature. He did not do that until he had read Lobachevskii's *Geometrische Untersuchungen*, and even then the connection between the differential geometry and the trigonometry rests on the choice of the same symbol for a constant of integration as for the (Gaussian) curvature. This symbol, k, is real for spherical trigonometry, and purely imaginary for non-Euclidean trigonometry.

We are returned, empty-handed, to where we began. In 1816 Gauss possessed a form of trigonometry applicable to more than just spherical geometry. He had proposed a surface which was a sphere of infinite radius and was not a plane. By 1831 he knew that the circumference of a semi-circle of radius r is $\frac{1}{2}\pi k\left(e^{r/k} - e^{-r/k}\right)$. This formula, dropped so casually in a letter to Schumacher, together with the one in his reply to Schweikart in 1818 for the maximum area of a triangle in terms of Schweikart's Constant (the maximum altitude of an isosceles right-angled triangle), are the only evidence we have that Gauss knew anything about non-Euclidean geometry in detail.

As has already been remarked, these formulae are not difficult to obtain if one simply assumes that non-Euclidean geometry is described by the formulae of hyperbolic trigonometry. Could Gauss have defended that assumption? There is no evidence either way. Given Gauss's very high standards, he might have felt confident of the validity of such an assumption but not been able to defend it to his own satisfaction. To introduce hyperbolic trigonometry into the study of non-Euclidean geometry properly is, as Bolyai and Lobachevskii found, a considerable labour of which no trace remains in Gauss's work. It would seem reasonable to assume that he made the assumption, but not the derivation from basic principles.

Another factor that must be considered is Gauss's aversion to the new geometry, expressed in the references to howling Boetians, wasp's nests, and the like. The absence of evidence is consistent with Gauss not giving this truly difficult topic sufficient attention. There are many occasions when Gauss worked, over a period of years, to bring topics to his satisfaction, and others, the theory of elliptic functions, for example, when he pushed hard to achieve the right levels of insight and of generality, to obtain the most appropriate proofs. In these cases the result is often books, book length memoirs, and many pages of nearly publishable notes. When we turn to the problem of parallels, there is nothing like so much material, and one is driven to wonder if it ever existed.

5. The need to study three-dimensional space

Let us grant that, perhaps by 1816, Gauss was convinced that there could be a non-Euclidean geometry, a geometry in which the angle sum of triangles is less than π, and the area of triangles is proportional to their angular defect and is bounded by a finite amount. The trigonometric formulae for this geometry are those of hyperbolic trigonometry, and the analogy with spherical geometry and trigonometry extends to formulae for the circumference and area of circles. Such a position is unsatisfactory, to Gauss and to us, because it is purely and simply an analogy. What is lacking is any argument that starts from an idea of geometry and derives the formulae according to convincing rules. In the case of spherical geometry, there is a basis: the Euclidean geometry of three-dimensional Space. Ordinary arguments about how geodesics and angles may be defined on a sphere (a surface sitting in three-dimensional Space) then lead to the familiar formulae. In the non-Euclidean case, the basis of Euclidean three-dimensional geometry is in question, and there is no surface.

Why, after all, does talk of two-dimensional geometry matter when the nature of three-dimensional Space is at stake? Because one believes that there is no essential obstacle to going up one dimension. In the Euclidean case, there seems to be no reason to suspect a problem, and there is an abundance of results about three dimensions even in Euclid's *Elements*. The methods of Cartesian coordinate geometry seem adaptable. Plus, if you believe that Space is actually Euclidean, you *know* there is no obstacle. But in the non-Euclidean case, there may be no reason to suspect a problem but equally there is no reason to believe success is guaranteed. There are no pre-existing results about three-dimensional non-Euclidean geometry, no coordinate methods to adapt. Even if one found a surface in three-dimensional Euclidean space with non-Euclidean geometry as its induced geometry, that would not licence the inference that three-dimensional Space was non-Euclidean, any more than the existence of spheres in three-dimensional Space forces the conclusion that Space is a three-dimensional sphere.

A more plausible way forward would be to start with non-Euclidean three-dimensional Space, and to derive a rich theory of non-Euclidean two-dimensional space from it. This would be better evidence that the assumption that there could be a non-Euclidean three-dimensional space made sense (although not convincing, to be sure). This is what Bolyai and Lobachevskii did, but not Gauss. Gauss did believe that such an assumption made sense, but he equivocated about it. Its consequences were paradoxical, but not self-contradictory; as a joke he might even wish it were true; its existence implies that the laws of geometry cannot be prescribed a priori.

The only hint that Gauss explored the non-Euclidean three-dimensional case in order to obtain new, suggestive, results about non-Euclidean two-dimensional geometry is the remark by Wachter about the sphere of infinite radius. This is a remark by Wachter, a mathematician of whom Gauss had a good opinion, and one must therefore wonder what Wachter himself brought to the discussion. There is no other reference to it at all in the surviving Gaussian *Nachlass*. What Wachter says is not encouraging: "Now the inconvenience arises that the parts of this surface are merely symmetrical, not, as in the plane, congruent; or, that the radius on one side is infinite and on the other imaginary[.]" and more of the same. This is a long way from saying, what enthusiasts for Gauss's grasp of non-Euclidean geometry suggest, that this is the Lobachevskian horosphere, a surface in non-Euclidean three-dimensional Space on which the induced geometry is Euclidean.

If the conclusion is that Gauss possessed tantalising hints of a new, non-Euclidean, geometry, but never worked his ideas up into a systematic theory, then his conviction is no less, and no greater, than that of Schweikart or Bessel. The grounds for his conviction are greater, but still insubstantial (we should not be too swayed by the fact that he turned out to be right). He did not possess almost all the substantial body of argument that gives Bolyai and Lobachevskii their genuine claim to be the discoverers of non-Euclidean geometry.

6. Another question about geometry

There is, however, a way of looking at what Gauss did that makes more sense of the available evidence. Our access to it starts with the letter to Bessel of January 1829, the problem of the definition of a plane as a surface that contains the line joining any two of its points, a definition, Gauss said, that contains more than is necessary for the determination of the surface, and tacitly involves a theorem which must first be proved. That this was a concern of Gauss's as early as 1797 is documented in his mathematical diary, indeed it is the only entry on a geometrical topic in the entire diary. Gauss's insight is that at one or another time in elementary geometry different properties of the plane are being used, and if one is taken as basic the others must be deduced as theorems. Moreover, it might be that unsuspected properties are being used, and these novelties should also be made explicit.

The plane has a problematic relationship to three-dimensional Space because it need not be a primitive given term but can be defined (Gauss proposed to obtain it by rotating a line about a perpendicular axis, other definitions are possible). In the same way, 'parallel' might be a primitive term, or a reducible one, and if primitive, capable of generating only one geometry, or more than one. Good housekeeping requires that one sort through these possibilities, and perhaps there will be no surprises, perhaps there will be. This locates the problem of parallels in the family of problems about the classical formulation of geometry, which is overwhelmingly how authors treated it. Only Legendre, among those who sought to defend the parallel postulate, used methods lying outside the classical formulation.

When Gauss wrote up his ideas about parallels in 1831, which Stäckel implied might be the ideas Gauss regarded as almost 40 years old, they were of this classical kind. One might argue that they had, for Gauss, a biographical aspect, but he could have gone on to say that these were the ideas of his youth and now he thinks something else. He did not, most likely because he thought these were the still good ideas. Even when he writes to Wolfgang Bolyai to praise Johann for knowing what he has known for some time, what he said by way of mathematical detail is about the area of triangles, a matter belonging to the classical formulation. He did not engage with the trigonometric aspect of Bolyai's work.

Gauss's investigations into the classical formulation of geometry were inconclusive. They remind us, as every attempt before and after Hilbert's does, of how slippery some elementary geometry is, and how hard it is to get it into a rigorous order. They do not put matters right, and they do not, in themselves, merit an account here (Stäckel's suffices). They are, however, the context for almost all of Gauss's investigations of the parallel postulate and non-Euclidean geometry. In this context, he concluded that there is more to the concept of the plane than meets the eye, but it could, with some effort, be spelled out properly. On the other hand, there was a lot more to parallelism than meets the eye. The fundamental intuition of lines that never meet no matter how far they are extended is literally ambiguous; it can be made to yield two theories. How much confidence Gauss placed in the new theory has been discussed already, but the bulk of the evidence is of the psychological kind: repeated failure to defend the parallel postulate gave way steadily to a feeling that the parallel postulate was indefensible and an alternative geometry therefore possible.

7. Gauss's letter to János Bolyai

If indeed János Bolyai's analysis of non-Euclidean geometry and the nature of space was more profound and wide-ranging than Gauss's, then the question of why Gauss answered as he did is raised with a new force. There is no evidence to suggest that Gauss was lying. On the contrary, everything we know about Gauss suggests that he was scrupulously honest, honest, indeed, to a fault. His own stated reason for writing as he did was that he 'could do no other'. He was not one to dissemble, and people who knew found him straight-forward and plain-speaking. But there is a wide gulf between saying something that is false, and telling the truth. One can be sincere, but mistaken, or wrong but innocent of any attempt to deceive.

In Gauss's case, several possibilities suggest themselves. To someone like Gauss, to return to a topic was surely to see it afresh. He need not have noticed what ideas were occurring to him for the first time, and which were recalled from earlier investigations (still less did he have to leave an accurate paper trail for subsequent historians). The facts of non-Euclidean geometry, as Bolyai presented them, could have been absorbed quickly by Gauss, much as if he had thought of them himself. The best mathematicians often have the habit, on hearing of a new result, of thinking how they could prove it themselves (Poincaré was said to have read a paper only if a proof of its results did not quickly occur to him). If Gauss was like that, he may not have read Bolyai's *Appendix* at all carefully.

It is not certain, for that matter, that Gauss read the *Appendix* properly at all. To this day there are stories (always apocryphal, to protect the people involved) of referees not reading papers carefully. Since János

Bolyai was a particularly concise writer, even Gauss may have taken the easy route of working his own way through the material by dipping in and out of the text. It is impossible to interpret the disparity between Gauss's stout claim that he knew all this already, and the two points he then discussed at length in his letter. These were János's unappealing choice of names for what are nowadays called, using Lobachevskii's terms, the horocycle and the horosphere; and an elementary proof of the relationship between area and angle sum of a triangle. But this response provides no evidence that Gauss engaged with the genuine novelties of the work: the systematic introduction of hyperbolic trigonometry, and the fact that the new geometry was introduced in three dimensions.

If Gauss read the *Appendix* his way, assimilating some of it to what he already knew and discarding the rest, then his reply is sincere, but not, without further corroboration, evidence that Gauss already new what Bolyai had published. And, as this paper has noted, such confirmatory evidence is lacking. What it is evidence for, of course, is what no-one who knows about Gauss can doubt, that he was not a charmer or a flatterer. Throughout his life he formed few friendships. His closest relationships were with astronomers such as Bessel, Olbers, Schumacher and others, with whom he enjoyed a long correspondence, but with whom he was not in daily contact. Such people could treat Gauss more nearly as an equal, something which most mathematicians, even János Bolyai, could not. Outside this limited circle, Gauss lacked the ability to respond warmly, to come over in the right way, to win people to his side. What he undoubtedly saw as the only honest way to proceed, if he was not to lapse into vanity, inevitably came over as arrogance, and permanently damaged János's enthusiasm for publishing.

8. Conclusion

It becomes clear that a mathematician persuaded of the truth of non-Euclidean geometry and seeking to convince others is almost driven to start by looking for, or creating, non-Euclidean three-dimensional Space, and to derive a rich theory of non-Euclidean two-dimensional Space from it – as Bolyai and Lobachevskii did, but not Gauss. The only hint that he explored the non-Euclidean three-dimensional case is the remark by Wachter, but what Wachter said was not encouraging: "Now the inconvenience arises that the parts of this surface are merely symmetrical, not, as in the plane, congruent; or, that the radius on one side is infinite and on the other imaginary" and more of the same. This is a long way from saying, what enthusiasts for Gauss's grasp of non-Euclidean geometry suggest, that this is the Lobachevskian horosphere, a surface in non-Euclidean three-dimensional Space on which the induced geometry is Euclidean. In particular, there is no three-dimensional differential geometry leading to an account of non-Euclidean space.

Gauss, by contrast, possessed a scientist's conviction in the possibility of a non-Euclidean geometry which was no less, and no greater, than that of Schweikart or Bessel. The grounds for his conviction are greater, but still insubstantial, because he lacks almost entirely the substantial body of argument that gives Bolyai and Lobachevskii their genuine claim to be the discoverers of non-Euclidean geometry.

Bibliography

[1] Bolyai, F. 1832, 1833 *Tentamen juventutem studiosam in Elementa Matheosis purae*, etc. Marosvásárhely

[2] Bolyai, J. 1832, *Appendix scientiam spatii absolute veram exhibens*, in Bolyai, F. [1832], tr. J. Houël, La Science Absolue de l'Espace, *Mémoires de la Société des Sciences physiques et naturelles de Bordeaux* 5, 1867, 189-248, tr. G. Battaglini, *Sulla scienza della spazio assolutamente vera*, Giornale di matematiche 6 1868, 97-115, tr. G.B. Halsted, *Science Absolute of Space*, Appendix in Bonola [1912], *János Bolyai Appendix: The theory of space, with introduction, comments, and addenda.* Edited by Prof. Ferenc Kárteszi, Doctor of the Mathematical Sciences. Supplement by Prof. Barna Szénássy. Akadémia Kiadó, Budapest, 1987, and North-Holland, Amsterdam, North-Holland Mathematics Studies. Number 138

[3] Bonola, R. 1906, *La geometria non-Euclidea*, Zanichelli, Bologna

[4] Bonola, R. 1912, *History of non-Euclidean geometry*, tr, H.S. Carslaw, preface by F. Enriques, Open Court, Chicago, Dover reprint, New York, 1955

[5] Breitenberger, E. 1984 *Gauss's Geodesy and the Axiom of Parallels*, Archive for History of Exact Sciences 29 273-289

[6] Coxeter, H.S.M. 1977, *Gauss as a geometer*, Historia Mathematica 4.4, 379-396.

[7] Dombrowski, P. 1979, *150 Years after Gauss' Disquisitiones generales circa superficies curvas, astérisque*, 62 (with the original text of Gauss and an English translation by A. Hiltebeitel and J. Morehead).

[8] Gauss, C.F. 1880, *Werke*, Vierter band, herausgegebenen von der Königlichen Gesellschaft der Wissenschaften zu Göttingen

[9] Goe, G. and van der Waerden, B. L. 1972, *Comments on Miller's 'The myth of Gauss's experiment on the Euclidean nature of physical space'* Isis 63, 345-348. With a reply by Arthur I. Miller Isis 65 (1974), 83-87.

[10] Gray, J.J. 1984, *A commentary on Gauss's mathematical diary*, 1796-1814, with an English translation, Expositiones Mathematicae, 2, 97-130 (reproduced with minor alterations in the new edition of G.W. Dunnington *Gauss – Titan of Science* 1956 published by the Mathematical Association of America in 2002)

[11] Gray, J.J. 1989, Ideas of space: *Euclidean, non-Euclidean, and relativistic* Oxford University Press, 2nd edition, Romanian edition, Idei Spatiu, Editura All Educational, 1998.

[12] Hilbert, D. 1899, *Grundlagen der Geometrie*, many subsequent editions.

[13] Hilbert, D. 1971, *Foundations of geometry*, 10th English edition, translation of the second German edition by L. Unger.

[14] Lambert, J.H. 1786, *Theorie der Parallellinien*, in F. Engel and P. Stäckel (eds.) *Theorie der Parallellinien von Euklid bis auf Gauss*, Teubner, Leipzig, 1899

[15] Legendre, A.M. 1794, *Éléments de Géométrie*, Paris, with numerous subsequent editions, e.g. 12th ed. 1823.

[16] Lobachevskii N. I. 1840, *Geometrische Untersuchungen etc.* Berlin, tr. G. B. Halsted as Geometrical researches on the theory of parallels, appendix in Bonola [1912].

[17] Miller, A. I. 1972, *The myth of Gauss' experiment on the Euclidean nature of physical space.* Isis 63, 345-348.

[18] Minding, F. 1839, *Wie sich entscheiden lässt, ob zwei gegebener Krummen Flächen, etc*, Journal für die reine und angewandte Mathematik 19, 370-387.

[19] Neumann, O. 1981, ed. *Mathematisches Tagebuch*, 1796-1814. With a historical introduction by Kurt-R. Biermann, translated from the Latin by Elisabeth Schuhmann, with notes by Hans Wussing. Third edition. Ostwalds Klassiker der Exakten Wissenschaften, 256. Leipzig, Akademische Verlagsgesellschaft Geest and Portig K.-G.

[20] Pasch, M. 1882, *Vorlesungen über neuere Geometrie*, Teubner, Leipzig

[21] Reichardt, H. 1976, *Gauss und die nicht-euklidische Geometrie*, Teubner, Leipzig

[22] Scholz, E. 1992, *Gauss und die Begründung der "hoheren" Geodäsie*, S.S. Demidov, M. Folkerts, D.E. Rowe, C.J. Scriba, (eds.) Amphora, Birkhäuser Verlag, Basel, Boston and Berlin, 631-648

[23] Stäckel, P. 1917, *Gauss als Geometer*, in Gauss Werke, X.2, Abh. 4, separately paginated

[24] Szénássy, B. 1980, *Remarks on Gauss's work on non-Euclidean geometry* (Hungarian), Mat. Lapok 28 1-3, 133-140.

[25] Waltershausen, W.S. von 1856, *Gauss zum Gedächtnis* Hirzel, Stuttgart, reprinted Dr. Martin Sändig, Wiesbaden, 1965, and translated from the German by Helen W. Gauss, Gauss, a Memorial (Colorado Springs, Colorado) 1966

[26] Zormbala, K. 1996, *Gauss and the definition of the plane concept in Euclidean elementary geometry*, Historia Mathematica, 23.4, 418-436

JÁNOS BOLYAI'S NEW FACE

Elemér Kiss
Sapientia University,
Marosvásárhely (Târgu-Mureş), Romania

1. Introduction

János Bolyai's (1802-1860) absolute and hyperbolic geometry made his name immortal in the history of mathematics. His life and scientific activity has been the topic of several studies. Bolyai's impact on modern thinking has been presented by several scholarly works, which also show its various facets. As a mathematician, he is almost exclusively known for his achievements in geometry and the theory of complex numbers. The personal image of Bolyai, most people were familiar with earlier, had mostly been influenced by some literary works, dramatizing his irreconcilable relationship with his father and his disharmony with the people around him.

Bolyai's many thousand-page legacy of manuscripts that has been brought to light by the research of recent years calls for a re-evaluation of his achievements as a scientist and his personality.

The Appendix [2] is not the only work János Bolyai left to posterity. Even after the completion of his exceptional work, he continued his research and put down his ideas. This resulted in his vast legacy of manuscripts, fourteen thousand pages of which are in the Teleki-Bolyai Library of Marosvásárhely. The notes contain those mathematical theorems, which prove not only that Bolyai brought a significant contribution to geometry, but also that he reached valuable conclusions in other mathematical fields.

First of all, we can assert that in the past our knowledge of his results in mathematics was rather superficial. Bolyai excels not only with his achievements in geometry, but also with his research in several other mathematical fields. His accomplishments in number theory and algebra are most astounding.

2. János Bolyai's Research on Number Theory

All Bolyai-monographs unanimously assert: although János Bolyai tried his hand at a few problems in number theory, his investigations were not particularly successful. However, his manuscripts attest the opposite of all this. Bolyai had a keen interest in questions of number theory and he had several original ideas, with which he preceded many other mathematicians of later ages.

Among the very first theorems found in Bolyai's legacy is the following: If p and q are prime numbers, and a is an integer divisible neither by p nor by q, and if $a^{p-1} \equiv 1 \pmod{q}$ and $a^{q-1} \equiv 1 \pmod{p}$, then

$$a^{pq-1} \equiv 1 \pmod{pq} \ . \tag{1}$$

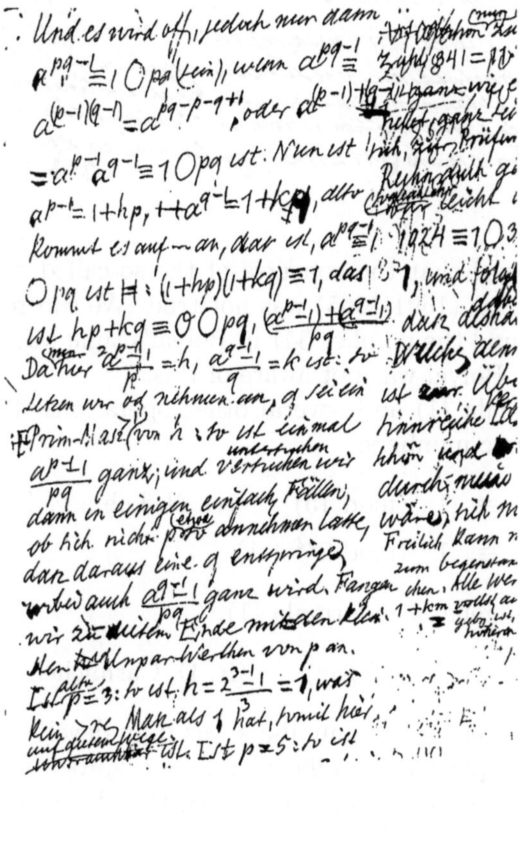

We can readily observe that it is the same theorem which James Hopwood Jeans (1877-1946) published decades later in 1898. Since we can definitely affirm that this relation was first recognized and demonstrated by János Bolyai, I propose that in the future it should be called Bolyai-Jeans Theorem.

At a time János Bolyai thought that he could find the formula of prime numbers by means of Fermat's little theorem. This is why he tried to prove its converse. As a result, he reached conclusion (1). By substituting $a = 2$ in this relation, with repeated attempts he got the numbers $p = 11$, $q = 31$, and thus Bolyai found the smallest pseudoprime number with respect to 2, that is, 341, for which

$$2^{340} \equiv 1 \pmod{341} .$$

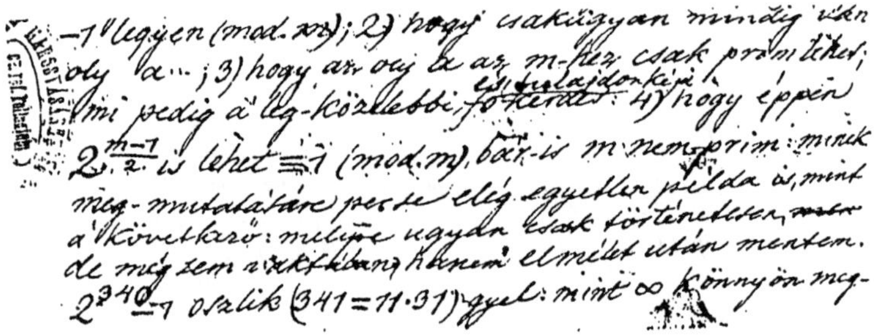

Thus, he showed that the converse of the so-called "little theorem" of Pierre de Fermat (1601-1665) was not true. We can remark that the only one who found the number 341 in Bolyai's time was an anonymous scientist [1], but Bolyai was not aware of this.

By using formula (1) and several other operations, Bolyai invented many pseudoprime numbers:

$$4^{14} \equiv 1 \pmod{15} \qquad \text{(D. N. Lehmer, 1927)}$$
$$2^{2^{32}} \equiv 1 \pmod{2^{32}+1}$$

He attempted to generalize formula (1):

$$a^{pqr-1} \equiv 1 \pmod{pqr} \qquad \text{(R. D. Carmichel, 1912)}$$

The research of pseudoprime numbers was fully developed in the twentieth century and played a significant role in cryptography.

Bolyai's further observations concerning the "little theorem" of Fermat can be found in [10], [11].

János Bolyai's inquiries concerning Fermat's two-square theorem are very valuable. The theorem goes as follows: every prime of the form $4k+1 (k \in \mathbf{N})$ can be written as a sum of two squares. The theorem had been formulated by Fermat but it was demonstrated by Leonhard Euler (1707-1783) about one hundred years later. Euler's proof reached the Teleki Téka of Marosvásárhely, where Farkas Bolyai read it. The proof seemed too long and complicated to him. Therefore, he asked his son to provide him with the "simplest" demonstration of the theorem. Within a short period of time János sent his father a two-page letter with four possible solutions to the problem. His solutions were so simple, because he deployed his achievements regarding the question of complex integers. One of his solutions is especially simple [12]. We feel, that nobody has provided a more brilliant solution to this theorem than Bolyai.

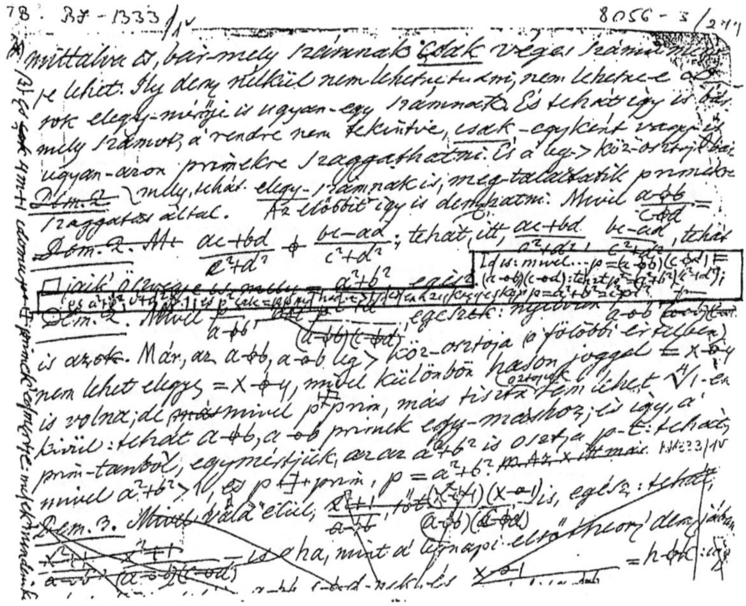

We note, that in 1844 Gotthold Eisenstein (1823-1852) also proved Fermat's two-square theorem using complex integers [4], but in a different way from János Bolyai.

In addition to these theorems János Bolyai dealt with several other number theoretical problems too, namely, with the Pell equation. Since he did not know that the converse of the Wilson's theorem – if $(n-1)! \equiv -1 \pmod{n}$, then n is prime – had already been demonstrated by Lagrange, he, just as his father, sought a solution to this problem. In his own words: "both my father and I solved the converse of the so beautiful and important Wilson's theorem" [3, 1193/16].

János Bolyai was also dealing with the $F_n = 2^{2^n} + 1$ (where $n \in \mathbf{N}$) type so called Fermat numbers. More than 20 years before Fortune

Landry (1799-?) he tried to determine whether the $F_6 = 2^{64} + 1$ is a prime or a composite number. He couldn't solve it, but he stated that every Fermat number F_n can be written in the form of $6n - 1$.

Regarding the so-called Comma of Didymos fraction of $\frac{81}{80}$, which plays a significant role in musical theory, he formulates the following problematic in 1840: let us determine that fraction $\frac{a}{b}(a, b \in \mathbf{N} \setminus \{0\})$ whose numerator is greater by 1 than its denominator and among the factors of the a and b natural numbers we can find only the prime numbers 2, 3, and 5. The answer to this question, which is by the way negative (that is, there is no such fraction [13]), can be found by resorting to the exponential diophantine equations proposed by Bolyai:

$$2^x - 3^y = 1,\ 3^x - 2^y = 1,\ 2^x - 5^y = 1,\ 2^x + 1 = 3^y \cdot 5^z, \ldots$$

etc. .

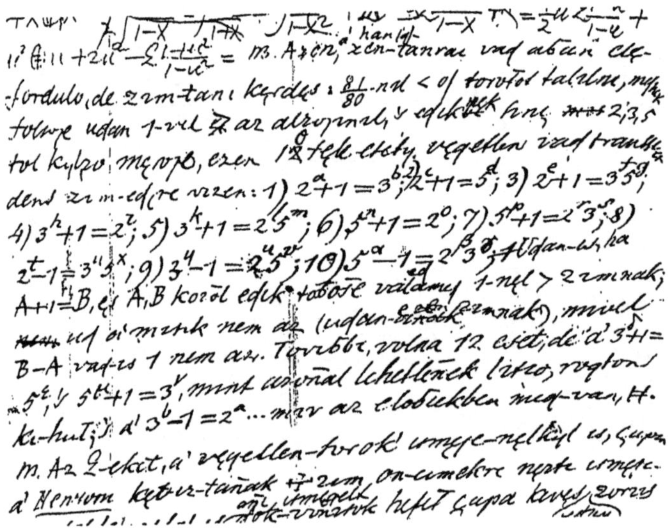

Although Bolyai did not solve these equations, his achievement was worth mentioning, because similar equations of the type $3^x - 2^y = 1$ can be found shortly after in the history of mathematics, proposed by Eugène Charles Catalan (1814-1894) in 1843 (he couldn't find the general solution to the problem either). Hence the so-called Catalan conjecture, which caused so many problems to twentieth century number theorists. We can say that János Bolyai was one of the first founders of the theory of "Exponential diophantine equations" [14].

Bolyai also engaged in the problem of the diophantine equations. He did not reach significant conclusions in this regard, but in Mathias

Hausser's book [9], for example, one can find that he solves the following equation in three different ways:

$$27x + 49y + 70z = 1000 \ .$$

He shows that the equation

$$x^2 + y^2 = 3 \ ,$$

does not have a solution within the ring of integers, but at the same time he observes that

$$i^2 + 2^2 = 3, \qquad i = \sqrt{-1} \ .$$

Finally, it is worth mentioning that Bolyai constructed a magic square:

x	y	3b-x-y
4b-2x-2y	B	2x-y-2b
x+y-b	2b-y	2b-x

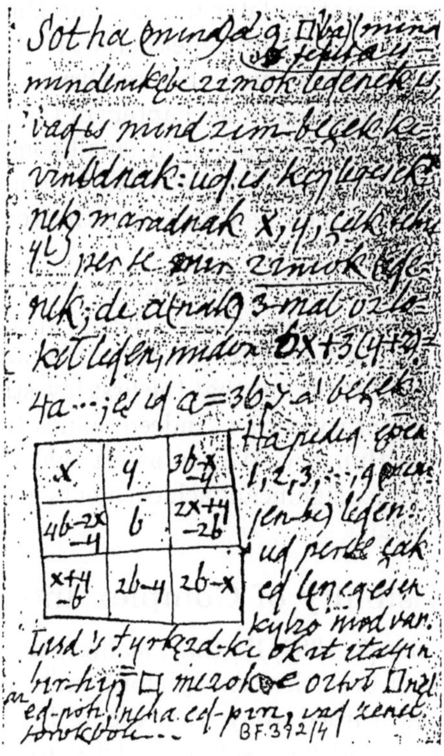

Bolyai got the bulk of his knowledge concerning number theory from Carl Friedrich Gauss's (1777-1855) *Disquisitiones Arithmeticae* (1801).

Including the most important and most recent problems, Gauss had sent a copy of his *Disquisitiones*, dedicated to his friend, Farkas Bolyai (1775-1856) (Amico suo de Bolyai per curam Pauli Vada, auctor) as early as 1803. János could have read it in his early childhood, because at the age of 13 he already spoke Latin. But later, he definitely bought some of Gauss's work in Vienna. The latter copy, accompanied by Bolyai's glosses can be found in the Library of the Hungarian Academy of Sciences. Farkas Bolyai's book is in the Teleki-Bolyai Library of Marosvásárhely.

János Bolyai called the *Disquisitiones Arithmeticae* "colossal". He often refers to it in his manuscript containing his inquiries into number theory and algebra. He used its theorems, but often made also critical observations.

At first, few mathematicians were able to penetrate the deep abstractions of the *Disquisitiones*, which was definitely ahead of its time. Gauss himself complained to Farkas Bolyai on September 2, 1808 [15, 93-94] that only six mathematicians had been able to understand his work in all of Europe. Only a generaration later did Lejeune Dirichlet (1805-1859) succeed in popularizing Gauss's arithmetical ideas. It is not mentioned in the history of mathematics that – before Dirichlet – it was the two Bolyais who had fully understood the chapters of the *Disquisitiones* and fathomed its secrets. It was, in fact, their small companion, handbook and constant book of reference. They studied it thoroughly. Nevertheless, due to their isolation, their new ideas and theorems did not become common property.

We used to think [16, 243, 287] that Hungarian mathematics did not have significant achievements in number theory up to the last quarter of the 19th century. If we take into consideration János Bolyai's ideas mentioned above, then it turns out that the first investigations in number theory started half a century earlier. Indeed, János Bolyai was the first to do such inquiries in Hungary. We can assert that the first Hungarian mathematician to produce significant achievements in the field of number theory was János Bolyai.

3. The Divisibility of Complex Integers

János Bolyai admits [3, 1389/1^v] that, at first, the concept of imaginary quantities had not been clear enough for him, he had regarded them as useless, and he had turned away from them. But later he glimpsed the formula $i^i = e^{-\frac{\pi}{2}}$, and he assiduously tried to create a clear idea about these numbers. He exchanged several letters with his father on this topic. When he learned that the Jablonowski Association of Leipzig commissioned a competition on the topic of the possiblity of constructing complex numbers, Bolyai wrote a brief paper, entitled *Responsio*, and he sent it to Leipzig in 1837.

However, János Bolyai's investigations of complex numbers cannot be restricted to his *Responsio*. If one meticulously gathers all the information left behind in his scattered manuscripts, one can find his theory regarding the divisibility of complex integers.

It is known that this theory was introduced and first published by Gauss in the years 1831 and 1832 [6], [7]. Still, Bolyai's notes prove that independent of Gauss and at about the same time he also conceived the divisibility properties of complex integers [11]. It had already been ready in 1837, but he did not include it in his *Responsio*.

In many of his notes, Bolyai sets the 1820s as starting point of his inquiry on complex integers. He even mentions that from the very beginning he had felt the need to extend number theory on the field of complex numbers [3, 982/8^v].

Prime numbers were in the centre of János Bolyai's attention. His notes on prime numbers can be found on various pages of his writings. In the case of complex primes he tries to distinguish them from rational primes. Thus, rational integer primes, that is, the numbers $\pm 2, \pm 3 \pm 5, \ldots$ are called principal numbers, those numbers, which are prime numbers in the ring of complex integers, such as, for example, $1+i, 1+2i, 2-3i, 6-i, K$ are called perfect prime numbers, and those which are prime numbers in both the ring of rational and in that of complex integers, such as, 3, 7, 11 ... (the prime numbers having the form $4m+3$) are called absolute prime numbers.

Various pages of his manuscripts show that Bolyai could clearly distinguish complex primes. Just as Gauss [8, 543], he ascertains that complex prime numbers are:

a. The numbers $1+i, 1-i, -1+i, -1-i$ [3, 1265/40],

b. The rational prime numbers of the form $4m+3$ [3, 1446, 10^v],

c. The complex factors of rational prime numbers of the form $4m+1$ [3, 1466/10, 10^v].

The manuscripts reveal the demonstration of all these, as well as the theorem of unique prime factorization [3, 1407/3^v, 1431/1].

János Bolyai is also concerned with the theory of congruences within the ring of complex integers. He determines – just as Gauss [8, 552], but by different means – those complex integers $e+fi$, for which – with given $a+bi$ and $c+di$ – we have $c+di \equiv e+fi \pmod{a+bi}$ [3, 1190/18^v,19].

4. On the Algebraic Investigations by János Bolyai

Fewer people know that Bolyai was also engaged in solving problems of algebra and that he achieved important results in this field, too. Several monographs dealing with the life and work of Bolyai mention that Bolyai was engaged in solving higher-order algebraic equation, but nowhere can we read about his actual achievements in the field.

Study of Bolyai's manuscripts revealed two stages in his research. In the first stage he, like many others before him, chose the wrong way of trying to solve algebraic equations of order 5. Moreover he thought to have found a way to solve algebraic equations of arbitrary order. After several unsuccessful attempts, in the second stage, he arrived at the so-called Ruffini-Abel theorem.

János Bolyai began to study this problem in 1837; on page [3, 143/3] of the manuscript he says: "*As far back as 1837 I already paid attention to this subject*". By this time the mathematicians already knew the Ruffini-Abel theorem, but Bolyai lived far away from the centers of mathematical scientific life of his age, working in isolation and seclusion. The current specialty publications, mathematical periodicals, recent information never reached him, which is why he began to work on problems that had already been solved by others before him. The same thing happened with the problem of solvability of algebraic equations: although he was a contemporary of Abel (1802-1829), he didn't know about his work, neither did he learn later of Galois' results that were published by Liouville in 1846.

Still, he had two important sources of information. One of them was his father, Farkas Bolyai, who gave him an excellent mathematical education in his childhood, and with whom he maintained a very close scientific relationship throughout his life. We know that Farkas Bolyai had a regular correspondence with Gauss [15]. The other interesting source is the public library established by count Sámuel Teleki (1739-1822) in Marosvásárhely in the early 1800s, the one called nowadays the "Teleki-Bolyai Library". János Bolyai was a regular reader in this library, and here he found several mathematical works that he studied with the utmost interest.

Here he read Gauss's book entitled "Disquisitiones aritmeticae" and his doctoral dissertation dated from 1799, in which he proves for the first time the so-called principle of classical algebra. János Bolyai's interest in solving the higher-degree equations was probably aroused by these works, since in them Gauss expresses his doubts about the solvability of higher degree equations. Bolyai being a great admirer of Gauss, his curiosity was stimulated. On page [3, 710/2^v] of the manuscript he writes: "*The very fact that even a notability like Gauss considers the algebraic solution of algebraic equations of degree higher than four impossible, comes out from his remark on page 645 of the Disquisitiones arithmeticae, where he even promises to prove the impossibility for equations of order five*". And in another passage: "*In chapter 9 of his excellent work* 'Demonstration nova', *published in Leipzig in 1799, Gauss says: Little hope was left for the general (algebraical) solution of the algebraic equations*". These lines had a great effect upon him, he quotes them in other passages later on.

As we have already mentioned, Bolyai did not know Abel's work, but in the Teleki Library he read about Ruffini's (1765-1822) proof, published in 1799, in which it is stated the algebraic equations of order higher than five cannot be solved by radicals. He also came across the work of Andreas von Ettingshausen (1796-1887), which discusses through a whole chapter Ruffini's theorem, but obviously does not mention Abel, since his valid proof concerning insolvability by radicals of general equations of degree higher than four appeared only a year before the publication of this book [5]. On page [3, 547/5] of the manuscript we read: "*Ruffini tried to prove the impossibility, and this has also been included in the 29th paragraph of the Wiener mathematician Ettingshausen. But several mistakes crept into this experiment, and later on I shall publish its interesting disclaimer, following it step by step.*"

So, on one hand, he realized that Ruffini's proof is not correct, on the other hand, he did not know about Abel's work. He concluded that the theorem is not true, and, as we have seen, he intended to disprove it. That is why, for a long time, he devoted all his efforts to solving the equations. First he sought the solution for the equations of order five. I have found two different attempts related to it: one of the manuscripts seemingly meant to be part of a larger paper, but of which only the first part is to be found. The title itself reveals that Bolyai thought he succeeded in solving the problem: "*The algebraical solution of the quintic algebraic equation.*" (on page [3, 710/1] of the manuscript). All we can make out is that he would have liked to apply Lagrange's method to solve the problem.

For a time he even thought he had managed to solve not only the equation of order five but also the algebraic equation of arbitrary order. On [3, 710/1] he writes: "*...here I speak only about the equation of order five, otherwise I also know the algebraic solution of algebraic equation of any degree, ...that I intend to publish.*" This fact is ascertained on [3, 396/5v]: "*Algebraische General-Auflösungs-Methoden aller* *algebraischen Gleichungen (jeder beliebigen Ordnung),*" or "*General algebraic solving methods for any algebraic equation of arbitrary order.*"

Later on he realized that these attempts were wrong. Among the manuscripts I have found several pieces in which he already wrote about the proof of the insolvability of algebraic equations of order higher than four. We can read statements and titles like:

"*As Ruffini's statement still needs proof, I will try to show the impossibility by another way.*" [3, 544/23^v]. "*Correction of Ruffini's theorem.*" [3, 544/24]. "*The most simple and shortest proof of the impossibility of algebraic solution of the general equation of order five ...*" [3, 395/12]. "*Algebraical solvability of the general equation of order five or the peculiar way or method for the impossibility of general solution of the equation of order five.*" [3, 1094/26].

Even from these sentences it is obvious that he was on the right track. On the following page [3, 395/1] we can read clearly the formulation of the Ruffini-Abel theorem as given by Bolyai: "*Theorem: Finding an algebraic solution of general equations of order higher than four, that is, at least of order five, is impossible.*"

The history of mathematics did not keep in evidence that there was a Hungarian mathematician in the 19th century who solved one of the greatest theorems in mathematics.

5. János Bolyai's Relationship with his Father

During the course of their lives the two Bolyais – father and son – exchanged several letters. All these recently found letters present their relationship in a wholly new light. Reading the letters one feels that the so often mentioned father-son relationship needs to be subjected to reconsiderations.

The letters are replete with mutual feelings of faithfulness, worry and respect. In addition to professional topics, almost every letter contains small advices or a few personal remarks. We can read into them that these two lonely men sought to communicate and meet as much as possible. It is worth quoting a few fragments from these letters:

"*Thanks for your inquiry about my herpes and the invitation for tomorrow*" – János starts off one of his letters. "*Many thanks for your wanting to visit me*" – he writes in another fragment. "*I would visit you, but I am afraid of the long journey*" – complains the elderly Farkas.

János always respected his father and admitted his professional merits:

"*The spirit of the Tentamen rises high*" – he writes about Farks's main work. He writes about Farkas's segmentation theorem: "*My father's wonderful conception*" or "*My father's clever conjecture is so beautiful and valuable*".

But Farkas in his turn also has an appreciative opinion about his son's work: "*It is an original work, which is unprecedented among Hungarian mathematicians*", "*The author of the Appendix solved the problem with wonderful acumen*".

We could go on quoting the opinion the two great scientists held about each other. We hope that the misconceptions blemishing the relationship of the two Bolyais will slowly fade away and they will be verified by the poet's words: "*Its clear light is increasing as it outstrips in space and time*".

6. Conclusions

There are very few scientists whose work is appreciated in their lifetime. The inventors of non-Euclidean geometry did not have the opportunity to enjoy the triumph of their discovery. Above all, János Bolyai deserved a better lot. His demonstration that the Euclidean axiom of parallelism was independent of other axioms ended a period of development of two millennia. He solved one of the most lasting problems of geometry and thus created modern geometry. At the same time, he also obtained significant results in other branches of mathematics.

Now, at his bicentennial it is important to evoke his course of life, his ideas and his achivements in mathematics in the light of the most recent research. Thus, we can get a more detailed and colourful portrait of the great inventor.

Acknowledgments

I thank the Domus Hungarica Scientarium et Artium and Arany János Public- Fundation for their support, which made the writing of this paper possible.

Bibliography

[1] Anonymous, *Théorèmes et problèmes sur les nombres*, Journal für die reine und angewandte Mathematik, 6 (1830), 100-106.

[2] János Bolyai, *Appendix prima scientia spatii ...*, Marosvásárhely, 1831 április.

[3] *János Bolyai's manuscripts*, Târgu-Mureş: Teleki-Bolyai Library.

[4] Leonard E. Dickson, *History of the theory of numbers*, I-III, Reprinted from the first edition, Chelsea Publ. Comp., New York, 1971

[5] Andreas von Ettingshausen, *Vorlesungen über die höhere Mathematik*, Wien, 1827.

[6] C.F. Gauss, *Theoria residuorum biquadraticorum: Commentatio secunda,* Göttingische gelehrte Anzeigen, Stück 64, 1831, 625-638.

[7] C.F. Gauss, *Theoria residuorum biquadraticorum: Commentatis secunda,* Commentationes Societatis Regiae Scientiarum *Göttingensis Recentiores,* 7 (1832), 89-198.

[8] C.F. Gauss, *Untersuchungen über höhere Arithmetik,* trans. Hans Maser, Berlin: Springer Verlag, 1889; reprint ed. New York: Chelsea 1965.

[9] Mathias Hausser, *Analitische Abhandlung der Anfangsgründe der Mathematik,* Wien, 1816.

[10] Elemér Kiss, *Mathematical Gems from the Bolyai Chests,* Budapest: Akadémiai Kiadó, 1999.

[11] Elemér Kiss, *Notes on János Bolyai's Researches in Number Theory,* Historia Mathematica 26, (1999), 68-76.

[12] Elemér Kiss, *A Short Proof of Fermat's Two-Square Theorem Given by János Bolyai,* Mathematica Pannonica, 8 (2) (1997), 293-295.

[13] Elemér Kiss, József Sándor, *János Bolyai's Arithmetic Problem,* MatLap, 5 (2001), nr.9, 321-325 (in Hungarian).

[14] Elemér Kiss, József Sándor, *János Bolyai's Arithmetic Problem and its Extension,* Octogon, Mathematical Magazin, 10 (2002), No. 2., 575-578.

[15] Franz Schmidt and Paul Stäckel, *Briefwechsel zwischen Carl Friderich Gauss und Wolfgang Bolyai,* Leipzig: Teubner, 1899.

[16] Barna Szénássy, *History of Mathematics in Hungary until the* 20th *Century,* Springer-Verlag, Berlin Heidelberg New York London Paris Tokyo Hong Kong Barcelona Budapest, 1992.

II

AXIOMATICAL AND LOGICAL ASPECTS

HYPERBOLIC GEOMETRY, DIMENSION-FREE

Walter Benz
Mathematisches Seminar
der Universität
Bundesstr. 55
20146 Hamburg

1. Introduction

If a geometry Γ is based on an arbitrary real inner product space X (as, for instance, Möbius Sphere Geometry, [5], or Lie Sphere Geometry, [3]), then we will understand by "Γ, Dimension-Free" a theory of Γ which applies to every described X, no matter whether finite- or infinite-dimensional, so, for instance, in the same way to $\mathbb{R}^2$ as well as to $C\,[0,1]$ with $fg = \int_0^1 t^2 f\,(t)\,g\,(t)\,dt$ for real-valued functions f, g defined and continuous in $[0,1]$. In the present paper Hyperbolic Geometry will play the central role. The basis of our theory will be an arbitrary real inner product space X of finite or infinite dimension containing at least two linearly independent elements. We will describe different classical, but isomorphic models of Hyperbolic Geometry in this general framework. Later on we will concentrate ourselves on the characterization of hyperbolic mappings as mappings preserving single hyperbolic distances. As a new result a large class of infinite-dimensional hyperbolic geometries will be presented, to which the Theorem of B. Farrahi – A.V. Kuz'minyh which concerns the finite-dimensional case, cannot be extended.

In [1] we presented a natural characterization of Hyperbolic Geometry, dimension-free, based on non-standard translations.

As far as the classical or foundational background of Hyperbolic Geometry is concerned we refer, for instance, to [8, 10–12].

2. Real Distance Spaces

$\Sigma = (S, d)$ is called a *real distance space* if $S \neq \emptyset$ is a set and d a mapping from $S \times S$ into $\mathbb{R}$. If $\Sigma = (S, d)$ and $\Sigma' = (S', d')$ are such spaces, then $\varphi : S \to S'$ is called a *homometry* from Σ into Σ' provided

$$d(x,y) = d'\left(\varphi(x), \varphi(y)\right) \tag{1}$$

holds true for all $x, y \in S$. An *isometry* is a bijective homometry. If there exists an isometry $\varphi : S \to S'$, then Σ, Σ' are said to be *isomorphic*. The set of isometries $\varphi : S \to S$ of Σ constitutes, under the permutation product, the *group of isometries Isom* (Σ) of Σ. Isomorphic spaces Σ, Σ' have isomorphic groups of isometries.

The elements of S are called *points* of $\Sigma = (S, d)$. If a, b are distinct points of Σ, we will call

$$\{x \in S \mid d(a,b) = d(a,x) + d(x,b)\}$$

the *Menger interval* $[a, b]$ and

$$[a,b] \cup \{x \in S \backslash \{a\} \mid b \in [a,x]\} \cup \{y \in S \backslash \{b\} \mid a \in [y,b]\}$$

the *m-line (Menger line)* $m(a,b)$. The set of all $x \in S$ such that $y \in S$, $d(a,x) = d(a,y)$ and $d(b,x) = d(b,y)$ imply $x = y$ is said to be the *l-line* $l(a,b)$. Observe $l(a,b) = l(b,a)$. The set $\{x \in S \mid d(c,x) = \varrho\}$ for $c \in S$ and $\varrho \in \mathbb{R}$ is called the *ball* $B(c, \varrho)$ with *center* c and *radius* ϱ. Suppose that $\Sigma = (S, d)$ and $\Sigma' = (S', d')$ are isomorphic real distance spaces and that $\varphi : S \to S'$ is an isometry. Then the following holds true.

Proposition 2.1. *Isom* $(\Sigma') = \varphi$ *Isom* $(\Sigma)\,\varphi^{-1}$ *and, moreover,*

(1) $\varphi([a,b]) = [\varphi(a), \varphi(b)]$,

(2) $\varphi\left(m(a,b)\right) = m\left(\varphi(a), \varphi(b)\right)$,

(3) $\varphi\left(l(a,b)\right) = l\left(\varphi(a), \varphi(b)\right)$,

(4) $\varphi\left(B(c,\varrho)\right) = B\left(\varphi(c), \varrho\right)$

for all points $a \neq b$ *of* Σ *and all* $c \in S$. *Furthermore, if* Σ *is a metric space in the sense of Fréchet, then also* Σ'.

If Σ satisfies the *axiom of coincidence*,

$$\forall_{x,y \in S}\; d(x,y) = 0 \Leftrightarrow x = y, \tag{2}$$

and also Σ', then, obviously, every homometry φ from Σ into Σ' must be injective. In this case, furthermore, $a, b \in [a,b]$ holds true. If Σ satisfies the *axiom of symmetry*,

$$\forall_{x,y \in S}\; d(x,y) = d(y,x), \tag{3}$$

then $[a,b] = [b,a]$ and $m(a,b) = m(b,a)$.

3. The Weierstraß model

Let X be a real inner product space, i.e. a real vector space equipped with an inner product

$$\delta : X \times X \to \mathbb{R},\ \delta(x, y) =: xy,$$

satisfying $x^2 > 0$ for all $x \neq 0$ in X. We assume that the dimension of X is at least 2. But, of course, the dimension of X may be infinite. The following real distance space Σ will be now of interest:

$$S = \{(x, \sqrt{1+x^2}) \in X \oplus \mathbb{R} \mid x \in X\}$$

with (see, for instance, G. Nöbeling [11, formula (3) on page 141])

$$\cosh d\left((x, \sqrt{1+x^2}), (y, \sqrt{1+y^2})\right) = \sqrt{1+x^2}\,\sqrt{1+y^2} - xy$$

and $d\left((x, \sqrt{1+x^2}), (y, \sqrt{1+y^2})\right) \geq 0$. Of course, $(x, \sqrt{1+x^2})$ may be identified with x, and so S with X (see T. Wiegand [14]). Rewriting the situation, the distance space Σ is given by (X, d) with

$$\cosh d(x, y) = \sqrt{1+x^2}\,\sqrt{1+y^2} - xy,\ d(x, y) \geq 0.$$

The distance space Σ is a metric space (see [2, Proposition 1 on page 22]). For distinct points $a, b \in X$ we have $m(a, b) = l(a, b)$. These sets are called *hyperbolic lines* of the *Weierstraß model* (X, d). All hyperbolic lines are given by all

$$\{p \cosh \xi + q \sinh \xi \mid \xi \in \mathbb{R}\}$$

with $p, q \in X$ satisfying $pq = 0$ and $q^2 = 1$. Through two distinct points a, b there is exactly one hyperbolic line l. If

$$l = \{x(\xi) = p \cos \xi + q \sinh \xi \mid \xi \in \mathbb{R}\}$$

we get $a = x(\alpha)$ and $b = x(\beta)$ for some real α, β. Eventually replacing q by $-q$ we may assume $\alpha < \beta$. The Menger interval $[a, b]$ can then be written in the form

$$[a, b] = \{x(\xi) \mid \alpha \leq \xi \leq \beta\}.$$

Remark 3.1. *The subset k of S of a real distance space $\Sigma = (S, d)$ is called a metric line of Σ if, and only if, there exists a bijection $f : k \to \mathbb{R}$ such that*

$$d(x, y) = |f(x) - f(y)|$$

holds true for all $x, y \in k$ (L.M. Blumenthal, K. Menger, Studies in Geometry, W.H. Freeman and Comp., San Francisco, 1970, p. 238). In a paper "Metric and Periodic Lines in Real Inner Product Space Geometries (Monatshefte Math., 2004, p. 1–10.)" we show that the metric lines of $\Sigma = (X, d)$ are exactly the hyperbolic lines of the Weierstraß model (X, d).

The balls $B(c, \varrho)$ are given as follows (see [2, p. 24]):

(1) $B(c, \varrho) = \emptyset$ for $\varrho < 0$,

(2) $B(c, 0) = c$,

(3) $B(c, \varrho) = \{x \in X \mid ||f - x|| + ||g - x|| = 2\alpha\}$ for $\varrho > 0$,

where $||a|| = \sqrt{a^2}$ for $a \in X$ and where $f := ce^{-\varrho}$, $g := ce^{\varrho}$ and $\alpha := \sinh \varrho \cdot \sqrt{1 + c^2}$ with the exponential function $e^t = \exp(t)$ for $t \in \mathbb{R}$.

In order to describe the homometries of (X, d) into itself and, furthermore, its isometries we proceed as follows. We define an orthogonal mapping $\omega : X \to X$ of X by $\omega(0) = 0$ and by

$$||\omega(x) - \omega(y)|| = ||x - y|| \tag{4}$$

for all $x, y \in X$ satisfying $||x - y|| = 1$ or $||x - y|| = 2$. Then ω must be linear and it must satisfy (4) for all $x, y \in X$ (E. Schröder [13]). Denote by $\tilde{O}(X)$ the set of all orthogonal mappings of X, and by $O(X)$ the set of all surjective orthogonal mappings, so it is the orthogonal group of X. Let, moreover, p be an element of X with $p^2 = 1$. For $t \in \mathbb{R}$ define

$$T_t(x) = x + \big((xp)(\cosh t - 1) + \sqrt{1 + x^2} \sinh t\big)\, p$$

for all $x \in X$. These $T_t : X \to X$ are the *hyperbolic translations* of X with *axis* p. They are isometries of (X, d) (see [2, pages 17, 18]). Denote by $\mathfrak{T}(p)$ the group of all hyperbolic translations of X with axis p. Then the following holds true ([2, p. 19], and [4, Remark 3, p. 9]).

Theorem 3.2. *All homometries of (X, d) into itself are given by*

$$O(X) \cdot \mathfrak{T}(p) \cdot \tilde{O}(X),$$

and all isometries by

$$H(X) := O(X) \cdot \mathfrak{T}(p) \cdot O(X).$$

$H(X)$ is called the hyperbolic group of X.

4. The Cayley – Klein model

Let X again be an arbitrary real inner product space of finite or infinite dimension ≥ 2. The real distance space (S, g) we are now interested in is given by

$$S := \{x \in X \mid x^2 < 1\}$$

and by

$$g(x, y) := \frac{1}{2} |ln(a, b; x, y)|$$

for $x, y \in S$, where the cross-ratio of a, b, x, y is involved, and where a, b are the points of intersection of the euclidean line

$$\{x + \xi (y - x) \mid \xi \in \mathbb{R}, \ \ x \neq y\}$$

with $\{x \in X \mid x^2 = 1\}$. We will show that this distance space (S, g) is isomorphic to the distance space (X, d) of section 3. In fact, the Weierstraß map $\mu : S \to X$,

$$\mu(x) = \frac{x}{\sqrt{1 - x^2}},$$

is a bijection, and it satisfies

$$g(x, y) = d(\mu(x), \mu(y))$$

for all $x, y \in S$. Hence, by section 2, the distance spaces (S, g) and (X, d) have isomorphic groups of isometries. In view of the Proposition of section 2 it is now easy to determine the corresponding objects to those of the Weierstraß model. For instance,

$$\mu^{-1}(\{p \cosh \xi + q \sinh \xi \mid \xi \in \mathbb{R}\}) = \left\{ \frac{p}{\sqrt{p^2 + 1}} + \frac{q}{\sqrt{p^2 + 1}} tanh \xi \mid \xi \in \mathbb{R} \right\}$$

is a segment of S, and suitable p, q lead to any segment of S. The Menger intervals are usual euclidean intervals. The group of *hyperbolic transformations* of the Cayley–Klein model, i.e. *Isom* (S, g), is given by

$$\mu^{-1} H(X) \mu = \mu^{-1} O(X) \mu \cdot \mu^{-1} \mathfrak{T}(p) \mu \cdot \mu^{-1} O(X) \mu \tag{5}$$

for a fixed $p \in X$ with $p^2 = 1$. Obviously,

$$\mu^{-1} \omega \mu(x) = \omega(x)$$

for all $\omega \in \tilde{O}(X)$ and $x \in S$, in view of $x \cdot x = \omega(x) \cdot \omega(x)$, so that $\mu^{-1} O(X) \mu$ is again $O(X)$, but now acting on S. In order to give (5) a final form, we have to determine $\mu^{-1} \mathfrak{T}(p) \mu$. So observe

$$\mu^{-1} T_t \mu(x) = \frac{x + [(xp)(\cosh t - 1) + \sinh t] p}{\cosh t + (xp) \sinh t}$$

for all $t \in \mathbb{R}$, in view of $\cosh t + (xp)\sinh t > 0$ since $(xp)^2 \leq x^2p^2 = x^2 < 1$. Put

$$R_t := \mu^{-1}T_t\mu.$$

Because of $T_{t_1+t_2} = T_{t_1}T_{t_2}$ (see [2, p. 18]) for all $t_1, t_2 \in \mathbb{R}$, we hence get

$$R_{t_1+t_2} = \mu^{-1}T_{t_1+t_2}\mu = \mu^{-1}T_{t_1}\mu \cdot \mu^{-1}T_{t_2}\mu = R_{t_1}R_{t_2}.$$

5. Poincaré's models

The results of this section can be found in [5]. Let X be an arbitrary real inner product space of finite or infinite dimension ≥ 2. The elements of

$$\overline{X} := X \cup \{\infty\}$$

are called M-points (*Möbius points*). The M-balls are defined as follows. These are the sets of M-points

$$\text{(i)} \quad S(m, \varrho) := \{x \in X \mid ||m - x|| = \varrho\}$$

with $0 < \varrho \in \mathbb{R}$ and $m \in X$,

$$\text{(ii)} \quad T(a, \alpha) := \{x \in X \mid ax = \alpha\} \cup \{\infty\}$$

with $\alpha \in \mathbb{R}$ and $a \neq 0$ in X.
$S(m, \varrho) \subseteq S(n, \sigma)$ implies $m = n$ and $\varrho = \sigma$. Moreover, $T(a, \alpha) \subseteq T(b, \beta)$ implies the existence of a real $\gamma \neq 0$ with $b = \gamma a$ and $\beta = \gamma\alpha$.

The bijections of $\overline{X}$ such that images and inverse images of M-balls are M-balls are called the *M-transformations* of X. The set of all these M-transformations is a group under the usual product of bijections. It will be called the *Möbius group* $\mathbb{M}(X)$ of X.

If $\omega \in O(X)$, $\gamma \in \mathbb{R}$ with $\gamma \neq 0$, and $a \in X$, then

$$f(x) := \gamma\omega(x) + a \text{ for } x \in X, \text{ and } f(\infty) = \infty,$$

is an M-transformation, called the *similitude* $\sigma(\omega, \gamma, a)$. The *inversion* in the M-ball $S(m, \varrho)$ is defined by

$$f(x) := m + \varrho^2 \cdot \frac{x - m}{||x - m||^2}$$

for $x \in X \backslash \{m\}$, and by $f(m) = \infty$, $f(\infty) = m$. It is also an M-transformation. The *inversion* in $T(a, \alpha)$ is defined by $f(\infty) = \infty$ and

$$f(x) := x + 2\frac{\alpha - ax}{a^2}a$$

for $x \in X$. Also this f is an M-transformation. If k is the inversion in the M-ball b, then sks^{-1} must be the inversion in the M-ball $s(b)$ for all $s \in \mathbb{M}(X)$. The group $\mathbb{M}(X)$ acts transitively on the set $\overline{X}$ of M-points, and it acts transitively on the set of M-balls. Every M-transformation which is not a similitude can be written in the form

$$\sigma_1 \iota \sigma_2,$$

where σ_1, σ_2 are suitable similitudes, and where ι denotes the inversion in $S(0,1)$. The two *sides* of the M-ball $S(m, \varrho)$ are defined by

$$S^+(m,\varrho) = \{\infty\} \cup \{x \in X \mid \|m - x\| > \varrho\},$$
$$S^-(m,\varrho) = \{x \in X \mid \|m - x\| < \varrho\},$$

and the two *sides* of $T(a, \alpha)$ by

$$\{x \in X \mid ax > \alpha\},\ \{x \in X \mid ax < \alpha\}.$$

If $\lambda \in \mathbb{M}(X)$ and if Π is a side, then also $\lambda(\Pi)$ must be a side. Moreover, $\mathbb{M}(X)$ acts transitively on the set of sides.

If a, b are M-balls, a is called *orthogonal* to b, $a \perp b$, in the case $\mu(b) = b$, where μ designates the inversion in a. If $a \perp b$ holds true, then also $b \perp a$.

Suppose that p, q, r are three distinct M-points. The intersection of all M-balls through these M-points is then called an M-*circle.* All M-circles are given by

$$\text{(i)} \quad \{u + \xi v \mid \xi \in \mathbb{R}\} \cup \{\infty\} \tag{6}$$

with $u, v \in X$ such that $v \neq 0$, and

$$\text{(ii)} \quad \{m + r\cos\varphi + s\sin\varphi \mid 0 \leq \varphi < 2\pi\} \tag{7}$$

with $m, r, s \in X$ such that $r^2 = s^2 \neq 0$ and $rs = 0$ hold true.

An M-circle c is said to be *orthogonal to an M-ball* a if all M-balls b containing c are orthogonal to a. In this case $\#(c \cap a) = 2$ holds true.

If p_1, p_2, p_3, p_4 are four distinct points of an M–circle their *cross ratio* is defined as usual in the case (6), and in the case (7) by

$$(p_1, p_2; p_3, p_4) = \frac{\sin\frac{1}{2}(\varphi_1 - \varphi_3)}{\sin\frac{1}{2}(\varphi_1 - \varphi_4)} : \frac{\sin\frac{1}{2}(\varphi_2 - \varphi_3)}{\sin\frac{1}{2}(\varphi_2 - \varphi_4)}$$

provided $p_i = m + r\cos\varphi_i + s\sin\varphi_i$, $0 \leq \varphi_i < 2\pi$, for $i \in \{1,2,3,4\}$.

If $\lambda \in \mathbb{M}(X)$ and if c is an M-circle, then also $\lambda(c)$ is an M-circle. If the M-circle c is orthogonal to the M-ball a, then $\lambda(c) \perp \lambda(a)$ holds true. If p_1, p_2, p_3, p_4 are four distinct points of c, then

$$(p_1, p_2; p_3, p_4) = \big(\lambda(p_1),\ \lambda(p_2);\ \lambda(p_3),\ \lambda(p_4)\big).$$

Remark 5.1. *For far-reaching generalizations of Möbius Geometry see A. Blunck and H. Havlicek [6].*

Suppose that a is an M-ball and that Δ is one of its sides. We then define the real distance space

$$\Sigma = (\Delta, \delta)$$

as follows. If $x \neq y$ are in Δ, there exists exactly one M-circle c through x, y with $c \perp a$. Since $\#(c \cap a) = 2$, put $c \cap a =: \{p, q\}$. Then define $\delta\,(x, x) = 0$ and

$$\delta\,(x, y) = |ln\,(p, q; x, y)|. \tag{8}$$

Theorem 5.2. *The group of all isometries $\varphi : \Delta \to \Delta$ of Σ consists exactly of all M-transformations λ with $\lambda\,(\Delta) = \Delta$.*

The hyperbolic lines, i.e. the sets $m\,(x, y) = l\,(x, y)$ for all $x \neq y$ in Δ, are given by $c \cap \Delta$, where c is an M-circle orthogonal to a.

All these distance spaces (Δ, δ) are isomorphic since $\mathbb{M}\,(X)$ acts transitively on the set of sides and since every M-transformation leaves invariant cross ratios. So up to isomorphism we may speak of exactly one *Poincaré model* over X.

Take a fixed $t \in X$ with $t^2 = 1$. With

$$t^{\perp} = \{x \in X \mid xt = 0\}$$

we get the direct sum $X = t^{\perp} \oplus \mathbb{R}t$. We shall write

$$x =: \overline{x} + x_0 t$$

for $x \in X$ with $\overline{x} \in t^{\perp}$ and $x_0 \in \mathbb{R}$. Observe

$$T\,(t, 0) = t^{\perp} \cup \{\infty\}.$$

Let Δ be the side

$$\Delta := \{x \in X \mid x_0 > 0\}.$$

Here we get with (8) for all $x, y \in \Delta$,

$$2 \sinh \frac{\delta\,(x, y)}{2} = \frac{||x - y||}{\sqrt{x_0 y_0}}.$$

It is now possible to establish the isomorphism between (Δ, δ) and the Weierstraß model over X (see [5]).

6. Mild-hypotheses characterizations of homometries

The result we prove in [4] is the following

Theorem 6.1. *Let $\varrho > 0$ be a fixed real number and $N > 1$ be a fixed integer. If $f : X \to X$ is a mapping satisfying*

$$\forall_{x,y\in X}\ d(x,y) = \varrho \Rightarrow d\big((f(x),\ f(y)\big) \leq \varrho,$$

$$\forall_{x,y\in X}\ d(x,y) = N\varrho \Rightarrow d\big(f(x),\ f(y)\big) \geq N\varrho,$$

then f must be a homometry of X, i.e.

$$\forall_{x,y\in X}\ d(x,y) = d\big(f(x),\ f(y)\big)$$

holds true.

Here again X designates a real inner product space of dimension at least 2. The underlying distance space is the Weierstraß model (X, d) of section 3.

The following result was proved for $\dim X = 2$ by B. Farrahi [7] and for $\dim X < \infty$ by A.V. Kuz'minyh [9].

Theorem 6.2. *If $\dim X < \infty$, then Theorem 6.1 holds also true in the case $N = 1$.*

In [4] we presented an example of an infinite-dimensional real inner product space such that Theorem 6.1 cannot be extended to $N = 1$. We now would like to prove that there exists a large class of such spaces with the same behaviour.

Let M be an infinite set and define X to be the set of all $f : M \to \mathbb{R}$ such that

$$\{m \in M \mid f(m) \neq 0\}$$

is finite. Put

$$(f+g)(m) := f(m) + g(m)$$

for $f, g \in X$ and $m \in M$, and $(\alpha f)(m) := \alpha f(m)$ for $f \in X$, $\alpha \in \mathbb{R}$, $m \in M$. Finally set

$$f \cdot g := \sum_{m\in M} f(m)\, g(m).$$

Define

$$\mathbb{Q}(M) := \{f \in X \mid \forall_{m\in M}\ f(m) \in \mathbb{Q}\}.$$

If $x \in X$ and $x(m) = 0$ for all $m \in M\backslash\{m_1, \ldots, m_n\}$, where $m_1, \ldots, m_n$ are pairwise distinct elements of M, we shall write

$$x = \xi_1 m_1 + \cdots + \xi_n m_n$$

with $x(m_i) =: \xi_i$, $i \in \{1, \dots, n\}$.

Suppose now that $\varrho > 0$ is a fixed given real number. For every $x \in X$ choose an element $\varphi(x) \in \mathbb{Q}(M)$ such that

$$d(x, \varphi(x)) < \frac{\varrho}{2} \tag{9}$$

with the distance d of section 3. This is possible: if

$$x = \xi_1 m_1 + \dots + \xi_n m_n,$$

put $\varphi(x) = \eta_1 m_1 + \dots + \eta_n m_n$ with $\eta_i \in \mathbb{Q}$ sufficiently close to $\xi_i \in \mathbb{R}$, $i \in \{1, \dots, n\}$, such that

$$1 \leq \cosh d(x, \varphi(x)) = \sqrt{1 + \Sigma\xi_i^2}\sqrt{1 + \Sigma\eta_i^2} - \Sigma\xi_i\eta_i$$

is sufficiently close to 1.

Since $\mathbb{Q}(M)$ has the same cardinality as M, let $\omega : M \to \mathbb{Q}(M)$ be a bijection. Put

$$\psi(\omega(m)) := \left(\sqrt{2}\sinh\frac{\varrho}{2}\right) m \tag{10}$$

for all $m \in M$. Hence $\psi : \mathbb{Q}(M) \to X$ and $f : X \to X$ with

$$f(x) := \psi(\varphi(x))$$

for all $x \in X$. We now will prove

$$d(f(x), f(y)) = \varrho \text{ for } d(x, y) = \varrho \text{ with } x, y \in X$$

and $d(f(x), f(y)) \in \{0, \varrho\}$ for $d(x, y) \neq \varrho$. In fact! $\varphi(x) = \varphi(y)$ implies $d(f(x), f(y)) = 0$, and $\varphi(x) \neq \varphi(y)$ obviously $d(f(x), f(y)) = \varrho$, in view of (10). What we finally have to show is that $d(x, y) = \varrho$ implies $\varphi(x) \neq \varphi(y)$. But $\varphi(x) = \varphi(y)$ would lead, in the case $d(x, y) = \varrho$, to the contradiction

$$\varrho = d(x, y) \leq d(x, \varphi(x)) + d(\varphi(y), y) < \frac{\varrho}{2} + \frac{\varrho}{2},$$

in view of (9).

Bibliography

[1] Benz, W.: A common characterization of euclidean and hyperbolic geometry by functional equations. Publ. Math. Debrecen 63 (2003) 495–510.

[2] Benz, W.: Hyperbolic distances in Hilbert Spaces. Aequat. Math. 58 (1999) 16–30.

[3] Benz, W.: Lie Sphere Geometry in Hilbert Spaces. Results Math. 40 (2001) 9–36.

[4] Benz, W.: Mappings preserving two hyperbolic distances. J. Geom. 70 (2001) 8–16.

[5] Benz, W.: Möbius Sphere Geometry in Inner Product Spaces. Preprint: Aequat. Math. 66 (2003) 284–320.

[6] Blunck, A., Havlicek, H.: Projective Representations II. Generalized chain geometries. Abh. Math. Sem. Univ. Hamburg 70 (2000) 301–313.

[7] Farrahi, B.: A characterization of isometries of absolute planes. Result. Math. 4 (1981) 34–38.

[8] Klotzek, B., Quaisser, E.: Nichteuklidische Geometrie. Deutscher Verlag der Wissenschaften. Berlin 1978.

[9] Kuz'minyh, A.V.: Mappings preserving the distance 1. Sibirsk. Mat. Z. 20 (1979) 597–602.

[10] Molnár, E.: Kreisgeometrie und konforme Interpretation des mehrdimensionalen metrischen Raumes. Periodica Math. Hung. 10 (1979) 237–259.

[11] Nöbeling, G.: Einführung in die nichteuklidische Geometrie. de Gruyter. Berlin 1976.

[12] Rédei, L.: Begründung der euklidischen und nichteuklidischen Geometrien. Akad. Kiadó. Budapest 1965.

[13] Schröder, E.M.: Eine Ergänzung zum Satz von Beckman und Quarles. Aequat. Math. 19 (1979) 89–92.

[14] Wiegand, T.: A polar-coordinate model of the hyperbolic plane. Publ. Math. 41 (1992) 161–171.

AN ABSOLUTE PROPERTY OF FOUR MUTUALLY TANGENT CIRCLES

H.S.M. Coxeter
University of Toronto
Toronto M5S 3G3, Canada

Abstract When Bolyai János was forty years old, Philip Beecroft discovered that any tetrad of mutually tangent circles determines a complementary tetrad such that each circle of either tetrad intersects three circles of the other tetrad orthogonally. By careful examination of a new proof of this theorem, one can see that it is absolute in Bolyai's sense. Beecroft's double-four of circles is seen to resemble Schläfli's double-six of lines.

Introduction

The absolute property of four mutually tangent circles that I am describing seems to have been discovered by Mr. Philip Beecroft (of Hyde Academy, Cheshire, England) and published in *The Lady's and Gentleman's Diary* "for the year of our Lord 1842, being the second after Bissextile, designed principally for the amusement and instruction of Students in Mathematics: comprising many useful and entertaining particulars, interesting to all persons engaged in that delightful pursuit." [3, p. 92].

In Beecroft's own words,

"If any four circles be described to touch each other mutually, another set of four circles of mutual contact may be described whose points of contact shall coincide with those of the first four."

I like to name this Beecroft's theorem and to express it as follows.

1. Beecroft's Theorem

Four circles, mutually tangent at six distinct points, determine four other circles, mutually tangent at the same six points, such that each circle of either tetrad intersects three circles of the other orthogonally at the points of mutual contact of those three.

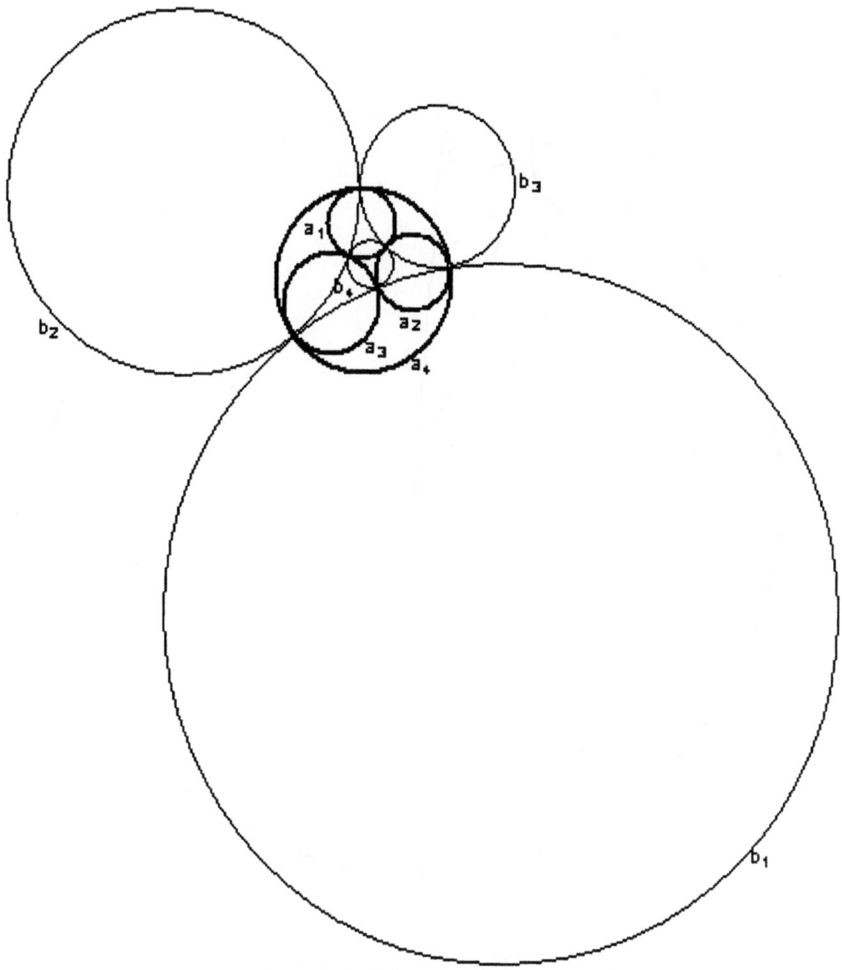

Figure 1. Beecroft's double-four of circles.

In Figure 1 we see four mutually tangent circles a_1, a_2, a_3, a_4 (dark) and another such set of four b_1, b_2, b_3, b_4 (light), such that b_1 passes through the points of mutual contact of a_2, a_3, a_4 and b_2 through the points of mutual contact of a_1, a_3, a_4 and so on. In other words, a_m and b_n intersect each other orthogonally whenever $m \neq n$. This figure makes the theorem almost obvious, but for the sake of completeness it seems desirable to consider further details.

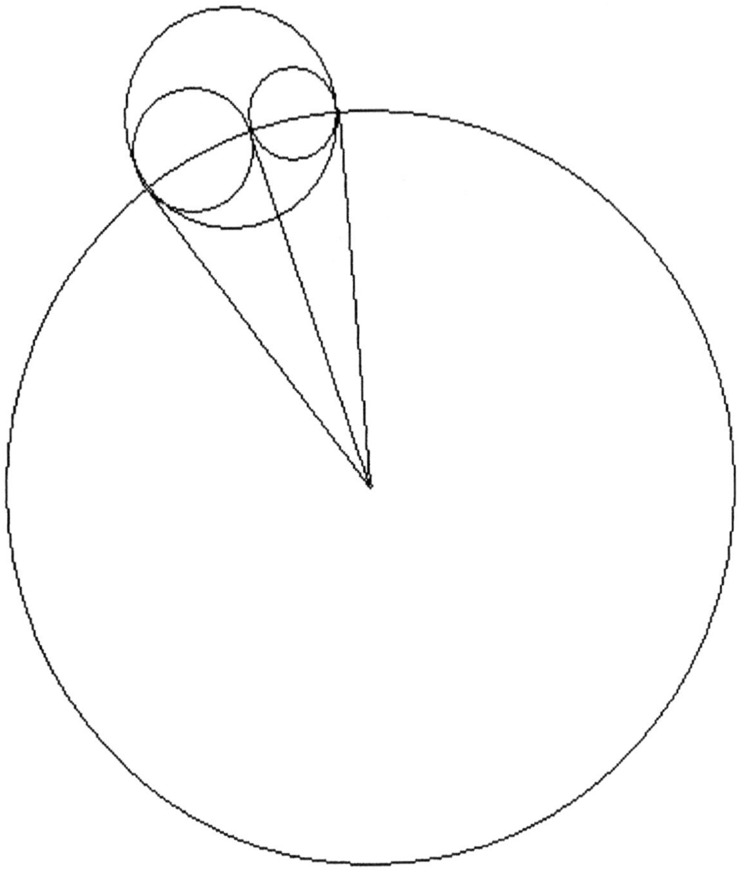

Figure 2. How a_2, a_3, a_4 determine b_1.

2. A new proof

How do we know that the common tangents of the three circles touching one another are concurrent? [6, pp. 311, 316]. It is because these common tangents are radical axes of pairs and all pass through the radical center of these three circles. These three tangents, drawn from the radical center to the points of contact, all have the same length and thus are radii of a new circle intersecting each of the three circles orthogonally (see Figure 2).

Since both b_1 and b_2 intersect a_3 and a_4 at their point of contact, the four b-circles yield the four a-circles by the same procedure that led from the a-circles to the b-circles.

Since no step in this proof uses Euclid's "parallel" postulate, directly or indirectly, Beecroft's theorem is indeed an absolute property of four mutually tangent circles; it holds not only in the Euclidean plane but also on a sphere and in the hyperbolic plane [4, pp. 27-32]. In the hyperbolic case one or more of the four circles may be replaced by a horocycle or a hypercycle (i.e., an equidistant curve). But Poincaré's circular model for the hyperbolic plane rules out the possibility of four mutually tangent horocycles!

3. Beecroft's theorem on a sphere

A spherical version of Beecroft's theorem is provided by two tetrads of mutually tangent circles which lie on the in-sphere of a cube and are the in-circles of the faces of two regular tetrahedra inscribed in the cube. This compound of two tetrahedra is often called *stella octangula* [5, p. 158; 7, p. 166]. The face-centres of the cube, which are the common midpoints of pairs of crossing edges of the two tetrahedra, are the six points at which corresponding circles of Beecroft's two tetrads intersect orthogonally. Reciprocation with respect to the sphere transforms the vertices of each tetrahedron into the face-planes of the other. Thus corresponding edges are polar lines.

In terms of Cartesian coordinates, the eight vertices of the cube are naturally taken to be $(\pm 1, \pm 1, \pm 1)$, with an even number of minus signs for one tetrahedron, an odd number for the other.

The six face-planes of the cube have the equations

$$x = \pm 1, \quad y = \pm 1, \quad z = \pm 1,$$

and the eight face-planes of the two regular tetrahedra are

$$\pm x \pm y \pm z = 1$$

with an odd number of minus signs for one tetrahedron, whose vertices include $(-1, -1, -1)$ in the plane

$$x + y - z = 1$$

and an even number of minus signs in the other, whose vertices include $(-1, -1, -1)$ in the plane

$$x - y - z = 1 \ .$$

In other words, Beecroft's $4 + 4$ circles are the sections of the sphere

$$x^2 + y^2 + z^2 = 1$$

by those eight planes.

4. The double-four of circles and the double-six of lines

It is, perhaps, not too fanciful to recognize some analogy relating Beecroft's double-four of circles

$$\begin{pmatrix} a_1, a_2, a_3, a_4 \\ b_1, b_2, b_3, b_4 \end{pmatrix}$$

in the plane, and Schläfli's double-six of lines

$$\begin{pmatrix} a_1, a_2, a_3, a_4, a_5, a_6 \\ b_1, b_2, b_3, b_4, b_5, b_6 \end{pmatrix}$$

in the projective space [13, p. 213]. (Schläfli was a Swiss contemporary of Bolyai and Beecroft.)

In Beecroft's double-four, two circles a_m and b_n intersect orthogonally whenever $m \neq n$. In Schläfli's double-six, two lines a_m and b_n meet whenever $m \neq n$.

5. Schäfli's Theorem

In Schläfli's own words [13, p. 214] (slightly altered because he abandoned his $a_m b_n$ notation in favour of $A, B, C, D, E, a, b, c, d, e, f$).

The double-sixes give rise to the remark that there is here exposed to view an apparently very elementary theorem which may be thus enunciated:

Draw at pleasure five lines a_2, a_3, a_4, a_5, a_6 which meet just one line b_1. Then (since any four mutually skew lines usually have just two transversals), any four of the five lines may be intersected by another line besides b_1. In this way we have the five tetrads

$$a_3a_4a_5a_6, \quad a_2a_4a_5a_6, \quad a_2a_3a_5a_6, \quad a_2a_3a_4a_6 \quad a_2a_3a_4a_5$$

intersected by

$$b_2 \qquad b_3 \qquad b_4 \qquad b_5 \qquad b_6$$

respectively. The apparently elementary theorem states that the five lines b_2, b_3, b_4, b_5, b_6 have a transversal, which we naturally name a_1, this completing the double-six.

Is there, for this elementary theorem, a demonstration more simple than the one derived from the theory of cubic forms?

Schläfli's challenging question has been answered by a number of geometers, as one can see in the list of References.

Bibliography

[1] H. F. Baker, *A Geometrical Proof of the Theorem of a Double Six of Straight Lines*, Proc. Royal Soc. A 84 (1911), p. 597.

[2] H. F. Baker, *The General Cubic Surface*, Principles of Geometry, Vol 3, Solid Geometry, Cambridge University Press (1934), pp. 159 and 225.

[3] P. Beecroft, *The Concordent Circles*, The Lady's and Gentleman's Diary, The Company of Stationers, London (1843).

[4] H. S. Carslaw, *The Elements of Non-Euclidean Plane Geometry and Trigonometry*, Longmans, London (1916).

[5] H. S. M. Coxeter, *Introduction to Geometry (2nd ed.)*, Wiley, New York (1969).

[6] H. S. M. Coxeter, *Inversive Geometry*, in Educational Studies in Mathematics, Vol 3 (1971), pp. 310-321.

[7] H. S. M. Coxeter, *A Geometriák Alapjai*, Műszaki könyvkiadó, Budapest (1973).

[8] H. L. Dorwart, *The Schäfli Double-Six Configurations*, C.R. Math Rep. Acad. Sci. Canada, Vol 15 (1993), pp. 54-58.

[9] J. Dougall, *The Double-Six of Lines and a Theorem*, in Euclidean Plane Geometry, Proc. Glasgow Math. Assoc., Vol 1 (1952), pp. 1-7.

[10] A. Ichida, *A Simple Proof of the Double-Six Theorem*, Tohoku Math. Journ., Vol 32 (1929), pp. 52-53.

[11] R. J. Lyons, *A Proof of the Theorem of the Double-Six*, Proc. Cambridge Philos. Society, Vol 37 (1941), pp. 433-434.

[12] L. Schäfli, *Theorie der vielfachen Kontinuität*, Gesammelte Mathematische Abhandlungen, Band I, Verlag Birhäuser, Basel (1953).

[13] L. Schäfli, *An attempt to determine the twenty-seven lines upon a surface of the third order, and to devide such surfaces into species in reference to the reality of the lines upon the surface*, Gesammelte Mathematische Abhandlungen, Band II, Verlag Birhäuser, Basel (1953).

[14] B. Segre, *Sulla costruzione delle bisestuple di nette*, Rend. Acad. Naz. Lincei (6), Vol II (1930), pp. 448-449.

[15] C. Yamashita, *An Elementary and Purely Synthetic Proof for the Double-Six Theorem of Schäfli*, Tohoku Math. Journ. (2), Vol 5 (1954), pp. 215-219.

REMEMBERING DONALD COXETER

Asia Ivic Weiss
Professor
Department of Mathematics and Statistics
York University
Phone: 416-736-2100 ext. 22554
Fax: 416-736-5757

William Weiss
Professor
Department of Mathematics
University of Toronto
Phone: 416-978-4150
Fax: 416-978-4107

Donald Coxeter was the leading figure, indeed the architect, of geometry in the 20th century. He passed away peacefully at his home in Toronto on March 31, 2003 at the age of 96. His daughter Susan was at his side as she has been since the death of his beloved wife Rien.

Over a span of eighty years Coxeter made remarkable contributions to mathematics, and to geometry in particular. Written legacy of Coxeter includes over 200 research and survey papers as well as thirteen monographs translated to eight languages. He wrote numerous reviews of other publications where his insights and reflections on the mathematical work of others was much appreciated. In 1974 Coxeter published what he considered his masterpiece, Regular Complex Polytopes. He worked on it for almost twenty years with the aim "...to construct it like a Bruckner symphony, with crescendos and climaxes, little foretastes of pleasure to come..." The exposition is beautiful, the design and presentation extraordinary.

Regular Complex Polytopes is a sequel to his famous Regular Polytopes, which for many of us had profound influence on both our research and our view of geometry. This book, first published in 1948, has been credited with reviving, some would say resurrecting, geometry in the twentieth century. It has drawn to geometry many mathematicians, among them John Conway and Peter McMullen who attest that the book had a major impact on their careers. Through Regular Polytopes, Coxeter established himself as a leading geometer. His work has had great influence outside geometry as well. With the publication of Generators and Relations for Discrete Groups, written jointly with his student Willy Moser and published in 1957, he attracted attention of other mathematicians.

Undoubtedly, Coxeter will be best remembered for his work on regular polytopes and reflection groups. Around 1930, examining symmetries of regular and semi-regular polytopes led him to investigate properties of a group generated by reflections. These are special cases of what is now known as the famous Coxeter groups. In his approach he combined geometry and group theory resulting in the first comprehensive treatment of finite reflection groups. In 1933 he completely classified spherical and Euclidean reflection groups. The corresponding classification for the hyperbolic groups is yet to be completed.

Among the numerous original ideas and solutions to problems in geometry, some results of Coxeter deserve a special mention. The construction of all uniform polyhedra, that is, of polyhedra with regular faces whose symmetry group is transitive on the vertices, was achieved jointly with M.S. Longuet-Higgins and J.C.P. Miller in 1954 (The completeness of the enumeration was proved in 1970s by Sopov and Skilling.) Also in 1954 he presented, as an invited speaker at the International Congress of Mathematicians in Amsterdam, the complete enumeration of hyperbolic tessellations. Coxeter will also be remembered for his many contributions to hyperbolic geometry and sphere packings (and hence to extreme forms and lattices), as well as graph theory and discrete and combinatorial geometry in general.

Coxeter's geometric intuition is legendary and he often disdained formal proof for those things his insight told him were obvious. When Asia was his graduate student she was given a paper in which Donald studied a geometric pattern of four mutually tangent circles first investigated by Renè Descartes. He extended this to an infinite sequence of circles in the plane each one tangent to the previous three and proved that the point of tangency of consecutive circles all lie upon a loxodromic (sometimes called equiangular or logarithmic) spiral. At the end of the paper he casually stated that a similar situation holds in three dimensions: the consecutive points of tangency of an infinite sequence of spheres, each tangent to the proceeding four, belong to an equiangular spiral on a cone. Asia found this statement impossible to verify. This may have been the only instance in which Coxeter's intuition failed him. In fact, she was able to show that these points belong to a loxodromic spiral on a sphere (a curve beginning at a pole and meeting each meridian at the same angle). Donald was surprised and pleased. To commemorate the discovery, Donald and Rien gave to us on our wedding day a crystal bowl in the shape of a sphere with loxodromes on its surface.

Later in life Donald revisited the subject. On his 90th birthday, the famous sculptor John Robinson presented Donald with Firmament, a sculpture of tangent spheres. This sculpture is now housed in the collection at York University. The collection also includes twenty-three miniature models of the stellated icosahedra. There are fifty-nine possible stellations of icosahedra; Coxeter was the first to complete the enumeration. These models were made by H.T. Flather, prior to the Second World War and shipped to Coxeter for safekeeping just before the war. After the end of the war Flather decided to leave them with Coxeter. He made another (complete) set of the icosahedra, which are housed at Cambridge University. Others, including M.C. Escher and Buckminster Fuller also gave original works of art to Donald Coxeter.

Reminiscing about Donald, we often recall the time Asia told him that they might have to postpone their next regular Friday meeting because she was soon due to give birth to our first child. When he realized that she might have to spend several hours in the labour room waiting for the birth, he immediately went to his filing cabinet. Withdrawing his latest preprint, he handed it to Asia, "This is something you can read while waiting." Asia tried to explain that it may be difficult to concentrate upon mathematics at such a time, but Donald reassured her "Don't worry. It is easy to read, mostly a survey." After a successful birth we telephoned him from the hospital room. He congratulated us and inquired about Asia. Being told that she was in good health, he replied "Wonderful, I shall be in my office on Friday." We shall never forget hearing Rien's voice over the telephone screaming "Donald!" Immediately he corrected himself, "I shall also be there next Friday."

Donald was very fond of animals and enjoyed fast cars. His life was filled with music and art. But his passion for geometry was above all. He put finishing touches on his last paper the day before his death. It will appear in the proceedings of the conference in Budapest in 2002 celebrating the 200th anniversary of the birth of the famous Hungarian geometer János Bolyai. He was honoured as the first plenary speaker at the conference.

Our mentor, friend and colleague, Donald Coxeter will be deeply missed; the memory of him, cherished.

(This essay originally appeared in Notes of the Canadian Mathematical Society.)

AXIOMATIZATIONS OF HYPERBOLIC AND ABSOLUTE GEOMETRIES

Victor Pambuccian
Department of Integrative Studies
Arizona State University West, Phoenix, AZ 85069-7100, USA
pamb@math.west.asu.edu

The two of us, the two of us, without return, in this world,
live, exist, wherever we'd go, we'd meet the same faraway point.
—Yeghishe Charents, *To a chance passerby.*

Abstract A survey of finite first-order axiomatizations for hyperbolic and absolute geometries.

1. Hyperbolic Geometry

Elementary Hyperbolic Geometry as conceived by Hilbert

To axiomatize a geometry one needs a language in which to write the axioms, and a logic by means of which to deduce consequences from those axioms. Based on the work of Skolem, Hilbert and Ackermann, Gödel, and Tarski, a consensus had been reached by the end of the first half of the 20th century that, as Skolem had emphasized since 1923, "*if* we are interested in producing an axiomatic system, we can *only* use first-order logic" ([21, p. 472]).

The language of first-order logic consists of the logical symbols $\wedge$, $\vee$, $\rightarrow$, $\neg$, $\leftrightarrow$, a denumerable list of symbols called *individual variables*, as well as denumerable lists of n-ary *predicate (relation)* and *function (operation)* symbols for all natural numbers n, as well as *individual constants* (which may be thought of as 0-ary function symbols), together with two quantifiers, $\forall$ and $\exists$ which can bind only individual variables, but not sets of individual variables nor predicate or function symbols. Its axioms and rules of deduction are those of classical logic.

Axiomatizations in first-order logic preclude the categoricity of the axiomatized models. That is, one cannot provide an axiom system in first-order logic which admits as its only model a geometry over the field of real numbers, as Hilbert [31] had done (in a very strong logic) in his *Grundlagen der Geometrie.* By the Löwenheim-Skolem theorem, if such an axiom system admits an infinite model, then it will admit models of any given infinite cardinality.

Axiomatizations in first-order logic, which will be the only ones surveyed, produce what is called an *elementary* version of the geometry to be axiomatized, and in which fewer theorems are true than in the standard versions over the real (or complex) field. It makes, for example, no sense to ask what the perimeter of a given circle is in elementary Euclidean or hyperbolic geometry, since the question cannot be formulated at all within a first-order language.

This does, however, not mean that the axiom systems surveyed here were presented inside a logical formalism by the authors themselves. In fact, those working in the foundations of geometry, unless connected to Tarski's work, even when they had worked in both logic and the foundations of geometry (such as Hilbert, Bachmann, and Schütte), avoided any reference to the former in their work on the latter. Some of the varied reasons for this reluctance are: (1) given that the majority of 20th century mathematicians nurtured a strong dislike for and a deep ignorance of symbolic logic, it was prudent to stay on territory familiar to the audience addressed; (2) logical formalism is of no help in achieving the crucial foundational aim of proving a representation theorem for the axiom system presented, i. e. for showing that every model of that axiom system is isomorphic to a certain algebraic structure; (3) logical formalism is quite often detrimental to the readability of the axiom system.

The main aim of our survey is the presentation of the axiom systems themselves, and we are primarily concerned with formal aspects of possible axiomatizations of well-established theories for which the representation theorem, arguably one of the most difficult and imaginative part of the foundational enterprise, has been already worked out. It is this emphasis on the manner of narrating a known story which makes the use of the logical formalism indispensable.

We shall survey only finite axiomatizations, i. e. all our axiom systems will consist of finitely many axioms. The infinite ones are interesting for their metamathematical and not their synthetically geometric properties, and were comprehensively surveyed by Schwabhäuser in the second part of [71]. All of the theories discussed in this paper are undecidable, as proved by Ziegler [88], and are consistent, given that they have consistent, complete and decidable extensions. The consistency proof can be carried out inside a weak fragment of arithmetic (as shown by H. Friedman (1999)).

Elementary hyperbolic geometry was born in 1903 when Hilbert [32] provided, using the end-calculus to introduce coordinates, a first-order axiomatization for it by adding to the axioms for plane absolute geometry (the plane axioms contained in groups I (Incidence), II (Betweenness), III (Congruence)) a *hyperbolic parallel axiom* stating that

HPA. *From any point P not lying on a line l there are two rays r_1 and r_2 through P, not belonging to the same line, which do not intersect l, and such that every ray through P contained in the angle formed by r_1 and r_2 does intersect l.*

Hilbert left out many details. The gaps were filled by Gerretsen (1942) and Szász [82, 83] (cf. also Hartshorne [26, Ch. 7, §41-43]), after initial attempts by Liebmann (1904), [49] and Schur (1904). Gerretsen, Szász, and Hartshorne succeeded in showing how a hyperbolic trigonometry could be developed in the absence of continuity, and in providing full details of the coordinatization. Different coordinatizations were proposed by de Kerékjártó (1940/41), Szmielew [85] and Doraczińska [18] (cf. also [71, II.2]).

Tarski's language and axiom system Given that Hilbert's language is a two-sorted language, with individual variables standing for *points* and *lines*, containing point-line incidence, betweenness, segment congruence, and angle congruence as primitive notions, there have been various attempts at simplifying it. The first steps were made by Veblen (1904, 1914) and Mollerup (1904). The former provided in 1904 an axiom system with *points* as the only individuals and with betweenness as the only primitive notion, arguing that segment and angle congruence may be defined in Cayley's manner in the projective extension, and thus, in the absence of a precise notion of elementary (first-order) definability, deemed them superfluous. In 1914 he provided an axiom system with *points* as individual variables and betweenness and equidistance as the only primitive notions. Mollerup (1904) showed that one does not need the concept of angle-congruence, as it can be defined by means of the concept of segment congruence. This was followed by Tarski's [86] most remarkable simplification of the language and of the axioms, a process started in 1926-1927, when he delivered his first lectures on the subject at the University of Warsaw, by both turning, in the manner of Veblen, to a one-sorted language, with *points* as the only individual variables — which enables the axiomatization of geometries of arbitrary dimension, without having to add a new type of variable for every dimension, as well as that of dimension-free geometry (in which there is only a lower-dimension axiom, but no axiom bounding the dimension from above) — and two relation symbols, the same used by Veblen in 1914, namely betweenness and equidistance.

We shall denote Tarski's first-order language by $\mathcal{L}(B, \equiv)$: there is one sort of individual variables, to be referred to as *points*, and two relation symbols, a ternary one, B, with $B(abc)$ to be read as 'point b lies between a and c', and a quaternary one, $\equiv$, with $ab \equiv cd$ to be read as 'a is as distant from b as c is from d', or equivalently 'segment ab is congruent to segment cd'. For improved readability, we shall use the following abbreviation for the concept of collinearity (we shall use the sign $:\Leftrightarrow$ whenever we introduce abbreviations, i. e. defined notions):

$$L(abc) :\Leftrightarrow B(abc) \vee B(bca) \vee B(cab). \tag{1}$$

In its most polished form (to be found in [71] (cf. also [87] for the history of the axiom system)), the axioms corresponding to the plane axioms of Hilbert's groups I, II, III, read as follows (we shall omit to write the universal quantifiers for universal axioms):

A 1.1. $ab \equiv ba$,

A 1.2. $ab \equiv pq \wedge ab \equiv rs \rightarrow pq \equiv rs$,

A 1.3. $ab \equiv cc \rightarrow a = b$,

A 1.4. $(\forall abcq)(\exists x)\, B(qax) \wedge ax \equiv bc$,

A 1.5. $a \neq b \wedge B(abc) \wedge B(a'b'c') \wedge ab \equiv a'b' \wedge bc \equiv b'c' \wedge ad \equiv a'd'$
$\wedge bd \equiv b'd' \rightarrow cd \equiv c'd'$,

A 1.6. $B(aba) \rightarrow a = b$,

A 1.7. $(\forall abcpq)(\exists x)\, B(apc) \wedge B(bqc) \rightarrow B(pxb) \wedge B(qxa)$,

A 1.8. $(\exists abc)\, \neg L(abc)$,

A 1.9. $p \neq q \wedge ap \equiv aq \wedge bp \equiv bq \wedge cp \equiv cq \rightarrow L(abc)$.

A1.4 is a segment transport axiom, stating that we can transport any segment on any given line from any given point; A1.5 is the five-segment

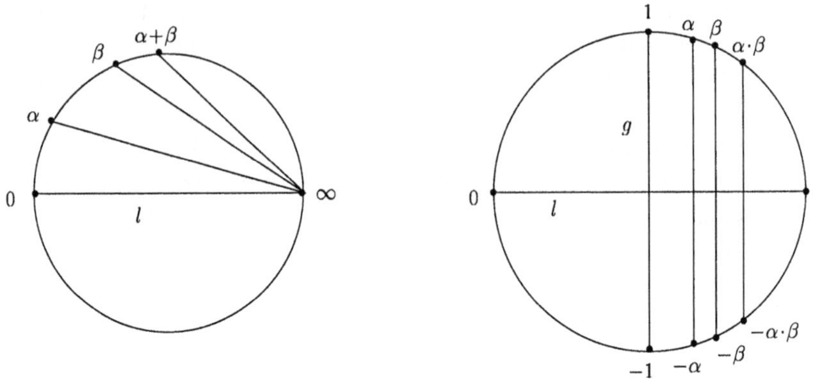

Figure 1. The end-calculus

axiom, whose statement is close to the statement of the side-angle-side congruence theorem for triangles; A1.7 is the Pasch axiom (in its *inner form*); A1.8 is a lower dimension axiom stating that the dimension is ≥ 2; A1.9 is an upper-dimension axiom, stating that the dimension is ≤ 2. We denote by $\mathcal{A}_2$ the $\mathcal{L}(B, \equiv)$-theory axiomatized by A1.1-A1.9, and by $\mathcal{A}$ the one axiomatized by A1.1-A1.8, i. e. $\mathcal{A}_2 := Cn(A1.1{-}A1.9)$, where $Cn(\Sigma)$ stands for the set of logical consequences of Σ.

The following axiom does not follow from Hilbert's axioms of groups I, II, III, but it nevertheless states a property common to Euclidean and hyperbolic geometry, usually called the Circle Axiom, which states that a circle intersects any line passing through a point which lies inside the circle.

CA. $(\forall abcpqr)(\exists x)\, B(cqp) \wedge B(cpr) \wedge ca \equiv cq \wedge cb \equiv cr \rightarrow cx \equiv cp \wedge B(axb)$.

The exact statement **CA** makes is: "If a is inside and b is outside a circle (with centre c and radius cp), then the segment ab intersects that circle."

All those involved in the coordinatization of elementary plane hyperbolic geometry proved a version of the following

Representation Theorem 1. *$\mathfrak{M}$ is a model of $\mathcal{H}_2 :=$ Cn($\mathcal{A}_2 \cup \{$**HPA**$\}$) if and only if $\mathfrak{M}$ is isomorphic to the Klein plane over a Euclidean ordered field (or, historically more accurate, the Beltrami-Cayley-Klein plane). These are the planes described in Pejas's classification of models of Hilbert's absolute geometry as planes of type III with $J = (0)$ and K a Euclidean ordered field.*

The end-calculus, which is the method developed by Hilbert [32] to prove the above theorem, uses the notion of limiting parallel ray, defined on the basis of **HPA**, to introduce the notion of an *end*, which is an equivalence class of limiting parallel rays. When we say that a line l has two ends α and β, we are saying that the two opposite rays in which the line l can be split, belong to the equivalence classes α and β. On the set of all ends, from which one end, denoted ∞, has been removed, one defines, with the help of the three-reflection theorem (which allows one to conclude that the composition of certain three reflections in lines is a reflection in a line), an addition and a multiplication operation, as well as an ordering, which turn the set of ends without ∞ into a Euclidean ordered field.

One starts by fixing a line l, and labeling its ends 0 and ∞ (see Fig. 1). Given ends α, β not equal to ∞, we let σ_a, σ_b denote the reflections in the lines having the ends ∞, α, and ∞, β respectively. We define $\alpha + \beta$ to be the end ξ, which is the end different from ∞ on the line x, for which $\sigma_x = \sigma_a \sigma_l \sigma_b$ (the fact that the composition of the three reflections $\sigma_a \sigma_l \sigma_b$ is a reflection in a line can be proved from the axioms in [4, §11,1], significantly weaker than those assumed here). To define multiplication, let h be a line perpendicular to l, and let 1 and -1 denote its two ends. Given two ends α and β different from both 0 and ∞, we let a, b denote the perpendiculars to l with ends α, respectively β. Then $\alpha \cdot \beta$ is defined to be that end of the line c for which $\sigma_c = \sigma_a \sigma_h \sigma_b$ which lies (i) on the same side of l in which 1 lies, provided that α and β lie on the same side of l; (ii) on the same side of l in which -1 lies, provided that α and β lie on different sides of l (by "an end ϵ lies on the side **s** of a line l" we mean to state that "there is a ray belonging to ϵ which lies completely in **s**"). The existence of the line c for which $\sigma_c = \sigma_a \sigma_h \sigma_b$ follows from the three-reflection theorem for three lines with a common perpendicular, i. e. A2.18. An end is positive if it lies on the same side of l as 1, and negative if it lies on the same side of l as -1, and zero if it is 0.

We can now extend the set of points of the hyperbolic plane by first adding all ends to it (see Fig. 2). The set of all perpendiculars to a line g of the hyperbolic plane will be called a *pole* of g, and will be denoted by $P(g)$. We shall treat poles as points, and add these new points to our plane, calling them *exterior* points. The extended plane thus consists of the points of the hyperbolic plane, to be referred to as *interior* points, of ends, to be referred to as *absolute* points, and of exterior points. We also extend the set of lines with two kinds of lines: *absolute* lines and *exterior* lines, in one-to-one correspondence with absolute respectively interior points. The absolute point uniquely determined by two different

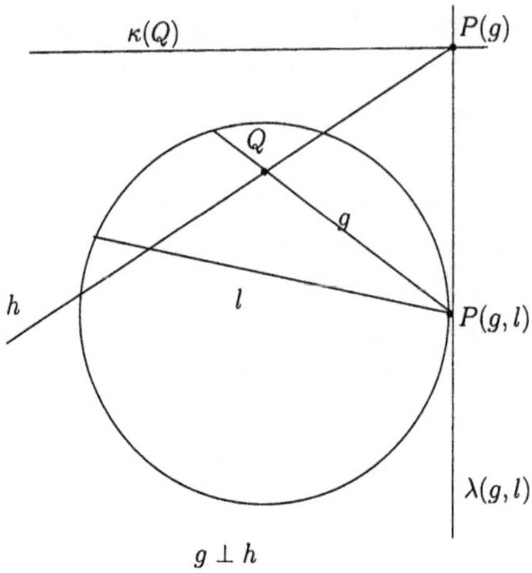

Figure 2. The projective extension

(interior) lines g and h passing through it will be denoted by $P(g,h)$, and its associated absolute line by $\lambda(g,h)$; the exterior line associated with the interior point P will be denoted by $\kappa(P)$. The incidence structure of the extended plane is given by the following rules: (i) interior points are not on absolute or exterior lines; (ii) absolute points are not on exterior lines; (iii) an absolute point $P(g,h)$ is on an interior line l if and only if l,g,h go through the same end; (iv) the absolute point $P(a,b)$ is on the absolute line $\lambda(g,h)$ if and only if $P(a,b) = P(g,h)$; (v) an exterior point $P(l)$ is on an interior line g if and only if g is perpendicular to l; (vi) an exterior point $P(l)$ is on an exterior line $\kappa(Q)$ if and only if $Q \in l$; (vii) an exterior point $P(l)$ is on an absolute line $\lambda(g,h)$ if and only if l passes through $P(g,h)$. The extended plane turns out to be a projective plane coordinatized by the ordered field of ends, and the correspondence between points and lines we have just defined turns out to be a hyperbolic projective polarity, with the set of ends as its *absolute conic.* The hyperbolic plane is thus the interior of the absolute conic, in other words, the Klein plane over the field of ends.

Synthetic proofs that **CA**, as well as the Two Circle Axiom (**TCA**), stating that two circles, one of which has points both inside and outside the other circle, intersect, follow from $\mathcal{H}_2$, were provided by Schur (1904), Szász (1958) and Strommer [80,81]. That $\mathcal{A}_2 \vdash$ **TCA** $\rightarrow$ **CA** was proved in [42, p. 168f], and the fact that **CA** $\rightarrow$ **TCA** can also be proved based on $\mathcal{A}_2$ was shown by Strommer (1973).

That **HPA** can be replaced by the weaker requirement that

$\mathbf{HPA}_0$. *There is a point P and a line l, with P not incident with l, and there are two rays r_1 and r_2 through P, not belonging to the same line, which do not intersect l, and such that every ray through P contained in the angle formed by r_1 and r_2 does intersect l.*

has been shown independently by Strommer (1962), Piesyk (1961), and Baumann and Schwabhäuser (1970). These authors were probably aware that at the time of their writing the problem of finding equivalents of **HPA** had been reduced to that of checking whether a certain statement holds in a certain algebraically described coordinate geometry, since Pejas [66] had succeeded in describing algebraically all models of $\mathcal{A}_2$, so their aim was to provide meaningful synthetic proofs for the equivalence. The same holds for the synthetic proofs of **CA** and **TCA**, and the proof of their equivalence in $\mathcal{A}_2$.

Among the weaker versions of **HPA**, there are a number of axioms which have been of interest. The first is the axiom characterizing the metric, and not the behaviour of parallels, as non-Euclidean. Among the many equivalent statements, "There are no rectangle" ($\neg\mathbf{R}$) is the most suggestive. A strengthening of $\neg\mathbf{R}$, stating that the metric is hyperbolic, can be expressed as "The midline of a triangle is less than half of that side of the triangle whose midpoint is not one of the endpoints of the midline" (**HM**). A weakening of **HPA**, which is stronger than **HM**, and was introduced by Bachmann [5], is the negation of his *Lotschnittaxiom* (**A**), stating that "There exists a quadrilateral with three right angles which does not close" (or, put differently, "There is a right angle and two perpendiculars on the sides of it which do not intersect"). Axiom $\neg\mathbf{A}$ is equivalent to an existential statement (cf. [57]). We have the following chain of implications, with no reverse implication holding $\mathcal{A}_2 \vdash \mathbf{HPA} \to \neg\mathbf{A} \to \mathbf{HM} \to \neg\mathbf{R}$ and none of the reverse implications holds in $\mathcal{A}_2 \vdash \mathbf{HPA} \to \neg\mathbf{A} \wedge \mathbf{CA} \to \mathbf{HM} \wedge \mathbf{CA} \to \neg\mathbf{R} \wedge \mathbf{CA}$ either (cf. [23]).

It is natural to ask what the missing link is, that one would need to add to $\mathcal{A}_2$, $\neg\mathbf{R}$, and **CA** to obtain an axiom system for hyperbolic geometry. As proved by Greenberg [24], it is from Aristotle's axiom **Ar**, stating that "The lengths of the perpendiculars from from one side to the other of a given angle increase indefinitely, i. e. can be made longer than any given segment", that we have

$$\mathcal{A}_2 \vdash \mathbf{HPA} \leftrightarrow \neg\mathbf{R} \wedge \mathbf{CA} \wedge \mathbf{Ar}. \tag{2}$$

It follows from (2), and has been pointed out in [46], that $\mathcal{H}_2$ admits a $\forall\exists$-axiom system, i. e. one in which for all axioms, when written in prenex form, all universal quantifiers (if any) precede all existential quantifiers (if any). It was shown by Kusak (1979) that one could replace **CA** in (2) by Liebmann's [49] axiom **L**, best expressed by using perpendicularity, defined by $\perp (abc) :\Leftrightarrow (\exists c')B(cac') \wedge ac \equiv ac' \wedge bc \equiv bc'$, as

L. $(\forall abcd)(\exists x) \perp (bac) \wedge \perp (cbd) \wedge \perp (dca) \rightarrow B(axd) \wedge ab \equiv cx,$

i. e. "In a quadrangle $abcd$ with three right angles b, c, d, the circle with centre c and radius ab intersects the side ad". Thus (2) becomes $\mathcal{A}_2 \vdash \mathbf{HPA} \leftrightarrow \neg\mathbf{R} \wedge \mathbf{L} \wedge \mathbf{Ar}$.

The Menger-Skala axiom system

Menger (1938) has shown that in hyperbolic geometry the concepts of betweenness and equidistance can be defined in terms of the single notion of point-line incidence, and thus that plane hyperbolic geometry can be axiomatized in terms of this notion alone by rephrasing a traditional axiom system in terms of incidence alone. This was one of the most important discoveries, for it shed light on the true nature of hyperbolic geometry, which moved nearer to projective geometry than to its one-time sister Euclidean geometry, in partial opposition to which it had been born. Menger even claimed that this fact alone proved — pace Poincaré — that hyperbolic geometry was actually *simpler* than Euclidean geometry.

Since an axiom system obtained by replacing all occurrences of betweenness and equidistance with their definitions in terms of incidence would look highly unnatural, its axioms long and un-intuitive, expressing properties of incidence in a roundabout manner, Menger and his students have looked for a more natural axiom system, that should not be derived from a traditional one, but should isolate some fundamental properties of incidence in plane hyperbolic geometry from which all others derive. This task was carried out by Menger, whose last word on the subject was [52], and by his students Abbott, DeBaggis and Jenks, but even their most polished axiom system contained a statement on projectivities that was not reducible to a first-order statement. Skala [73] showed that that axiom can be replaced by the axioms of Pappus and Desargues for the hyperbolic plane, thus accomplishing the task of producing the first elegant first-order axiom system for hyperbolic geometry based on incidence alone.

This axiom system is formulated in a two-sorted first-order language, with individual variables for *points* (upper-case) and *lines* (lower-case), and a single binary relation $|$ as primitive notion, with $P|l$ to be read 'point P is incident with line l'. To shorten the statement of some of the axioms we define: (1) the notion of betweenness β, with $\beta(A, B, C)$ ('B lies between A and C') to denote 'the points A, B, and C are three distinct collinear points and every line through B intersects at least one line of each pair of intersecting lines which pass through A and C; (2) the notions of ray and segment in the usual way, i. e. a point X is on (incident with) a ray $\overrightarrow{AB}$ (with $A \neq B$) if and only if $X = A$ or $X = B$ or $\beta(A, X, B)$ or $\beta(A, B, X)$, and a point X is incident with the segment AB if and only if $X = A$ or $X = B$ or $\beta(A, X, B)$; (3) the notion of ray parallelism, for two rays $\overrightarrow{AB}$ and $\overrightarrow{CD}$ not part of the same line by the condition that every line that meets one of the two rays meets the other ray or the segment AC; two lines or a line and a ray are said to be parallel if they contain parallel rays; (4) the notion of a rimpoint as a pair (a, b) of parallel lines, which is said to be incident with a line l if $l = a$, $l = b$ or l is parallel to both a and b, and there exists a line that intersects a, b, and l; a rimpoint (a, b) is identical with a rimpoint (c, d) if both c and d are incident with (a, b). Rimpoints will be denoted in the sequel by capital Greek characters. A point of the closed hyperbolic plane (i. e. a point or a rimpoint) will be denoted in the sequel by capital Latin characters with a bar on top, and will be referred to as a *Point.* If Π_1 and Π_2 are rimpoints, then $\Pi_1\Pi_2$ denotes the line l incident with both Π_1 and Π_2, and $\Pi_1 P$ denotes the line l incident with Π_1 and P.

The axioms, which we present in informal language, their formalization being straightforward, are:

A 1.10. *Any two distinct points are on exactly one line.*

A 1.11. *Each line is on at least one point.*

A 1.12. *There exist three collinear and three non-collinear points.*

A 1.13. *Of three collinear points, at least one has the property that every line through it intersects at least one of each pair of intersecting lines through the other two.*

A 1.14. *If P is not on l, then there exist two distinct lines on P not meeting l and such that each line meeting l meets at least one of those two lines.*

A 1.15. *Any two non-collinear rays have a common parallel line.*

A 1.16. *(Pascal's theorem on hexagons inscribed in conics) If Π_i $(i = 1, \ldots, 6)$ are rimpoints and M, N, P are the intersection points of the lines $\Pi_1\Pi_2$ and $\Pi_4\Pi_5$, $\Pi_2\Pi_3$ and $\Pi_5\Pi_6$, $\Pi_3\Pi_4$ and $\Pi_6\Pi_1$, then M, N, and P are collinear.*

A 1.17. *(Pappus) Let a and b be different lines containing Points $\overline{A}_1$, $\overline{A}_2$, $\overline{A}_3$ and $\overline{B}_1$, $\overline{B}_2$, $\overline{B}_3$ respectively, with $\overline{A}_i \neq \overline{B}_j$ for all $i, j \in \{1, 2, 3\}$ and $\overline{A}_i \neq \overline{A}_j$, $\overline{B}_i \neq \overline{B}_j$ for $i \neq j$. If $\overline{M}$ lies on the lines $\overline{A}_1\overline{B}_2$ and $\overline{A}_2\overline{B}_1$,*

$\overline{N}$ lies on the lines $\overline{A}_1\overline{B}_3$ and $\overline{A}_3\overline{B}_1$, and $\overline{P}$ lies on the lines $\overline{A}_2\overline{B}_3$ and $\overline{A}_3\overline{B}_2$, then $\overline{M}$, $\overline{N}$, and $\overline{P}$ are collinear.

A 1.18. *(Desargues) Let a, b, c be three different lines, $\overline{O}$ a Point incident with each of them, each containing pairs of distinct Points $(\overline{A}_1, \overline{A}_2)$, $(\overline{B}_1, \overline{B}_2)$, and $(\overline{C}_1, \overline{C}_2)$ respectively. If $\overline{M}$ lies on the lines $\overline{A}_1\overline{B}_1$ and $\overline{A}_2\overline{B}_2$, $\overline{N}$ lies on the lines $\overline{A}_1\overline{C}_1$ and $\overline{A}_2\overline{C}_2$, and $\overline{P}$ lies on the lines $\overline{B}_1\overline{C}_1$ and $\overline{B}_2\overline{C}_2$, then $\overline{M}$, $\overline{N}$, and $\overline{P}$ are collinear.*

Although this axiomatization is simpler than any possible one for Euclidean geometry, by being based on point-line incidence alone, there is no $\forall\exists\forall$-axiom system for hyperbolic geometry formulated only in terms of incidence or collinearity (L), as noticed by Pambuccian (2004). All the axioms of the Menger-Skala axiom system can be formulated as $\forall\exists\forall\exists$-axioms, this being the simplest possible one for hyperbolic geometry expressed by means of L alone, as far as quantifier-complexity is concerned.

Since in hyperbolic geometry of any dimension three points a, b, c are collinear if and only if there is a point d such that, for all x we have $\sigma(a\sigma(b\sigma(cx))) = \sigma(dx)$, where $\sigma(xy)$ stands for the point obtained by reflecting y in x (i. e. three points are collinear if and only if the composition of the reflections in those points is a reflection in a point, cf. [39, 41]), we can rephrase any axiom system of hyperbolic geometry expressed in terms of collinearity into one expressed in terms of the binary operation of point reflection σ (or equivalently in terms of the midpoint operation). However, one can provide an axiom system for hyperbolic geometry of any finite dimension in terms of σ of lower quantifier complexity than that of any axiom system based on collinearity. A $\forall\exists$-axiom system for it has been provided in [62].

In the above two theories for plane hyperbolic geometry presented so far, the one expressed in $\mathcal{L}(B, \equiv)$ and the one expressed in terms of collinearity (L) alone, the primitive notions are different, although the intended interpretation of the individual variables is the same (*points* in both cases). In this case, the equivalence or synonymity of two theories can be defined by stipulating that there exists a definition of each predicate and operation symbol of one theory in terms of the primitive notions of the other, and that the theory expressed in the language that contains all the primitive notions of the two theories, containing all the sentences true in both theories, as well as all the definitions referred to above is consistent (in other words, if the two theories have a common definitional extension).

In case the individual variables of two theories do not necessarily have the same intended interpretation (and such a case would be hyperbolic

geometry expressed in $\mathcal{L}(B,\equiv)$ and expressed in a two-sorted language by means of the single binary relation $\in$ of point-line incidence), we say that two theories $\mathcal{T}_1$ and $\mathcal{T}_2$, in languages $\mathcal{L}_1$ and $\mathcal{L}_2$, which we shall, for simplicity's sake (the general case being a straightforward extension), consider to be one-sorted (they each have only one sort of individual variables), axiomatize the 'same geometry', if the following conditions hold:
There are natural numbers k_i for $i = 1, 2$ such that:
(i) one can identify the individuals X of L_1 with any $k = k_1$-tuple $(x_1, \ldots, x_k)$ of individuals from L_2 which satisfies a certain formula with k free variables $\varphi(x_1, \ldots, x_k)$ of L_2;
(ii) there is a definition for the equivalence of two k-tuples, in terms of an L_2 formula ψ with $2k$ free variables, such that $(x_1, \ldots, x_k) \equiv (y_1, \ldots, y_k)$ if and only if $\psi(x_1, \ldots, x_k, y_1, \ldots, y_k)$ holds;
(iii) for every n-ary relation symbol π of L_1, there is an L_2-formula δ_π with kn free variables, such that $\pi(X_1, \ldots, X_n)$ holds if and only if $\delta_\pi(x_{1,1}, \ldots, x_{1,k}, \ldots, x_{n,1}, \ldots, x_{n,k})$ holds, where $(x_{i,1}, \ldots, x_{i,k})$ is a k-tuple associated via (i) to X_i; analogously for operation symbols;
(iv) For every formula θ, if $\mathcal{T}_1 \vdash \theta$ then $\mathcal{T}_2 \vdash \overline{\theta}$, where $\overline{\theta}$ is the L_2-formula obtained by replacing all of its individual variables with k-tuples satisfying φ, all equality symbols with the $\equiv$-relation, and all occurring relation and function symbols with the L_2-formulas that correspond to them by (iii);
(v) these conditions must also hold with 1 and 2 interchanged (k thus becomes k_2).

Constructive axiomatizations

Constructive axiomatizations of geometry were introduced in [53], and the constructive theme was continued with axiomatizations in infinitary logic by Engeler (1968) and Seeland (1978). In the finitary case, they can be characterized as being formulated in first-order languages without predicate symbols, and consisting entirely of universal axioms (a purely existential axiom eliminates the need for individual constants in the language, and will be allowed in constructive axiomatizations). These languages contain only function symbols and individual constants as primitive notions, and the axioms contain, with the possible exception of a single purely existential axiom (in case the language contains no individual constants), *no existential quantifiers.*

Such universal axiomatizations in languages without relation symbols capture the essentially constructive nature of geometry, that was the trademark of Greek geometry. For Proclus, who relates a view held by Geminus, "a postulate prescribes that we construct or provide some simple or easily grasped object for the exhibition of a character, while an axiom asserts some inherent attribute that is known at once to one's auditors". And "just as a problem differs from a theorem, so a postulate differs from an axiom, even though both of them are undemonstrated; the one is assumed because it is easy to construct, the other accepted because it is easy to know." That is, *postulates* ask for the production, the ποίησις of something not yet given, of a τι, whereas *axioms* refer to the γνῶσις of a given, to insight into the validity of certain relationships that hold between given notions. In traditional axiomatizations, that contain relation symbols, and where axioms are not universal statements, such as Tarski's, this ancient distinction no longer exists. The constructive axiomatics preserves this ancient distinction, as the ancient postulates are the primitive notions of the language, namely the individual constants and the geometric operation symbols themselves, whereas what Geminus would refer to as "axioms" are precisely the axioms of the constructive axiom system.

In a certain sense, one may think of a constructive axiomatization as one in which all the existence claims have been replaced by the existence of certain operation symbols, and where there is no need for the usual predicate symbols since they may be defined in a quantifier-free manner in terms of the operations of the constructive language.

A constructive axiom system for plane hyperbolic geometry was provided by Pambuccian (2004). It is expressed in the language $\mathcal{L}_{con} := \mathcal{L}(T', C_1, C_2, K_1, K_2, P, A', H_1, H_2)$, with *points* as variables, which contains only ternary operation symbols having the following intended interpretations: $T'(abc)$ is the point d on the ray opposite to ray $\overrightarrow{ac}$ with $ad \equiv ab$, provided that $a \neq c$ or $a = b$, and arbitrary otherwise; $C_i(abc)$, for $i = 1, 2$, stand for the two points d for which $da \equiv db$ and $da \equiv ac$, provided that $a \neq b$ and b lies between a and c, arbitrary points otherwise; $K_i(abc)$, for $i = 1, 2$, stand for the two points d for which $ad \equiv ab$ and $bd \equiv bc$, provided that (i) c lies between b and a, and is different from b or (ii) c lies strictly between a and the reflection of b in a, two arbitrary points, otherwise; $P(abc)$ stands for the point d on the side ac or bc of triangle abc, for which $da \equiv db$ and $B(adc) \vee B(bdc)$, provided that a, b, c are three non-collinear points, an arbitrary point, otherwise; $A'(abc)$ stands for the point d on the ray $\overrightarrow{ac}$ for which $dd' \equiv ab$, where d' is the reflection of d in the line ab, provided that a, b, c are three non-collinear points, arbitrary, otherwise; $H_i(abc)$, for $i = 1, 2$, stand for the two points d for which $db \perp ba$ and $ad \perp dc$, provided that a, b, c are three different points with $B(abc)$, arbitrary, otherwise (if one does not like the fact that some operations may take "arbitrary" values for some arguments — which means that there is no axiom fixing the value of that operation for certain arguments, for which the operation is geometrically meaningless — one may add axioms stipulating a particular value in cases with no geometric significance (such as $T'(aba) = a$)). Its axioms are all universal, with one exception, a purely existential axiom, which states that there exist two different points. All the operations used are absolute, and by replacing $\neg\mathbf{R}$ (expressed in $\mathcal{L}_{con}$) with $\mathbf{R}$ we obtain an axiom system for plane Euclidean geometry.

That axiom system is the simplest possible axiom system for plane hyperbolic geometry among all axiom systems expressed in languages with only one sort of variables, to be interpreted as points, and without individual constants. The simplicity it displays is twofold. If, from the many possible ways to look at simplicity we choose the syntactic criterion which declares that axiom system to be simplest for which the maximum number of variables which occur in any of its axioms, written in prenex form, is minimal, then the axiom system referred to above is the simplest possible, regardless of language. Each axiom is a prenex statement containing no more than 4 variables. By a theorem of Scott [72] for axiom systems for Euclidean geometry which is valid in the hyperbolic case as well, there is no axiom system with individuals to be interpreted as points for plane hyperbolic geometry, consisting of at most 3-variable sentences, since all the at most 3-variable sentences which hold in plane hyperbolic geometry hold in all higher-dimensional hyperbolic geometries as well. The quantifier-complexity of its axioms is the simplest possible, as it consists of universal and existential axioms, so there are no quantifier-alternations at all in any of its axioms. It is also simplest among all constructive axiomatizations in that it uses only ternary operations, and one cannot axiomatize plane hyperbolic geometry by means of universal and existential axioms solely in terms of binary operations.

Pambuccian (2001) has also shown that plane hyperbolic geometry can be axiomatized by universal axioms in a two-sorted first-order language $\mathcal{L}$, with variables for both *points* and *lines*, to be denoted by upper-case and lower-case Latin alphabet letters respectively, three individual constants A_0, A_1, A_2, standing for three non-collinear points, and the binary operation symbols φ, ι, π_1, π_2 as primitive notions, where $\varphi(A, B) = l$, $\iota(g, h) = P$, $\pi_1(P, l) = g_1$, $\pi_2(P, l) = g_2$ may be read as: 'l is the line joining A and B' (provided that $A \neq B$, an arbitrary line, otherwise), 'P is the point of intersection of g and h' (provided that g and h are distinct and have a point of intersection, an arbitrary point, otherwise), 'g_1 and g_2 are the two limiting parallel lines from P to l' (provided that P is not on l, arbitrary lines, otherwise).

Another constructive axiomatization with *points* as variables, containing three individual constants a_0, a_1, a_2, standing for three non-collinear points, with $\Pi(a_0a_1) = \pi/3$ ($\Pi(xy)$ stands for the Lobachevsky function associating the angle of parallelism to the segment xy), one quaternary operation symbol $\tilde{\iota}$, with $\tilde{\iota}(abcd) = p$ to be interpreted as 'p is the point of intersection of lines $\overline{ab}$ and $\overline{cd}$, provided that lines $\overline{ab}$ and $\overline{cd}$ are distinct and have a point of intersection, an arbitrary point, otherwise', and two ternary operation symbols, $\varepsilon_1(abc)$ and $\varepsilon_2(abc)$, with $\varepsilon_i(abc) = d_i$ (for $i = 1, 2$) to be interpreted as 'd_1 and d_2 are two distinct points on line $\overline{ac}$ such that $ad_1 \equiv ad_2 \equiv ab$, provided that $a \neq c$, an arbitrary point, otherwise', was provided by Klawitter (2003).

Constructive axiomatizations also serve as a means to show that certain elementary hyperbolic geometries are in a precise sense "naturally occurring". If we were to explore the land of plane hyperbolic geometry in one of its models over the real numbers, and all the notes we can take of all the wonders we see have to be written down without the use of quantifiers, being allowed to use only the three constants A_0, A_1, A_2, which are marked in the model we are visiting and represent three generic fixed non-collinear points, as well as the joining, intersection, and hyperbolic parallels operations φ, ι, π_1, π_2, then all the notes we take will be theorems of plane elementary hyperbolic geometry as conceived by Hilbert. If we are as thorough as possible in our observations, then we would have written down an axiom system for that geometry. This may not be so surprising, given that we have the operations π_1 and π_2 as part of our language, so it may be said that we have built into our language the element of surprise we claim to have obtained.

However, even if the language had been perfectly neutral vis-à-vis **HPA**, such as $\mathcal{L}_{con}$, to which we add three individual constants a_0, a_1, a_2, standing for three non-collinear points, the story we could possibly tell of our visit to the land of plane hyperbolic geometry over the reals is that of plane elementary hyperbolic geometry as conceived by Hilbert. Since all the operations in $\mathcal{L}_{con}$ are absolute, in the sense of having a perfectly meaningful interpretation should we have landed in the Euclidean kingdom, the proof that Hilbert's elementary hyperbolic geometry *is* a most natural fragment of full plane hyperbolic geometry over the reals no longer suffers from the shortcoming of the previous one.

Another constructive axiomatization with a similar property of being formulated in a language containing only absolute operations will be referred to in the survey of H-planes.

Other languages with simple axiom systems

The simplest language in which n-dimensional hyperbolic (as well as Euclidean) geometry, i. e. of a theory synonymous with $\mathcal{H}_n$, can be axiomatized, with individuals to be interpreted as *points*, is one containing a single ternary relation such as Pieri's P ($P(abc)$ standing for $ab \equiv ac$) or the perpendicularity predicate $\perp$ (cf. [71]). That no finite set of binary relations with points as variables can axiomatize hyperbolic (or Euclidean) geometry was shown by Robinson (1959) (cf. [71]).

There are several results related to the axiomatizability of hyperbolic geometry in languages in which the individual variables have interpretations other than points or points and lines. In the two-dimensional case, Prażmowski (1986) has shown that hyperbolic geometry can be axiomatized with individual variables to be interpreted as *equidistant lines*, or *circles*, or *horocycles*, in languages containing a single ternary relation, or several binary relations. Prażmowski (1984, 1986) also showed that *horocycles* or *equidistant lines* or *circles* and *points*, with point-horocycle or point-equidistant line, or point-circle incidence can serve as a language in which to axiomatize plane hyperbolic geometry.

Unlike the Euclidean 2-dimensional case, for $n = 2$ and for $n \geq 4$, n-dimensional hyperbolic geometry over Euclidean ordered fields can be (as shown in [62], using a result from [69]) axiomatized by means of $\forall\exists$-axioms with *lines* as individual variables by using only the binary relation of line orthogonality (with intersection) as primitive notion. Just like Euclidean three-space, hyperbolic three-space cannot be axiomatized with line perpendicularity alone (as shown by List [51]), but it can be axiomatized — as noticed by Pambuccian (2000) — with *planes* as individual variables and plane-perpendicularity as the only primitive notion. More remarkable, as noticed in [61], both n-dimensional hyperbolic and Euclidean geometry (coordinatized by Euclidean fields) can be axiomatized with *spheres* as individuals and the single binary predicate of sphere tangency for all $n \geq 2$.

Generalized hyperbolic geometries

Order based generalizations of hyperbolic geometry A generalization of the ordered structure of hyperbolic planes was provided by Prażmowski [68, §2.1] under the name *quasihyperbolic plane*, in a language with points, lines, point-line incidence, and a quaternary relation among points, with $ab \upharpoonleft\upharpoonright cd$ to be interpreted as 'ray $\overrightarrow{ab}$ is parallel to $\overrightarrow{cd}$ or $a = b$ or $c = d$'. Another (dimension-free) generalization was proposed by Karzel and Konrad [38] (cf. also [36, 45]). It is not known how the two generalizations, the quasihyperbolic and that of [38], are related.

Plane geometries Klingenberg [43] introduced the most important fragment of hyperbolic geometry, the generalized hyperbolic geometry over arbitrary ordered fields. Its axiom system, consists of the axioms for metric planes, i. e. A2.13-A2.22, and the two axioms $(\exists ab)\, U(ab)$ and $\bigwedge_{i=1}^{3}(pq|a_i \wedge U(a_i g)) \rightarrow \bigvee_{i=1}^{3} a_i = a_{i+1}$ (addition in the indices being mod 3), with $U(ab)$, to be read 'the lines a and b have neither a point nor a perpendicular in common', being defined by $U(ab) :\Leftrightarrow (\forall xy)\, \neg(xy|a \wedge xy|b) \wedge \neg(x|a \wedge x|b)$ (a different axiom system, based on that of semi-absolute planes can be found in [8]). The axiom system could also have been expressed by means of universal axioms in the bi-sorted first-order language $\mathsf{L}(A_0, A_1, A_2, A_3, \varphi, F, \pi, \iota, \zeta)$, with points and lines as individual variables, where the A_i are point constants such that $\varphi(A_0, A_1)$ and $\varphi(A_2, A_3)$ have neither a common point nor a common perpendicular, ζ is a binary operation with lines as arguments and a line as value, with $\zeta(g, h)$ to be interpreted as 'the common perpendicular to g and h, provided that $g \neq h$ and that the common perpendicular exists, an arbitrary line otherwise'; F is a ternary operation, $F(abc)$ standing for the footpoint of the perpendicular from c to the line ab, provided that $a \neq b$, an arbitrary point, otherwise; and π a ternary operation symbol, $\pi(abc)$ being interpreted as the fourth reflection point whenever a, b, c are collinear points with $a \neq b$ and $b \neq c$, and arbitrary otherwise. By *fourth reflection point* we mean the following: if we designate by σ_x the mapping defined by $\sigma_x(y) = \sigma(xy)$, i. e. the reflection of y in the point x, then, if a, b, c are three collinear points, by [4, §3,9, Satz 24b], the composition $\sigma_c\sigma_b\sigma_a$, is the reflection in a point, which lies on the same line as a, b, c. That point is designated by $\pi(abc)$. If we denote by Ξ the axiom system expressed in this language, then we can prove that Ξ is the axiom system for the universal $\mathsf{L}(A_0, A_1, A_2, A_3, \varphi, F, \pi, \iota, \zeta)$-theory of the standard Kleinian model of the hyperbolic plane over the real numbers. In other words, that if we are allowed to express ourselves only by using the above operations, and none of our sentences is allowed to have existential quantifiers, then all we could say about the phenomena taking place in the hyperbolic plane over the reals in which there are four fixed points A_0, A_1, A_2, A_3, such that the lines $\varphi(A_0, A_1)$ and $\varphi(A_2, A_3)$ are hyperbolically parallel, is precisely the theory of Klingenberg's hyperbolic planes. In this sense, Klingenberg's hyperbolic geometry is a naturally occurring fragment of full hyperbolic geometry. As shown in [43] and [4], all models of Klingenberg's axiom system are isomorphic to the generalized Kleinian models over ordered fields K. Their point-set consists of the points of a hyperbolic projective-metric plane over K that lie inside the absolute, the lines being all the lines of the hyperbolic projective-metric plane that pass through points that are interior to the absolute, and the operations have the intended interpretation. The difference between generalized Kleinian models and Kleinian models over Euclidean ordered fields is that in the former neither midpoints of segments nor hyperbolic (limiting) parallels from a point to a line (in other words intersection points of lines with the absolute) need to exist. In fact, if we add to the $\mathcal{L}(\equiv)$ axiom system for Klingenberg's generalized hyperbolic planes an axiom stating the existence of the midpoint of every segment, we obtain an axiom system for $\mathcal{H}_2$.

A hyperbolic projective polarity ω defined in a Pappian projective plane induces a notion of perpendicularity: two lines are perpendicular if each passes through the pole of the other. If the projective plane is orderable, then one obtains a hyperbolic geometry by the process described earlier. If the original projective plane is not orderable, then one cannot define hyperbolic geometry in this way, but the whole projective plane, with the exception of those lines which pass through their poles, with its perpendicularity relation defined by ω may be considered as a geometry of hyperbolic-type, whose models are called *hyperbolic projective-metric* planes. They were axiomatized by Lingenberg, who also provided an axiomatization for these planes over quadratically closed fields (see [50] and the literature cited therein). As shown in [63], they can be axiomatized in terms of *lines* and *orthogonality*.

Treffgeradenebenen have been introduced by Bachmann [4, §18,6], as models where lines are all the *Treffgeraden*, i. e. lines in $\mathfrak{A}\ (K, -1)$, with K a Pythagorean field, which intersect the unit circle (the set of all points (x, y) with $x^2 + y^2 = 1$) in two points, and where points are all points of $\mathfrak{A}\ (K, -1)$ for which all lines of $\mathfrak{A}\ (K, -1)$ which pass through them are *Treffgeraden*. For a constructive axiom system see [59].

Further generalizations of hyperbolic planes, with a poorly understood class of models, have been put forward by Baer [9] (and formally axiomatized by Pambuccian (2001)) and Artzy [3]. All models of the former must have infinitely many points and lines, whereas the models of the latter may be finite as well.

Higher-dimensional geometries Axiom systems for both three-dimensional hyperbolic geometry over Euclidean ordered fields, and for more general hyperbolic geometries in terms of planes and reflections in planes, obtained from that of Ahrens by adding certain axioms to it, were presented by Scherf (1961). Hübner [35] described algebraically the models of Kinder's (1965) axiom system for n-dimensional absolute metric geometry to which certain additional axioms, in particular axioms of hyperbolic type, have been added, thus generalizing Scherf's work to the n-dimensional case with $n \geq 2$. Kroll and Sörensen [48] have axiomatized a dimension-free hyperbolic geometry, all of whose planes are generalized hyperbolic planes in the sense of Klingenberg.

2. Absolute Geometry

The concept of *absolute* geometry was introduced by Bolyai in §15 of his *Appendix*, its theorems being those that do not depend upon the assumptions of the existence of no more than one or of several parallels. It turned out that this concept has even farther-reaching consequences than that of hyperbolic geometry, for it provides, for an era which knows that geometry cannot be equated with Euclidean geometry, a framework for a definition of what one means by geometry. In a first approximation one would think that the body of theorems common to Euclidean and hyperbolic geometry would form *geometry per se*, and it is this geometry that was first thoroughly studied. One can also conceive of any body of theorems common to Euclidean and hyperbolic geometry as forming an absolute geometry, and it is this more liberal view that we take in this section.

Order-based absolute geometries

Ordered Geometry A most natural choice for a geometry to be called *absolute* would be one based on the groups I and II of Hilbert's axioms, i. e. a geometry of incidence and order, with no mention of parallels. We could call this geometry with Coxeter [14], who follows Artin [2], *ordered geometry*, or the geometry of *convexity spaces*, which were axiomatized by Bryant [11] and shown to be equivalent to ordered geometry by Precup (1980). Its axioms, expressed in $\mathcal{L}(B)$, with L defined by (1), are: A1.8, A1.6, and

A 2.1. $B(aab)$,

A 2.2. $L(abc) \to L(cba) \wedge L(bac)$,

A 2.3. $a \neq b \wedge L(abc) \wedge L(abd) \to L(acd)$,

A 2.4. $B(abc) \to B(cba)$,

A 2.5. $B(abc) \wedge B(acd) \to B(bcd)$,

A 2.6. $(\forall ab)(\exists c)\, B(cab) \wedge c \neq a$,

A 2.7. $(\forall abcxy)(\exists z)\, \neg L(abc) \wedge B(bxc) \wedge x \neq b \wedge x \neq c \wedge B(ayx) \wedge y \neq a \wedge y \neq x \to L(cyz) \wedge B(azb) \wedge z \neq a \wedge z \neq b$.

In the presence of Pasch's axiom A2.7, one can define in a first-order manner the concept of dimension in ordered geometry. Pasch [64] has shown that models of 3-dimensional ordered geometries can be embedded in ordered projective spaces of the same dimension if they satisfy additional conditions. Kahn (1980) has shown that the Desargues theorem need not be postulated in at least 3-dimensional spaces, provided that a condition which ordered spaces do satisfy, holds. Sörensen (1986), Kreuzer (1989), [47] and Frank (1988) offered shorter or more easy to follow presentations of these results, which rely on minimal sets of assumptions. The extent to which convex geometry can be developed within a very weak ordered geometry, which may also be formulated in $\mathcal{L}(B)$ can be read from [13].

Grochowska-Prażmowska [25] has provided an axiom system for an ordered geometry based on the quaternary relation of *oriented parallelism* $\upharpoonleft\!\upharpoonright$. That axiom system is equivalent to that of ordered geometry to which an axiom on ray parallels has been added. That axiom states the existence, for every triple a, b, c of non-collinear points, of a point d, different from c, such that the rays $\overrightarrow{ab}$ and $\overrightarrow{cd}$ are parallel, ray parallelism being defined in the Menger-Skala manner presented earlier. It has both affine ordered planes and hyperbolic planes as models.

Weak general affine geometry Szczerba [84] has shown that if one adds to A1.6, A1.8, A2.6, and the projective form of the Desargues axiom, the following axioms, of which A2.11 is an upper dimension axiom, A2.10 the outer form of the Pasch axiom, and A2.12 an axiom stating that there is a line in the projective closure of the plane which lies outside the plane,

A 2.8. $B(xyz) \wedge B(yzu) \wedge y \neq z \rightarrow B(xyu)$,

A 2.9. $B(xyz) \wedge B(xyu) \wedge x \neq y \rightarrow B(yzu) \vee B(yuz)$,

A 2.10. $(\forall txyzu)(\exists v)\, B(xtu) \wedge B(yuz) \rightarrow B(xvy) \wedge B(ztv)$,

A 2.11. $(\forall xyzt)(\exists u)\, (L(yuz) \wedge L(xtu)) \vee (L(xuy) \wedge L(ztu)) \vee (L(xuz) \wedge L(yut))$,

A 2.12. $(\exists abcd)(\forall xyzuvs)\, L(axb) \wedge L(byc) \wedge L(czd) \wedge L(dua) \wedge L(avc) \wedge L(xvz) \wedge L(yvu) \wedge L(xus) \wedge L(yzs) \rightarrow x = u \vee y = z$,

one obtains an axiom system Ψ for a general affine geometry, all of whose models are isomorphic to open, convex sets in affine betweenness planes over ordered skew fields, an affine betweenness plane $\mathfrak{B}(F)$ over the ordered skew field F being the structure $\langle F \times F, \mathbf{B}_F \rangle$, with $\mathbf{B}_F(\mathbf{abc})$ if and only if $\mathbf{b} = \mathbf{a}(1 - \xi) + \mathbf{c}\xi$ for some $\xi \in F$. If one adds a projective form of Pappus' axiom to Ψ, an extension giving rise to an axiom system Ψ_a, then the skew fields in the representation theorem for Ψ become commutative. This is another naturally occurring theory. Pappian general affine geometry is the $\mathsf{L}(B)$-theory common to ordered affine Pappian geometry and to Klingenberg's generalized Kleinian models, i.e. it contains precisely those $\mathsf{L}(B)$-sentences which are true in both of these geometries. Szczerba [84] also proves a representation theorem for the weaker theory axiomatized by $\Psi \setminus \{\mathrm{A2.12}\}$.

Plane metric ordered geometries

H-planes Following Bolyai, a vast literature on absolute geometries has come into being. The main aim of the authors of this literature is that of establishing systems of axioms that are on the one hand *weak* enough to be common to various geometries, among which the Euclidean and the hyperbolic, and on the other hand, *strong* enough to allow the proof of a substantial part of the theorems of elementary (Euclidean) geometry, and to allow an algebraic description as subgeometries of some projective geometry with a metric defined in the manner of Cayley [12].

The first three groups of Hilbert's [31] axioms provided the first elementary axiomatization of an absolute geometry, which we have already encountered, expressed in $\mathcal{L}(B, \equiv)$ as $\mathcal{A}_3$ (which is $\mathcal{A}$ together with an axiom fixing the dimension to 3).

Of great importance, since it facilitates absolute proofs of theorems, was Pejas's [66] algebraic description of all models of $\mathcal{A}_2$ (also referred to in [29] as Hilbert planes, or H-planes). It reads:

Let K be a field of characteristic $\neq 2$, and k an element of K. By the *affine-metric plane* $\mathfrak{A}\ (K, k)$ (cf. [29, p.215]) over the field K with the *metric constant* k we mean the projective plane $\mathfrak{P}\ (K)$ over the field K from which the line $[0, 0, 1]$, as well as all the points on it have been removed (and we write $\mathfrak{A}\ (K)$ for the remaining point-set), for whose points of the form $(x, y, 1)$ we shall write (x, y) (which is incident with a line $[u, v, w]$ if and only if $xu + yv + w = 0$), together with a notion of orthogonality, the lines $[u, v, w]$ and $[u', v', w']$ being orthogonal if and only if $uu' + vv' + kww' = 0$. If K is an ordered field, then one can order $\mathfrak{A}\ (K)$ in the usual way.

The algebraic characterization of the H-planes consists in specifying a point-set E of an affine-metric plane $\mathfrak{A}\ (K, k)$, which is the universe of the H-plane. Since E will always lie in $\mathfrak{A}\ (K)$, the H-plane will inherit the order relation B from $\mathfrak{A}\ (K)$. The congruence of two segments **ab** and **cd** will be given by the usual Euclidean formula $(a_1 - b_1)^2 + (a_2 - b_2)^2 = (c_1 - d_1)^2 + (c_2 - d_2)^2$ if $E \subset \mathfrak{A}\ (K, 0)$, and by

$$\frac{F(\mathbf{a},\mathbf{b})^2}{Q(\mathbf{a})Q(\mathbf{b})} = \frac{F(\mathbf{c},\mathbf{d})^2}{Q(\mathbf{c})Q(\mathbf{d})}, \tag{3}$$

if $E \subset \mathfrak{A}\ (K,k)$ with $k \neq 0$, where $F(\mathbf{x},\mathbf{y}) = k(x_1y_1+x_2y_2)+1,\ Q(\mathbf{x}) = F(\mathbf{x},\mathbf{x})$, and $\mathbf{x} = (x_1,x_2), \mathbf{y} = (y_1,y_2)$.

Let now K be an ordered Pythagorean field (i. e. the sum of any two squares of elements of K is the square of an element of K), R the ring of *finite* elements, i. e. $R = \{x \in K : (\exists n \in \mathbf{N})\, |x| < n\}$ and P the ideal of *infinitely small* elements of K, i. e. $P = \{0\} \cup \{x \in K : x^{-1} \notin R\}$. All H-planes are isomorphic to a plane of one of the following three types:

Type I. $E = \{(a,b) \mid a,b \in M\} \subset \mathfrak{A}\ (K,0)$, *where M is an R-module $\neq (0)$;*

Type II. $E = \{(a,b) \mid a,b \in M\} \subset \mathfrak{A}\ (K,k)$ *with $k \neq 0$, where M is an R-module $\neq (0)$ included in $\{a \in K \mid ka^2 \in P\}$, that satisfies the condition*

$$a \in M \Rightarrow ka^2 + 1 \in K^2;$$

Type III. $E = \{\mathbf{x} \mid Q(\mathbf{x}) > 0, Q(\mathbf{x}) \notin J\} \subset \mathfrak{A}\ (K,k)$ *with $k < 0$, where $J \subseteq P$ is a prime ideal of R that satisfies the condition*

$$ka^2 + 1 > 0, ka^2 + 1 \notin J \Rightarrow ka^2 + 1 \in K^2,$$

with K satisfying

$$\{a \in K \mid ka^2 \in R \setminus P\} \neq \emptyset.$$

In planes of type I there exist rectangles, so their metric is Euclidean, and we may think of them as 'finite' neighborhoods of the origin inside a Cartesian plane. Those of type II can be thought of as infinitesimally small neighborhoods of the origin in a non-Archimedean ordered affine-metric plane. There is no rectangle in them, and their metric may be of hyperbolic type (should $k < 0$) or of elliptic type (should $k > 0$) — in the latter the sum of the angles of a triangle can exceed two right angles only by an infinitesimal amount. Planes of type III are generalizations of the Klein inner-disc model of hyperbolic geometry. A certain infinitesimal collar around the boundary may be deleted from the inside of a disc, and the metric constant k, although negative, may not be normalizable to -1, as the coordinate field is only Pythagorean and not necessarily Euclidean. In case K is a Euclidean field (every positive element has a square root) and $J = (0)$, we can normalize the metric constant k to -1 and we have Klein's inner-disc model of plane hyperbolic geometry with K as coordinate field.

The axioms for absolute geometry, in particular the *five-segment axiom* A1.5, have been the subject of intensive research. Significant simplifications, which allow the formulation of all plane axioms as prenex sentences with at most six variables, have been achieved by Rigby (1968, 1975), building up on the results obtained by Mollerup (1904), R. L. Moore (1908) Dorroh (1928,1930), Piesyk (1965), Forder (1947), Szász (1961). As shown by Pambuccian (1997), it is not possible to axiomatize $\mathcal{A}_2$ in $\mathcal{L}(B, \equiv)$ by means of prenex axioms containing at most 4 variables, raising the question whether it is possible to do so with at most 5 variables. The axiomatization of Hilbert planes can also be achieved by completely separating the axioms of order from those of congruence and collinearity, i. e. if one expresses the axiom system in $\mathcal{L}(L, B, \equiv)$, then the symbol B does not occur in any axiom in which the symbol $\equiv$ occurs. This was shown in polished form by Sörensen [78]. A constructive axiomatization for a theory $\mathcal{CA}_2$ synonymous with $\mathcal{A}_2$ was provided in [58]. It is expressed in $\mathsf{L}(a_0, a_1, a_2, T', J')$, where the a_i stand for three non-collinear points, T' being the segment transport operation encountered earlier, and J' is a quaternary segment-intersection predicate, $J'(abcd)$ being interpreted as the point of intersection of the segments ab and cd, provided that a and b are two distinct points that lie on different sides of the line cd, and c and d are two distinct points that lie on different sides of the line ab, and arbitrary otherwise. This shows the remarkable fact that plane absolute geometry is a theory of two geometric instruments: segment-transporter and segment-intersector. If we enlarge the language by adding a ternary operation A — with $A(abc)$ representing the point on the ray $\overrightarrow{ac}$, whose distance from the line ab is congruent to the segment ab, provided that a, b, c are three non-collinear points, and an arbitrary point otherwise — we can express constructively a strengthened version of Aristotle's axiom, to be denoted by **Ars**, as: $\neg L(abc) \rightarrow (B(aA(abc)c) \vee B(acA(abc))) \wedge ab \equiv A(abc)F(abA(abc))$. If we add **Ars** and **R** to the axiom system for $\mathcal{CA}_2$, we obtain a constructive axiom system for plane Euclidean geometry over Pythagorean ordered fields, whereas if we add **Ars** and **HM** to the axiom system for $\mathcal{CA}_2$ we obtain a constructive axiom system for plane hyperbolic geometry over Euclidean ordered fields, i. e. a theory synonymous with $\mathcal{H}_2$. Thus both Euclidean and hyperbolic plane geometry may be axiomatized in the same language $\mathsf{L}(a_0, a_1, a_2, T', J', A)$, the only difference consisting in the axiom specifying the metric, no specifically Euclidean or specifically hyperbolic operation symbol being needed to constructively axiomatize the two geometries.

An interesting constructive axiomatization of H-planes over Euclidean ordered fields can be obtained by translating into constructive axiomatizability results the theorems of Strommer (1977) (proved independently by Katzarova (1981) as well), or their generalization in [15], where it is shown that Steiner's theorem on constructions with the ruler, given a circle and its centre C, can be generalized to the absolute setting by having a few additional fixed points in the plane (such as two points P and Q together with the midpoint M of the segment PQ, provided that $CP \not\equiv CQ$).

A very interesting, but never cited, absolute geometry weaker than that of H-planes, is the one considered by Smid [74], who also characterized it algebraically by showing that it can be embedded in a projective metric plane. That geometry is obtained from Hilbert's axiom system of H-planes by replacing the axiom of segment transport with the weaker version which asks, given two point-pairs a, b and c, d, that one of the following two statements hold: (i) the segment ab can be transported on ray $\vec{cd}$ or (ii) the segment cd can be transported on ray $\vec{ab}$. It holds in all open convex subsets of H-planes.

Metric planes

Schmidt-Bachmann planes The geometry of metric planes can be thought of as the metric geometry common to the three classical plane geometries (Euclidean, hyperbolic, and elliptic). Neither order nor free mobility is assumed.

It originates in the observation that a significant amount of geometric theorems can be proved with the help of the three-reflection theorem, which proved for the first time its usefulness in [32]. The axiomatization of metric planes grew out of the work of Hessenberg, Hjelmslev, A. Schmidt, and Bachmann, whose life and students' work have been devoted to their study. Metric planes have never been presented as models of an axiom system in the logical sense of the word, but as a description of a subset satisfying certain conditions inside a group. Given that most mathematicians were familiar with groups, but not with logic — which is the area in which Bachmann had worked before embarking on the reflection-geometric journey — and that the group-theoretical presentation flows smoothly and gracefully, which cannot be said of the formal-logical one, it is perfectly reasonable to present it the way Bachmann did, when writing a book on the subject. Since we are interested here only in the axiom systems themselves, and not in the development of a theory based on them, it is natural to present these structures as axiomatized in first-order logic.

There are two main problems for these purely metric plane geometries: (i) that of their embeddability in a Pappian projective plane, where line-perpendicularity and line-reflection are represented in the usual manner by means of a quadratic form, and (ii) that of characterizing algebraically those subsets of lines in the projective plane in which the metric plane has been embedded, which are the lines of the metric plane. While the first problem has been successfully solved whenever it had a solution, there are only partial results concerning the second one, complete representation theorems being known only for metric planes satisfying additional requirements (such as free mobility or orderability).

In its most polished form, to be found in [4], the axiom system can be understood as being expressed with one sort of variables for *lines*, and a binary operation ϱ, with $\varrho(a,b)$ to be interpreted as 'the reflection of line b in line a'. To improve the readability of the axioms, we shall use the following abbreviations ($p|q$ may be read 'p is orthogonal to q' (i. e., given A2.20, $\varrho(p,q)=q$) — and we may think of the pair (p,q) with $p|q$ as a 'point', namely the intersection point of p and q — and $pq|a$ may be read 'a passes through the intersection point of p and q, two orthogonal lines' or 'the point pq lies on a'):

$$g_1 \dots g_n = h_1 \dots h_m :\Leftrightarrow (\forall x)\, \varrho(g_1, \dots \varrho(g_n, x)\dots) = \varrho(h_1, \dots \varrho(h_m, x)\dots),$$
$$(a_1 \dots a_n)^2 = 1 :\Leftrightarrow (\forall x)\, \varrho(a_1, \dots \varrho(a_n, \varrho(a_1, \dots \varrho(a_n, x)\dots) = x,$$
$$a|b :\Leftrightarrow a \neq b \wedge (ab)^2 = 1, J(abc) :\Leftrightarrow abc \neq 1 \wedge (abc)^2 = 1,$$
$$pq|a :\Leftrightarrow p|q \wedge J(pqa).$$

A 2.13. $a^2 = 1,$

A 2.14. $(\forall ab)(\exists c)\, aba = c,$

A 2.15. $(\forall abcd)(\exists g)\, a|b \wedge c|d \rightarrow J(abg) \wedge J(cdg),$

A 2.16. $ab|g \wedge cd|g \wedge ab|h \wedge cd|h \rightarrow (g = h \vee ab = cd),$

A 2.17. $(\forall a_1a_2a_3pq)(\exists d)\, \bigwedge_{i=1}^3 pq|a_i \rightarrow a_1a_2a_3 = d,$

A 2.18. $(\forall a_1a_2a_3g)(\exists d)\, \bigwedge_{i=1}^3 a_i|g \rightarrow a_1a_2a_3 = d,$

A 2.19. $(\exists ghj)\, g|h \wedge \neg j|g \wedge \neg j|h \wedge \neg J(jgh),$

A 2.20. $a|b \rightarrow \varrho(a,b) = b,$

A 2.21. $a \neq b \wedge \varrho(a,b) = b \rightarrow a|b,$

A 2.22. $\varrho(a,b) = a \rightarrow a = b.$

A2.13 states that reflections in lines are involutions; A2.14 that for all line-reflections a and b, aba is a line-reflection as well; A2.15 that for any two points (a,b) and (c,d) there is a line g joining them; A2.16 states that the joining line of two different points is unique; A2.17 and A2.18 that the composition of three reflections in lines with a point or with a perpendicular in common is a reflection in a line; A2.19 that there are three lines forming a right triangle; A2.20-A2.22 ensure that ϱ has the desired interpretation.

The Euclideanity of a Euclidean plane may be considered as being determined by its affine structure (i. e. by the fact that an Euclidean plane is an affine plane), or as being determined by its Euclidean metric, i. e. by the fact that there are rectangles in that plane. On the basis of orthogonality, one may define in the usual manner a notion of parallelism, and ask whether having a Euclidean metric implies the affine structure, i. e. the intersection of non-parallel lines). It was shown by Dehn [16] that the latter is not the case, i. e. that there are planes with a Euclidean metric, to be called *metric-Euclidean planes*, that are not Euclidean planes (i. e. where the parallel axiom does not hold). Such planes, which are precisely the planes of type I in Pejas's classification for which $E \neq \mathfrak{A}\ (K, 0)$, must be non-Archimedean.

Metric-Euclidean planes were introduced by Bachmann (1948) (see also [4]), as metric planes in which the rectangle axiom, i. e. $(\exists abcd)\, a \neq b \wedge c \neq d \wedge a|c \wedge a|d \wedge b|c \wedge b|d$, holds. The point-set of a metric-Euclidean plane of characteristic $\neq 2$ is a subset of the Gaussian plane over (L, K), where L is a quadratic extension of K (a generalization of the standard complex numbers plane), which contains 0, 1, is closed under translations and rotations around 0, and contains the midpoints of any point-pair consisting of an arbitrary point and its image under a rotation around 0.

Non-elliptic metric planes (i. e. those satisfying the axiom $(\forall abc)\, abc \neq 1$, which will be referred to as $\neg\mathbf{P}$) can also be axiomatized in $\mathcal{L}(\equiv)$, as proposed in [77] (see [59] for a formalization of that axiom system).

By an *ordinary metric-projective* plane $\mathfrak{P}\ (K, \mathfrak{f}\ )$ over a field K of characteristic $\neq 2$, with $\mathfrak{f}$ a symmetric bilinear form, which may be chosen to be defined by $\mathfrak{f}\ (\mathbf{x}, \mathbf{y}) = \alpha x_1 y_1 + \beta x_2 y_2 + \gamma x_3 y_3$, with $\alpha\beta\gamma \neq 0$, for $\mathbf{x}, \mathbf{y} \in K^3$ (where $\mathbf{u}$ always denotes the triple (u_1, u_2, u_3), line or point, according to context), we understand a set of points and lines, the former to be denoted by (x, y, z) the latter by $[u, v, w]$ (determined up to multiplication by a non-zero scalar, not all coordinates being allowed to be 0), endowed with a notion of incidence, point (x, y, z) being incident with line $[u, v, w]$ if and only if $xu + yv + zw = 0$, an orthogonality of lines defined by $\mathfrak{f}$, under which lines $\mathbf{g}$ and $\mathbf{g}'$ are orthogonal if and only if $\mathfrak{f}\ (\mathbf{g}, \mathbf{g}') = 0$, and a segment congruence relation defined by (3) for points $\mathbf{a}$, $\mathbf{b}$, $\mathbf{c}$, $\mathbf{d}$ for which $Q(\mathbf{a}), Q(\mathbf{b}), Q(\mathbf{c}), Q(\mathbf{d})$ are all $\neq 0$, with $F(\mathbf{x}, \mathbf{y}) = \beta\gamma x_1 y_1 + \alpha\gamma x_2 y_2 + \alpha\beta x_3 y_3$. An ordinary projective metric plane is called *hyperbolic* if $F(\mathbf{a}, \mathbf{a}) = 0$ has non- zero ($\mathbf{a} \neq \mathbf{0}$) solutions, in which case the set of solutions forms a conic section, the *absolute* of that projective-metric plane.

The algebraic characterization of non-elliptic metric planes is given by

Representation Theorem 2. *Every model of a non-elliptic metric plane is either a metric-Euclidean plane, or else it can be represented as an embedded subplane (i. e. containing with every point all the lines of the projective-metric plane that are incident with it) that contains the point* $(0,0,1)$ *of a projective-metric plane* $\mathfrak{P}\,(K,\mathfrak{f})$ *over a field* K *of characteristic* $\neq 2$, *in which no point lies on the line* $[0,0,1]$, *from which it inherits the collinearity and segment congruence relations.*

The proof of this most important representation theorem follows, to some extent, in the non-elliptic (and most difficult) case, the pattern of the proof of Representation Theorem 1. One defines, for any two lines a and b, the *pencil* of lines defined by a and b to be the set $G(ab) := \{c \mid (\exists d)\, abc = d\})$. One can extend the set of lines and points of the metric plane to an *ideal plane* by letting the set of all line pencils be the set of points (pencils $G(ab)$ which contain two orthogonal lines m and n (in other words, for which there is a point mn on both a and b) can be thought of as representing points of the metric plane (the point mn in our example), whereas those which do not contain two orthogonal lines are *ideal points*, i. e. points that have been added to the metric plane). In analogy to the hyperbolic case, among the ideal points $G(ab)$, one may think of those for which there is a line g with $a|g$ and $b|g$ (i. e. a common perpendicular to a and b), as representing what used to be the exterior points, and of those for which there is neither a point on both a and b nor a common perpendicular (should such pencils exist), as representing what used to be absolute points. Given that two distinct pencils have at most one common line, one can extend the set of lines with *ideal lines* to make sure that any two points are joined by a line. That this is possible and that the ideal plane turns out to be a Pappian projective plane is due to several crucial ideas and theorems due to Hjelmslev, such as the concept of semi-rotation, and his *Lotensatz* (cf. [4, §3,6; §6], [29, §42] for full details of the proof of this representation theorem).

Non-elliptic metric planes have also been constructively axiomatized by Pambuccian [60] in the language $\mathcal{L}(a_0, a_1, a_2, F, \pi)$ with individual variables to be interpreted as *points*, with a_0, a_1, a_2 standing for three non-collinear points, and with F and π as described earlier. A representation theorem for ordered metric planes with non-Euclidean metric (i. e. in which there are no rectangles) has been provided by Pejas [67]. Metric planes with various kinds of degrees of free mobility (all of which are expressed inside $\mathcal{L}(\varrho)$) have been studied by Diller [17], who provides an algebraic description for their models.

Among generalizations of metric planes, which can be expressed in the same language $\mathcal{L}(\varrho)$, one that would encompass Minkowski planes as well was proposed by Wolff (1967) (cf. [4]). It arises from the axiom system for metric planes by weakening A2.15 and changing A2.19. In it two "points" do not always have a joining line.

Sperner planes and their generalizations Asking for the weakest axiom system in terms of line-reflections from which one could prove the projective form of Desargues' theorem — expressed in a particular manner in the language of line-reflections — Sperner [79] provided an axiom system for a large class of planes, to be referred to as *Sperner planes*, and opened up a vast area of research into axiom system weaker than the Schmidt-Bachmann one. Unlike the axiom system for metric planes, Sperner's axiom systems also allows geometries over fields of characteristic 2. It follows from [19] that Sperner planes are embeddable in Pappian projective planes, and there is a quadratic form $\mathbf{Q}$ such that line-reflection has the usual algebraic expression relative to $\mathbf{Q}$.

Building upon the work of Sperner [79], Lingenberg (1959-1965) has introduced in a long series of papers a class of planes considerably larger than that of metric planes, and later presented his results in book form in [50].

Hjelmslev and semiabsolute planes Hjelmslev [34] considered geometries in which the uniqueness of the line joining two points, A2.16 in Bachmann's axiom system for metric planes, is no longer required, being replaced by that of the uniqueness of the perpendicular raised from a point of a line to that line. Such geometries were axiomatized in $\mathcal{L}(\varrho)$ by Bachmann [7].

More general structures, called pre-Hjelmslev planes, were axiomatized in [44]. Representation theorems for ordered Hjelmslev planes with or without free mobility, similar to those of Pejas, have been provided by Kunze (1981).

Semiabsolute planes, which were defined and studied in [8] as models of A2.13-A2.16, A2.18-A2.22, and $(\forall abg)(\exists c)\, ab|g \rightarrow abg = c$, cannot, in general, be embedded in projective planes.

Three- and higher-dimensional absolute geometries

The first three-dimensional generalization of Bachmann's plane absolute geometry was provided by Ahrens [1]. A weakening of Ahrens's axiom system in the non-elliptic case, which may be understood as a three-dimensional variant of Lingenberg's planes (axiomatized by **S, EB***, and **D**) was provided by Nolte [55]. Nolte's axiom system can be formulated in a one-sorted language $\mathcal{L}(\varrho)$, with *planes* as individual variables and the binary operation ϱ, with $\varrho(g,h)$ to be interpreted as the reflection of the plane h in the plane g.

Dimension-free ordered spaces with congruence and free mobility with all subplanes H-planes, have been considered in Karzel and König [37], where a representation theorem for the models of these spaces, amounting to their embeddability in a Pappian affine plane, is proved. A like-minded, more general geometry was studied in [45] (see also [36]).

An n-dimensional generalization of Bachmann's [4] axiom system for metric planes, with *hyperplanes* as individual variables, and a binary operation ϱ_n with $\varrho_n(ab)$ to be interpreted as "the reflection of b in a", which is equivalent to that of [1] for $n = 3$, was proposed by Kinder (1965). A first algebraic description of Kinder's axiom system was provided in [35]. An algebraic description of some classes of the *ordered* n-dimensional absolute geometries axiomatized by Kinder, generalizing the results of [67], was provided in great detail in [28], after one for the ordered ones with free mobility had been provided by Klopsch (1985). The situation in higher dimensions is significantly more complex than in the two-dimensional case. Just as Kinder's axiom system generalizes Ahrens's axiom system to finite dimensions, [56] generalizes the axiom system from [55] to finite dimensions. A like-minded, but less researched, axiom system was introduced by Lenz (1974).

A dimension-free absolute metric geometry based on incidence and orthogonality was first proposed by Lenz (1962). It has been weakened to admit elliptic models as well in [75], [76]. The axiom system from [75] can be formalized in $\mathcal{L}(L,\kappa,\perp)$, with individuals to be interpreted as *points*, L a ternary relation standing for collinearity, κ a quaternary relation standing for coplanarity, and $\perp$ a ternary relation standing for orthogonality, with $\perp(abc)$ to be interpreted as ab is perpendicular to ac.

An axiom system for these spaces, but excluding the elliptic case, formulated in a language containing two sorts of variables, for points and hyperplanes, the binary relation of point-hyperplane incidence, and the binary notion of hyperplane orthogonality was presented in [76]. A like-minded axiom system, for dimension-free absolute metric geometry, with points and lines as individual variables, was presented in [20], simplified in [27], and reformulated in a different language by J. T. Smith (1985). If one adds to that axiom system axioms implying that the space has finite dimension $n \geq 3$, then the axiom system is equivalent to that proposed by Kinder (1965), and all finite-dimensional models can be embedded in a projective geometry over a field of characteristic different from 2 of elliptic, hyperbolic or Euclidean type, the orthogonality being given by a symmetric bilinear form. A weaker version of the above geometry was considered by J. T. Smith (1974). The most general axiom system for dimension-free absolute metric geometries, with points and lines as variables, was provided in [70], the last paper on this subject.

Axiomatizations of a large class of dimension-free absolute geometries with *points* as variables and the binary operation of point-reflection were proposed in [39] and [22]. In an earlier paper, Karzel (1971) had shown that axiom systems for line-reflections (formulated in $\mathcal{L}(\varrho)$) satisfying Sperner-Lingenberg-type axioms, can be interpreted not only as axiomatizing plane geometries, but also as axiomatizing geometries of dimension ≥ 3.

A very general absolute geometry, axiomatized in $\mathcal{L}(B, \equiv)$, whose betweenness and equidistance can be represented by a special class of generalized metrics, with values in ordered Abelian groups, was proposed by Moszyńska [54].

Reverse geometry

One source of motivations for research in absolute geometries has been mentioned already. It is the quest for a theory common to Euclidean and hyperbolic geometry, which would be rich enough to allow its models to be embedded in projective spaces coordinatized by commutative fields (Pappian projective spaces), and, in case a notion of orthogonality is definable inside that theory, that it may be represented in the usual manner by means of a quadratic form.

A second source can be discerned in questions of what one might call *reverse geometry*: Which axioms are needed to prove a particular theorem? The programme of reverse geometry was stated by Hilbert [30, p. 50], with his characteristic eloquence (cf. also [65] for the history of the regressive method):

> Unter der axiomatischen Erforschung einer mathematischen Wahrheit verstehe ich eine Untersuchung, welche nicht dahin zielt, im Zusammenhange mit jener Wahrheit neue oder allgemeinere Sätze zu entdecken, sondern die vielmehr die Stellung jenes Satzes innerhalb des Systems der bekannten Wahrheiten und ihren logischen Zusammenhang in der Weise klarzulegen sucht, daß sich sicher angeben läßt, welche Voraussetzungen zur Begründung jener Wahrheit notwendig und hinreichend sind.

An early example of an explicitly reverse geometric undertaking can be found in Barbilian [10], where it is analyzed on which of Hilbert's axioms the truth of a theorem due to Pompeiu, on the distances from a point to the vertices of an equilateral triangle, depends.

Much more common are investigations on the possibility of proving inside $\mathcal{A}$ theorems first established in Euclidean geometry, among the latest being that of Pambuccian (1998) on the existence of equilateral triangles over any given segment, or of the existence of even one equilateral triangle in H-planes.

Bachmann [6], following an earlier investigation by Toepken (1941), embarked on a reverse study of the altitude theorem, which states that the three altitudes of a triangle are concurrent. The axiom system for metric planes itself can be seen as an answer to the question: Which axioms are needed to prove all four important line concurrency theorems in triangles (concurrency of altitudes, angle bisectors, medians, and perpendicular bisectors)? All of these theorems figure prominently in both Hjelmslev [33, 34] and Bachmann [4, §4].

Explicitly reverse are also the proof in ordered geometry of the Sylvester-Gallai theorem in [14], as well as those of Pambuccian (2001, 2003).

Bibliography

[1] Ahrens, J. (1959). Begründung der absoluten Geometrie des Raumes aus dem Spiegelungsbegriff. *Math. Z.*, 71: 154–185.

[2] Artin, E. (1957). *Geometric algebra.* New York: Interscience.

[3] Artzy, R. (1966). Non-Euclidean incidence planes. *Israel J. Math.*, 4: 43–53.

[4] Bachmann, F. (1973 (1959)). *Aufbau der Geometrie aus dem Spiegelungsbegriff*, 2. Auflage. Berlin: Springer-Verlag.

[5] Bachmann, F. (1964). Zur Parallelenfrage. *Abh. Math. Sem. Univ. Hamburg*, 27: 173–192.

[6] Bachmann, F. (1967). Der Höhensatz in der Geometrie involutorischer Gruppenelemente. *Canad. J. Math.*, 19: 895–903.

[7] Bachmann, F. (1989). *Ebene Spiegelungsgeometrie.* Mannheim: B.I.-Wissenschaftsverlag.

[8] Bachmann, O. (1976). Zur spiegelungsgeometrischen Begründung von Geometrien. *Geom. Dedicata*, 5: 497–516.

[9] Baer, R. (1948). The infinity of generalized hyperbolic planes. *Studies and Essays Presented to R. Courant on his 60th Birthday, January 8, 1948*, pages 21–27. New York: Interscience Publishers.

[10] Barbilian, D. (1936). Exkurs über die Dreiecke. *Bull. Math. Soc. Roum. Sci.*, 38: 3–62.

[11] Bryant, V. (1974). Independent axioms for convexity. *J. Geom.*, 5: 95–99.

[12] Cayley, A. (1859). A sixth memoir upon quantics. *Collected mathematical papers*, vol. 2, pages 561–606. Cambridge, 1889.

[13] Coppel, W. A. (1998). *Foundations of convex geometry.* Cambridge University Press.

[14] Coxeter, H. S. M. (1969). *Introduction to geometry.* New York: J. Wiley.

[15] Csorba, F., Molnár, E. (1982/86) Steiner constructions on the projective-metric plane (Hungarian). *Mat. Lapok*, 33: 99–122.

[16] Dehn, M. (1900). Die Legendre'schen Sätze über die Winkelsumme im Dreieck. *Math. Ann.*, 53: 404–439.

[17] Diller, J. (1970). Eine algebraische Beschreibung der metrischen Ebenen mit ineinander beweglichen Geraden. *Abh. Math. Sem. Univ. Hamburg*, 34: 184–202.

[18] Doraczyńska, E. (1977). On constructing a field in hyperbolic geometry. *Bull. Acad. Polon. Sci. Sér. Sci. Math. Astronom. Phys.*, 25: 1109–1114.

[19] Ellers, E., Sperner, E. (1962). Einbettung eines desarguesschen Ebenenkeims in einer projektiven Ebene. *Abh. Math. Sem. Univ. Hamburg*, 25: 206–230.

[20] Ewald, G. (1974). Spiegelungsgeometrische Kennzeichnung euklidischer und nichteuklidischer Räume beliebiger Dimension. *Abh. Math. Sem. Univ. Hamburg*, 41: 224–251.

[21] Ferreirós, J. (2001). The road to modern logic — an interpretation. *Bull. Symbol. Logic*, 7: 441–484.

[22] Gabrieli, E. (1999). Loops with reflection germ: a characterization of absolute planes. *Discrete Math.*, 208/209: 285–298.

[23] Greenberg, M. J. (1979). On J. Bolyai's parallel construction. *J. Geom.*, 12: 45–64.

[24] Greenberg, M. J. (1988). Aristotle's axiom in the foundations of geometry. *J. Geom.*, 33: 53–57.

[25] Grochowska-Prażmowska, M. (1993). A proof of Pasch's axiom in the absolute theory of oriented parallelity. *J. Geom.*, 46: 66–81.

[26] Hartshorne, R. (2000). Geometry: Euclid and Beyond. New York: Springer-Verlag.

[27] Heimbeck, G. (1980). Zum Aufbau der absoluten Geometrie nach Ewald. *Abh. Math. Sem. Univ. Hamburg*, 50: 70–88.

[28] Hellwig, G. (1991). *Modelle der absoluten Geometrie*. Dissertation, Universität Kiel.

[29] Hessenberg, G., Diller, J. (1967). *Grundlagen der Geometrie*. Berlin: W. de Gruyter.

[30] Hilbert, D. (1902/1903). Über den Satz von der Gleichheit der Basiswinkel im gleichschenkligen Dreieck. *Proc. London Math. Soc.* 35: 50-68.

[31] Hilbert, D. (1899). *Grundlagen der Geometrie*. Stuttgart: B. G. Teubner, 12. Auflage, 1977.

[32] Hilbert, D. (1903). Neue Begründung der Bolyai-Lobatschefskyschen Geometrie. *Math. Ann.*, 57: 137–150.

[33] Hjelmslev, J. (1907). Neue Begründung der ebenen Geometrie. *Math. Ann.*, 64: 449–474.

[34] Hjelmslev, J. (1929, 1929, 1942, 1945, 1949). Einleitung in die allgemeine Kongruenzlehre. *Danske Vid. Selsk., mat.-fys. Medd.*, 8, Nr. 11; 10, Nr. 1; 19, Nr. 12; 22, Nr. 6; Nr. 13; 25, Nr. 10.

[35] G. Hübner, (1969). Klassifikation n-dimensionaler absoluter Geometrien, *Abh. Math. Sem. Univ. Hamburg*, 33: 165–182.

[36] Karzel, H. (1999). Recent developments on absolute geometries and algebraization by K-loops. *Discrete Math.*, 208/209: 387–409.

[37] Karzel, H. , König, M. (1981). Affine Einbettung absoluter Räume beliebiger Dimension. In Butzer, P. L. and Fehér, F., editors, *E. B. Christoffel. The influence of his work on mathematics and the physical sciences*, pages 657–670. Basel: Birkhäuser Verlag.

[38] Karzel, H., Konrad, A. (1994). Eigenschaften angeordneter Räume mit hyperbolischer Inzidenzstruktur. *Beiträge Geom. Algebra TU München*, 28: 27–36.

[39] Karzel, H., Konrad, A. (1995). Reflection groups and K-loops. *J. Geom.*, 52: 120–129.

[40] Karzel, H., Kroll, H.-J. (1988). *Geschichte der Geometrie seit Hilbert.* Darmstadt: Wissenschaftliche Buchgesellscahft.

[41] Karzel, H., Wefelscheid, H. (1995). A geometric construction of the K-loop of a hyperbolic space. *Geom. Dedicata* 58: 227–236.

[42] de Kerékjártó, B. (1955). *Les fondements de la géométrie I.* Budapest: Akademiai Kiádo.

[43] Klingenberg, W. (1954). Eine Begründung der hyperbolischen Geometrie. *Math. Ann.*, 127: 340–356.

[44] Knüppel, F. (1991). On pre-Hjelmslevgroups and related topics. In Barlotti, A., Ellers, E. W., Plaumann, P. and Strambach, K., editors, *Generators and relations in groups and geometries*, pages 125–177. Kluwer Academic Publishers.

[45] Konrad, A. (1995). *Nichteuklidische Geometrie und K-loops.* Dissertation, Technische Universität München.

[46] Kordos, M., Szczerba, L. W. (1969). On the $\Pi\Sigma$-axiom systems for hyperbolic and some related geometries. *Bull. Acad. Polon. Sci. Sér. Sci. Math. Astronom. Phys.*, 17: 175–180.

[47] Kreuzer, A. (1996). Locally projective spaces which satisfy the bundle theorem. *J. Geom.*, 56: 87–98.

[48] Kroll, H.- J., Sörensen, K. (1998). Hyperbolische Räume. *J. Geom.*, 61: 141–149.

[49] Liebmann, H. (1905). Elementargeometrischer Beweis der Parallelenkonstruktion und neue Begründung der trigonometrischen Formeln der hyperbolischen Geometrie. *Math. Ann.*, 61: 185–199.

[50] Lingenberg, R. (1979). Metric planes and metric vector spaces. New York: J. Wiley.

[51] List, K. (2001). Harmonic mappings and hyperbolic Plücker transformations. *J. Geom.*, 70: 108–116.

[52] Menger, K. (1971). The new foundation of hyperbolic geometry. In Butcher, J. C., editor, *Spectrum of Mathematics*, pages 86–97. Auckland University Press.

[53] Moler N. , Suppes, P. (1968). Quantifier-free axioms for constructive plane geometry. *Compositio Math.*, 20: 143–152.

[54] Moszyńska, M. (1977). Theory of equidistance and betweenness relations in regular metric spaces. *Fund. Math.*, 96: 17–29.

[55] Nolte, W. (1966). Zur Begründung der absoluten Geometrie des Raumes. *Math. Z.*, 94: 32–60.

[56] Nolte, W. (1974). Metrische Räume mit dreiseitverbindbaren Teilräumen I,II. *J. Geom.*, 4: 53–90, 91–117.

[57] Pambuccian, V. (1994). Zum Stufenaufbau des Parallelenaxioms. *J. Geom.*, 51: 79–88.

[58] Pambuccian, V. (2001). Constructive axiomatization of plane absolute, Euclidean and hyperbolic geometry, *Math. Logic Quart.*, 47: 129–135.

[59] Pambuccian, V. (2001). Fragments of Euclidean and hyperbolic geometry. *Sci. Math. Jpn.*, 53: 361–400.

[60] Pambuccian, V. (2003). Constructive axiomatization of non-elliptic metric planes. *Bull. Pol. Acad. Math.*, 51: 49–57.

[61] Pambuccian, V. (2003). Sphere tangency as single primitive notion for hyperbolic and Euclidean geometry. *Forum Math.*, 15: 943–947.

[62] Pambuccian, V. (2004). Hyperbolic geometry in terms of point-reflections or of line-orthogonality, *Math. Pannon.*, to appear.

[63] Pambuccian, V. (2004). An axiomatics of hyperbolic projective-metric planes in terms of lines and orthogonality, *Bull. Pol. Acad. Sci. Math.*, 52: 297–302.

[64] Pasch, M., Dehn, M. (1926 (1882)). Vorlesungen über neuere Geometrie. 2. Auflage. Berlin: Springer-Verlag.

[65] Peckhaus, V. (2002). Regressive Analysis. *Log. Anal. Hist. Philos.*, 5: 97–110.

[66] Pejas, W. (1961). Die Modelle des Hilbertschen Axiomensystems der absoluten Geometrie. *Math. Ann.*, 143: 212–235.

[67] Pejas, W. (1964). Eine algebraische Beschreibung der angeordneten Ebenen mit nichteuklidischer Metrik. *Math. Z.*, 83: 434–457.

[68] Prażmowski, K. (1989). An axiomatic description of the Strambach plane. *Geom. Dedicata*, 32: 125–156.

[69] Sakowicz, K. (1988). *Axiomatizability of the dual external hyperbolic plane* (Polish), MS Thesis, University of Warsaw, Białystok Section.

[70] Schröder, E. M. (1984). Aufbau metrischer Geometrie aus der Hexagrammbedingung. *Atti Sem. Mat. Fis. Univ. Modena*, 33: 183–217.

[71] Schwabhäuser, W., Szmielew, W., Tarski, A. (1983). *Metamathematische Methoden in der Geometrie.* Berlin: Springer-Verlag.

[72] Scott, D. (1959). Dimension in elementary Euclidean geometry. In Henkin, L., Suppes P. and Tarski, A., editors, *The axiomatic method*, pages 53–67. North-Holland, Amsterdam.

[73] Skala, H. L. (1992). Projective-type axioms for the hyperbolic plane. *Geom. Dedicata*, 44: 255–272.

[74] Smid, L. J. (1936). Eine absolute Axiomatik der Geometrie in einer begrenzten Ebene. *Math. Ann.*, 112: 125–138.

[75] Smith, J. T. (1973). Orthogonal geometries, I, II. *Geom. Dedicata*, 1: 221–235, 334–339.

[76] Smith, J. T. (1985). Point-hyperplane axioms for orthogonal geometries, *Geom. Dedicata*, 17: 279–286.

[77] Sörensen, K. (1984). Ebenen mit Kongruenz. *J. Geom.*, 22: 15–30.

[78] Sörensen, K. (1999). Eine Bemerkung über absolute Ebenen. *J. Geom.*, 64: 160–166.

[79] Sperner, E. (1954). Ein gruppentheoretischer Beweis des Satzes von Desargues in der absoluten Axiomatik. *Arch. Math. (Basel)*, 5: 458–468.

[80] Strommer, J. (1961). Ein elementarer Beweis der Kreisaxiome der hyperbolischen Geometrie. *Acta. Sci. Math.*, 22: 190–195.

[81] Strommer, J. (1990). Ein neuer elementarer Beweis der Kreisaxiome der hyperbolischen Geometrie. *Acta Sci. Math.*, 54: 269–271.

[82] Szász, P. (1958). Unmittelbare Einführung Weierstrassscher homogenen Koordinaten in der hyperbolischen Ebene auf Grund der Hilbertschen Endenrechnung. *Acta Math. Acad. Sci. Hungar.*, 9: 1–28.

[83] Szász, P. (1962). Einfache Herstellung der hyperbolischen Trigonometrie in der Ebene auf Grund der Hilbertschen Endenrechnung. *Ann. Univ. Sci. Budapest. Eötvös Sect. Math.*, 5: 79–85.

[84] Szczerba, L. W. (1972). Weak general affine geometry. *Bull. Acad. Polon. Sci. Sér. Sci. Math. Astronom. Phys.*, 20: 753–761.

[85] Szmielew, W. (1961/1962). A new analytic approach to hyperbolic geometry. *Fund. Math.*, 50: 129–158.

[86] Tarski, A. (1959). What is elementary geometry? In Henkin, L., Suppes P. and Tarski, A., editors, *The axiomatic method*, pages 16–29. North-Holland, Amsterdam.

[87] Tarski A., Givant, S. (1999). Tarski's system of geometry. *Bull. Symb. Log.*, 5: 175–214.

[88] Ziegler, M. (1982). Einige unentscheidbare Körpertheorien. *Enseign. Math.*, 28: 269–280.

LOGICAL AXIOMATIZATIONS OF SPACE-TIME. SAMPLES FROM THE LITERATURE

Hajnal Andréka, Judit X. Madarász and István Németi
Rényi Institute of Mathematics Budapest
andreka@renyi.hu,madarasz@renyi.hu,nemeti@renyi.hu

Abstract We study relativity theory as a theory in the sense of mathematical logic. We use first-order logic (FOL) as a framework to do so. We aim at an "analysis of the logical structure of relativity theories". First we build up (the kinematics of) special relativity in FOL, then analyze it, and then we experiment with generalizations in the direction of general relativity. The present paper gives samples from an ongoing broader research project which in turn is part of a research direction going back to Reichenbach and others in the 1920's. We also try to give some perspective on the literature related in a broader sense. In the perspective of the present work, axiomatization is not a final goal. Axiomatization is only a first step, a tool. The goal is something like a *conceptual analysis of relativity* in the framework of logic.

In section 1 we recall a complete FOL-axiomatization **Specrel** of special relativity from [5, 32]. In section 2 we answer questions from papers by Ax and Mundy concerning the logical status of faster than light motion (FTL) in relativity. We claim that already very small/weak fragments of **Specrel** prove "*No FTL*". In section 3 we give a sketchy outlook for the possibility of generalizing **Specrel** to theories permitting accelerated observers (gravity). In section 4 we continue generalizing **Specrel** in the direction of general relativity by localizing it, i.e. by replacing it with a version still in first-order logic but now local (in the sense of general relativity theory). In section 5 we give samples from the broader literature.

Keywords: logic, first-order logic, axiomatization, completeness theorem, foundation of space-time, special and general relativity, faster-than-light motion, reverse geometry, reverse spacetime theory, Alexandrov-Zeeman type theorem, accelerated observers, local theory

Introduction

The interplay between logic and relativity theory goes back to around 1920 and has been playing a non-negligible role in works of researchers like Reichenbach, Carnap, Suppes, Ax, Szekeres, Malament, Walker, and of many other contemporaries. For example, the logical theory of definability can find its roots in this interplay, cf. Reichenbach [45](1924). On the other side, definability has always been an important topic in the foundation of relativity from Reichenbach (1924) to e.g. Malament (1977) [38], Friedman (1983), [17] (1991), and Hogarth (2003) [27].

The present paper intends to give samples from the area called *analysis of the logical structure of relativity theories.* The first step in this analysis is building up relativity theory as a theory in the sense of first-order logic (FOL). The reason why we chose FOL and not e.g. second-order logic is presented in detail in [5, App.1] as well as in Ax [9], Pambuccian [44], but the reasons in Väänänen [59], Ferreirós [16], or Woleński [61] also apply.[1] Axiomatizations of special relativity have been extensively studied in the literature, cf. section 5. These works usually stop with a kind of completeness theorem for their axiomatizations. What we call the analysis of the logical structure of relativity theory begins with proving such a completeness theorem but the real work comes afterwards, during which one often concludes that we have to change the axioms. Very roughly, one could phrase this as "we start off where the others stopped (namely, at completeness)". In the present work, especially in section 2, we try to illustrate what we understand by this kind of analysis of logical structure.

In sections 1 and 2 we recall (from [5, 32]) and study a, basically, complete FOL-axiomatization **Specrel** of special relativity. In sections 3, 4 we give an outlook on generalizing the logic based approach towards general relativity. We will make "two steps" towards general relativity: In section 3 we extend **Specrel** by permitting accelerated observers and this way we become able to study some aspects of gravity via Einstein's equivalence principle. The other step is in section 4 where we make our theory local in the sense of general relativity. We call this process localization of a theory and we note that it is related to the process of relativization which already proved successful in algebraic logic and in the area known as "Logic, Language and Information" (LLI), cf. e.g. [7]. We do all this in the framework of first-order logic. Our intention is to

[1] Being in the framework of FOL is the same as being *elementary*, according to the standard terminology, cf. e.g. Tarski [58].

use these "two steps" towards general relativity together, i.e. to combine the theories obtained by the methods of sections 3 and 4.

Axiomatization of a subject matter (like e.g. geometry) usually evolves through three stages. The first to appear is an axiomatization without formal logic, followed by an axiomatization in the framework of formal (but perhaps higher-order) logic, and finally by an axiomatization in the framework of first-order logic (which might be many-sorted). In the case of geometry, Euclid gave the first kind of axiomatization, Hilbert gave the second, and finally Tarski and his school developed the third, cf. e.g. [58, 44]. In the case of set theory, it was Cantor, Russell-Zermelo, modern set theory. The history of the "nonlogical", "higher-order logic", "first-order logic" stages of evolution of axiomatizations for some other theories (e.g. Peano's Arithmetic) can be found e.g. in Ferreirós [16]. This is a natural temporal sequence of development, since each one of these stages "prepares the ground" for the next one. For the purposes of conceptual analysis the first-order-logic-based axiomatizations are the most suitable, e.g. because FOL forces us to be explicit about many things.

As we will see in section 5, the *interplay between logic and relativity* has been extensively studied in the literature. Turning to axiomatizations as special parts of this interplay, a large part of the axiomatizations of relativity in the literature are of the first stage (nonlogical) and concern special relativity. There is a growing number of works which are in the framework of formal logic, e.g. the book Schutz [50] (second stage). Some are already in the framework of FOL; these will be further reflected on in section 5 from the point of view of e.g. amenability for conceptual analysis or other purposes of insight-seeking. Motivation for the research direction surveyed/reported here is nicely summarized in Ax [9], Suppes [52]; cf. also the introduction of [5].

1. An axiomatization of special relativity, in FOL

In this paper we deal with the kinematics of relativity only, i.e. we deal with motion of *bodies* (or *test-particles*). The motivation for our choice of vocabulary (for special relativity) is summarized as follows. We will represent motion as changing spatial location in time. To do so, we will have reference-frames for co-ordinatizing events and, for simplicity, we will associate reference-frames with special bodies which we will call *observers*. We visualize an observer-as-a-body as "sitting" in the origin of its reference frame, or equivalently, "living" on the time-axis of the reference frame. There will be another special kind of bodies which we will call *photons*. For co-ordinatizing events we will use an arbitrary *ordered field* in place of the field of the real numbers. Thus the elements of this field will be the "*quantities*" which we will use for marking time and space.

Let us fix a natural number $n > 1$. n will be the number of space-time dimensions. In most works $n = 4$, i.e. one has three space-dimensions and one time-dimension.[2]

Motivated by the above, our language contains the following symbols:

unary relation symbols $\mathsf{B}, \mathsf{Ob}, \mathsf{Ph}, \mathsf{F}$ (for bodies, observers, photons, and quantities, i.e. elements of the field, respectively),

binary function symbols $+, -, \cdot, /$, constants $0, 1$ and a binary relation symbol $<$ (for the field-operations and ordering on F),

a $2+n$-ary relation symbol W (for co-ordinatizing events, i.e. for the world-view relation).

We will read "$\mathsf{B}(x), \mathsf{Ob}(x), \mathsf{Ph}(x), \mathsf{F}(x)$" as "$x$ is a body", "x is an observer", "x is a photon", "x is a field-element", and we will read "$\mathsf{W}(x, y, z_1, z_2, \ldots, z_n)$" as "observer x sees (or observes) the body y at time z_1 at location $(z_2, \ldots z_n)$". This "seeing" or "observing" has nothing to do with seeing via photons or observing via experiments, it simply means that, according to x's co-ordinate system or reference frame, y is present at co-ordinates $(z_1, \ldots, z_n)$.

The above, together with statements of the form $x = y$, $x < y$ are the so-called *atomic formulas* of our first-order language, where $x, y, \ldots, z_n$ can be arbitrary variables or terms built up from variables by using the field-operations $+, -, \cdot, /, 0, 1$. The *formulas* of the first-order language are built up from these atomic formulas by using the logical connectives *not* ($\neg$), *and* ($\wedge$), *or* ($\vee$), *implies* ($\rightarrow$), *if-and-only-if* ($\leftrightarrow$) and the quantifiers *exists* ($\exists$) and *for all* ($\forall$).

Usually we will use the variables m, k to denote observers, b to denote bodies, ph to denote photons and $p_1, \ldots, q_1, \ldots$ to denote field-elements. We will write p and q in place of $p_1, \ldots, p_n$ and $q_1, \ldots, q_n$, e.g. we will write $\mathsf{W}(m, b, p)$ in place of $\mathsf{W}(m, b, p_1, \ldots, p_n)$, and we will write $\forall p$ in place of $\forall p_1, \ldots, p_n$ etc.

The models of this language are of the form $\mathsf{M} = \langle U, \mathsf{B}, \mathsf{Ob}, \mathsf{Ph}, \mathsf{F}, \ldots, \mathsf{W}\rangle$ where U is a nonempty set, $\mathsf{B}, \mathsf{Ob}, \mathsf{Ph}, \mathsf{F}$ are unary relations on U, etc.

[2]Recent generalizations of general relativity in the literature (e.g. M-theory) indicate that it might be useful to leave n as a variable.

A unary relation on U is just a subset of U, so we will use B, Ob etc as sets as well, e.g. we will write $x \in \mathsf{B}$ in place of $\mathsf{B}(x)$. In our investigations, we will always assume that $\mathsf{Ob}, \mathsf{Ph} \subseteq \mathsf{B}$, and that $U = \mathsf{B} \cup \mathsf{F}$ is the universe of M. Therefore we will write our models in the form $\mathsf{M} = \langle \mathsf{B}, \mathsf{Ob}, \mathsf{Ph}, \underline{\mathsf{F}}, \mathsf{W} \rangle$ where $\underline{\mathsf{F}} = \langle \mathsf{F}, +, \cdot, -, /, 0, 1, < \rangle$ will be an ordered field with universe F, and $\mathsf{B} \cup \mathsf{F}$ plays the role of the "universe" of M.

Having fixed our language, we now turn to formulating an axiom system for special relativity in this language.

We will formulate each axiom on three levels. First we give a very intuitive formulation, then we give a precise formalization using notions that will be useful later as well (like life-line), and finally, for completeness, we give a concrete first-order formula without using the introduced notions and abbreviations, at the beginning, at least. Readers familiar with first-order logic can read the first-order formulas right away.

${}^n\mathsf{F}$ denotes the set of all n-tuples of elements of F. If a is an n-tuple, then we will assume that $a = \langle a_1, \ldots, a_n \rangle$, i.e. a_i denotes the i-th member of the n-tuple a (for $0 < i \leq n$). We will use the *vector-space* structure of ${}^n\mathsf{F}$. I.e. if $p, q \in {}^n\mathsf{F}$ and $\lambda \in \mathsf{F}$, then $p + q, p - q, \lambda p \in {}^n\mathsf{F}$, and $\overline{0} = \langle 0, \ldots, 0 \rangle$ is the *origin*. The *slope* of a vector $p \in {}^n\mathsf{F}$ is defined as $\mathsf{slope}(p) = (p_2^2 + \cdots + p_n^2)/p_1^2$ if $p_1 \neq 0$, and $\mathsf{slope}(p) = \infty$ otherwise, where $\infty \notin \mathsf{F}$ is fixed, as usual, to denote a kind of formal infinity.

Let $q, v \in {}^n\mathsf{F}, v \neq \overline{0}$. The (straight) *line* going through q and with *squared speed* [3] or *slope* $(v_2^2 + \cdots + v_n^2)/v_1^2$ and in the *spatial direction* $\langle v_2, \ldots, v_n \rangle$ is

$\{q + \lambda v : \lambda \in \mathsf{F}\}$. The set of straight lines is then

$\mathsf{Lines} \stackrel{d}{=} \{\{q + \lambda v : \lambda \in \mathsf{F}\} : q, v \in {}^n\mathsf{F}, v \neq \overline{0}\}$. If $\ell \in \mathsf{Lines}$, then

$\mathsf{slope}(\ell) \stackrel{d}{=} \mathsf{slope}(p - q)$ for some (and then for all) $p, q \in \ell, p \neq q$.

The *life-line*, or *trace* of a body b in observer m's world-view, or as seen by m, is the set of co-ordinate points at which m sees b, and the set of bodies m sees at a given co-ordinate point p is the *event* happening for m at p:

$\mathsf{tr}_m(b) \stackrel{d}{=} \{p \in {}^n\mathsf{F} : \mathsf{W}(m, b, p)\}$ and

$\mathsf{ev}_m(p) \stackrel{d}{=} \{b \in B : \mathsf{W}(m, b, p)\}$.

[3]For technical reasons, we use the square of the speed instead of speed, throughout.

The *speed* of a body b as seen by m is

$\mathsf{speed}_m(b) \stackrel{d}{=} \mathsf{slope}(\mathsf{tr}_m(b))$, this is meaningful when $\mathsf{tr}_m(b) \in \mathsf{Lines}$.

We are ready now for formulating the axioms on all three levels.

AxSelf Each observer sees himself standing still at the origin, i.e. the life-line of m in his own world-view is the *time-axis*

$\mathsf{tr}_m(m) = \bar{t} = \{\langle t, 0, \ldots, 0\rangle : t \in \mathsf{F}\}$, if $m \in \mathsf{Ob}$. A first-order formula expressing this without abbreviations is

$$(\forall m, p)\Big((\mathsf{Ob}(m) \wedge \mathsf{F}(p_1)) \rightarrow [\mathsf{W}(m, m, p) \leftrightarrow (p_2 = 0 \wedge \cdots \wedge p_n = 0)]\Big).$$

AxPh The photon-traces are exactly the lines with slope 1 (in each observer's world-view). In other words, all photons move with speed 1, and it is possible to send out a photon with speed 1 at each point and in each direction,

$\{\mathsf{tr}_m(\mathsf{ph}) : \mathsf{ph} \in \mathsf{Ph}\} = \{\ell \in \mathsf{Lines} : \mathsf{slope}(\ell) = 1\}$ for all $m \in \mathsf{Ob}$. A first-order formula expressing this without abbreviations is

$$\begin{aligned}&(\forall m, p, q)(\mathsf{Ob}(m) \rightarrow\\ &\quad [(\exists \mathsf{ph})(\mathsf{Ph}(\mathsf{ph}) \wedge \mathsf{W}(m, \mathsf{ph}, p) \wedge \mathsf{W}(m, \mathsf{ph}, q))\\ &\quad \leftrightarrow (p_1 - q_1)^2 = (p_2 - q_2)^2 + \cdots + (p_n - q_n)^2]).\end{aligned}$$

AxEvent All observers see the same events,

$\{\mathsf{ev}_m(p) : p \in {}^n\mathsf{F}\} = \{\mathsf{ev}_k(p) : p \in {}^n\mathsf{F}\}$ for all $m, k \in \mathsf{Ob}$. A first-order formula expressing this without abbreviations is

$$(\forall m, k, p)(\exists q)\Big((\mathsf{Ob}(m) \wedge \mathsf{Ob}(k)) \rightarrow (\forall b)[\mathsf{W}(m, b, p) \leftrightarrow \mathsf{W}(k, b, q)]\Big).$$

AxField The usual first-order axioms saying that $\langle \mathsf{F}, +, -, \cdot, /, 0, 1, < \rangle$ is a *Euclidean* ordered field, i.e. an ordered field in which positive elements have square-roots; together with the formula expressing $\mathsf{Ob}, \mathsf{Ph} \subseteq \mathsf{B}$ and $\mathsf{W} \subseteq \mathsf{Ob} \times \mathsf{B} \times {}^n\mathsf{F}$.

$$\mathbf{Specrel}_0 \stackrel{d}{=} \{\mathbf{AxSelf}, \mathbf{AxPh}, \mathbf{AxEvent}, \mathsf{AxField}\}.$$

Since, in some sense, AxField is only an "auxiliary" axiom about the "mathematical frame" of our reasoning, the heart of $\mathbf{Specrel}_0$ consists of three very natural axioms, **AxSelf**, **AxPh**, **AxEvent**. The reader is invited to check that these are really intuitively convincing, natural and simple assumptions.

From these four axioms already one can prove the most characteristic predictions of special relativity theory. What the average layperson usually knows about relativity is that "moving clocks slow down", "moving spaceships shrink", and "moving pairs of clocks get out of synchronism". We call these the *paradigmatic effects* of special relativity. All these can be proven from the above four axioms, in some form. E.g. one can prove that "if m, k are any two observers not at rest relative to each other, then one of m, k will "see" or "think" that the clock of the other runs slow". Careful analysis, statement, and proofs (from our axioms) of the various paradigmatic effects of special relativity can be found in [5, §§2.5, 2.6, 2.8, 4.8] and in [32].

As an example here we formulate that time cannot be taken as absolute[4], i.e. it is impossible to construct a model for the above four axioms in which all observers agree about what events are happening simultaneously (assuming not all the observers are at rest w.r.t. each other). From now on we do not spell out the concrete, abbreviation-free first-order formulas, it will always be straightforward to construct them (if someone wanted to see them). Let

NoAbstime $(\exists m, k, p, q, p', q')[p_1 = q_1 \land p'_1 \neq q'_1 \land \mathsf{ev}_m(p) = \mathsf{ev}_k(p') \land \mathsf{ev}_m(q) = \mathsf{ev}_k(q')]$.

Intuitively, events $\mathsf{ev}_m(p)$ and $\mathsf{ev}_m(q)$ are simultaneous for m but not simultaneous for k.

NotAllrest

$(\exists m, k, p, q)[\mathsf{W}(m, k, p) \land \mathsf{W}(m, k, q) \land (p_2 \neq q_2 \lor \cdots \lor p_n \neq q_n) \land \mathsf{Ob}(k)]$.

Intuitively, k is not at rest relative to m.

If Σ is a set of formulas and M a model, then $\mathsf{M} \models \Sigma$ denotes that all formulas in Σ are true in the model M, in this case we say that M is *a*

[4]In the more general framework towards which we will be working in section 3, stronger theorems will be provable about time not being absolute (cf. e.g. Gödel's rotating universes [21], [32, Fig.135, p.366]). However, making this precise has to be postponed.

model of Σ. If φ is a formula, then $\Sigma \models \varphi$ denotes that φ is true in all models of Σ. In this case we say that φ *follows from* (or, *is a semantical consequence of*) Σ. By Gödel's completeness theorem, this implies that φ is *provable* from Σ in the usual, syntactical sense of first-order logic's proof theory.

Theorem 1.1. $\mathbf{Specrel_0} \models (\mathbf{NotAllrest} \rightarrow \mathbf{NoAbstime})$, *assuming* $n > 2$.

Theorem 1.1 follows from Theorem 1.2 way below.

Adding the next axiom **AxSym** to $\mathbf{Specrel_0}$ will imply all the paradigmatic effects of relativity in their strongest form (and even the so-called Twin Paradox in a form), cf. Thm.1.2(ii). The *world-view transformation* f_{mk} between two observers m, k is defined as

$$\mathsf{f}_{mk} \stackrel{d}{=} \{\langle p, q\rangle : \mathsf{ev}_m(p) = \mathsf{ev}_k(q), \mathsf{ev}_k(q) \neq \emptyset \text{ and } p, q \in {}^n\mathsf{F}\} \ .$$

From the axioms so far we can see that the world-view transformations play a central role in relativity. From our previous axioms it follows that f_{mk} is a transformation of ${}^n\mathsf{F}$ (and not only an arbitrary binary relation) if m, k are observers.[5] Therefore we will use f_{mk} as a function. Then $\mathsf{f}_{mk}(p)$ is the "place" where k sees the same event that m sees at p, i.e.

$$\mathsf{ev}_m(p) = \mathsf{ev}_k(\mathsf{f}_{mk}(p)) \ .$$

Let $p, q \in {}^n\mathsf{F}$. Then $p_1 - q_1$ is the time passed between the events $\mathsf{ev}_m(p)$ and $\mathsf{ev}_m(q)$ as seen by m and $\mathsf{f}_{mk}(p)_1 - \mathsf{f}_{mk}(q)_1$ is the time passed between the same two events as seen by k. Hence $\|(\mathsf{f}_{mk}(p)_1 - \mathsf{f}_{mk}(q)_1)/(p_1 - q_1)\|$ is the rate with which k's clock runs slow as seen by m. Here, $\|a\|$ denotes the *absolute value* of a when $a \in \mathsf{F}$, i.e. $\|a\| \in \{a, -a\}$ and $\|a\| \geq 0$.

AxSym All observers see each other's clocks run slow to the same extent,

$$\|\mathsf{f}_{mk}(p)_1 - \mathsf{f}_{mk}(q)_1\| = \|\mathsf{f}_{km}(p)_1 - \mathsf{f}_{km}(q)_1\|, \text{ when } m, k \in \mathsf{Ob} \text{ and } p, q \in \bar{t}.$$

$$\mathbf{Specrel} \stackrel{d}{=} \mathbf{Specrel_0} \cup \{\mathbf{AxSym}\} = \{\mathbf{AxSelf}, \mathbf{AxPh}, \mathbf{AxEvent}, \mathbf{AxSym}, \mathsf{AxField}\}.$$

[5]This is a typical example of a property of special relativity which will be relaxed in the process of localization (towards general relativity) in sections 3,4. Namely, the axioms of our local theories in sections 3,4 will not imply that the function f_{mk} is everywhere defined in ${}^n\mathsf{F}$. This is an essential generalization towards general relativity.

The (square of the) so-called *Minkowski-length* of $p \in {}^n\mathsf{F}$ is defined as

$$\mu(p) \stackrel{d}{=} p_1^2 - p_2^2 - \cdots - p_n^2, \quad \text{while} \quad |p| \stackrel{d}{=} p_1^2 + p_2^2 + \cdots + p_n^2$$

denotes the (square of the) usual *Euclidean length*. Let $f : {}^n\mathsf{F} \to {}^n\mathsf{F}$ be a mapping. Recall from the literature that f is called an *affine* transformation if it is the composition of a bijective linear transformation with a translation. f is called a *Poincaré-transformation* (or an *inhomogeneous Lorentz-transformation*) if f is an affine transformation which preserves Minkowski-distance, i.e. if $\mu(f(p) - f(q)) = \mu(p - q)$ for all $p, q \in {}^n\mathsf{F}$. [6] f is called a *dilation* if there is a positive $\delta \in \mathsf{F}$ such that $f(p) = \delta p$ for all $p \in {}^n\mathsf{F}$ and f is called *field-automorphism-induced* if there is an automorphism π of the field $\langle \mathsf{F}, +, \cdot \rangle$ such that $f(p) = \langle \pi p_1, \ldots, \pi p_n \rangle$ for all $p \in {}^n\mathsf{F}$.

The following is proved in [5, 2.9.4, 2.9.5] and in [32, 2.9.4–2.9.7].

Theorem 1.2. *Let $n > 2$, let* M *be a model of our language and let m, k be observers in* M. *Then (i)-(ii) below hold.*

(i) f_{mk} *is a Poincaré-transformation composed with a dilation and a field-automorphism-induced mapping, if* M $\models$ **Specrel**$_0$.

(ii) f_{mk} *is a Poincaré-transformation, if* M $\models$ **Specrel**.

Assume $n > 2$. Theorem 1.2 (i) and (ii) above are best possible in the sense that e.g. for every Poincaré-transformation f over an arbitrary Euclidean ordered field there are a model M $\models$ **Specrel** and observers m, k in M such that the world-view transformation f_{mk} between m's and k's world-views in M is f, see [5, 2.9.4(iii), 2.9.5(iii)]. Similarly for (i). Hence, Thm.1.2 can be refined to a pair of completeness theorems, cf. [5, 3.6.13, p.271].

It follows from Thm.1.2 that the paradigmatic effects hold in **Specrel** in their strongest form, e.g. if m and k are observers not in rest w.r.t. each other, then both will "think" that the clock of the other runs slow. **Specrel** also implies the "inertial approximation" of the Twin Paradox, see e.g. [5, 2.8.18], and [55]. For the Twin Paradox see also section 3 herein.

[6] A concrete description of Poincaré-transformations in Euclidean fields is known, as follows. f is called an *isometry* if it preserves Euclidean length, and f is called a *Newtonian* transformation (or sometimes "trivial") if it is an isometry which is the identity on the time axis $\bar{t}$, composed with a translation. (Indeed, Newtonian, or Galilean, re-coordinatizations are such.) f is called a *Lorentz-boost* if there is $0 \leq v < 1$ such that $f(p_1, \ldots, p_n) = \langle (p_1 - vp_2)/\sqrt{1-v^2}, (p_2 - vp_1)/\sqrt{1-v^2}, p_3, \ldots, p_n \rangle$ for all $p \in {}^n\mathsf{F}$. Now, it is known that f is a Poincaré-transformation iff it is a composition of a Newtonian transformation, a Lorentz-boost, and another Newtonian transformation (cf. e.g. [5, §2.9]).

By Theorem 1.2(ii), **Specrel** implies that the f_{mk}'s are affine transformations (i.e. linear + translation). In models of **Specrel**$_0$, the world-view-transformations are not necessarily affine transformations, but (by Thm.1.2(i)) they are still line-preserving bijections (or, in other words, *collineations*), i.e. the image of any line under a world-view transformation is a line again. That the f_{mk}'s are collineations follows from the axioms in **Specrel**$_0$ by a suitable generalization of the following theorem. Let $\Re$ denote the ordered field of real numbers, also called the *real line.*

Theorem 1.3. (Alexandrov and Zeeman) *Let $n > 2$. If a bijection $f : {}^n\Re \to {}^n\Re$ maps lines of slope 1 onto lines of slope 1, then f maps any line onto a line, i.e. f is a collineation.*

Theorem 1.3 is very useful and has many strengthened versions and variants in the literature. For a survey of these see Guts [24] and Lester [30]. In particular, Theorem 1.3 is true if we replace $\Re$ with any Euclidean ordered field in it. (Thm.s 4.1,4.2 in section 4 herein are also Alexandrov-Zeeman type theorems.)

The condition $n > 2$ is necessary in Thm.1.2, because the Alexandrov-Zeeman theorem is not true for $n = 2$ and for this same reason, Theorem 1.2 is not true, either, for $n = 2$. However, a version of Theorem 1.2 becomes true for $n = 2$ as well, if we add the following two axioms to **Specrel**$_0$.

It has been a basic principle in physics that if no external force is acting on a body b (i.e. if it is *inertial*) then it changes neither the speed nor the direction of its movement, and this means that b's life-line is a straight line; this is expressed by the axiom

AxLine The life-line of an observer is a straight line as seen by any other observer,

$\mathsf{tr}_m(k) \in \mathsf{Lines}$ if $m, k \in \mathsf{Ob}$.

AxOb It is possible to move with any given speed less than 1, everywhere and in every direction, i.e. for all $\ell \in \mathsf{Lines}$ and $m \in \mathsf{Ob}$

$\mathsf{slope}(\ell) < 1 \to (\exists k \in \mathsf{Ob})\ell = \mathsf{tr}_m(k)$.

Let $\rho : {}^2\mathsf{F} \longrightarrow {}^2\mathsf{F}$ denote "*reflection to the $x = y$-line*", i.e. $\rho(x, y) = \langle y, x\rangle$ for all $x, y \in \mathsf{F}$. Notice that ρ is a linear transformation but it does not preserve the Minkowski-distance (since $\mu(\rho(x, y)) = -\mu(x, y)$).

Theorem 1.4. *Let $n = 2$, let M be a model of our language and let m, k be observers in M. Then (i)-(ii) below hold.*

(i) f_{mk} *is a Poincaré-transformation composed with a dilation and a field-automorphism-induced mapping, and composed perhaps with the reflection* ρ*, if* $M \models$ $\mathbf{Specrel}_0 \cup \{\mathbf{AxLine}, \mathbf{AxOb}\}$.

(ii) f_{mk} *is a Poincaré-transformation composed perhaps with the reflection* ρ*, if* $M \models$ $\mathbf{Specrel} \cup \{\mathbf{AxLine}, \mathbf{AxOb}\}$.

Theorems 1.2, 1.4 say that $\mathbf{Specrel}_0$ and **Specrel** are adequate axiomatizations for the "metric-free" and "metric" parts of special relativity, respectively[7], because the decisive part of the completeness theorem for special relativity is the description of the world-view transformations. In [5, Thm.3.8.14, p.301], [3, Thm.4, p.15] and [4, §§6,7] these theorems are extended to complete first-order axiomatizations of the standard models of special relativity (by adding some more, "book-keeping", axioms, which are called "book-keeping" axioms because they are of a trivial nature, in some sense).

2. A piece of conceptual analysis: faster than light motion

Our main aim is more ambitious than providing a complete axiomatization for the kinematics of special relativity. Our complete axiomatization is only a byproduct. Our aim is to provide an analysis of the logical structure of special relativity (or in other words, giving a conceptual analysis for relativity in a precise, explicit and transparent logical framework). To this end we have to start with a list of axioms and a completeness theorem but the emphasis is on what comes beyond these.

Many of the efforts reported here are strongly connected to what is called "reverse geometry" in Pambuccian [44]. At least many of the goals of the presently reported approaches can be interpreted as doing reverse geometry for the geometry of space-time.

To illustrate what we mean by the logical analysis of relativity, we choose the much debated topic of *faster than light motion.* We refer to this as the *No FTL* conjecture where "*No FTL*" abbreviates the conjecture that faster than light motion is impossible (in relativity). The issue whether faster than light (*FTL*) observers or bodies can, in principle, exist is being seriously debated even today, cf. e.g. [20] or Matolcsi-Rodriguez [39]. A large part of the literature claims that *No FTL* is an

[7]This is for $n > 2$, for $n = 2$ we have to add two axioms as in Thm.1.4. Roughly, by results like Thm.1.2, $\mathbf{Specrel}_0$ and **Specrel** correspond to what in Malament [38], Hogarth [27] are called "*causal space-time*" and "*space-time*" respectively, while the time-oriented version $\mathbf{Specrel} \cup \{f_{mk}(\langle 1, 0, \ldots, 0\rangle)_1 > 0\}$ of **Specrel** corresponds to "*time-oriented space-time*". Cf. [4].

axiom (of relativity) while another substantial part wants to get rid of *No FTL* and they maintain that *No FTL* is *not* true (i.e. that *FTL is* possible).

At the same time, in the area connecting relativity with quantum mechanics, the *No FTL* conjecture plays an essential role, e.g. the so-called Einstein-Podolsky-Rosen paradox is based on assuming *No FTL*. Many authors think that *No FTL* is the essence of relativity as is suggested by the subtitle "Physics according to Newton - A world with no speed limit" in Kogut's book on relativity.

This controversy in the literature[8] makes *No FTL* an ideal testing-ground for logic in relativity. E.g. we can study which potential axioms of (special) relativity imply *No FTL*, what is implied by *No FTL*, how to obtain the most "conservative" modification of the axiom system such that it will not imply *No FTL* etc. We will return to "*No FTL*" under conditions more general than special relativity in section 4.

Let "No FTL" abbreviate the formula saying that no observer k can move *faster than light* relative to any other observer m, formally it abbreviates

$$\mathsf{speed}_m(k) \leq 1 \text{ when } m, k \in \mathsf{Ob}.$$

The following is a corollary of Theorem 1.2(i):

Corollary 2.1. $\mathbf{Specrel}_0 \models \mathsf{No\ FTL}$, *if* $n > 2$.

Corollary 2.1 implies that if we do not want to have No FTL as a theorem in special relativity, then we have to give up or weaken at least one of the axioms in $\mathbf{Specrel}_0$. However, even those authors who debate the status of "*No FTL*" accept all axioms of $\mathbf{Specrel}_0$.

Corollary 2.1 above can be considerably improved. Namely, if we derive "No FTL" from a weaker subsystem of $\mathbf{Specrel}_0$, we get a stronger, more interesting theorem. In particular, we get information about what other usually accepted axioms we also have to give up if we want to permit "*FTL*" travel in special relativity. Indeed, $\mathbf{Specrel}_0$ can be replaced by the much weaker subsystem **Locrel** in Cor.2.1 above, see Theorem 4.4 herein. Similar results are proved in Madarász-Tőke [36, Thm.4], [35, 37, Thm.s3-5], [32, Thm.3.2.13, p.118.], [5, Thm.4.3.24, p.497].

In [32, pp.89-90] it is also shown that No FTL does not follow from Einstein's *Special Principle of Relativity* (SPR) going back to Galileo, in 2-dimensional space-time, contrary to what Einstein claims in [14,

[8] The above-mentioned controversy is documented in [32, §2.7]. Cf. also Lewis [31, pp.67-80] and Novikov [43].

pp.126-127]. In [5, §3.4.2, Thm.3.4.22, Thm.4.3.25] it is also studied how to refine the axioms of special relativity in order to make No FTL independent from the rest of the axioms. I.e. variants of special relativity are elaborated there in which *FTL is* possible (i.e. where No FTL is no more a theorem). In this way the approaches reported here intend to also contribute to the kind of research that wants to experiment with admitting *FTL* motion in relativity. For a thorough discussion of the status and literature of FTL cf. e.g. [32, §2.7, pp.70-73 (especially footnote 163)], and Lewis [31, pp.67-80, 212-213].

Ax [9] adds No FTL to his list of axioms for special relativity and he raises the question whether his axiom system is redundant. Here we will give an answer to Ax's question which at the same time will refute a claim of Mundy's for space-times of more than two dimensions. Namely, we will prove that No FTL is a logical consequence of Ax's remaining axioms (hence Ax's system *is* redundant), and will prove that Mundy's axioms in [42] do entail No FTL when $n > 2$. Intuitively, we conclude that No FTL is not an axiom but a theorem of (special) relativity.[9]

Now we turn to applying Cor.2.1 to answering questions in Ax [9], Mundy [42]. Ax [9] contains a finite first-order axiomatization Σ of special relativity. In this axiom system, one of the axioms, namely **AxC4**, states explicitly that all observers move slower than light. We recall **AxC4** in section 5.

$\Sigma - \{\mathbf{AxC4}\}$ denotes the set of axioms that remain of Σ after **AxC4** is deleted from it.

Theorem 2.1. *In the axiom-system* Σ *in Ax* [9], *the axiom* **AxC4** *stating that all observers move slower than light is superfluous, i.e.*

$$\Sigma - \{\mathbf{AxC4}\} \models \mathbf{AxC4}.$$

Proof-outline: Let $\mathsf{U} = \langle U, \mathsf{P}, \mathsf{S}, \mathsf{T}, \mathsf{R} \rangle$ be a model of $\Sigma - \{\mathbf{AxC4}\}$. Here, P denotes the set of "particles" and S denotes the set of "light-signals". These correspond, roughly, to our Ob and Ph, cf. section 5. Using Tarski's first-order axiomatization of Euclidean geometry, on pp. 531-532 Ax constructs a Euclidean ordered field F and to any $a \in \mathsf{P}$ a bijection $\sigma_a : {}^4\mathsf{F} \longrightarrow \mathsf{Events}$, where Events corresponds roughly to our set of events. In this construction, Ax uses **AxC4** three times. With some ingenuity, one can replace the first and last uses of **AxC4** by the uses of the axioms **T1** and **AxC1** of Σ respectively. The second

[9] A similar statement can be proved about general relativity, too, but there more work is needed for formalizing the *No FTL* conjecture in a precise, purely logical form; cf. sections 3,4 and Thm.4.4.

use of **AxC4** is not needed for the construction itself, only to ensure a specific property. Now, using the above σ_a, one can construct a model $\mathsf{M} = \langle \mathsf{B}, \mathsf{Ob}, \mathsf{Ph}, \mathsf{F}, \mathsf{W} \rangle$ of our language and check that $\mathsf{M} \models$ $\mathbf{Specrel}_0$. In Ax's axiomatization, $n = 4$. By Corollary 2.1 then $\mathsf{M} \models$ No FTL. Using the definition of M, this means that $\mathsf{U} \models$ **AxC4**. QED

On the other hand, the axiomatization in Mundy[10] [42] does not contain an axiom explicitly stating "*No FTL*". But then, Mundy claims that there are models of his axiom system $\mathcal{T}$ in which there exist FTL observers. On p.43 he writes: "Therefore the only line-type information left open by the theories $\mathcal{T}$ and $\mathcal{T}'$ is which if any of the lines on or outside of the light cone are of type T, i.e. are possible paths of inertial motion. In physical terms this amounts to asking whether inertial motion can proceed at a speed equal to or greater than that of light. My contention is that nothing in either the classical or the special relativistic space-time theories provides any answer to this question. The evidence for this is that the theories $\mathcal{T}$ and $\mathcal{T}'$ seem to formalise adequately the physical content of those space-time theories, and yet do not fix an answer to this question."

Theorem 2.2. *Let $\mathcal{T}$ be the axiom system in Mundy* [42]. *In any model of $\mathcal{T}$ all the lines in T (i.e. the so-called time-like lines) are within the light-cones. When $c \neq 0$, they are strictly within the light-cones. Hence $\mathcal{T} \models$ "No FTL".*

Proof-outline: The proof-idea is very similar to the previous one. Let M be a model of $\mathcal{T}$. (Assume that $c \neq 0$. Mundy does not seem to be aware of the fact that $\mathcal{T}$ allows $c = 0$. But one can show that if $c = 0$ in a model of $\mathcal{T}$, then the time-like and light-like lines coincide, hence we are done.) In [42, pp. 40-42], Mundy constructs co-ordinate systems for each time-like line in T, based on which one can construct a model M' of our language. One can check then that $\mathsf{M}' \models$ $\mathbf{Specrel}_0$, and so $\mathsf{M}' \models$ No FTL by our Corollary 2.1 (since $n = 4$ in [42]). This means that there are no time-like lines outside the light-cones. That there are no time-like lines on the light-cone follows from [3, Prop.1]. QED

We hope the above example (of analyzing *No FTL*) illustrates how analyzing the logical structure works in the *theory of inertial observers* (special relativity proper).

A relatively comprehensive analysis of the theories **Specrel**, $\mathbf{Specrel}_0$ and their fragments can be found in [5], [32]. An investigation of the so-called "why-type" questions is also found there. Below we turn to going

[10] Mundy's work "comes" from the seminars of Suppes on the topic of logic and relativity. Suppes (1974) credits his motivation for this line of research to interactions with Tarski.

beyond special relativity. We will make two steps in this direction: (i) admitting accelerated observers (section 3), and (ii) making the theory local (section 4).

3. Axioms for accelerated observers, in FOL

In **Specrel** we restricted attention to inertial observers. It is a natural idea to generalize the theory to including accelerated observers as well. We will refer to such a generalized theory as a theory of accelerated observers. It is explained in the classic textbook [40, pp.163-165] that the study of accelerated observers can be regarded as a natural first step (from special relativity) towards general relativity.

The most important axiom for accelerated observers will state that at each moment of his life-time, the accelerated observer "sees" the world near him and for a short while like some inertial observer does, co-moving with him. I.e.: at every point p on the life-line of an accelerated observer k, there is an inertial observer m such that in a small enough neighborhood of p the observers k and m agree, i.e. they "see" the world roughly the same way there. We now begin to formalize this. The *sphere* with center p and radius $\varepsilon \in \mathsf{F}$ is $S(p,\varepsilon) \stackrel{d}{=} \{q \in {}^n\mathsf{F} : |q-p| \le \varepsilon\}$ and F^+ denotes the set of strictly positive members of F.

We say that m, k are *co-moving observers at* $q \in {}^n\mathsf{F}$, in symbols $\mathsf{comove}(m,k,q)$, if (i)-(iii) below hold for m, k as well as for k, m.

(i) $q \in \mathsf{tr}_m(k) \cap \mathsf{tr}_m(m)$,

(ii) f_{mk} is an injective function on $S(q,\eta)$ for some $\eta \in \mathsf{F}^+$, and

(iii) $\forall \varepsilon \in \mathsf{F}^+ \quad \exists \delta \in \mathsf{F}^+ \quad \forall p \in S(q,\delta) \quad |p - \mathsf{f}_{mk}(p)| \le \varepsilon\, |p-q|$.

This notion of co-moving observers matches our intuition when $\mathsf{tr}_m(m)$ agrees with $\mathsf{tr}_k(k)$ in a neighborhood of q. **AxSelf$^-$** below will ensure this.

To speak about accelerated observers, we introduce a new unary relation symbol Ib into our language. We will read "$\mathsf{Ib}(x)$" as "x is an *inertial body*", and then "accelerated observer" will mean "not (necessarily) inertial observer".

We are ready now for stating our main axiom for accelerated observers.

AxAcc At any point on the life-line of any observer k there is a co-moving inertial observer, i.e.

$$(\forall k \in \mathsf{Ob})(\forall q \in \mathsf{tr}_k(k))(\exists m \in \mathsf{Ob} \cap \mathsf{Ib})\mathsf{comove}(k,m,q)\ .$$

In models of **Specrel**, the world-view transformations f_{mk} are bijections everywhere defined on ${}^n\mathsf{F}$. Experimenting with constructing world-views for accelerated observers suggests that it is not natural to assume that the domain of f_{mk} is the whole ${}^n\mathsf{F}$ when k is truly accelerated and m is inertial, cf. e.g. [40, §6.6, p.173 (Fig.6.4)].[11] Therefore we relax the assumption that the accelerated observer k sees some body in each point of ${}^n\mathsf{F}$, i.e. we relax the condition that the *co-ordinate domain* of observer m is the whole ${}^n\mathsf{F}$, for every observer m. (This assumption is implicit in **Specrel**, cf. e.g. **AxPh**.) The next section is devoted to this aspect of generalizing our theory towards general relativity (where f_{mk} is only a partial function on ${}^n\mathsf{F}$).

AxSelf⁻ A (perhaps accelerated) observer sees himself in his own co-ordinate system on that part of the time-axis where he sees a body at all, and this part is nonempty,

$$\mathsf{tr}_m(m) = \bar{t} \cap \{p \in {}^n\mathsf{F} : \exists b \mathsf{W}(m,b,p)\} \neq \emptyset.$$

In sections 1,2 we had only inertial observers (see e.g. **AxLine**), therefore, for brevity, we said "observer" in place of "inertial observer". For inertial observers we continue to use the axiom system **Specrel**, but since we now have accelerated observers as well, we will have to restrict **Specrel** to inertial observers. If **Ax** is any axiom, **Ax^in** will denote the axiom we get by restricting it to inertial observers. E.g.

AxSelf^in $(\forall m,p)\Big((\mathsf{Ib}(m) \wedge \mathsf{Ob}(m) \wedge \mathsf{F}(p_1)) \rightarrow [\mathsf{W}(m,m,p)$
$\leftrightarrow (p_2 = 0 \wedge \cdots \wedge p_n = 0)]\Big).$

We define

Specrel^in $\stackrel{d}{=}$ {**AxSelf^in**, **AxPh^in**, **AxEvent^in**, **AxSym^in**, **AxLine^in**, **AxOb^in**, AxField}.

To be able to use **AxAcc** in models where the field-reduct is different from the real line $\Re$, we will need a kind of "induction" axiom schema. It will state that every parametrically definable subset of the field-reduct has a supremum if bounded and nonempty; as follows.

Let φ be a formula in the first-order language of our models $\mathsf{M} = \langle \mathsf{B}, \mathsf{Ob}, \mathsf{Ph}, \underline{\mathsf{F}}, \mathsf{W} \rangle$, and let $t, a_1, \ldots, a_m$ be all the free variables of φ.

[11] Actually, if we assume that "radar distance" between vertical coordinate-lines is constant and we use 1/2-radar simultaneity for setting up the co-ordinate system for each accelerated observer k, then we cannot have the whole ${}^n\mathsf{F}$ as the co-ordinate domain of all accelerated observers k, assuming enough of them exist.

Sup_φ The subset $\{t \in \mathsf{F} : \varphi(t, a_1, \ldots, a_m)\}$ of the field-reduct defined by φ when using $a_1, \ldots, a_m$ as fixed parameters has a supremum if it is bounded and nonempty, i.e.

$$(\forall a_1, \ldots a_m)[\exists t\varphi \wedge (\exists b \in \mathsf{F})(\forall t \in F)(\varphi \rightarrow t \leq b) \rightarrow \\ (\exists d \in \mathsf{F})(\forall b \in \mathsf{F})\left((\forall t \in \mathsf{F})(\varphi \rightarrow t \leq b) \leftrightarrow d \leq b\right)].$$

Note that having $a_i \in \mathsf{F}$ and $a_j \in \mathsf{B}$ in the *same* formula φ is permitted here.

$\mathsf{IND} \overset{d}{=} \{\mathsf{Sup}_\varphi : \varphi$ is a first-order formula in our language$\}$.

Accrel $\overset{d}{=}$ **Specrel**$^{\mathbf{in}} \cup \{$**AxAcc**, **AxSelf**$^{-}\} \cup \mathsf{IND}$.

We note that if the field-reduct of a model M is the real line $\Re$, then $\mathsf{M} \models \mathsf{IND}$.

Accrel is at the heart of the theory of accelerated observers, in some sense, cf. e.g. [5, §8] together with [40, §6]. Some interesting statements of relativity can already be derived from it at the present point. As an example we show that the so-called *Twin Paradox* can be naturally formulated and proved in **Accrel**. See Figure 1. More importantly, the details of the Twin Paradox (e.g. who sees what, when) can be analyzed with the clarity of logic, cf. [5, pp.139-150] for part of such an analysis.

Intuitively, the *Twin Paradox* says that if one of two twin brothers leaves the other (accelerating) and returns to him later, the brother who stayed behind will be *older* at the time of their reunion. That is, more time has passed for the "non-accelerating" brother than for the traveling one.

Twinp $(\forall m, k \in \mathsf{Ob})(\forall p, q, p', q' \in \bar{t})[(\mathsf{Ib}(m) \wedge p_1 < q_1 \wedge p'_1 < q'_1$ $\wedge\, \mathsf{meet}(m, k, p, p', q, q') \wedge \mathsf{depart}(m, k, p, q) \wedge \mathsf{live}(k, p', q'))$ $\rightarrow q_1 - p_1 > q'_1 - p'_1]$, where

$\mathsf{meet}(m, k, p, p', q, q')$ denotes $\mathsf{ev}_m(p) = \mathsf{ev}_k(p') \wedge \mathsf{ev}_m(q) = \mathsf{ev}_k(q')$,

$\mathsf{depart}(m, k, p, q)$ denotes $(\exists r \in \bar{t})(p_1 < r_1 < q_1 \wedge r \notin \mathsf{tr}_m(k))$, and

$\mathsf{live}(k, p', q')$ denotes $\{r \in \bar{t} : p'_1 < r_1 < q'_1\} \subseteq \mathsf{tr}_k(k)$.

The following is from Székely [54].

Theorem 3.1.

(i) **Accrel** $\models$ **Twinp**.

(ii) For any Euclidean ordered field F *different from the real line there is* $\mathsf{M} \models$ (**Accrel** $- \mathsf{IND}$) *with field-reduct* F *such that* $\mathsf{M} \not\models$ **Twinp**.

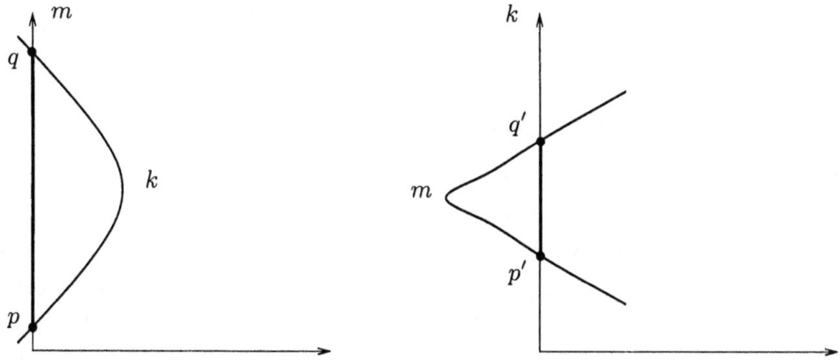

m: "non-moving" (inertial) brother

k: traveling (accelerated) brother

Figure 1. The "twin paradox".

Similar results and investigation apply to the effect of gravity on clocks (also known as the *Tower Paradox*). For the time being, more concrete formulations, details etc. are available in [5] using some axioms extra to our present **Accrel**. We conjecture that the axioms extra to **Accrel** in [5] can be either eliminated or significantly simplified. (Concerning the above effect of gravity on clocks, we note that this effect is at the heart of the "science-fiction"-like behaviour of certain black holes.)

Adding the axiom schema IND to our axioms in **Accrel** represents a first step in the direction pursued in the so-called nonstandard-time logics of time represented by [48], [2] and based on Feferman [15]. Feferman's paper is devoted to reformulating everyday mathematical practice in the framework of many-sorted first-order logic.[12] All this enables us to "import" just as much of e.g. mathematical analysis into our first-order theory **Accrel** of accelerated observers as we need. Explicit and detailed elaborations of these ideas to situations similar to our present one (theory of accelerated observers) can be found in the above quoted [48, 2] and in the works quoted therein. In developing **Accrel** further, and in related work, we will adopt the methods of the works just quoted to the framework of **Accrel** (extended with a pseudo-second-order sort).

[12] Other works in this direction carry the adjective "weak-second-order" or "Henkin-style" to indicate that in reality they use only many-sorted first-order logic. An example is weak second-order analysis. In [Takeuti, G. Proof theory, North-Holland, 1987] (Chap.3) the adjective "weak or Henkin-style" is dropped from "weak-second-order" theories.

The logical analysis of the relativity theory **Accrel** is an intensively pursued research topic, cf. e.g. [5], [54]. For lack of space, here we cannot present this subject to the extent it deserves.

4. Localizing relativity

We mentioned in the previous section that it is not natural to assume that the co-ordinate domain of a truly accelerated (i.e. not inertial) observer would always be the whole ${}^n\mathsf{F}$. Motivated by this, and by other considerations that come up in developing general relativity, it is useful to investigate generalizations of **Specrel** and **Accrel** in which the co-ordinate domains of the observers are only subsets of ${}^n\mathsf{F}$. These theories are called local versions of **Specrel** and **Accrel**, respectively (where "local" is used in the same sense as in general relativity). Here we illustrate these investigations by a few examples taken from Madarász-Tőke [37, 35] and from [5, §4.9]. Localization as a step towards general relativity theory is used also in Latzer [29] and Busemann [12].

To describe the new, localized theories, first we describe what their typical models are like. Let $\mathsf{M} = \langle \mathsf{B}, \mathsf{Ob}, \mathsf{Ph}, \underline{\mathsf{F}}, \mathsf{W} \rangle$ be a model and let O be a set of subsets of ${}^n\mathsf{F}$; we define the model M^- "localized" or "relativized" to O as follows. Intuitively, for each $D \in O$ and $m \in \mathsf{Ob}$ we will have a new observer $\langle m, D \rangle$ such that the world-view of $\langle m, D \rangle$ is that of m restricted to the "co-ordinate domain" D. Formally, let $\mathsf{M}^- = \langle \mathsf{B}^-, \mathsf{Ob}^-, \mathsf{Ph}^-, \underline{\mathsf{F}}^-, \mathsf{W}^- \rangle$ where

$$\mathsf{B}^- \stackrel{d}{=} \mathsf{B} \times O, \quad \mathsf{Ob}^- \stackrel{d}{=} \mathsf{Ob} \times O, \quad \mathsf{Ph}^- \stackrel{d}{=} \mathsf{Ph} \times O, \quad \underline{\mathsf{F}}^- \stackrel{d}{=} \underline{\mathsf{F}},$$

$$\mathsf{W}^-(\langle m, D \rangle, \langle b, D' \rangle, p) \stackrel{d}{\Leftrightarrow} (p \in D \wedge \mathsf{W}(m, b, p)).$$

We now begin to give axioms suitable for these localized models. Throughout this section , $n \geq 2$ is arbitrary. We will always assume AxField, and we will not indicate AxField explicitly in the theorems.

For simplicity, temporarily, we will talk about "inertial" observers only, so we will not use the new relation symbol Ib and "in our mind" all observers will be inertial as in the axiom system **Specrel**. Later, of course, we will extend our localization procedure from **Specrel** to **Accrel**, since it is the localized version **LocAccrel** of **Accrel** which brings us closer to general relativity.

If **Ax** is an axiom, then $\mathbf{Ax}^-$ will denote the axiom we get from **Ax** by "localizing", or "relativizing" it, and then usually $\mathbf{Ax}^{--}$ will denote an even weaker axiom. In particular, $\mathbf{Ax}^-$ will be such that if $\mathsf{M} \models \mathbf{Ax}$ then $\mathsf{M}^- \models \mathbf{Ax}^-$. (We will always assume that the elements of O are open and that they intersect the time-axis. We call a set $D \subseteq {}^n\mathsf{F}$ *open* if $(\forall p \in D)(\exists \varepsilon \in \mathsf{F}^+) S(p, \varepsilon) \subseteq D$.)

When f is a function, $\mathsf{Dom}(f)$ denotes its domain. The *co-ordinate domain* $\mathsf{CD}(m)$ of an observer m is defined as $\mathsf{Dom}(\mathsf{f}_{mm})$, i.e.

$$\mathsf{CD}(m) \stackrel{d}{=} \{p \in {}^n\mathsf{F} \,:\, (\exists b \in \mathsf{B})\mathsf{W}(m,b,p)\}\,.$$

AxLine^{--} The traces of observers and photons are subsets of straight lines, but they must be restrictions of lines to the co-ordinate domain (or empty),

$$(\forall m \in \mathsf{Ob})(\forall h \in \mathsf{Ob} \cup \mathsf{Ph})(\exists \ell \in \mathsf{Lines})(\mathsf{tr}_m(h) = \ell \cap \mathsf{CD}(m) \text{ or } \mathsf{tr}_m(h) = \emptyset).$$

AxOb^{--} Each point in the co-ordinate domain has a neighborhood and a "speed threshold" λ such that within this neighborhood each line slower than λ is the life-line of an observer,

$$(\forall m \in \mathsf{Ob})(\forall p \in \mathsf{CD}(m))(\exists \varepsilon, \lambda \in \mathsf{F}^+)(\forall \ell \in \mathsf{Lines}) \Big((\mathsf{slope}(\ell) < \lambda \,\wedge\, \ell \cap \mathsf{S}(p,\varepsilon) \neq \emptyset) \to \;(\exists k \in \mathsf{Ob})\emptyset \neq \mathsf{tr}_m(k) \subseteq \ell\Big).$$

AxOpen $(\forall m, k \in \mathsf{Ob})(\mathsf{Dom}(\mathsf{f}_{mk})$ is an *open* set$)$.

Notice that the "speed threshold" λ in **AxOb^{--}** can vary from point to point.

If F is a field then coll denotes the ternary *collinearity* relation on ${}^n\mathsf{F}$, i.e. $\mathsf{coll}(p,q,r)$ iff $(\exists \ell \in \mathsf{Lines})\{p,q,r\} \subseteq \ell$. The *affine structure* $\mathcal{A} = \langle {}^n\mathsf{F}, \mathsf{coll}\rangle$ can be extended (uniquely) to an n-dimensional *projective structure* $\mathcal{P} = \langle \overline{{}^n\mathsf{F}}, \mathsf{coll}\rangle$ the natural way (${}^n\mathsf{F} \subseteq \overline{{}^n\mathsf{F}}$ etc.), see any textbook on projective geometry (or e.g. [23]). Throughout, by a $\mathcal{P}$*-collineation* we understand an automorphism of the projective structure $\mathcal{P}$ and by an $\mathcal{A}$*-collineation* we understand an automorphism of the affine structure $\mathcal{A}$. In other words, an $\mathcal{A}$-collineation is a permutation of ${}^n\mathsf{F}$ which preserves lines. It is known that these are exactly the bijective affine transformations composed with field-automorphism-induced mappings.

The next theorem (which is Thm.1 in [35]) says that, locally, the world-view transformations are $\mathcal{P}$-collineations in models of **AxLine^{--}**, **AxOb^{--}**, **AxOpen**. This is a rather strong Alexandrov-Zeeman type theorem.

Theorem 4.1. *Assume* **AxLine^{--}**, **AxOb^{--}**, **AxOpen**. *Let* $m, k \in \mathsf{Ob}$ *and* $p \in \mathsf{Dom}(\mathsf{f}_{mk})$. *Then there is a unique* $\mathcal{P}$*-collineation* g *such that* f_{mk} *agrees with* g *on* $S(p,\varepsilon)$ *for some* $\varepsilon \in \mathsf{F}^+$. *In particular, the* f_{mk}*'s preserve* coll *and* $\neg\mathsf{coll}$ *locally.*

In the above theorem, $\mathcal{P}$-collineation cannot be replaced by $\mathcal{A}$-collineation. We get $\mathcal{A}$-collineation, however, if we add axioms about photons.

Notice that so far, nothing has been used about photons. We will assume that the photon-traces that cross a given point form a cone, called a *light-cone.* We are going to formalize this.

The set of *spatial directions* dir is defined as

$$\mathsf{dir} :\stackrel{\mathsf{d}}{=} \{\mathsf{d} \in {}^{\mathsf{n}-1}\mathsf{F} \ : \ \mathsf{d} \neq \bar{0}\}.$$

Assume $m \in \mathsf{Ob}$, $b \in \mathsf{B}$, $d \in \mathsf{dir}$. We say that b *moves in direction* d *as seen by* m iff $(\forall p, q \in \mathsf{tr}_m(b))(\exists \lambda \in \mathsf{F})$
$[\langle q_2, \ldots, q_n\rangle - \langle p_2, \ldots, p_n\rangle = \lambda d \ \wedge \ (q_1 > p_1 \ \rightarrow \ \lambda \geq 0)]$.
We note that $\mathsf{speed}_m(b)$ is meaningful also when $\mathsf{tr}_m(b)$ is an at least two-element subset of a line. In the following we will use $\mathsf{speed}_m(\mathsf{ph})$ in this sense. Recall that $\mathsf{speed}_m(\mathsf{ph}) = \infty$ abbreviates that $\mathsf{tr}_m(\mathsf{ph})$ is "horizontal" (as introduced at the beginning of section 1).

Ax∃Ph From any point $p \in \mathsf{CD}(m)$ in any direction there is a photon moving forwards in that direction,

$$(\forall m \in \mathsf{Ob})(\forall p \in \mathsf{CD}(m))(\forall d \in \mathsf{dir})(\exists \mathsf{ph} \in \mathsf{Ph})$$
$$[p \in \mathsf{tr}_m(\mathsf{ph}) \wedge \ (\mathsf{ph} \text{ moves in direction } d \text{ as seen by } m)].$$

AxItr The speed of light is direction-independent locally,

$$(\forall \mathsf{ph}, \mathsf{ph}' \in \mathsf{Ph})\left(\mathsf{tr}_m(\mathsf{ph}) \cap \mathsf{tr}_m(\mathsf{ph}') \neq \emptyset \rightarrow \mathsf{speed}_m(\mathsf{ph}) = \mathsf{speed}_m(\mathsf{ph}')\right).$$

In effect, the photon-traces that cross a given $p \in \mathsf{CD}(m)$ form a cone, called a *light-cone.* Notice that the speed of light—the angle (or "openness" or "width") of the light-cone—may differ from point to point. Here, *Itr* abbreviates "isotropy".

AxFin The speed of each photon is nonzero and finite,

$$(\forall m \in \mathsf{Ob})(\forall \mathsf{ph} \in \mathsf{Ph})\,(0 < \mathsf{speed}_m(\mathsf{ph}) \neq \infty \text{ or } \mathsf{tr}_m(\mathsf{ph}) = \emptyset).$$

Theorem 4.2. *Assume* **AxLine^{--}**, **AxOb^{--}**, **AxOpen**, **Ax∃Ph**, **AxItr**, **AxFin**. *Assume* $m, k \in \mathsf{Ob}$ *and* $p \in \mathsf{Dom}(\mathsf{f}_{mk})$. *Then there is a unique* $\mathcal{A}$*-collineation* g *such that* f_{mk} *agrees with* g *on* $S(p, \varepsilon)$, *for some* $\varepsilon \in \mathsf{F}^+$. *Hence the* f_{mk}*'s preserve parallelism,* coll *and* ¬coll *locally.*

Theorem 4.3 below, taken from Lester [30, p.929], shows that the assumption **AxOb^{--}** cannot be omitted from Theorem 4.2 above, even if we replace **AxItr** with the stronger **AxE** defined below. It also shows that the localized version of the Alexandrov-Zeeman theorem Thm.1.3 fails to hold.

AxE $(\forall m \in \mathsf{Ob})(\forall \mathsf{ph} \in \mathsf{Ph})\mathsf{speed}_m(\mathsf{ph}) = 1$.

Theorem 4.3.
For any $n \geq 2$ *there is a model of* $\{$**AxLine**$^{--}$, **AxOpen**, **Ax∃Ph**, **AxE**, **AxSelf**$\}$ *with* $m, k \in \mathsf{Ob}$ *such that* $\mathsf{Dom}(\mathsf{f}_{mk}) \neq \emptyset$ *and for any* $p \in \mathsf{Dom}(\mathsf{f}_{mk})$ *and any* $\varepsilon \in \mathsf{F}^+$ *the world-view transformation* f_{mk} *does not preserve* coll *on* $\mathsf{S}(p, \varepsilon)$, *i.e. there are* $a, b, c \in \mathsf{S}(p, \varepsilon)$ *such that* $\mathsf{coll}(a, b, c)$ *but* $\neg\mathsf{coll}(\mathsf{f}_{mk}(a), \mathsf{f}_{mk}(b), \mathsf{f}_{mk}(c))$. *I.e., the collinear points* a, b, c *the collinearity of which is not preserved are densely located everywhere in* $\mathsf{Dom}(\mathsf{f}_{mk})$.

It might be interesting to notice that while studying **Specrel** and its fragments (cf. section 1) assuming or omitting axiom **AxOb** for $n > 2$ did not matter too much, in the present, local theory (or localized version of **Specrel**) **AxOb**$^-$ does matter by Theorems 4.1–4.3.

We now turn to proving a "no FTL" theorem in the local setting. We will replace **AxItr** with a much weaker assumption, **AxP1**, but at the same time we will weaken-and-strengthen **AxOb**$^{--}$ to requiring that the observer-traces "fill" the light-cones.

AxP1 The speed of light is unique and well-defined in each direction at each location,

$(\forall \mathsf{ph}, \mathsf{ph}' \in \mathsf{Ph})[\mathsf{ph}$ and ph' move in the same direction as seen by $m \rightarrow (\mathsf{tr}_m(\mathsf{ph}) = \mathsf{tr}_m(\mathsf{ph}')$ or $\mathsf{tr}_m(\mathsf{ph}) \cap \mathsf{tr}_m(\mathsf{ph}') = \emptyset)]$.

Basically, this is the first-order logic formalization of Friedman's principle (P1) in [18, p.159].

AxPh^{--} $\stackrel{d}{=}$ **Ax∃Ph** $\wedge$ **AxP1** $\wedge$ **AxFin**.

AxOb$^-$ There are observers on lines which are slower than light locally,

$(\forall m \in \mathsf{Ob})(\forall \mathsf{ph} \in \mathsf{Ph}, p \in \mathsf{tr}_m(\mathsf{ph}))(\forall 0 \leq \lambda < \mathsf{speed}_m(\mathsf{ph}))(\exists k \in \mathsf{Ob})$ $[p \in \mathsf{tr}_m(k),\ \mathsf{speed}_m(k) = \lambda,$ and ph, k move in the same direction as seen by $m]$.

We will also use a weakened version of the local version of **AxEvent** formalized as follows. Assume $k, h \in \mathsf{Ob}$. Then we say that k is a *brother* of h iff $(\forall m \in \mathsf{Ob})\ \mathsf{tr}_m(k) = \mathsf{tr}_m(h)$. In a model M^- relativized to a set O of subsets of $^n\mathsf{F}$, if $k \in \mathsf{Ob}$, then the new observers $\langle k, D\rangle, \langle k, D'\rangle$ are brothers. They may "see" different events.

AxEvent^{--} If m sees an event happening to k, some brother of k sees it, too,

$(\forall m, k \in \mathsf{Ob}, p \in \mathsf{tr}_m(k))(\exists h \in \mathsf{Ob}, q \in {}^n\mathsf{F})[h$ is a brother of k and $\mathsf{ev}_m(p) = \mathsf{ev}_h(q)]$.

Locrel $\stackrel{d}{=}$ {**AxSelf**$^-$, **AxLine**$^{--}$, **AxPh**$^{--}$, **AxOb**$^-$, **AxEvent**$^{--}$, **AxOpen**, AxField}.

Now, **Locrel** is a localized version of **Specrel**$_0$. FTL abbreviates faster than light. Let $k, m \in \mathsf{Ob}$. We call k FTL w.r.t. m iff there is a $\mathsf{ph} \in \mathsf{Ph}$ such that k and ph move in the same direction as seen by m, they meet, i.e. $\mathsf{tr}_m(k) \cap \mathsf{tr}_m(\mathsf{ph}) \neq \emptyset$, and $\mathsf{speed}_m(k) > \mathsf{speed}_m(\mathsf{ph})$. **NoFTL** abbreviates the formula saying that no observer k can move *faster than light*, in the present *local* sense, relative to any other observer, i.e. it abbreviates the formula $\neg(\exists m, k \in \mathsf{Ob})[k$ FTL w.r.t. $m]$.

Theorem 4.4. *Assume* $n > 2$. *Then* **Locrel** $\models$ **NoFTL**.

For $n = 2$, or if we omit any assumption from **Locrel**, FTL observers do become possible.

The local version **LocAccrel** of the theory **Accrel** of accelerated observers can be obtained by adding **AxAcc**, **AxSelf**$^-$, IND to **Locrel**$^{\mathbf{in}}$ (where the latter is defined analogously to **Specrel**$^{\mathbf{in}}$).

5. Samples from the literature

Einstein named his theories of relativity after the *principle of relativity* (PR) a special case of which is SPR mentioned in section 2. Einstein derives the theory of special relativity as a consequence of SPR. (PR is analogously related to general relativity.) The principle of relativity has been playing a central role in European thinking for almost 2400 years by now, very roughly as follows. Primitive versions of PR were proposed by Heraclides of Ponticus (388-315 BC) and Aristarchus (310-230 BC). PR was "abolished" by Ptolemy (c. 140 AD), and restored by Nicole d'Oresme (c. 1350), Copernicus (c. 1510) and Galileo (c. 1600). It came again in doubt during the 1800's because the theory of electromagnetism seemed to contradict PR. Finally, Einstein restored PR (and even generalized it).[13] Physicist Max Born said about Einstein's theory: "The theory appeared to me, and it still does, the greatest feat of human thinking ...". This brief history indicates that it is worthwhile to apply the machinery of modern logic for obtaining a refined analysis of the logical structure of relativity theory, or theory of space-time.

Indeed, one could say that logic was started by the effort of axiomatizing geometry (by Euclid). This axiomatization was brought to its

[13]Some ideas about the principle of relativity remain open for future work to clarify, in the form of Mach's principle, cf. Barbour [10] and Gödel [21]. The first "approximately modern" formulation of PR is due probably to Oresme, cf. [10, §4.4 (.. the early ideas about relativity), pp.206-207] and §2.5 in [Mlodinow, L., Euclid's window. Touchstone, New York, 2002].

modern form by Tarski, who replaced Hilbert's second-order logic axiomatization by one based on purely first-order logic, cf. Tarski [58], Pambuccian [44]. But the geometry of space-time[14] is at the center of relativity, hence it is natural to base the logical analysis of relativity on (a continuation of) the Tarskian tradition of logicizing geometry.

Axiomatizations of special relativity are abundant in the literature. To mention some: axiomatizations of special relativity have been studied in e.g. Robb 1914 [46], Reichenbach 1924 [45], Carathéodory 1924 [13], Alexandrov and his school starting with 1950 [1, 24], Suppes and his school starting with 1959 [51, 52, 53], Basri 1966 [11], Szekeres 1968 [56], Ax 1978 [9], Friedman 1983 [18], Mundy 1986 [42], Goldblatt 1987 [23], Schutz 1997 [50]. Latzer [29], Busemann [12], Walker [60] contain experiments in the direction of extending the logic based, axiomatic approach towards general relativity. All this is only a small sample. There are more works listed in our bibliography and the bibliographies of these. In this section we briefly recall a few samples from the rich literature mentioned earlier. For lack of space, we cannot even come close to do justice to the works deserving attention. A rich source of further references is [49].

Guts [24] is a survey of work done in the field of axiomatizing relativity theory in and around Russia. Their axiomatizations are not yet logic-based, but they go beyond just giving completeness theorems. We already saw a theorem of this school, the Alexandrov-Zeeman theorem. Here we cite another theorem of Alexandrov to get a flavor of these results. For a more thorough survey we refer to Guts [24].

First we give an axiomatization and then the completeness theorem. A *space-time* in [24] is defined to be a system $\langle V, \langle P_a : a \in V\rangle, T\rangle$ which satisfies **A1** below.

A1 V is a connected, simply-connected locally compact 4-dimensional Hausdorff space, $\langle P_a : a \in V\rangle$ is a family of subsets of V and T is a *transitive, commutative* group of homeomorphisms of V onto itself satisfying

$$\{a\} \subset P_a,$$

$$P_x \subseteq P_y \text{ if } x \in P_y,$$

$$P_{t(a)} = t[P_a] \text{ for all } a \in V, t \in T.$$

[14]The space-time of special relativity was geometrized by Minkowski, Minkowski-geometry is one of the so-called pseudo-Euclidean geometries. The geometry of general relativity is locally pseudo-Euclidean.

We note that this implies that V is homeomorphic to ${}^n\Re$ where $\Re$ is the real line.

The survey [24] emphasizes the "*causal-future-cone*" structure $\langle V, \langle P_x : x \in V\rangle\rangle$ and in this respect is related to Robb's approach (cf. [46, 23]) emphasizing the partial order structure $\langle V, \ll\rangle$ of space-time, where $x \ll y$ iff $y \in P_x$. Many authors call $\ll$ the *causality relation* and call P_x the causal future (cone) of x. The general relativistic approach of Busemann [12] also starts out with the structure $\langle V, \ll\rangle$ (then localizes it in a spirit similar to our section 4). The topology on V can usually be derived from $\ll$.

We are now going to state further axioms. If $X \subseteq V$ then ∂X and $intX$ denote the *boundary* and *interior* of X respectively and if $a \in V$ then G_a denotes the set of all permutations g of V for which $g(a) = a$ and $g[P_x] = P_{g(x)}$ for all $x \in V$.

A2 $P_x \cap \{z \in V : y \in P_z\}$ is bounded if $y \in P_x$,

A3 $(\forall x, y \in \partial P_a - \{a\})(\exists g \in G_a)g(x) = y$,

A4 $(\forall x, y \in intP_a)(\exists g \in G_a)g(x) = y$,

A5 $intP_e \neq \emptyset$ for some $e \in V$.

Motivation for this definition of space-time and for some of the axioms is given in Guts [24, pp.44-46]. The philosophy is that "space-time is a form of existence of matter". $\langle V, T\rangle$ is a 4-dimensional affine space with group T of translations and "$x \in P_y$" intends to represent the so-called causality relation $x \ll y$ with the intended meaning "action (energy-momentum) is transmitted from x to y". So P_x can be interpreted as the causal future cone of event x. Further, ∂P_x is the "light-cone" starting at event x. G_x serves to represent those world-view transformations which leave event x fixed. Axiom **A2** is interpreted as saying that the velocity of transmission of energy is finite.

Theorem 5.1. (Alexandrov [1]) *Assume* **A1** – **A5** *and let* $e \in V$. *Then* P_e *is a closed or an open elliptic cone, and* G_e *is the homogeneous Lorentz-group with dilations. I.e. there are Cartesian coordinates* $x_1 \ldots$, x_4 *such that if* e *has co-ordinates* $\langle 0, 0, 0, 0\rangle$ *then*

$$P_e = \{\langle x_1, \ldots, x_4\rangle \in V : x_4^2 \geq x_1^2 + x_2^2 + x_3^2,\ x_4 \geq 0\} \text{ or}$$

$$P_e = \{\langle x_1, \ldots, x_4\rangle \in V : x_4^2 > x_1^2 + x_2^2 + x_3^2,\ x_4 > 0\}, \text{ and}$$

$$G_e = \{f : f \text{ is a linear transformation preserving Minkowski-distance composed with a dilation}\}.$$

The other axiomatizations in the list we gave at the beginning of this section are in the framework of logic more or less, though not always in a very formalized manner. Robb [46] is the earliest of these, and his work is the model or starting point of many later axiomatizations. E.g. both Ax and Mundy quote his work explicitly as the inspiration of their own axiom systems. Robb had only one binary relation in his logical language, that of the causality relation $\ll$, called often also the relation "after". He had 21 axioms.

Suppes' 1959 axiom system in [51] is in second-order logic, but later in [53] (1972) he outlined how to modify this system to be a first-order one. Here we quote the axiom system from [51].

The pair $\langle M, \mathsf{Ob}\rangle$ is called a *collection of relativistic frames* in [51] if **S1, S2** below hold.

S1 M is a set and Ob is a set of bijections between M and the set ${}^4\Re$ of four-tuples of real numbers.

If $x, y \in M$ and $m \in \mathsf{Ob}$, then define $\mu_m(x,y) \stackrel{d}{=} \mu(m(x) - m(y))$ where μ denotes the Minkowski-length (introduced earlier, in section 2).

S2 $(\forall x, y \in M)[(\exists m \in \mathsf{Ob})\mu_m(x,y) > 0 \rightarrow (\forall m, k \in \mathsf{Ob})\mu_m(x,y) = \mu_k(x,y)]$.

Theorem 5.2. (Suppes [51]) *Let $\langle M, \mathsf{Ob}\rangle$ be a collection of relativistic frames. Then $m^{-1} \circ k$ is a Poincaré-transformation for all $m, k \in \mathsf{Ob}$.*

Reichenbach [45] emphasized the importance of *observation-oriented* axioms. The idea is that we start out from observation-oriented axioms, then we work with these axiom systems, and during this work useful *theoretical* notions arise. Based on these theoretical notions we get a more thorough understanding, and we develop another, theoretical-notion oriented axiom system. We can call the first approach "bottom up" while the second one "top down". The first kind of axiom systems are more intuitive, while the second kind more compact mathematically. We consider Suppes' system as top down, since there is no intuition given why to think in terms of exactly the Minkowski-distance of events (as opposed to e.g. the usual Euclidean distance). The Minkowski-distance of events in the "bottom up" approaches (like ours) emerges naturally as "theoretical", defined concepts, see e.g. Theorem 1.2.

The idea is that the theoretical concepts should be first-order definable in terms of the observational-oriented concepts. Definability theory of first-order logic is very important in this process. Indeed, definability theory was started by Reichenbach (for the purposes of relativity theory), and later worked out for first-order logic by Tarski. Cf. also [17, 25], and the related literature of identifiability [26]. Madarász [32, §4.3] extended the theory of definability to defining new sorts in many-sorted logic, i.e. to defining new universes. This theory could be fruitfully used in comparing the various axiomatizations present in the literature. More on this and on the equivalence of "bottom up" and "top down" approaches can be read in [32, §4].

The axiom system of Ax [9] is the first truly first-order and observation oriented axiom system. His system is very elegant and intuitively clear. His language contains two unary relation symbols P, S for "particles" and "signals", and two binary relation symbols T, R for "transmitting a signal" and "receiving a signal". Particles are "imagined" as life-lines of bodies and signals are "imagined" as finite segments of life-lines of photons. Signals of length zero are conceived as "events". Ax has three groups of axioms, altogether 23 ones. Here we quote two typical axioms from his paper. The first axiom corresponds to our "No FTL" formula, while the second axiom corresponds to our axiom **AxOb**.

AxC4 [maximality of signal speed] For all particle a and event γ there is a unique signal β which begins at γ and reaches a,

$(\forall a \in \mathsf{P}, \gamma \in \mathsf{S})[\mathsf{Ev}(\gamma) \rightarrow (\exists! \beta)[\mathsf{Beg}(\beta) = \gamma \wedge a\mathsf{R}\beta]]$, where

$\mathsf{Ev}(\gamma)$ abbreviates $(\forall a)(a\mathsf{T}\gamma \rightarrow a\mathsf{R}\gamma)$ and
$\mathsf{Beg}(\beta) = \gamma$ abbreviates $\forall a[a\mathsf{T}\beta \rightarrow (a\mathsf{T}\gamma \wedge a\mathsf{R}\gamma)]$.

AxC5 [axiom of the limiting value of speed of light] Assume that the particle a receives the signal β in the event γ_1 and the event γ_2 occurs later on a's life-line. Then there is a particle b which receives both the beginning of β and γ_2,

$(\forall a, \beta, \gamma_1, \gamma_2, \gamma_3)[(a\mathsf{R}\beta \wedge \mathsf{End}(\beta) = \gamma_1 \wedge \mathsf{Ev}(\gamma_2) \wedge a\mathsf{R}\gamma_2 \wedge \gamma_1 \ll \gamma_2$
$\wedge\, \mathsf{Beg}(\beta) = \gamma_3) \rightarrow (\exists b)(b\mathsf{R}\gamma_3 \wedge b\mathsf{R}\gamma_2)]$, where

$\mathsf{End}(\beta) = \gamma_3$ abbreviates $(\forall a)[a\mathsf{R}\beta \rightarrow (a\mathsf{R}\gamma_3 \wedge a\mathsf{T}\gamma_3)]$ and

$\gamma_1 \ll \gamma_2$ abbreviates $\mathsf{Ev}(\gamma_1) \wedge \mathsf{Ev}(\gamma_2) \wedge \gamma_1 \neq \gamma_2 \wedge (\exists a, \alpha, \beta)[a\mathsf{T}\gamma_1$
$\wedge\, \mathsf{Beg}(\alpha) = \gamma_1 \wedge \mathsf{End}(\alpha) = \mathsf{Beg}(\beta) \wedge \mathsf{End}(\beta) = \gamma_2 \wedge a\mathsf{R}\gamma_2)$.

This $\ll$ is the same causality relation as in the earlier systems.

Ax [9] thoroughly presents the intuitive, physical motivation for his approach, his choice of framework, his axioms etc. which are ambitious and satisfying even on the level of abstraction of philosophy of physics (and philosophy of science). The motivation followed by the present team coincides with the motivation given by Ax (and is very close to the one in Suppes [52]).

Goldblatt [23] contains another first-order axiomatization of Minkowski space-time, i.e. of special relativity. While Ax's language is observational-oriented, and he emphasizes this, Goldblatt's system is clearly theoretical-oriented. One of his basic notions is that of orthogonality. Both Ax and Goldblatt build on Tarski's first-order axiomatization of Euclidean geometry. In fact, Goldblatt's system is a very harmonious extension/modification to Minkowski-geometry of Tarski's system for Euclidean geometry.

The structures axiomatized by Ax and Goldblatt do not contain the (relativistic) information carried by the so-called relativistic distance or Minkowski-metric. So it seems that the first, observation-oriented axiomatization in first-order logic of special relativity with all the relevant data (including relativistic distance) was obtained in the works of the present team.

Acknowledgments

We would like to thank János Makowsky, Ildikó Sain, and Jesse Alama for helpful comments on an earlier version. Heartfelt thanks go to Victor Pambuccian for helpful suggestions and pointers to the literature. This research was supported by the Hungarian National Fund for basic research grants No's T30314, T43242, T35192, and by COST project 274.

Bibliography

[1] Alexandrov, A. D.: *On foundations of space-time geometry. I.*, Soviet Math. Dokl., **15**, 1974, 1543-1547

[2] Andréka, H. and Goranko, V. and Mikulás, Sz. and Németi, I. and Sain, I.: *Effective first order temporal logics*, in Time and Logic, a computational approach, Ed. by L. Bolc and A. Szałas, UCL Press Ltd., London, 1995, 51-129

[3] Andréka, H. and Madarász, J. X. and Németi, I.: *Logical analysis of special relativity*, in JFAK. Essays Dedicated to Johan van Benthem on the Occasion of his 50th Birthday, Gerbrandy,J. and Marx,M. and de Rijke,M. and Venema,Y., editor, Vossiuspers, Amsterdam University Press, http://www.wins.uva.nl/~j50/cdrom, 1999, CD-ROM, ISBN: 90 5629 104 1, 23pp

[4] Andréka, H. and Madarász, J. X. and Németi, I.: *Logical analysis of relativity theories*, In Hendricks et al., editor, First Order Logic Revisited, Logos Verlag Berlin, 2004, 7-36

[5] Andréka, H. and Madarász, J. X. and Németi, I.: *On the logical structure of relativity theories*, Alfréd Rényi Institute of Mathematics, Hungar. Acad. Sci., Budapest, Research Report, July 5, 2002, with contributions from A. Andai, G. Sági, I. Sain and Cs. Tőke. http://www.math-inst.hu/pub/algebraic-logic/Contents.html. 1312 pp.

[6] Andréka, H. and Madarász, J. X. and Székely, G.: *Seminar notes*, Alfréd Rényi Institute of Mathematics, Budapest, 2003

[7] Andréka, H. and van Benthem, J. and Németi, I.: *Modal languages and bounded fragments of predicate logic*, Journal of Philosophical Logic, **65**, 1998, 217-274

[8] Andréka, H. and Németi, I.: *Course Notes*, ELTE University Budapest, 2000

[9] Ax, J.: *The elementary foundations of spacetime*, Found. Phys., **8**, 7-8, 1978, 507-546

[10] Barbour, J. B.: *Absolute or relative motion?*, Cambridge University Press, 1989

[11] Basri, S.: *A deductive theory of space and time*, North-Holland Publishing Co, Amsterdam, 1966

[12] Busemann, H.: *Time-like spaces*, Dissertationes Math. (Rozprawy Math.), **53**, 1967

[13] Carathéodory, C.: *Zur Axiomatik der speziellen Relativitätstheorie*, Sitzungsber. phys. math., **14**. Febr., 1924, 12-27

[14] Einstein, A.: *On the special and the general theories of relativity*, Verlag von F. Vieweg & Son, Braunschweig, 1921, In German

[15] Feferman, S.: *Theories of finite type related to mathematical practice*, Handbook of Mathematical Logic, Barwise, J., editor, North-Holland, 1977, 913-971

[16] Ferreirós, J.: *The road to modern logic – an interpretation*, The Bulletin of Symbolic Logic, **7**, 4, 2001, 441-484

[17] Fetzer, J. H. and Shatz, D. and Schlesinger, G. N.: *Definitions and definability: philosophical perspectives*, Kluwer Academic Publishers, 1991

[18] Friedman, M.: *Foundations of Space-Time Theories. Relativistic Physics and Philosophy of Science*, Princeton University Press, 1983

[19] Friedman, M.: *Reconsidering logical positivism*, Cambridge University Press, 1999

[20] Gibbs, Ph.: *Is faster than light travel or communication possible?*, Internet, www.math.ucr.edu/home/baez/physics/Relativity/SpeedOfLight/FTL.html, 1997

[21] Gödel, K.: *An example of a new type of cosmological solutions of Einstein's field equations of gravitation*, Reviews of Modern Physics, **21**, 1949, 447-450

[22] Gödel, K.: *Lecture on rotating universes*, Kurt Gödel Collected Works, Vol. III., Ed. Feferman, S. Dawson, J.W. Goldfarb, W. Parsons, C. Solovay, R. N., Oxford University Press, New York Oxford, 1995, 261–289

[23] Goldblatt, R.: *Orthogonality and Spacetime Geometry*, Springer-Verlag, 1987

[24] Guts, A. K.: Axiomatic relativity theory, Russian Math. Survey, **37**, 2, 1982, 41-89

[25] Henkin, L. and Suppes, P. and Tarski, A.: *The axiomatic method with special reference to geometry and physics*, Studies in Logic and the Foundations of Mathematics, North-Holland, 1959

[26] Hintikka, J.: *Towards a general theory of identifiability*, in Definitions and definability: philosophical perspectives, Kluwer Academic Publishers, 1991, 161-183

[27] Hogarth, M.: *Conventionality of simultaneity: Malament's result revisited*, Cambridge University, 2003, Manuscript

[28] Hogarth, M.: *Deciding arithmetic in Malament-Hogarth spacetimes*, 2001, University of Cambridge (UK), preprint

[29] Latzer, R. W.: *Nondirected light signals and the structure of time*, Synthese, **24**, 1972, 236-280

[30] Lester, J. A.: *Distance preserving transformations*, in Handbook of Incidence Geometry, Buekenhout, F., editor, Elsevier Science B. V., 1995, 921-944

[31] Lewis, D.: *The paradoxes of time travel*, in David Lewis Philosophical Papers Volume II, Oxford University Press, New York, Oxford, 1986, 67-80

[32] Madarász, J. X.: *Logic and Relativity (in the light of definability theory)*, ELTE, Budapest, http://www.math-inst.hu/pub/algebraic-logic/Contents.html., 2002

[33] Madarász, J. X. and Németi, I.: *On faster than light observers, Part I.*, Manuscript, Budapest, 1998

[34] Madarász, J. X. and Németi, I.: *On FTL observers in a generalized version of first-order relativity (The generalization is in the direction of partiality coming from general relativity)*, 1999, Manuscript in preparation

[35] Madarász, J. X. and Németi, I. and Tőke, Cs.: *Local relativity theories*, in First-Order Logic'75, Logos Verlag, Berlin, 2004, 225-268

[36] Madarász, J. X. and Tőke, Cs.: *Duality between observational and theoretical concepts in relativity theory analyzed by FOL and definability theory*, Szeged, Hungary, Extended abstract, July, 2003, Proceedings of the Kalmár Workshop on Logic and Computer Science, 133-142

[37] Madarász, J. X. and Tőke, Cs.: *Partial relativity theories*, Alfréd Rényi Institute of Mathematics, Hungar. Acad. Sci., Budapest, Research Report, July, 2003

[38] Malament, D.: *Casual theories of time and the conventionallity of simultaneity*, Noûs, **11**, 1977, 293-300

[39] Matolcsi, T. and Rodrigues, W. A. Jr.: *Geometry of Spacetime with Superluminal Phenomena*, Algebr. Group Geom, **14**, 1997, 1-16

[40] Misner, C. W. and Thorne, K. S. and Wheeler, J. A.: *Gravitation*, W. H. Freeman, San Francisco, 1973

[41] Muis, R.: *Physical Structures*, Master Thesis, University of Utrecht, December 2002, www.phys.uu.nl/~ wwwgrnsl/muis/scriptie.pdf

[42] Mundy, J.: The philosophical content of Minkowski geometry, British J. Philos. Sci., **37**, 1, 1986, 25-54

[43] Novikov, I. D.: *The river of time*, Cambridge University Press, Cambridge, England, 1998

[44] Pambuccian, V.: *Axiomatizations of hyperbolic and absolute geometries*, Non-Euclidean Geometries, Kluwer, 2004

[45] Reichenbach, H.: *Axiomatization of the theory of relativity*, University of California Press, Berkeley, 1969, Translated by M. Reichenbach. Orignal German edition published in 1924.

[46] Robb, A. A.: *A Theory of Time and Space*, Cambridge University Press, 1914, Revised edition, *Geometry of Time and Space*, published in 1936

[47] Rudin, W.: *Principles of Mathematical analysis*, McGraw-Hill, New York, 1953

[48] Sain, I.: *Dynamic logic with nonstandard model theory*, Hungarian Academy of Sciences, Budapest, 1986, ix+180pp, (In Hungarian)

[49] Schelb, U.: *Zur physikalischen Begründung der Raum-Zeit-Geometrie*, Habilitationsschrift, Fachbereich Physik der Universität Paderborn, August 1997, 327pp, www.phys.uni-paderborn.de/ udo

[50] Schutz, J. W.: *Independent axioms for Minkowski space-time*, Longman, London, 1997

[51] Suppes, P.: *Axioms for relativistic kinematics with or without parity*, in Henkin, L. and Tarski, A. and Suppes, P.: Symposium on the Axiomatic Method with Special Reference to Physics, North-Holland, 1959, 291-307

[52] Suppes, P.: *The desirability of formalization in science*, The Journal of Philosophy, **27**, 1968, 651-664

[53] Suppes, P.: *Some open problems in the philosophy of space and time*, Synthese, **24**, 1972, 298-316

[54] Székely, G.: *Accelerated observers and the twin paradox*, Master Thesis, ELTE, Budapest, 2004

[55] Székely, G.: *A first-order logic investigation of the Twin Paradox*, Manuscript, ELTE, Budapest (in Hungarian), 23pp., 2003

[56] Szekeres, G.: *Kinematic geometry: an axiomatic system for Minkowski space-time*, Journal of the Australian Mathematical Society, **8**, 1968, 134-160

[57] Takeuti, G.: *Proof Theory*, North-Holland, 1987

[58] Tarski, A.: *What is elementary geometry?*, in The Axiomatic Method with Special Reference to Geometry and Physics, Henkin, L. and Suppes, P. and Tarski, A., editor, North-Holland, Amsterdam, 1959, 16-29

[59] Väänänen, J.: Second-order logic and foundations of mathematics, The Bulletin of Symbolic Logic, **7**, 4, 2001, 504-520

[60] Walker, A. G.: *Axioms for Cosmology*, in The Axiomatic Method with Special Reference to Geometry and Physics, Henkin, L. and Suppes, P. and Tarski, A., editor, North-Holland, Amsterdam, 1959, 308-321

[61] Woleński, J.: *First-order logic: (Philosophical) pro and contra*, in First-order Logic'75, Logos Verlag, Berlin, 2004

III

POLYHEDRA, VOLUMES, DISCRETE ARRANGEMENTS, FRACTALS

STRUCTURES IN HYPERBOLIC SPACE

Robert Connelly *
Department of Mathematics,
Malott Hall, Cornell University,
Ithaca, NY 14853
connelly@math.cornell.edu

Abstract This is an overview of some of the similarities and differences between structures such as frameworks and cabled tensegrities in the hyperbolic plane and hyperbolic space on the one hand and the Euclidean plane, the sphere and Euclidean space on the other hand.

MSC: 52C25, 52C35, 52Cxx, 52.50

Keywords: Hyperbolic plane, configurations, framework, tensegrity

1. Introduction

There is a strong analogy between the unit sphere in Euclidean space and the hyperbolic plane. One can regard the hyperbolic plane $\mathbb{H}^2$ as half of the "unit sphere" in Lorenz space, namely the following set:

$$\mathbb{H}^2 = \{(x, y, t) \in E^{1,2} \mid x^2 + y^2 - t^2 = -1,\ t > 0\}.$$

Of course, there is a similar definition for higher-dimensional Euclidean, spherical and hyperbolic spaces. So one can often "change the sign" in a formula in spherical geoemtry and create a correct formula in hyperbolic geometry. For example, in spherical trigonometry many of the standard formulas for a spherical triangle involve sines and cosines. To get an appropriate correct formula in the hyperbolic plane, when these functions are applied to edge lengths, replace the sine and cosine functions by the hyperbolic sine and the hyperbolic cosines function respectively. When they are applied to angles, leave the sine and cosine functions as they were.

In the following we show some similarities and then differences between hyperbolic and non-hyperbolic situations.

*Research supported in part by NSF Grant No. DMS-0209595

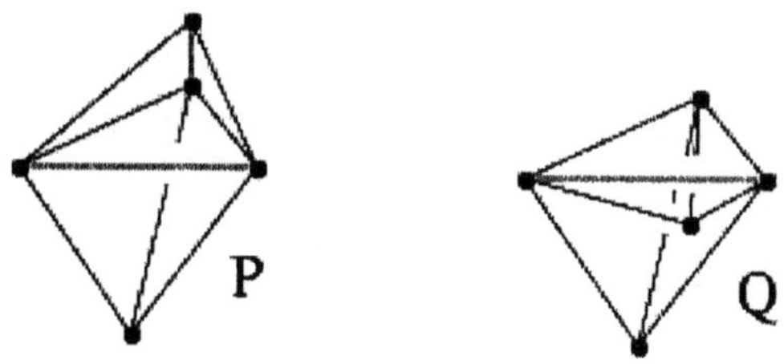

Figure 1. Non-congruent polyhedral surfaces

2. Polyhedra in dimension 3

In 1813 Cauchy [Cauchy, 1813] proved the following, which we state in modern language. A function f between sets in any space with a metric, such as Euclidean, spherical or hyperbolic space, is called a *congruence* if the distance between every pair of points p and q in one set is the same as the distance between $f(p)$ and $f(q)$ in the other set. Recall that a *face* of a convex polytope P (in Euclidean, spherical or hyperbolic space) is the intersection of a hyperplane with P that contains P on one side. (A *homeomorphism* is a function such that it and its inverse are one-to-one and continuous.)

Theorem 2.1 (Cauchy). *Suppose that $f : P \to Q$ is a homeomorphism between two convex polyhedral surfaces P and Q in $\mathbb{E}^3$ such that f restricted to each face of P is a congruence. Then f itself is a congruence.*

This is an example of a local but partly global result. It is not entirely global because of the restriction that both the source and target sets are assumed to be convex. But if the correspondence f is assumed to be close enough to the identity and the source P is convex, then the target Q will be convex as well and Cauchy's theorem applies. But if only P is convex, Figure 1 shows that Q may not be congruent to P even in the simplest case when P has just five vertices.

But what about hyperbolic space? Here it is instructive to look at Cauchy's proof. It has essentially two parts, a combinatorial topological lemma, and a local (and partly global) geometric lemma. The combinatorial lemma is the following:

Lemma 2.2. *If some subset of the edges of a convex polytope are assigned a + or − such that there are at least four changes in sign (or no labeled vertices) around each vertex, then no edges are labeled anywhere.*

It is instructive to try to find such a labeling for the first non-trivial case of a convex polyhedral surface, the (regular) octahedron. The proof

Figure 2. Polygonal arcs resembling opening arms

of this lemma is a nice application of the Euler characteristic. One can find a good description of this lemma and the next in [Lyusternik] or [Aigner-Ziegler] for example. This lemma clearly holds when the convex polyhedral surface is in $\mathbb{H}^3$, since convexity plays a minor role. It is only the topological and combinatorial structure of the graph of edges that matters.

The geometric lemma is the following. I like to call this Cauchy's "arm" lemma, because the polygon involved looks like an opening arm. (There was a problem with Cauchy's proof of this lemma. It was not corrected until some years after his 1813 publication. The message seems to be that one should be careful here. See [Connelly] for a discussion of this.)

Lemma 2.3 (Cauchy's arm). *Suppose that $f : A \to B \subset \mathbb{S}^2$ is a continuous map from a convex polygonal arc A in the unit 2-sphere $\mathbb{S}^2$ that is a congruence when restricted to each edge of A and the angle at each internal vertex of A does not decrease. Then the distance between the endpoints of A are not less than the distance between the endpoints of B.*

Figure 2 shows what this looks like in the Euclidean plane, a limiting case of the sphere.

Note that the target polygon B is not assumed to be convex. This lemma is then applied to small spheres centered at each vertex of the convex polyhedral surface P. Each edge incident to a vertex of P corresponds to a vertex on the corresponding sphere, and each dihedral angle corresponds to an internal angle of a spherical arc. The idea is that if the edges of the spherical polyhedron are labeled $+$ or $-$ depending on whether they increase or decrease, respectively, and there are only two changes in sign around a vertex, an arc with only $+$ followed by an arc with $-$ signs contradicts the Arm Lemma 2.3.

But for a hyperbolic polyhedral surface we see that the Arm Lemma 2.3 still applies since it is used for polygons in a sphere. So Cauchy's Theorem 2.1 is still true in hyperbolic space $\mathbb{H}^3$. Indeed, Cauchy's Theorem 2.1 generalizes to dimensions greater than 3, as well as to spherical and hyperbolic spaces of dimension greater than 3. See [Connelly] for more information and related results.

3. Static rigidity

A natural approach for rigidity questions is to linearize the conditions for the rigidity of constraints. In order to simplify the discussion, suppose that the polyhedral surfaces we consider all have only triangular faces. The rigidity constraints can be restricted only to pairs of vertices with an edge between them, where the edges act as bars. One version of the linear constraints can be described with forces acting on this framework of vertices and bars. For each vertex p_i in this framework there is associated a force vector F_i, the external load, and the question of static rigidity is whether each appropriate external load can be *resolved* by some internal stress in the framework. This means that there are scalars ω_{ij} associated with every edge between vertices p_i and p_j such that for each i,

$$F_i = \sum_j \omega_{ij}(p_j - p_i). \tag{1}$$

(The stress $\omega_{ij} = 0$ when there is no bar between p_i and p_j.) We say that an external force $F = (\dots, F_i, \dots)$ is an *equilibrium force* if it has 0 linear and angular momentum, which means that $\sum_i F_i = 0$, and $\sum_i F_i \wedge p_i = 0$, where $\wedge$ is the wedge product from linear algebra. There is no way that an external force $F = (\dots, F_i, \dots)$ can be resolved if it has any non-zero linear or angular momentum. So we define a bar framework to be *statically rigid* if every equilibrium force can be resolved.

Note that this definition only depends on the underlying linear algebra, and so it applies to the sphere and hyperbolic space, as well as Euclidean space, since we can regard them all as being in an appropriate ambient linear space. Note that in dimension 3 the dimension of the space of equilibrium forces is 6 (assuming the framework is not just one bar or a single vertex), and there is an equilibrium equation for each of the three coordinates. If there are n vertices in the framework, there are $3n$ equations to be satisfied. So there must be at least $3n - 6$ bars in the framework. (A similar calculation holds in other dimensions.)

We say that a framework is *rigid* if there is no continuous noncongruent motion of the points that preserves the lengths of the bars. The following is a basic result. See [Connelly] for a proof, for example.

Theorem 3.1. *If a framework is statically rigid in Euclidean, spherical or hyperbolic space, it is rigid.*

So static rigidity is a very local form of rigidity. It is easy to check that for a triangulated 2-dimensional spherical surface the number of edges in the triangulation is $3n-6$, and so there are just enough equations for static rigidity. So we have an infinitesimally local version of Cauchy's semi-global result due originally to Max Dehn. There is a clear proof in [Gluck].

Theorem 3.2 (Dehn). *If P is a convex polyhedral surface in Euclidean, spherical or hyperbolic space, where all the 2-dimensional faces of P are triangles, then P is statically rigid.*

Note that since there are just enough equations for static rigidity, it is enough to show that the homogenized version of the linear equilibrium equations (1) only have the 0 solution. In other words, in this case, it is enough to show that the 0 external force has only the 0 stress as a resolution. (A stress that resolves the 0 external force is called a *self stress.*) It is easy to check that there cannot be just two sign changes in the stresses in the edges adjacent to a vertex, and so Cauchy's combinatorial Lemma applies us to show that a self stress for P is 0. (This is the idea in [Gluck] following some ideas of A. D. Alexsandrov, but Dehn's proof is somewhat different.) Incidentally, W. Whiteley in [Connelly-Whiteley] showed that it is possible to get Dehn's Theorem 3.2 formally from Cauchy's Theorem 2.1 by a technique related to Pogorelov's correspondence mentioned below.

4. The Pogorelov correspondence

There is a very interesting formal corresondence from pairs of configurations of points in Euclidean space $\mathbb{E}^d$ to pairs of configurations of points in hyperbolic space $\mathbb{H}^d$ that has very useful properties with respect to distances. This is described in [Pogorelov].

Theorem 4.1 (Pogorelov). *There is a rational function $f : \mathbb{E}^d \times \mathbb{E}^d \to \mathbb{H}^d \times \mathbb{H}^d$ such that for all (p_1, q_1) and (p_2, q_2) in $\mathbb{E}^d \times \mathbb{E}^d$ we have*

$$\begin{aligned} |p_1 - p_2| &\leq |q_1 - q_2| \quad \textit{if and only if} \\ |f_1(p_1, q_1) - f_1(p_2, q_2)| &\leq |f_2(p_1, q_1) - f_2(p_2, q_2)|, \quad \textit{and} \\ |p_1 - p_2| &\geq |q_1 - q_2| \quad \textit{if and only if} \\ |f_1(p_1, q_1) - f_1(p_2, q_2)| &\geq |f_2(p_1, q_1) - f_2(p_2, q_2)|, \end{aligned}$$

where $f = (f_1, f_2)$, and $|\dots|$ is the usual Euclidean norm for $\mathbb{E}^d$ and the norm using the Lorentz inner product when the points are in $\mathbb{H}^d$.

This map can be regarded as taking the configurations $\mathbf{p} = (p_1, \ldots, p_n)$ and $\mathbf{q} = (q_1, \ldots, q_n)$ to the configurations

$$\begin{aligned} f_1(\mathbf{p}, \mathbf{q}) &= (f_1(p_1, q_1), \ldots, f_1(p_1, q_1)) \quad \text{and} \\ f_2(\mathbf{p}, \mathbf{q}) &= (f_2(p_1, q_1), \ldots, f_2(p_1, q_1)), \end{aligned}$$

where, in particular if some pair of vertices in the $\mathbf{p}$ configuration is the same distance apart as in the $\mathbf{q}$ configuration in Euclidean space, then the corresponding pair of vertices are also the same distance apart in hyperbolic space. But an important consideration is that the image pair of configurations in hyperbolic space each depend on both configurations $\mathbf{p}$ and $\mathbf{q}$. Nevertheless, the image of a point $f_1(p_i, q_i)$, say, only depends on p_i and q_i, and it turns out that it is possible to extend correspondences from $\mathbf{p}$ to $\mathbf{q}$ so the conditions of Theorem 4.1 hold. Furthermore, if both $\mathbf{p}$ and $\mathbf{q}$ are convex polyhdra in $\mathbb{E}^d$, so are $f_1(\mathbf{p})$ and $f_2(\mathbf{p})$ in $\mathbb{H}^d$. So this provides another proof of Cauchy's Theorem in hyperbolic space. But this still provides no direct way of transfering global results in Euclidean space to hyperbolic space.

5. Global motions

There are applications, where it is desirable to have a truly global result, with no restriction on the target configuration. But there can be problems when it comes to the case of hyperbolic geometry. For example, consider the following Lemma, which we called a "Leapfrog Lemma" and which was an essential component in the paper [Bezdek-Connelly].

Lemma 5.1 (Leapfrog).
Suppose that $\mathbf{p} = (p_1, \ldots, p_n)$ and $\mathbf{q} = (q_1, \ldots, q_n)$ are two configurations in $\mathbb{E}^d$. Then there is a motion $\mathbf{p}(t) = (\mathbf{p}_1(t), \ldots, \mathbf{p}_n(t))$ in $\mathbb{E}^{2d}$, which is analytic in t, such that $\mathbf{p}(0) = \mathbf{p}$, $\mathbf{p}(1) = \mathbf{q}$ and for $0 \le t \le 1$, $|\mathbf{p}_i(t) - \mathbf{p}_j(t)|$ is monotone.

The method in [Bezdek-Connelly] uses a specific formula, but we will discuss here another method. Both methods have difficulties in extending the results to hyperbolic space. For any configuration of points $\mathbf{p} = (p_1, \ldots, p_n)$ in $\mathbb{E}^d$ define the *Gramm matrix* Gr as an $(n-1)$-by-$(n-1)$ symmetric matrix, whose ij entry is the inner product $p_i \cdot p_j$ for $i = 1, \ldots, n-1$ and $j = 1, \ldots, n-1$, where we assume $p_n = 0$. Define the *configuration matrix* as

$$P = \begin{bmatrix} p_1 & p_2 & \ldots & p_{n-1} \end{bmatrix},$$

where each p_i for $i = 1, \ldots, n-1$ is regarded as a column vector. Then the Gramm matrix $Gr = P^T P$, where P^T is the transpose of P.

So any Gramm matrix coming from a configuration of points in $\mathbb{E}^d$ is positive semi-definite with a kernel of dimension at most d. Conversely, any positive semi-definite matrix Gr with a kernel of dimension at most d can be written as $Gr = P^T P$, where P is a d-by-n matrix and the columns of P serve as the coordinates of a configuration in $\mathbb{E}^d$ with $p_n = 0$. So the Leapfrog Lemma 5.1 can be viewed as coming from the convexity of positive semi-definite matrices.

But what about hyperbolic space? One can still define a Gramm matrix using the indefinite inner product, where

$$< p_i, p_j >= x_i x_j + y_i y_j + z_i z_j + \cdots - t_i t_j$$

and for each i

$$p_i = (x_i, y_i, z_i, \ldots, t_i),$$

and hyperbolic space itself is defined as in the first section:

$$\mathbb{H}^d = \{(x, y, z, \ldots, t) = p \in E^{1,d} \mid < p, p >= -1,\ t > 0\}.$$

We can assume that 0 is in the configuration as before. But now the Gramm matrix for this inner product is $Gr = P^T DP$, where P is the configuration matrix, as before, and D is a diagonal matrix with all 1's on the diagonal, except for the last which is -1. So Gr is indefinite with one negative eigenvalue while all the rest are positive. Convex combinations of these matrices could have more than one negative eigenvalue, and we have a quandry. Do we have a leapfrog lemma in hyperbolic space? We are left with the following rather unsettling questions and conjecture.

Conjecture 5.2 (Hyperbolic-Leapfrog).
Suppose that $\mathbf{p} = (p_1, \ldots, p_n)$ *and* $\mathbf{q} = (q_1, \ldots, q_n)$ *are two configurations in* $\mathbb{H}^d$. *Then there is a motion* $\mathbf{p}(t) = (\mathbf{p}_1(t), \ldots, \mathbf{p}_n(t))$ *in* $\mathbb{H}^{2d}$, *which is analytic in* t, *such that* $\mathbf{p}(0) = \mathbf{p}$, $\mathbf{p}(1) = \mathbf{q}$ *and for* $0 \leq t \leq 1$, $< \mathbf{p}_i(t) - \mathbf{p}_j(t), \mathbf{p}_i(t) - \mathbf{p}_j(t) >$ *is monotone.*

It is not clear if there is even a continuous monotone motion in any $\mathbb{H}^N$, for N sufficiently large, although this seems more likely to be true. On the positive side, we do have the following analogous result for the spherical case.

Lemma 5.3 (Spherical-Leapfrog). *Suppose that* $\mathbf{p} = (p_1, \ldots, p_n)$ *and* $\mathbf{q} = (q_1, \ldots, q_n)$ *are two configurations in* $\mathbb{S}^d$. *Then there is a motion* $\mathbf{p}(t) = (\mathbf{p}_1(t), \ldots, \mathbf{p}_n(t))$ *in* $\mathbb{S}^{2d+1}$, *which is analytic in* t, *such that* $\mathbf{p}(0) = \mathbf{p}$, $\mathbf{p}(1) = \mathbf{q}$ *and for* $0 \leq t \leq 1$, $|\mathbf{p}_i(t) - \mathbf{p}_j(t)|$ *is monotone.*

This follows easily from the Euclidean case Lemma 5.1, but notice that the ambient space has dimension $2d + 1$ instead of $2d$. If $2d + 1$

could be replaced with $2d$ even for $d = 2$, it would yield an extension of the main result (about unions or intersections of spherical disks) in [Bezdek-Connelly] to $\mathbb{S}^2$ following a method of Csikos. Similarly, there would be another extension to $\mathbb{H}^2$ if Conjecture 5.2 were true. Even if these conjectures are true, it is clear that there seems to be some particular asymmetry between spherical space and hyperbolic space.

Bibliography

[Aigner-Ziegler] Aigner, Martin; Ziegler, Günter M., Proofs from The Book. Including illustrations by Karl H. Hofmann. Second edition. Springer-Verlag, Berlin, 2001. viii+215 pp. ISBN: 3-540-67865-4 MR 2001j:00001

[Bezdek-Connelly] K. Bezdek and R. Connelly, Pushing disks apart - the Kneser-Poulsen conjecture in the plane, J. reine angew. Math. 553 (2002), 221-236.

[Cauchy] Sur les polygones et les polyédres, Second Mémoire, I. École Polythchn., 9(1813) 87 (= Oeuvres Complétes d'Augustin Cauchy, 2e série, Tome 1 (1905) 26-38).

[Connelly] Connelly, Robert, Rigidity. Handbook of convex geometry, Vol. A, B, 223–271, North-Holland, Amsterdam, 1993. MR 94j:52041

[Connelly-Whiteley] Connelly, Robert(1-CRNL); Whiteley, Walter(3-YORK-MS), Second-order rigidity and prestress stability for tensegrity frameworks. (English. English summary) SIAM J. Discrete Math. 9 (1996), no. 3, 453–491. MR 97e:52037

[Gluck] Gluck, Herman, Almost all simply connected closed surfaces are rigid. Geometric topology (Proc. Conf., Park City, Utah, 1974), pp. 225–239. Lecture Notes in Math., Vol. 438, Springer, Berlin, 1975. MR 53 #4074

[Lyusternik] Lyusternik, L. A. Convex figures and polyhedra. Translated and adapted from the first Russian edition (1956) by Donald L. Barnett D. C. Heath and Co., Boston, Mass. 1966 viii+191 pp. 52.00 MR 36 #4435

[Pogorelov] Pogorelov, A. V. (Russian) [Topics in the theory of surfaces in an elliptic space] Izdat. Harcprime kov. Gos. Univ., Kharkov 1960 92 pp. MR 22 #8460

THE SYMMETRY OF OPTIMALLY DENSE PACKINGS

Charles Radin*
Mathematics Department
University of Texas at Austin
radin@math.utexas.edu

Abstract This is a slightly expanded version of a talk given at the János Bolyai Conference on Hyperbolic Geometry, held in Budapest in July, 2002. The general subject of the talk was the densest packings of simple bodies, for instance spheres or polyhedra, in Euclidean or hyperbolic spaces, and describes recent joint work with Lewis Bowen. One of the main points was to report on our solution of the old problem of treating optimally dense packings of bodies in hyperbolic spaces. The other was to describe the general connection between aperiodicity and nonuniqueness in problems of optimal density.

1. Packings of Euclidean space

For motivation we begin with packings of regular pentagons in the Euclidean plane, $\mathbb{E}^2$. First we recall that by a "packing" P of pentagons in a square C, we mean a collection of congruent copies, pent_j, of such a body, all contained in C and with pairwise disjoint interiors. By the "density" of such a packing P we mean:

$$\frac{\sum_j \text{volume}(\text{pent}_j)}{\text{volume}(C)}. \tag{1}$$

It is clear that for given square C there exists some maximum possible value of this density, over all P. We are however more interested in an optimum density in regions of infinite volume rather than C, and therefore we need a more sophisticated definition.

To analyze the optimum density in the whole of $\mathbb{E}^2$ we proceed as follows. First, for each square C and packing P of the whole plane, we

*Research supported in part by NSF Grant No. DMS-0071643 and Texas ARP Grant 003658-158.

consider the relative density

$$d_C(P) \equiv \frac{\sum_j \text{volume}(\text{pent}_j \cap C)}{\text{volume}(C)} \tag{2}$$

and then obtain a density for P as

$$d(P) \equiv \lim_C \frac{\sum_j \text{volume}(\text{pent}_j \cap C)}{\text{volume}(C)}, \tag{3}$$

in which we allow C to grow so as to contain every point of $\mathbb{E}^2$.

It is not hard to construct packings P for which this limiting density $d(P)$ does not exist, for instance by constructing P to have arbitrarily large regions empty of pentagons, so that the relative density oscillates instead of having a limit. This is an essential feature of analyzing density in spaces of infinite volume. Density is inherently a *global* quantity, and fundamentally requires a formula somewhat like (3) for its definition [13]. As a consequence we are trying to optimize $d(P)$ over packings P even though $d(P)$ is undefined for some P. However this does not prevent us from showing the existence of a convincing optimal density for our pentagon problem, for instance as follows.

First consider a sequence of squares C_n with sides of length n. It is easy to show the existence of packings P_n of $\mathbb{E}^2$ which achieve a maximum for the relative density $d_{C_n}(\cdot)$. We then trim P_n by removing all pentagons from it which have nonempty intersection with the complement of C_n, and "periodize" the result by appropriate translations, obtaining a packing $\tilde{P}_n$ invariant under two perpendicular translations of length n. For $n >> 1$ the density of $\tilde{P}_n$ is still reasonably high relative to any square of edge length n since the only relevant loss is from those pentagons lying on the boundary of C_n (and its translates), and the volume of these is negligible for purposes of density, for large n. It is then easy to show that the density of $\tilde{P}_n$ has a well defined limit as $n \to \infty$, and it is reasonable to accept this limit as the optimum density of regular pentagons in $\mathbb{E}^2$. And finally, it is also easy to show the existence of a packing which has this value as a well defined density, in the sense of (3).

The above technique allows us not only to prove the *existence* of an optimal density (for regular pentagons or other bodies) but even to estimate its value – though not to actually determine the optimum value. In fact it is difficult to determine the optimal density for most simple bodies. One of the first interesting examples was that of unit disks in $\mathbb{E}^2$. The history of this, culiminating in the fully acceptable proof of L. Fejes Tóth in 1940, is interesting; see [24, 12].

Of particular relevance here are the optimal packings of $\mathbb{E}^2$ by congruent copies of the two bodies in Figure 1, known as the kite and dart, introduced by Roger Penrose in 1977 [15]. It is possible to construct tilings of the plane with these bodies (see Figure 2), which are evidently the densest packings. (See [23] for a general introduction to the mathematics of these sorts of tilings.) The relevant point however is that *every* such optimal packing/tiling of kites and darts has "low symmetry": the symmetry group of such a packing does not have a compact fundamental domain in $\mathbb{E}^2$. (A packing is called "periodic" (resp. "nonperiodic") if it has (resp. does not have) a symmetry group with compact fundamental domain, and we say an optimal packing problem is "aperiodic" if *all* its optimally dense packings are nonperiodic.)

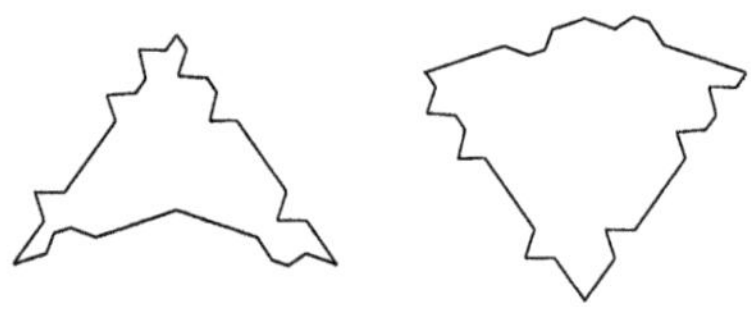

Figure 1. The kite and dart tiles

One of our main points is that aperiodicity is strongly connected to the uniqueness of the packing problem.

Theorem 1. *If there is only one optimally dense packing of* $\mathbb{E}^d$ *or* $\mathbb{H}^d$, *up to congruence, by congruent copies of the bodies from some fixed, finite collection, then that packing must have a symmetry group with compact fundamental domain.*

(We sketch the proof later, after introducing some notation.) So we may conclude for instance that there are many kite and dart tilings, that is, many equivalence classes modulo congruence. Another interesting feature of the kite and dart tilings, besides their low symmetry, is that they are all "locally identical": every bounded region of one such tiling has a congruent copy in every other such tiling. So the nonuniqueness in this case cannot be seen locally – it is an essentially global feature. Of course the optimally dense packing for a given collection of bodies may fail to be unique up to congruence in a simpler way: for unit spheres in $\mathbb{E}^3$ this follows from an accidental degeneracy wherein optimal packings can have different bounded regions.

We summarize some of the above as follows. If we consider the optimization problem in which we seek to optimize the density of packings of $\mathbb{E}^d$ by congruent copies bodies from some some fixed, finite collection, we see from the above that there always *exists* an optimally dense packing,

Figure 2. A kite and dart tiling of the Euclidean plane

but that the solution may not be unique (up to congruence) — not just because of accidental (local) degeneracy as in sphere packings of $\mathbb{E}^3$, but more fundamentally (globally), as in the kite and dart tilings. We will see later that other considerations suggest a small modification of the framework of this problem, which may eliminate the sort of nonuniquess associated with aperiodicity.

2. Packings of hyperbolic space

Roger Penrose also introduced another example of interest here, in 1978 [22]. Congruent copies of the body shown in Figure 3 can tile the hyperbolic upper-half plane, as in Figure 4, but *only nonperiodically*, the latter following from an elegant argument. If a tiling by copies of that body had a symmetry group with compact fundamental domain, that domain would have to contain as many dents as bumps, since they would be paired up. But the body under consideration has two (inward, triangular) dents and one (outward, triangular) bump, and since the compact domain can only contain finitely many bodies, there is an automatic imbalance! (This example led to the following interesting works: [1, 19, 18, 16].)

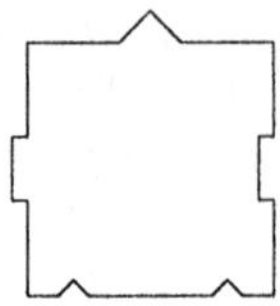

Figure 3. The hyperbolic Penrose tile

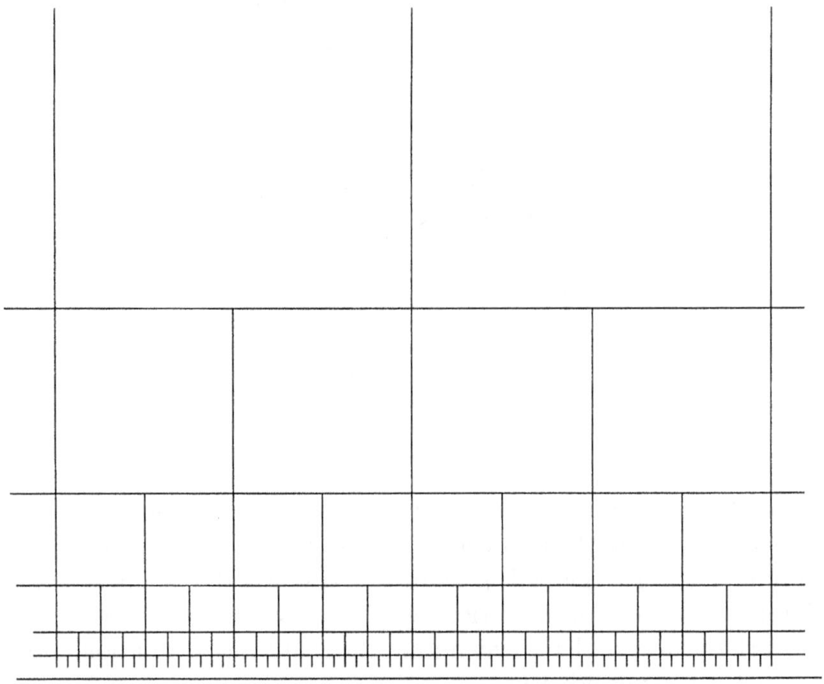

Figure 4. A hyperbolic Penrose tiling

We are bringing up this example in the context of optimally dense packings in part to show a connection with the packings of congruent disks published a few years earlier by Károly Böröczky [2]; see Figure 5. Those disk packings had been influential in convincing the discrete geometry community that *there could not in fact be a consistent theory of densest packings in hyperbolic spaces*; see [8, 9, 10, 11, 12, 2, 3, 4, 13, 14, 17]. Basically, there have been attempts for at least 50 years to deal with the global notion of density in hyperbolic space, in particular for packings of congruent spheres, and the effort has essentially died out, and turned to various local substitutes. We sketch the two-part argument against a theory of optimal density as follows.

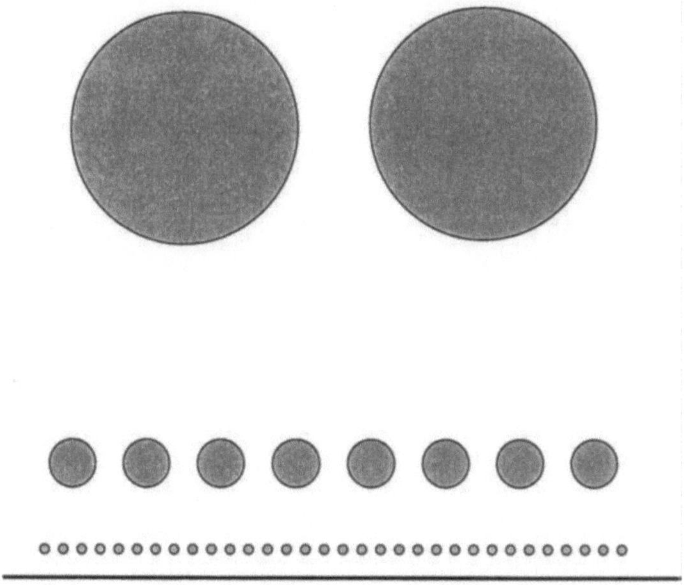

Figure 5. The Böröczky disk packing of the hyperbolic plane

First we note that the technique used above, to prove the *existence* of optimally dense packings of Euclidean spaces, does not have a simple analog for hyperbolic spaces; the weak point is where the bodies straddling the boundary of the square C_n are thrown out. In hyperbolic spaces the role of the C_n could naturally be played by compact fundamental domains, but the problem is that in a hyperbolic space a large fundamental domain has a finite fraction of its volume near its boundary, so the bodies straddling the boundary would not be negligible.

Of course this was only one approach to proving existence of optimally dense packings. However this boundary phenomenon underscores the intrinsic difficulty of proving that limits of the form (3) would exist for any but the simplest sorts of packings.

The other part of the argument, based on Böröczky's packing, is as follows. In Figure 6 the packing is displayed together with a copy of Penrose's tiling – a tiling by congruent bodies. Consider the two regions in dark outline in Figure 7, each made from three copies of the Penrose tile (and therefore the two regions have the same volume). From each of these we could, in an obvious way, make congruent copies to produce a tiling of the hyperbolic plane. But each of these tilings then suggests an obvious value for the density of the packing: the relative density in each of its tiles of the disks contained in the tile. But this would suggest one density twice the value of the other!

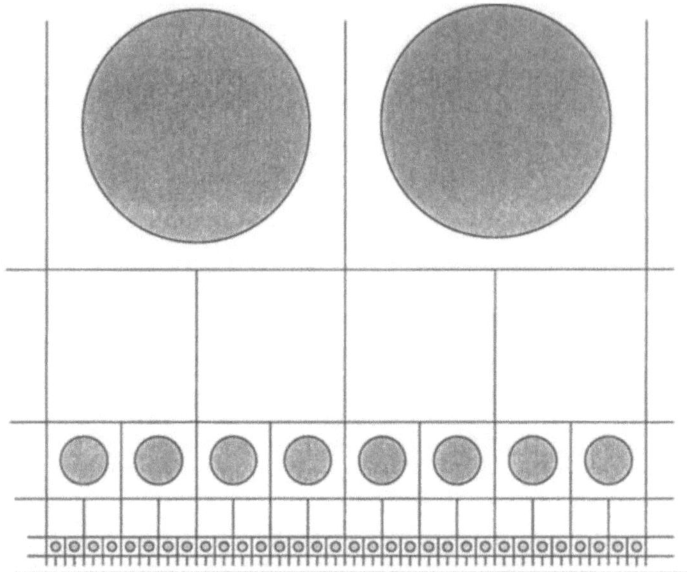

Figure 6. Böröczky's packing of disks, with tiling background

A similar inconsistency can be demonstrated between the densities based on Voronoi tilings and Dirichlet tilings associated with Böröczky's packing.

In summary, one part of the difficulty of dealing with optimally dense packings in hyperbolic space has been proving the existence of limiting densities of the form (3) for nonperiodic packings, and the other part was the inconsistencies that arise when trying to avoid the limit definition of density by appealing to densities relative to associated tilings.

The other of our main goals here, besides the connection between aperiodicity and nonuniqueness, is to show how a standard part of mathematics, ergodic theory, can in fact be used to prove the existence of limiting densities of the form (3) for complicated (nonperiodic) packings, enough to produce a useful theory of optimally dense packings. We now outline this approach, for the simple case of packings of hyperbolic space, $\mathbb{H}^d$, by balls of fixed radius R; see [6] and [7] for details.

Consider the space X_R of all possible "relatively dense" packings of hyperbolic space by balls of fixed radius R, and put a metric topology on X_R such that convergence of a sequence of packings corresponds to

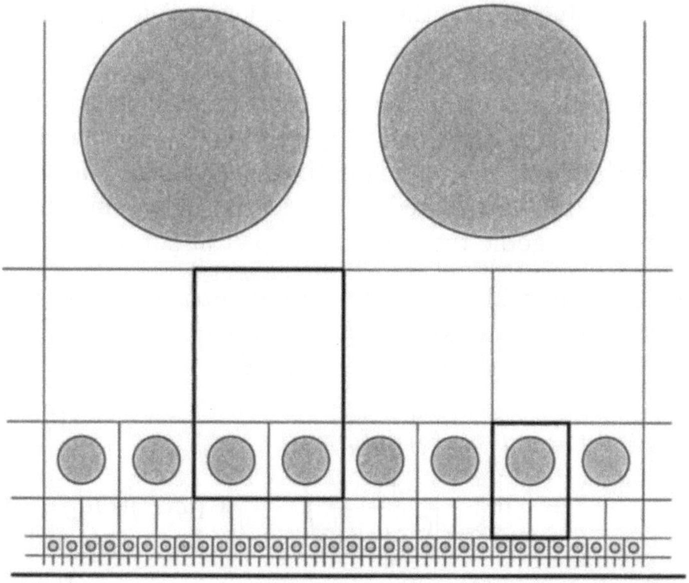

Figure 7. Böröczky's packing with two tiles in dark outline

uniform convergence on compact subsets of hyperbolic space. (A packing of R-balls is relatively dense if every sphere of radius R intersects a ball in the packing.) Such a metric makes X_R compact, and makes continuous the natural action on X_R of the group $\mathcal{G}^d$ of rigid motions of hyperbolic space. We then consider Borel probability measures on X_R which are invariant under $\mathcal{G}^d$.

As examples of such measures, one can identify the orbit $O(P)$ of a periodic packing P with the quotient of $\mathcal{G}^d$ by the symmetry group of P, and thus project Haar measure on the $\mathcal{G}^d$ to $O(P)$. This idea is easily exploited to prove Theorem 1, first for cofinite symmetry groups and then using results on complete saturation [Bow] to handle the cocompact case.

We define the density $d(\mu)$ of each invariant measure μ on X_R as $\mu(A)$, where A is the following set of packings:

$$A \equiv \{P \in X_R \,|\, \text{the origin of } \mathbb{H}^d \text{ is in a ball in } P\}. \tag{4}$$

(It is easy to see from the invariance of μ that this definition is independent of the choice of origin.) We may now introduce the notion of optimal density.

Definition 2. A probability measure $\bar{\mu}$ on the space X_R of packings, ergodic under rigid motions, is "optimally dense" if $d(\bar{\mu}) = \sup_\mu d(\mu) =$

$\sup_\mu \mu(A)$; the value $d(\bar{\mu})$ is the "optimal density" for packing balls of radius R.

(An invariant measure μ is "ergodic" if it cannot be expressed as an average: $\mu = a_1\mu_1 + a_2\mu_2$, with a_1, $a_2 > 0$ and μ_1, μ_2 invariant.) It is easy to show the existence of such optimal measures. The terminology is then justified – that is, related to the density of packings in the sense of (3) – by recent ergodic theorems of Nevo et al [20, 21], as follows.

Theorem 3 (Nevo et al). *Let μ be a Borel probability measure on the compact metric space X, ergodic with respect to an action of the isometry group $\mathcal{G}^d$ of $\mathbb{H}^d$. For any open subset A of X,*

$$\begin{aligned} \mu(A) &= \int_X \chi_A(q)\, d\mu(q) \\ &= \lim_{\rho\to\infty} \frac{1}{\nu[\mathcal{G}^d(\rho,p)]} \int_{\mathcal{G}^d(\rho,p)} \chi_A[g(P)]\, d\nu(g) \end{aligned} \tag{5}$$

for μ-a.e. P, where χ_A is the indicator function for A, ν is Haar measure on $\mathcal{G}^d$ and:

$$\mathcal{G}^d(\rho,p) = \{g \in \mathcal{G}^d \,:\, m_{\mathbb{H}}[g(p),p] < \rho\}, \tag{6}$$

where $m_{\mathbb{H}}$ is the metric on, and p any fixed point in, $\mathbb{H}^d$.

We use this ergodic theorem as follows. With A as in (4), the theorem shows the existence of a limiting density, in the sense of (3), (with expanding spheres instead of squares) for μ-a.e. packing P. (We improved Nevo's theorem somewhat in [7] to obtain existence relative to *every* p.)

Now that we have a powerful mechanism to prove the *existence* of limiting densities in packings, we define the key notion of "optimally dense packings" as those packings which reproduce an optimally dense measure.

Definition 4. A packing $P \in X_R$ is "optimally dense" if it is in the support of some optimally dense measure $\bar{\mu}$ and, for every p in $\mathbb{H}^d$,

$$\int_{X_R} f(q)\, d\bar{\mu}(q) = \lim_{\rho\to\infty} \frac{1}{\nu[\mathcal{G}^d(\rho,P)]} \int_{\mathcal{G}^d(\rho,P)} f[g(p)]\, d\nu(g) \tag{7}$$

for every continuous function f on X_R.

It follows easily from the above that for every optimally dense $\bar{\mu}$, $\bar{\mu}$-almost every packing is optimally dense (and in particular optimally dense packings exist!)

From the next result we see that optimally dense packings need not be periodic.

Theorem 5. *For most $R > 0$ (all but countably many), the densest packing of $\mathbb{H}^d$ by spheres of radius R is not unique (up to rigid motion) – in fact for most fixed radii R the sphere packing problem in $\mathbb{H}^d$ is aperiodic.*

3. Conclusion

It is appropriate to step back and see what this formalism, introduced to solve the old problem of densest packings of bodies in hyperbolic spaces, has to say about the general problem, for Euclidean as well as hyperbolic spaces. (Our definitions of optimal density are easily shown to agree with the standard ones for packings in Euclidean space – using Birkhoff's pointwise ergodic theorem instead of Nevo's.) As we saw, in Euclidean spaces the phenomenon of aperiodicity, as exemplified for instance by the Penrose kite and dart tilings, could be understood as a certain (global) form of nonuniqueness, up to congruence, of the optimally dense packings of some set of bodies. For natural reasons, aperiodicity was first noted when the bodies were polyhedra and the densest packings were tilings. (See [25], which was not sufficiently appreciated when first published.) We now see that in hyperbolic space this same phenomenon already appears in the simpler setting of densest packings of congruent spheres. And the ergodic theory formalism, which we introduced to overcome the conceptual difficulties of densest packings in hyperbolic space, also suggests that the nonuniqueness associated with aperiodicity could be eliminated by reformulating the optimization problem as having its solutions be invariant measures, rather than packings which reproduce such measures. Indeed, it is reasonable to conjecture that, with an appropriate notion of genericity, the problem of optimally dense packings of $\mathbb{E}^d$ or $\mathbb{H}^d$, by copies of a generic finite set of bodies, has a unique invariant measure as solution.

Bibliography

[1] J. Block and S. Weinberger, Aperiodic tilings, scalar curvature, and amenability of spaces, *J. Amer. Math. Soc.* 5(1992), 907–918.

[2] K. Böröczky, Gömbkitöltések állandó görbületű terekben I, *Mat. Lapok.* 25(1974), 265–306.

[3] K. Böröczky, Packing of spheres in spaces of constant curvature, *Acta Math. Acad. Sci. Hung.* 32(1978) 243–261.

[4] K. Böröczky and A. Florian, Über die dichteste Kugelpackung in hyperbolischen Raum, *Acta Math. Acad. Sci. Hung.* 15(1964) 237–245.

[5] L. Bowen, On the existence of completely saturated packings and completely reduced coverings, *Geometriae Dedicata* 98(2003), 211-226.

[6] L. Bowen and C. Radin, Densest packing of equal spheres in hyperbolic space, *Discrete Comput. Geom.* 29(2003), 23–39.

[7] L. Bowen and C. Radin, Optimally dense packings of hyperbolic space, *Geometriae Dedicata* 104(2004), 37–59.

[8] L. Fejes Tóth, Über einen geometrischen Satz, *Math. Z.* 46(1940) 79–83.

[9] L. Fejes Tóth, On close-packings of spheres in spaces of constant curvature, *Publ. Math. Debrecen* 3(1953) 158–167.

[10] L. Fejes Tóth, Kreisaufuellungen der hyperbolischen Ebene, *Acta Math. Acad. Sci. Hung.* 4(1953), 103–110.

[11] L. Fejes Tóth, Kreisuberdeckungender hyperbolischen Ebene, *Acta Math. Acad. Sci. Hung.* 4(1953), 111–114.

[12] L. Fejes Tóth, *Regular Figures*, Macmillan, New York, 1964.

[13] G. Fejes Tóth and W. Kuperberg, Packing and covering with convex sets, chapter 3.3, pp. 799-860, in Vol B of *Handbook of Convex Geometry*, ed. P. Gruber and J. Wills, North Holland, Amsterdam, 1993.

[14] G. Fejes Tóth, G. Kuperberg and W. Kuperberg, Highly saturated packings and reduced coverings, *Monatsh. Math.* 125(1998) 127–145.

[15] M. Gardner, Mathematical Games, in *Sci. Amer.* (January 1977) 110–121.

[16] C. Goodman-Strauss, A strongly aperiodic set of tiles in the hyperbolic plane, preprint, Univ. Arkansas, 2000.

[17] G. Kuperberg, Notions of denseness, *Geom. Topol.* 4(2000) 274–292.

[18] G.A. Margulis and S. Mozes, Aperiodic tilings of the hyperbolic plane by convex polygons, *Israel J. Math.* 107(1998) 319–332.

[19] S. Mozes, Aperiodic tilings, *Invent. Math.* 128(1997), 603–611.

[20] A. Nevo, Pointwise ergodic theorems for radial averages on simple Lie groups I, *Duke Math. J.* 76(1994) 113–140.

[21] A. Nevo and E. Stein, Analogs of Weiner's ergodic theorems for semisimple groups I, *Annals of Math.* 145(1997) 565–595.

[22] R. Penrose, Pentaplexity, *Eureka* 39(1978) 16–22.

[23] C. Radin, *Miles of Tiles*, Student Mathematical Library, Vol 1, Amer. Math. Soc., Providence, 1999.

[24] C.A. Rogers, *Packing and Covering*, University Press, Cambridge, 1964.

[25] P. Schmitt, Disks with special properties of densest packings, *Discrete Comput. Geom.* 6(1991) 181–190.

FLEXIBLE OCTAHEDRA IN THE HYPERBOLIC SPACE

Hellmuth Stachel
Institute of Geometry, Vienna University of Technology
stachel@geometrie.tuwien.ac.at

Abstract This paper treats flexible polyhedra in the hyperbolic 3-space $\mathbb{H}^3$. It is proved that the geometric characterization of octahedra being infinitesimally flexible of orders 1 or 2 is quite the same as in the Euclidean case. Also Euclidean results concerning continuously flexible octahedra remain valid in hyperbolic geometry: There are at least three types of continuously flexible octahedra in $\mathbb{H}^3$; the line-symmetric Type 1, Type 2 with planar symmetry, and the non-symmetric Type 3 with two flat positions. However, Type 3 can be subdivided into three subclasses according to the type of circles in hyperbolic geometry. The flexibility of Type 3 octahedra can again be argued with the aid of Ivory's Theorem.

Keywords: Flexible polyhedra, Bricard's octahedra, infinitesimal flexibility, hyperbolic geometry

Introduction

R. Bricard's *continuously flexible* octahedra ([1], compare also [9, 5]) play an essential role in the theory of flexible polyhedra. The first two types of flexible octahedra in the Euclidean 3-space $\mathbb{E}^3$ admit self-symmetries: All pairs of opposite vertices of *Type 1* are symmetric with respect to a line; at *Type 2*, two pairs of vertices are symmetric with respect to a plane which passes through the remaining two vertices. Octahedra of *Type 3* are unsymmetric and admit two flat positions which in a certain way are related to three concentric circles (see e.g. [7]). Bricard proved in [1] that these three types are the only octahedra in $\mathbb{E}^3$ which are continuously flexible — apart from two trivial cases which either have one equator aligned or two opposite vertices coinciding.

Beside the continuously flexible exemplars also the *infinitesimally flexible* octahedra deserve interest. They can be classified with respect to the *order* $n \geq 1$ of flexibility. In [6] geometric characterizations were given for octahedra which are flexible either of first or of second order (compare also [2, 9]).

The aim of this paper is to demonstrate that these characterizations remain valid in the hyperbolic space $\mathbb{H}^3$ (Theorem 1). Furthermore, it will be proved (Theorem 3) that the hyperbolic counterparts of Bricard's octahedra are again continuously flexible.

The flexibility of Types 1 and 2 in $\mathbb{H}^3$ can be proved as in $\mathbb{E}^3$: Let a skew isogram $B_1C_1B_2C_2$ be given, i.e., a quadrangle with the property that *opposite* sides have equal length. Then each pair (B_1, B_2) and (C_1, C_2) is symmetric with respect to an axis a.[1] Any arbitrary point A_1 can serve as a vertex for a pyramid with basis $B_1C_1B_2C_2$. This pyramid consisting of four triangles is flexible in $\mathbb{H}^3$. And this flexibility is not restricted when we add its mirror under reflection in the axis a. Of course, at each Bricard's octahedron we have to neglect self-intersections.

When a quadrangle $B_1C_1B_2C_2$ is given where two pairs of *neighboring* sides are of equal length, e.g., $d_h(B_1, C_i) = d_h(B_2, C_i)$, $i = 1, 2$, then the vertices B_1, B_2 are symmetric with respect to a plane through C_1 and C_2. In a similar way as before, two symmetric pyramids with the common basis $B_1C_1B_2C_2$ constitute a flexible octahedron which is of Type 2.

The description of octahedra of Type 3 (with flat positions) is more complicated and will be given in Section 2 below.

It is conjectured that these three types are the only nontrivial examples of flexible octahedra in $\mathbb{H}^3$. However, a complete proof is still open. For the Euclidean case Bricard's main result in [1] has been reproved in [5] with methods from projective geometry. The proof was based on a configuration theorem concerning bipartite frameworks (see [8]). The hyperbolic counterpart of this theorem has not yet been proved.

Most of the following statements are based on the projective model of $\mathbb{H}^3$ with the absolute quadric Ω. We use a coordinate system in the real projective 3-space $\mathbb{P}^3$ such that for any two points $X = \mathbf{x}\mathbb{R}$, $Y = \mathbf{y}\mathbb{R}$ conjugate position with respect to Ω is equivalent to

$$\langle \mathbf{x}, \mathbf{y} \rangle := x_0y_0 - x_1y_1 - x_2y_2 - x_3y_3 = 0. \qquad (1)$$

Then for points[2] $\mathbf{x}, \mathbf{y} \in \mathbb{H}^3$ the coordinates can be normalized to $\langle \mathbf{x}, \mathbf{x} \rangle = \langle \mathbf{y}, \mathbf{y} \rangle = 1$, and their hyperbolic distance $d_h(\mathbf{x}, \mathbf{y})$ obeys

$$\cosh d_h(\mathbf{x}, \mathbf{y}) = \langle \mathbf{x}, \mathbf{y} \rangle, \quad \text{provided} \quad \langle \mathbf{x}, \mathbf{x} \rangle = \langle \mathbf{y}, \mathbf{y} \rangle = 1. \qquad (2)$$

[1]The triangles $B_1B_2C_1$ and $B_2B_1C_2$ are congruent. Therefore there is a product of reflections in two perpendicular planes with $B_1 \mapsto B_2$, $C_1 \mapsto C_2$, and vice versa.

[2]In the sequel we often identify the point X with any of its coordinate vectors $\mathbf{x}$ when we briefly speak about 'point' $\mathbf{x}$.

This is the so-called Weierstraß model of $\mathbb{H}^3$. It is located on the unit sphere of the four-dimensional Minkowski space $\mathbb{M}^4$.

1. Infinitesimally flexible octahedra in $\mathbb{H}^3$

Theorem 1. *Let $\mathcal{O}$ be an octahedron in $\mathbb{H}^3$ with the non-coplanar 'equator' $B_1C_1B_2C_2$ and the 'poles' $A_1 \neq A_2$.*

1. *$\mathcal{O}$ is infinitesimally flexible of first order if and only if there is a second-order surface Φ_1 passing through the vertices A_1, A_2 and through the sides of the equator $B_1C_1B_2C_2$.*

2. *A first-order infinitesimally flexible octahedron $\mathcal{O}$ with surface Φ_1 according to 1. is infinitesimally flexible of order 2 if and only if there are second-order surfaces Φ_{2a} through the poles A_1, A_2 and Φ_{2b} through the sides of the equator $B_1C_1B_2C_2$ such that the pencil spanned by Φ_{2a} and Φ_{2b} includes the surface Ψ_2 which is polar to the absolut quadric Ω with respect to Φ_1.*

Proof. In analogy to the Euclidean definition (cf. [3, 4]) a framework $\mathcal{F}$ with vertices $\mathbf{v}_1, \ldots, \mathbf{v}_n$ and edges $e_{jk} = \mathbf{v}_j\,\mathbf{v}_k$, $(j,k) \in E$, is called *infinitesimally flexible of order* n (in the classical sense) in $\mathbb{H}^3$ if and only if for each $i \in \{1, \ldots, n\}$ there is a polynomial function

$$\mathbf{v}_i' := \mathbf{v}_i + \mathbf{v}_{i,1}t + \cdots + \mathbf{v}_{i,n}t^n, \quad n \geq 1, \tag{3}$$

such that

(a) the replacement of $\mathbf{v}_i$ by $\mathbf{v}_i' \in \mathbb{R}[t]^4$ in the formulas for the edge lengths gives stationary values of multiplicity $\geq n$ at $t = 0$, i.e., due to (2)

$$\begin{aligned} \langle\, \mathbf{v}_j',\, \mathbf{v}_k' \rangle - \langle\, \mathbf{v}_j,\, \mathbf{v}_k \rangle &= o(t^n) \quad \forall (j,k) \in E, \\ \text{while } \langle\, \mathbf{v}_i',\, \mathbf{v}_i' \rangle - 1 &= o(t^n) \quad \forall i \in \{1, \ldots, n\}. \end{aligned} \tag{4}$$

(b) In order to exclude *trivial* flexes, the vectors $\mathbf{v}_{1,1}, \ldots, \mathbf{v}_{n,1}$ do not originate from any motion of $\mathcal{F}$ as a rigid body.

The n-tupel $(\mathbf{v}_1', \ldots, \mathbf{v}_n')$ of polynomial vector functions is called a non-trivial *n-th-order flex* of $\mathcal{F}$. □

Conditions for n-th order flexibility of $\mathcal{O}$. The 12 edges of the octahedron $\mathcal{O}$ define a framework in $\mathbb{H}^3$ with 6 vertices $\mathbf{a}_1, \ldots, \mathbf{c}_2$. We change the notation of the equator slightly by setting

$$\mathbf{v}_1 := \mathbf{b}_1, \ \mathbf{v}_2 := \mathbf{c}_1, \ \mathbf{v}_3 := \mathbf{b}_2, \ \mathbf{v}_4 := \mathbf{c}_2\,.$$

Now in analogy to [6] we subdivide the edge set of $\mathcal{O}$ into the equator $\{\mathbf{v}_1\mathbf{v}_2, \ldots, \mathbf{v}_4\mathbf{v}_1\}$ and the eight sides $\mathbf{v}_j\mathbf{a}_k$, $j \in \{1, \ldots, 4\}$, $k \in \{1, 2\}$. The latter form a bipartite sub-framework $\mathcal{O}'$ of $\mathcal{O}$.

Let an n-th-order flex of $\mathcal{O}$ be given by

$$\mathbf{v}_j' = \mathbf{v}_j + \mathbf{v}_{j,1}t + \cdots + \mathbf{v}_{j,n}t^n, \ \mathbf{a}_k' = \mathbf{a}_k + \mathbf{a}_{k,1}t + \cdots + \mathbf{a}_{k,n}t^n \tag{5}$$

such that

$$\begin{array}{ll} \langle\, \mathbf{v}_j', \mathbf{a}_k' \rangle - \langle\, \mathbf{v}_j, \mathbf{a}_k \rangle = o(t^n), & \langle\, \mathbf{v}_j', \mathbf{v}_{j+1}' \rangle - \langle\, \mathbf{v}_j, \mathbf{v}_{j+1} \rangle = o(t^n), \\ \langle\, \mathbf{v}_j', \mathbf{v}_j' \rangle - \langle\, \mathbf{v}_j, \mathbf{v}_j \rangle = o(t^n), & \langle\, \mathbf{a}_k', \mathbf{a}_k' \rangle - \langle\, \mathbf{a}_k, \mathbf{a}_k \rangle = o(t^n) \end{array} \tag{6}$$

for all $j \in \{1, \ldots, 4\}$ and $k \in \{1, 2\}$. From now on we assume a *non-coplanar* equator $\mathbf{v}_1 \ldots \mathbf{v}_4$. Then at each $t \in \mathbb{R}$ there is a linear mapping

$$l(t)\colon\ \mathbb{R}^4 \to \mathbb{R}^4, \quad \mathbf{v}_j \mapsto \mathbf{v}_j'(t) \ \text{ for } \ j = 1, \ldots, 4.$$

For each $k \in \{1, 2\}$ the equations

$$\langle\, l(\mathbf{v}_j), \mathbf{a}_k' \rangle - \langle\, \mathbf{v}_j, \mathbf{a}_k \rangle = o(t^n)$$

define a system of four linearly independent equations for the unknown vector $\mathbf{a}_k$. Let the mapping $l^\star$ be adjoint to l, i.e., obeying $\langle\, l(\mathbf{x}), \mathbf{y} \rangle = \langle\, \mathbf{x}, l^\star(\mathbf{y}) \rangle$ for each $t \in \mathbb{R}$. Then the solution of the linear system can be expressed as

$$\mathbf{a}_k = l^\star(\mathbf{a}_k') + \mathbf{o}(t^n) \ \text{ for } \ k = 1, 2.$$

Thus the first equation of (6) holds true for all edges of $\mathcal{O}'$. For t sufficiently near to 0 the linear mapping l is bijective as well as $l^\star$. Here we introduce two bilinear forms over $\mathbb{R}[t]$,

$$\begin{array}{l} f(t; \mathbf{x}, \mathbf{y}) := \langle\, l(\mathbf{x}), l(\mathbf{y}) \rangle - \langle\, \mathbf{x}, \mathbf{y} \rangle, \\ g(t; \mathbf{x}, \mathbf{y}) := \langle\, \mathbf{x}, \mathbf{y} \rangle - \langle\, l^{\star-1}(\mathbf{x}), l^{\star-1}(\mathbf{y}) \rangle. \end{array} \tag{7}$$

Thus we can summarize the remaining conditions in (6) for an n-th order infinitesimal flex as

$$f(\mathbf{v}_j, \mathbf{v}_j) = o(t^n), \quad f(\mathbf{v}_j, \mathbf{v}_{j+1}) = o(t^n), \quad g(\mathbf{a}_k, \mathbf{a}_k) = o(t^n). \tag{8}$$

These equations are homogeneous in $\mathbf{v}_j$ and $\mathbf{a}_k$.

Now we turn to matrix notation. We write the coordinate vectors as columns and set up

$$l(t)\colon\ \mathbf{v}_j \ \mapsto\ \mathbf{v}_j' = A(t)\mathbf{v}_i \ \text{ with } \ A(0) = I_4\,, \tag{9}$$

where I_4 denotes the 4×4 unit matrix. The entries of A are polynomials of degree $\leq n$. Let H denote the diagonal matrix $\operatorname{diag}(1, -1, -1, -1)$

with $H^{-1} = H^T = H$. Then the fundamental bilinear form (1) can be expressed as

$$\langle \mathbf{x}, \mathbf{y} \rangle = \mathbf{x}^T H \mathbf{y},$$

and the mapping adjoint to l reads

$$l^\star(t)\colon\ \mathbf{a}'_k \mapsto \mathbf{a}_k = HA^T H\mathbf{a}'_k\,. \tag{10}$$

Therefore the coordinatizations of the bilinear forms defined in (7) are

$$\begin{aligned} f(t;\mathbf{x},\mathbf{y}) &:= \mathbf{x}^T(A^T HA - H)\mathbf{y},\\ g(t;\mathbf{x},\mathbf{y}) &:= \mathbf{x}^T(H - HA^{-1}HA^{T^{-1}}H)\mathbf{y}. \end{aligned} \tag{11}$$

Lemma 1. *The octahedron $\mathcal{O}$ with non-coplanar equator $\mathbf{v}_1 \dots \mathbf{v}_4$ is infinitesimally flexible of order n if and only if in a neighborhood of $t = 0$ there is a regular matrix $A(t)$ with entries of class C^n and $A(0) = I_4$ such that the vertices $\mathbf{v}_1, \dots, \mathbf{a}_2$ obey the equations (8) with bilinear forms f, g according to (11) — provided the flex corresponding to $\mathbf{v}'_j(t) = A\mathbf{v}_j$ and $\mathbf{a}'_k(t) = HA^{T^{-1}}H\mathbf{a}_k$ is not trivial.*

In order to obtain geometric characterizations for n-order flexible octahedra, $n \in \{1, 2, \dots\}$, we compare the coefficients of t^n in the equations (8). For this purpose we use the Taylor expansions (compare [7])

$$\begin{aligned} A(t) &= I_4 + A_1 t + A_2 t^2 + \cdots,\\ A^{-1}(t) &= I_4 - A_1 t + (-A_2 + A_1^2)t^2 + \cdots. \end{aligned} \tag{12}$$

This implies for the bilinear forms (11)

$$\begin{aligned} &A^T HA - H = (A_1^T H + HA_1)t + (A_2^T H + A_1^T HA_1 + HA_2)t^2 + \cdots,\\ &H - HA^{-1}HA^{T^{-1}}H = (A_1^T H + HA_1)t\\ &+ (A_2^T H - A_1^{T^2} H - HA_1 HA_1^T H - HA_1^2 + HA_2)t^2 + \cdots. \end{aligned}$$

We set

$$\begin{aligned} f(t;\mathbf{x},\mathbf{y}) &= f_1(\mathbf{x},\mathbf{y})t + f_2(\mathbf{x},\mathbf{y})t^2 + \dots,\\ g(t;\mathbf{x},\mathbf{y}) &= g_1(\mathbf{x},\mathbf{y})t + g_2(\mathbf{x},\mathbf{y})t^2 + \dots \end{aligned} \tag{13}$$

and obtain

$$f_1(\mathbf{x},\mathbf{y}) = g_1(\mathbf{x},\mathbf{y}) = \mathbf{x}^T(HA_1 + A_1^T H)\mathbf{y}, \tag{14}$$

$$\begin{aligned} f_2(\mathbf{x},\mathbf{y}) &= \mathbf{x}^T(A_2^T H + A_1^T HA_1 + HA_2)\mathbf{y},\\ g_2(\mathbf{x},\mathbf{y}) &= \mathbf{x}^T(A_2^T H - A_1^{T^2} H - HA_1 HA_1^T H - HA_1^2 + HA_2)\mathbf{y}. \end{aligned} \tag{15}$$

First-order flexibility. Suppose the bilinear form f_1 in (14) is not zero. Then $f_1(\mathbf{x},\mathbf{x}) = 0$ is the equation of a second-order surface Φ_1. Due to

$$f_1(\mathbf{v}_i,\mathbf{v}_i) = f_1(\mathbf{v}_i,\mathbf{v}_{i+1}) = f_1(\mathbf{a}_k,\mathbf{a}_k) = 0$$

this surface Φ_1 passes through the equator and through the poles.

According to Lemma 1 and (12) the 'velocity vectors' under this flex in the Minkowski space $\mathbb{M}^4$ are

$$\mathbf{v}_{i,1} = A_1\mathbf{v}_i, \quad \mathbf{a}_{k,1} = -HA_1^T H\mathbf{a}_k.$$

The velocity vector $\dot{\mathbf{x}}$ of point $\mathbf{x}$ under any hyperbolic motion obeys

$$\dot{\mathbf{x}} = B\mathbf{x} \text{ with } B^T = -HBH.$$

$B := -H_1 + HA_1^T H$ is an example for such a 'pseudo-skew-symmetric' matrix. We superimpose this instantaneous motion and obtain the new velocity vectors

$$\mathbf{v}_{i,1} = \tfrac{1}{2}(A_1 + HA_1^T H)\mathbf{v}_i, \quad \mathbf{a}_{k,1} = -\tfrac{1}{2}(A_1 + HA_1^T H)\mathbf{a}_k. \tag{16}$$

In analogy to the Euclidean case these vectors in $\mathbb{M}^4$ are *perpendicular* in the Minkowski sense to the surface Φ_1. This follows from the fact that the plane tangent to Φ_1 at $\mathbf{x}$ obeys the equation $(HA_1 + A_1^T H)\mathbf{x} = 0$. The vector $(A_1 + HA_1^T H)\mathbf{x}$ is perpendicular to this tangent plane.

Since $\mathbf{v}_1,\ldots,\mathbf{v}_4$ are supposed to be linearly independent, the velocity vectors of our flex are trivial if and only if the matrix $(A_1 + HA_1^T H)$ is pseudo-skew-symmetric. But then this must be the zero-matrix, and the bilinear form f_1 in (14) would be zero, too. Hence the existence of a second-order surface Φ_1 guarantees that the flex is non-trivial.

Second-order flexibility. The second-order terms in the bilinear forms $f(t;\mathbf{x},\mathbf{y})$, $g(t;\mathbf{x},\mathbf{y})$ are listed in (15). Their difference

$$\begin{aligned} h_2(\mathbf{x},\mathbf{y}) :=\ & f_2(\mathbf{x},\mathbf{y}) - g_2(\mathbf{x},\mathbf{y}) \\ =\ & \mathbf{x}^T(A_1^T HA_1 + A_1^{T^2} H + HA_1HA_1^T H + HA_1^2)\mathbf{y} \\ =\ & \mathbf{x}^T(A_1^T H + HA_1)H(HA_1 + A_1^T H)\mathbf{y} \end{aligned} \tag{17}$$

depends on the first-order terms only. The points of the second-order surface $\Psi_2\colon h_2(\mathbf{x},\mathbf{x}) = 0$ have polar planes with respect to Φ_1 which are tangent to the absolute quadric Ω. This means for regular Φ_1 that Ψ_2 is polar to Ω with respect to Φ_1.

Due to (8), for a 2nd-order flexible octahedron $\mathcal{O}$ there is a surface $\Phi_{2b} : f_2(\mathbf{x},\mathbf{x}) = 0$ passing through the equator. The second surface $\Phi_{2a}\colon g_2(\mathbf{x},\mathbf{x}) = 0$ passes through the poles $\mathbf{a}_1$ and $\mathbf{a}_2$. And in addition, Φ_{2a} and Φ_{2b} span a pencil which includes Ψ_2.

2. Flexible octahedra of Type 3 in $\mathbb{H}^3$

Any Bricard octahedron $\mathcal{O}$ of Type 3 in $\mathbb{E}^3$ admits flat positions which can be determined in the following way (see Fig. 1):

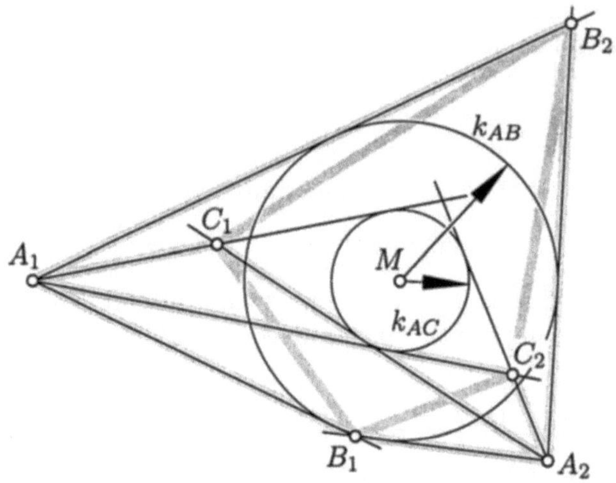

Figure 1. The flat position of Type 3, proper case

Let k_{AC}, k_{AB} be two different circles with the common center M, and let A_1, A_2 be two different points outside k_{AC} and k_{AB}. The tangent lines of k_{AB} passing through A_1 or A_2 define a quadrilateral. We specify (B_1, B_2) as a pair of opposite vertices.[3] Then $A_1B_1A_2B_2$ is a quadrangle with the four sides $A_1B_1, \ldots, B_2A_1$ tangent to k_{AB}. In the same way we specify a second quadrangle $A_1C_1A_2C_2$ tangent to the circle k_{AC}. Then (A_1, A_2), (B_1, B_2) and (C_1, C_2) are the pairs of opposite vertices of $\mathcal{O}$ in a flat position. We restrict to proper octahedra by the assumption that points $B_1, \ldots, C_2$ are finite. And we exclude self-symmetries by requiring that A_1, A_2 are not aligned with M and the distances $\overline{A_1M}$ and $\overline{A_2M}$ are different.

This definition can immediately be used in the hyperbolic plane $\mathbb{H}^2$, too. However, the demand for finite points $B_1, \ldots, C_2$ is of course much more restrictive. And we have to distinguish whether the concentric circles k_{AC}, k_{AB} are proper circles, hypercircles or horocircles in $\mathbb{H}^2$. This means that in the projective model which is used here the center M can be located in the interior, in the exterior or on the absolute conic u of $\mathbb{H}^2$.

[3]When k_{AB} happens to be tangent to the line A_1A_2 then the pair (B_1, B_2) is unique; one B-point is the point of contact.

We prove in several steps that in $\mathbb{H}^3$ an octahedron $\mathcal{O}$ with such a flat position is continuously flexible. Due to the projective model of $\mathbb{H}^3$ we can frequently follow the arguments given in [7] for the Euclidean case.

2.1 Properties of the flat positions

The pairs of line pencils with carriers (A_1, A_2), (B_1, B_2) and (C_1, C_2) span a two-parametric linear system $\mathcal{S}$ of second-class curves.

Any two different curves of this system span a one-parameter linear system ('range') which is completely included in $\mathcal{S}$. Therefore $\mathcal{S}$ contains the circles k_{AB} and k_{AC} and the spanned range, i.e., all circles centered at M, the multiplicity-two pencil with carrier M, and the absolute conic u, if seen as the set of isotropic lines.

Any two different ranges from $\mathcal{S}$ share one curve. This implies that also the quadrangle $B_1C_1B_2C_2$ is tangent to a circle k_{BC} centered at M. Hence at $\mathcal{O}$ no pair of opposite vertices can be distinguished among the others.

Furthermore, with any conic $c \in \mathcal{S}$ all conics confocal[4] to c are included in $\mathcal{S}$. And this range shares a curve c' with the range spanned by the pair of pencils with carriers (C_1, C_2) and the twofold pencil M. We therefore conclude for any position of M

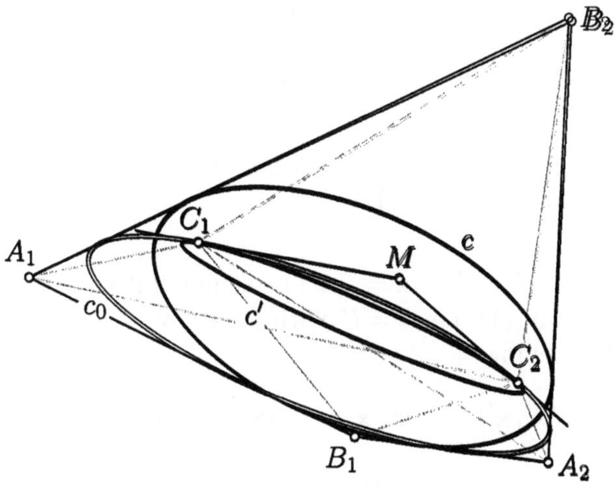

Figure 2. Illustration to Lemma 2

[4]Two conics in $\mathbb{H}^2$ are *confocal* if and only if these tangential conics span a linear system which contains the absolute conic u.

Lemma 2. *For any conic c tangent to the sides of $A_1B_1A_2B_2$ there is a confocal conic c' which passes simultanously through C_1 and C_2 such that the tangent lines at C_1 and C_2 intersect in M (Fig. 2).*

2.2 Particular case of Ivory's Theorem in $\mathbb{H}^3$

Let Φ' be a ruled quadric in $\mathbb{H}^3$ with a plane σ of symmetry. The principal section $\Phi' \cap \sigma$ is denoted by c'. Then the following Lemma 3 is a hyperbolic counterpart of Ivory's Theorem in $\mathbb{E}^3$. The affine transformation between 'corresponding points' of confocal surfaces in the Euclidean case is now replaced by a selfadjoint linear mapping l in the Weierstraß model or by a collinear transformation in the projective model of $\mathbb{H}^3$.

Lemma 3. *For any ruled quadric Φ' in $\mathbb{H}^3$ with a plane σ of symmetry and a real focal curve*[5] $c \subset \sigma$ *there is a selfadjoint linear mapping* $l\colon\ \mathbb{R}^4 \to \mathbb{R}^4$ *with the following properties:*

a) Points $\mathbf{x}' \in \Phi'$ are mapped on points $\mathbf{x} := l(\mathbf{x}') \in \sigma$ obeying $\langle \mathbf{x}, \mathbf{x}\rangle = \langle \mathbf{x}', \mathbf{x}'\rangle$. Therefore absolute points of Φ' remain on Ω.

b) The restriction of l to $c' := \Phi' \cap \sigma$ is a bijection onto the focal curve c which must be located in the exterior of c'. Conversely, for any pair (c', c) of confocal curves in $\mathbb{H}^2$ with c in the exterior of c' there is a ruled surface Φ' through c' with c as focal conic.

c) The restriction of l to any generator of Φ' is an isometry onto a tangent line of c.

d) For any $\mathbf{x}', \mathbf{y}' \in \Phi'$ we obtain equal hyperbolic distances $d_h(l(\mathbf{x}'), \mathbf{y}') = d_h(\mathbf{x}', l(\mathbf{y}'))$.

Proof. Let $l\colon \mathbb{R}^4 \to \mathbb{R}^4$, $\mathbf{x}' \mapsto \mathbf{x} = l(\mathbf{x}')$ be selfadjoint with respect to the bilinear form $\langle\ ,\ \rangle$ in (1), i.e., $l = l^\star$ with $\langle l(\mathbf{x}'), \mathbf{y}\rangle = \langle \mathbf{x}', l^\star(\mathbf{y})\rangle$. There is a quadric Φ' obeying

$$q(\mathbf{x}', \mathbf{x}') := \langle \mathbf{x}', \mathbf{x}'\rangle - \langle l(\mathbf{x}'), l(\mathbf{x}')\rangle = 0. \tag{18}$$

The points $\mathbf{x}' \in \Phi'$ are characterized by the property

$$\langle \mathbf{x}', \mathbf{x}'\rangle = \langle l(\mathbf{x}'), l(\mathbf{x}')\rangle = \langle \mathbf{x}, \mathbf{x}\rangle.$$

[5] Two second-order surfaces Φ, Φ' in $\mathbb{H}^3$ are *confocal* if and only if their dual surfaces span a linear system which includes the absolute surface Ω — seen as the set of absolute planes. If the dual of Φ is singular, i.e., consisting of the tangential planes of a conic c, then c is called a *focal conic* of Φ'.

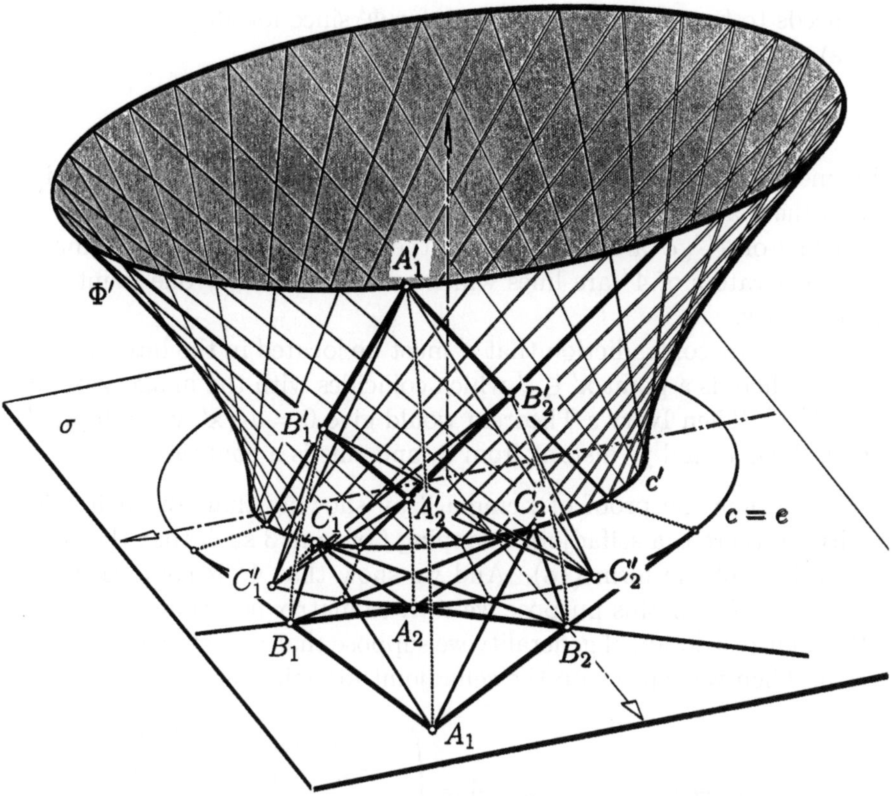

Figure 3. Proving the flexibility of Type 3 with Ivory's Theorem

Therefore according to (2) for any two points $\mathbf{x}', \mathbf{y}' \in \Phi'$ we have a statement of Ivory type

$$d_h(l(\mathbf{x}'), \mathbf{y}) = d_h(\mathbf{x}', l^\star(\mathbf{y})).$$

In addition, the distance between two points $\mathbf{x}', \mathbf{y}' \in \Phi'$ is preserved if and only if

$$q(\mathbf{x}', \mathbf{y}') = \langle\, \mathbf{x}', \mathbf{y}' \rangle - \langle\, l(\mathbf{x}'),\, l(\mathbf{y}') \rangle = 0.$$

This characterizes conjugate position of $\mathbf{x}', \mathbf{y}'$ with respect to Φ', i.e., the connecting line $\mathbf{x}'\mathbf{y}'$ is a generator of Φ'.

Suppose the selfadjoint l has rank 3 with $l(\mathbb{R}^4) = \sigma$. Let $\mathbf{s}$ denote the absolute pole of σ. Then we have

$$0 = \langle\, \mathbf{s},\, l(\mathbf{x}') \rangle = \langle\, l(\mathbf{s}),\, \mathbf{x}' \rangle$$

for all $\mathbf{x}' \in \mathbb{R}^4$. This implies $l(\mathbf{s}) = \mathbf{o}$, i.e., the fibres of l are perpendicular to σ.

σ needs to be a plane of symmetry for Φ' since for all $\mathbf{y}' \in \sigma$ we have $\langle \mathbf{y}', \mathbf{s} \rangle = 0$ and therefore according to (18)

$$q(\mathbf{y}', \mathbf{s}) := \langle \mathbf{y}', \mathbf{s} \rangle - \langle l(\mathbf{y}'), l(\mathbf{s}) \rangle = 0.$$

This means, $\mathbf{s}$ is the pole of σ also with respect to Φ'. The restriction of l to the plane σ is bijective and transfers $c' = \Phi' \cap \sigma$ onto a conic c. All points of $\Phi' \setminus c'$ are mapped into the exterior of c since the images of the generators of Φ' are lines which meet c in exactly one point, i.e., tangent lines.

We prove by contradiction that c' must be located in the interior of c: Suppose there is a point $\mathbf{x}' \in c'$ which coincides with the image $\mathbf{y}$ of any $\mathbf{y}' \in \Phi' \setminus c'$. Then Ivory's Theorem would give $0 = d_h(\mathbf{x}', \mathbf{y}) = d_h(\mathbf{x}, \mathbf{y}')$ with $\mathbf{x} \in c \subset \sigma$ and $\mathbf{y}'$ distant from σ, and this is a contradiction.

We continue the proof of Lemma 3 by showing that for each ruled quadric Φ' there is a selfadjoint mapping l of rank 3 such that the equation of Φ' is of the form (18). And we show that Φ' is confocal to its image under l. For this purpose we turn to matrix notation:

Without restriction of generality we suppose that σ obeys the equation $x_3 = 0$. Then we can set up the selfadjoint mapping as

$$l\colon\ \mathbf{x}' \mapsto \mathbf{x} = A\mathbf{x}' \quad \text{with} \quad A = \begin{pmatrix} a_{00} & a_{01} & a_{02} & 0 \\ -a_{01} & a_{11} & a_{12} & 0 \\ -a_{02} & a_{12} & a_{22} & 0 \\ 0 & 0 & 0 & 0 \end{pmatrix} \tag{19}$$

since $A^T = HAH$. Hence equation (18) gets the form

$$\Phi'\colon\ q(\mathbf{x}', \mathbf{x}') = \mathbf{x}'^T(H - A^THA)\mathbf{x}' = \mathbf{x}'^T H(I_4 - A^2)\mathbf{x}' = 0. \tag{20}$$

The symmetric coordinate matrix

$$Q := H - A^THA$$

of the quadratic form $q(\mathbf{x}', \mathbf{x}')$ is supposed to be regular, and from (20) we obtain

$$A^2 = I_4 - HQ. \tag{21}$$

Points $\mathbf{x}'$ of any plane with coordinate vector $\mathbf{u}' = A^T\mathbf{u}$ have their image $\mathbf{x} = l(\mathbf{x}')$ in the plane $\mathbf{u}^T\mathbf{x} = 0$. The mapping $\mathbf{u} \mapsto A^T\mathbf{u}$ is adjoint to l with respect to the standard scalar product. Under this mapping all planes obeying the quadratic form

$$\Phi\colon\ \mathbf{u}^TA(H - A^THA)^{-1}A^T\mathbf{u} = 0 \tag{22}$$

are transferred into planes tangent to Φ'. So, (22) is the dual representation of the image $l(\Phi')$, while $\mathbf{u}'^T(H - A^T HA)^{-1}\mathbf{u}' = 0$ is the dual equation of Φ'.

Now we state

$$H + A(H - A^T HA)^{-1}A^T = (H - A^T HA)^{-1}. \qquad (23)$$

This reveals that $l(\Phi')$ and the dual of the absolute quadric Ω with coordinate matrix H span a range which includes Φ', i.e., Φ' and $l(\Phi')$ are confocal — whether l has rank 4 or 3.

For proving eq. (23) we multiply both sides with $(H - A^T HA)$ and obtain

$$I_4 - HA^T HA + A(H - A^T HA)^{-1}A^T(H - A^T HA) = I_4,$$

and this is an identity because of $HA^T = AH$ and

$$A^T(H - A^T HA) = A^T H - A^T A^T HA = (H - A^T HA)A.$$

It remains to prove that each hyperboloid Φ' in $\mathbb{H}^3$ with σ as a plane of symmetry has an equation of type (20). This is equivalent to the existence of a matrix A, which is of type (19) and obeys (21).

There are three types of ruled quadrics Φ' to distinguish: The principal section $c' = \Phi' \cap \sigma$ has

(I) a center of symmetry,

(II) no center, but an axis of symmetry,

(III) neither a center, nor an axis of symmetry.

Ad (I): In this case we can set up the equation of Φ' as

$$q(\mathbf{x}', \mathbf{x}') = \gamma_{00}{x_0'}^2 + \gamma_{11}{x_1'}^2 + \gamma_{22}{x_2'}^2 - {x_3'}^2 = 0.$$

This is a ruled quadric with points in the interior of the absolute Ω and with a real focal conic c in σ if either

$$\gamma_{00} < 0, \quad \gamma_{11}, \gamma_{22} > 0, \quad \gamma_{00} + \gamma_{11} > 0 \text{ or } \gamma_{00} + \gamma_{11} > 0,$$

or

$$0 < \gamma_{00} < 1, \quad -1 < \gamma_{11} < 0, \quad \gamma_{22} > 0, \quad \gamma_{00} + \gamma_{11} > 0.$$

In both cases there is a matrix A obeying (21), namely

$$A^2 = \operatorname{diag}(1 - \gamma_{00},\ 1 + \gamma_{11},\ 1 + \gamma_{22},\ 0).$$

We get the solutions

$$A = \operatorname{diag}(\pm\sqrt{1-\gamma_{00}},\ \pm\sqrt{1+\gamma_{11}},\ \pm\sqrt{1+\gamma_{22}},\ 0),$$

and the corresponding focal curve c obeys

$$x_3 = 0 \ \text{ and } \ \frac{\gamma_{00}}{1-\gamma_{00}}\, x_0^2 + \frac{\gamma_{11}}{1+\gamma_{11}}\, x_1^2 + \frac{\gamma_{22}}{1+\gamma_{22}}\, x_0^2 = 0.$$

Ad (II): Without restriction of generality we can set up

$$q(\mathbf{x}',\mathbf{x}') = \gamma_{11}{x_1'}^2 + 2\gamma_{01}x_0'x_1' + \gamma_{22}{x_2'}^2 - {x_3'}^2 = 0$$

with

$$\gamma_{01} \neq 0, \quad \gamma_{11}^2 \leq 4\gamma_{01}^2, \quad \gamma_{22} > 0.$$

According to (21) there is a matrix A obeying

$$A^2 = \begin{pmatrix} 1 & -\gamma_{01} & 0 & 0 \\ \gamma_{01} & 1+\gamma_{11} & 0 & 0 \\ 0 & 0 & 1+\gamma_{22} & 0 \\ 0 & 0 & 0 & 0 \end{pmatrix}$$

since the upper-left 2×2-matrix has either conjugate complex eigenvalues or a twofold eigenvalue with a one-dimensional eigenspace. We obtain

$$A = \begin{pmatrix} (-\delta+\gamma_{11})/2\sqrt{\delta} & \gamma_{01}/\sqrt{\delta} & 0 & 0 \\ -\gamma_{01}/\sqrt{\delta} & -(\delta+\gamma_{11})/2\sqrt{\delta} & 0 & 0 \\ 0 & 0 & \sqrt{1+\gamma_{22}} & 0 \\ 0 & 0 & 0 & 0 \end{pmatrix}$$

for $\delta := 2 + \gamma_{11} + 2\sqrt{1+\gamma_{01}^2+\gamma_{11}}$. The focal curve $c \in \sigma$ obeys

$$(\gamma_{22}+1)\left[\gamma_{01}^2 x_0^2 - 2\gamma_{01}x_0x_1 - (\gamma_{01}^2+\gamma_{11})x_1^2\right] - \gamma_{22}(1+\gamma_{01}^2+\gamma_{11})x_2^2 = 0.$$

Ad (III): We can set up

$$q(\mathbf{x}',\mathbf{x}') = \gamma_{00}({x_0'}^2 - {x_1'}^2 - {x_2'}^2) + 2\gamma_{02}(x_0' - x_1')x_2' - {x_3'}^2 = 0$$

with

$$\gamma_{00} < 0, \quad \gamma_{02} \neq 0.$$

There is a matrix A obeying

$$A^2 = \begin{pmatrix} 1-\gamma_{00} & 0 & -\gamma_{02} & 0 \\ 0 & 1-\gamma_{00} & -\gamma_{02} & 0 \\ \gamma_{02} & -\gamma_{02} & 1-\gamma_{00} & 0 \\ 0 & 0 & 0 & 0 \end{pmatrix}$$

since the upper-left 3×3-matrix has a positive triple eigenvalue with a one-dimensional eigenspace. The solution is

$$A = \frac{1}{8\sqrt{\delta^3}}\begin{pmatrix} 8\delta^2+\gamma_{02}^2 & -\gamma_{02}^2 & -4\gamma_{02}\delta & 0 \\ \gamma_{02}^2 & 8\delta^2-\gamma_{02}^2 & -4\gamma_{02}\delta & 0 \\ 4\gamma_{02}\delta & -4\gamma_{02}\delta & 8\delta^2 & 0 \\ 0 & 0 & 0 & 0 \end{pmatrix}$$

for $\delta := 1-\gamma_{00}$; the focal curve $c \in \sigma$ obeys

$$\gamma_{00}\delta^2(x_0^2-x_1^2-x_2^2) - \gamma_{02}(x_0-x_1)\left[\gamma_{02}(x_0-x_1)-2\delta x_2\right] = 0.$$

The last remark of Lemma 3,b can be verified by comparing the equations of c with that of Φ' in the cases (I)–(III). □

2.3 Conclusions for Octahedra in $\mathbb{H}^3$

Now we combine the previous statements: We identify σ with the projective model of the hyperbolic plane $\mathbb{H}^2$ where the flat position of $\mathcal{O}$ is located. We see each conic c' of Lemma 2 (Fig. 2) as the principal section of a one-sheet hyperboloid Φ' and c as its focal curve (see Fig. 3).

Then Lemma 3,c reveals that there is a quadrangle $A_1'B_1'A_2'B_2'$ with sides on Φ' which is mapped by l onto $A_1B_1A_2B_2$ while all side-lengths are preserved.[6] Under l the vertices $C_1, C_2 \in c'$ are mapped onto $C_1', C_2' \in c$ (notation reversed!), and Ivory's Theorem in Lemma 3,d implies $d_h(A_i, C_j) = d_h(A_i', C_j')$ and $d_h(B_i, C_j) = d_h(B_i', C_j')$ (see Fig. 3). Hence the spatial octahedron $A_1' \dots C_2'$ is isometric to the flat position $A_1 \dots C_2$.

For completing the proof of the continuous flexibility of $\mathcal{O}$ two items remain to be checked:

(i) c' needs to be inside the focal curve c, to say, no tangent line of c may intersect c', and

(ii) c and c' must be of the same type with respect to u, i.e., both intersect u in the same way.

It is substantial that due to the properties of the linear system $\mathcal{S}$ there is a conic c_0 tangent to $A_1B_1A_2B_2$ and passing through both line elements (C_i, MC_i), $i = 1, 2$. So we can use continuity arguments:

Ad (i): Let t denote the side A_1B_1. While the 2nd-class curve c' with line elements (C_i, MC_i) varies, the pole T of t with respect to c' traces

[6]The quadrangle $A_1'B_1'A_2'B_2' \in \Phi$ is unique up to the reflection in the plane σ.

a line t'. For $c' = c_0$ we obtain the point T_0 of contact between t and c_0. We get $T = M$ when c degenerates into the pencil with carrier M. And $S = t' \cap C_1C_2$ is the pole of t with respect to the pair of line pencils (C_1, C_2). Conversely, any point T of line T_0M defines a unique curve c' of this contact range.

Now it depends on the choice of direction when starting from c_0: If T moves along t' torwards the interior of c_0, i.e., if the pair (T, M) separates (T_0, S), then the corresponding conic c' will not intersect t. This results from properties of the polarity with respect to c' and the involution of conjugate points on t'. So c' meets the necessary condition; it is included in the interior of the confocal c, which according to Lemma 2 is tangent to t.

Ad (ii): When starting from c_0, the types of c and c' with respect to the absolute conic u can only begin to differ at a position where c or c' contacts u. Since c and c' are confocal, this contact with u happens for both conics simultaneously at the same point U. So, it could only happen that — from this contact at U on — one conic has real points of intersection near U, the other has no intersection. But this is a contradiction with Lemma 3,a,b, which states that there is a bijection $c' \to c$ mapping absolute points again on absolute points, provided c' is in the interior of c.

All octahedra of Type 3 admit a *second flat position.* This results from the concentric circles k_{AB}, k_{AC}, k_{BC} in the given flat position (see Fig. 1) for the following reason:

At each of the six vertices, e.g. at A_i, the connecting lines with the other pairs (B_1, B_2) and (C_1, C_2) are symmetric with respect to the line through M: Suppose we keep the face $A_1B_1C_1$ fixed. Then for the second flat position it is necessary that each vertex $\overline{A}_2, \overline{B}_2, \overline{C}_2$ of the opposite face is obtained by reflecting the single points A_2, B_2 and C_2 in the sides B_1C_1, A_1C_1 and A_1B_1, respectively (see Fig. 4). In order to guarantee that the distances do not change, we must e.g. demonstrate that there is one isometry in $\mathbb{H}^2$ which maps simultaneously $B_2 \mapsto \overline{B}_2$ and $C_2 \mapsto \overline{C}_2$. The first can be carried out by the consecutive reflections in the lines A_1B_2 and A_1C_1. For the latter we use the reflections in A_1C_2 and A_1B_1. Now it results from the Three-Reflection-Theorem of absolute geometry that these products of reflections are equal because of the symmetry with respect to line A_1M.

It turns out that in the sense of Fig. 3 this second flat position is reached when c' degenerates into the pair of line pencils (C_1, C_2). The corresponding hyperboloid Φ degenerates into a focal conic of c_0.

Thus we end up with

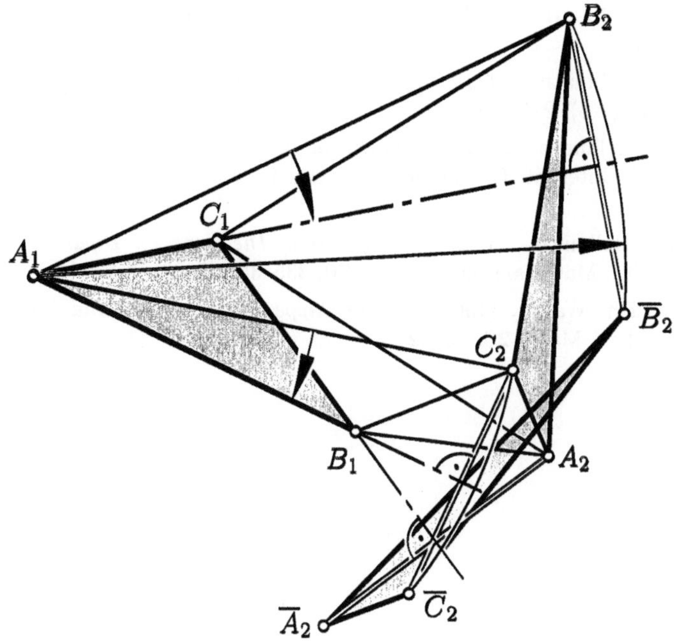

Figure 4. The two flat positions $A_1B_1C_1A_2B_2C_2$ and $A_1B_1C_1\overline{A}_2\overline{B}_2\overline{C}_2$ of $\mathcal{O}$

Theorem 2. *All three classes of Type 3 octahedra in $\mathbb{H}^3$ are continuously flexible and they admit a second flat position.*

Theorem 3. *There are at least three types of continuously flexible octahedra in $\mathbb{H}^3$. At Type 1 all pairs of opposite vertices are symmetric with respect to a line, at Type 2 two pairs of vertices are symmetric with respect to a plane which passes through the remaining two vertices. Flexible octahedra of Type 3 are unsymmetric with flat positions according to Fig. 1.*

Bibliography

[1] Bricard, Raoul. (1897). *Mémoire sur la théorie de l'octaèdre articulé.* J. math. pur. appl., Liouville **3**, 113–148.

[2] Blaschke, Wilhelm. (1920). *Über affine Geometrie XXVI: Wackelige Achtflache.* Math. Z. **6**, 85–93.

[3] Connelly, Robert, and Hermann Servatius. (1994). *Higher-order rigidity – What is the proper definition?* Discrete Comput. Geom. **11**, no. 2, 193–200.

[4] Sabitov, Idzhad Kh. (1992). *Local Theory of Bendings of Surfaces.* In Yu.D. Burago, V.A. Zalgaller (eds.): Geometry III, Theory of Surfaces. Encycl. of Math. Sciences, vol. 48, Springer-Verlag, pp. 179–250.

[5] Stachel, Hellmuth. (1987). *Zur Einzigkeit der Bricardschen Oktaeder.* J. Geom. **28**, 41-56.

[6] Stachel, Hellmuth. (1999). *Higher Order Flexibility of Octahedra.* Period. Math. Hung. **39** (1-3), 225–240.

[7] Stachel, Hellmuth. (2002). *Remarks on Bricard's Flexible Octahedra of Type 3.* Proc. 10th Internat. Conference on Geomety and Graphic, Kiev (Ukraine), Vol. 1, 8–12.

[8] Stachel, Hellmuth. (2002). *Configuration Theorems on Bipartite Frameworks.* Rend. Circ. Mat. Palermo, II. Ser., **70**, 335–351 (2002).

[9] Wunderlich, Walter. (1965). *Starre, kippende, wackelige und bewegliche Achtflache.* Elem. Math. **20**, 25–32.

FRACTAL GEOMETRY ON HYPERBOLIC MANIFOLDS

Bernd O. Stratmann
Mathematical Institute, University of St Andrews,
St Andrews KY16 9SS, Scotland.
bos@maths.st-and.ac.uk

... casting aside all scruples, we shall plunge wholeheartedly into the "new universe" which Bolyai "created from nothing".
—H.S.M. Coxeter (1907 - 2003), [20] page 287

Abstract In this survey we give a report on some recent results obtained in the studies of hyperbolic manifolds by means of fractal geometry. Emphasis has been put on results derived in the quantitative and qualitative fractal analysis of long term geodesic dynamics on hyperbolic manifolds.

Keywords: Conformal dynamics, fractal geometry, hyperbolic manifolds, Kleinian groups, ergodic theory.

1. Fractal geometry and hyperbolic manifolds

Hyperbolic manifolds are located at the junction of various different areas of modern mathematics. For instance, they play a central role in low-dimensional topology, complex dynamics, Teichmüller theory, harmonic analysis, spectral theory, analytic number theory, Diophantine approximations, and non-commutative algebra, to name a few.

On small scales the geometry of an $(n+1)$-dimensional hyperbolic manifold $\mathcal{G}$ coincides with the geometry of the surrounding $(n+1)$-dimensional hyperbolic space $\mathbb{D}$, whereas on large scales the topology of

$\mathcal{G}$ affects the global geometry of $\mathcal{G}$. This can then be analysed by a group of isometries, namely $\mathcal{G}$ can be represented by a Kleinian group, that is a discrete subgroup of the group of all isometries of $\mathbb{D}$. For a more detailed study of the interplay between the local and the global structures, it is a rather fruitful method to investigate $\mathcal{G}$ in terms of geodesic dynamical systems. Part of this type of investigations is to locate various different dynamical aspects of $\mathcal{G}$ and then to study these by means of concepts from fractal geometry. This is the main theme of this survey.

Throughout let G be a non-elementary Kleinian group acting discontinuously on hyperbolic $(n+1)$-space $\mathbb{D} = \{z \in \mathbb{R}^{n+1}; \|z\| < 1\}$ (we shall always use the Poincaré ball model $(\mathbb{D}, d)$ whose boundary at infinity is the unit sphere $\mathbb{S} = \{z \in \mathbb{R}^{n+1}; \|z\| = 1\}$) (cf. [5]). We shall always assume that G has no torsion. The limit set $L(G) \subset \mathbb{S}$ of G is the set of accumulation points of the G-orbit of some arbitrary point in $\mathbb{D}$. Since we are mainly interested in fractal properties of $L(G)$, we shall always assume that the set of ordinary points $\Omega(G) := \mathbb{S} \setminus L(G)$ is non-empty. Important subsets of $L(G)$ are the set $L_p(G)$ of bounded parabolic fixed points of G, the radial limit set $L_r(G)$ and the uniformly radial limit set $L_{ur}(G)$. Here s_ξ refers to the hyperbolic ray from the origin to ξ, and $b(x, r) \subset \mathbb{D}$ refers to the hyperbolic ball of radius r centred at x.

L_r: *A point $\xi \in L(G)$ is called radial limit point if there exists a positive constant $c = c(\xi)$ such that $s_\xi \cap b(g(0), c) \neq \emptyset$ for infinitely many different orbit points $g(0) \in G(0)$.*

L_{ur}: *A point $\xi \in L(G)$ is called uniformly radial limit point if for some positive $c = c(\xi)$ we have that $s_\xi \subset \bigcup_{g \in G} b(g(0), c)$.*

L_p: *A point $p \in L(G)$ is called a bounded parabolic point if it is a fixed point of some parabolic element of G and if its stabiliser G_p has the following properties. There exists a set $M \subset \mathbb{S}$ and a compact set $N \subset \mathbb{S} \setminus \{p\}$ such that $\bigcup_{g \in G_p} g(M) = \mathbb{S} \setminus \{p\}$ and $(M \setminus N) \cap g(M \setminus N) = \emptyset$ for all $g \in G \setminus \{id\}$. In particular, G_p is always isomorphic to some finite extension of $\mathbb{Z}^{k(p)}$ for some $k(p) \in \{1, \dots, n\}$, where $k(p)$ is referred to as the rank of p.*

It is well-known that, by fixing some base frame, we can associate to each Kleinian group G a hyperbolic $(n+1)$-manifold $\mathcal{G} = \mathbb{D}/G$ (we always assume that $\mathcal{G}$ is oriented) (cf. [34, 77]). In $\mathcal{G}$ the limit set $L(G)$ is recovered as follows. For $m \in \mathcal{G}$ fixed, let $\Lambda_m(\mathcal{G})$ denote the set of all geodesic loops[1] which start and terminate at m. A geodesic l in $\mathcal{G}$

[1]Note that these loops are not necessarily closed geodesics.

is called loop-approximable if each finite segment of l can be approximated with arbitrary accuracy by segments of elements in $\Lambda_m(\mathcal{G})$[2]. The geodesic core $C(\mathcal{G})$ of $\mathcal{G}$ is then defined by

$$C(\mathcal{G}) := \{l \text{ geodesic in } \mathcal{G} : l \text{ is loop-approximable}\}.$$

A limit direction is an element of the unit tangent space of $\mathcal{G}$ at m such that tracing this direction on $\mathcal{G}$ results in a geodesic ray which is eventually asymptotic to a geodesic in $C(\mathcal{G})$. The set of these so obtained limit directions is in 1-1-correspondence to the limit set $L(G)$. To demonstrate this transfer between $L(G)$ and the set of limit directions we remark that clearly every ray s_ξ for $\xi \in \mathbb{S}$ admits a projection via the universal covering map onto $\mathcal{G}$ where it becomes a geodesic ray emanating from the point corresponding to the origin. If now for instance $\xi \in L_r(G)$ then the projected ray is recurrent on $\mathcal{G}$, meaning that while travelling along this ray some bounded region in $\mathcal{G}$ gets visited infinitely often. Similarly, if $\xi \in L_{ur}(G)$ then the projected ray describes a bounded excursion, meaning that the whole ray is contained in some suitable bounded region of $\mathcal{G}$. Clearly, every uniformly radial point is radial (whereas the opposite is only true for convex cocompact Kleinian groups[3]).

Poincaré was presumably the first who realised the significance of a certain series which can be associated in a canonical way to any arbitrary hyperbolic manifold. This series is nowadays called the Poincaré series and its abcissa of convergence is usually referred to as the exponent of convergence or often also as the Poincaré exponent. More precisely, for $s \in \mathbb{R}$ the Poincaré series $\mathcal{P}_s(z,w)$ associated with G is given by

$$\mathcal{P}_s(z,w) := \sum_{g \in G} \exp(-sd(z, g(w))).$$

Clearly, convergence and divergence of this series does not depend on the choice of z and w. Hence, the exponent of convergence of G, that is the Poincaré exponent $\delta(G)$, is uniquely determined by

$$\delta(G) := \inf\{s \in \mathbb{R} : \mathcal{P}_s(0,0) \text{ converges}\}.$$

For various special types of Kleinian groups it had been known for some time that $\delta(G)$ quantifies the fractal nature of the uniformly radial limit set (see e.g. [31, 61, 52, 22, 63, 66, 26]). More recently,

[2]i.e. each finite segment of γ is contained in an arbitrarily small neighbourhood of some element of $\Lambda_m(\mathcal{G})$.

[3]A Kleinian group is convex cocompact if and only if $L(G) = L_r(G)$. Note that in this paper we exclusively consider groups for which $L(G) \neq \mathbb{S}$, i.e. G can not be cocompact

Bishop and Jones [8] found an astonishingly elementary method which allows specifying this relation between $\delta(G)$ and the Hausdorff dimension $\dim_H(L_{ur}(G))$ in its complete generality, that is for arbitrary non-elementary Kleinian groups. Consequently, their result gives an ultimate clarification of the Hausdorff-dimensional significance of the Poincaré exponent of any arbitrary Kleinian group. More precisely, the following result was obtained in [8] (cf. also [70]).

Theorem 1 (Theorem of Bishop and Jones (I)).
For every Kleinian group G we have that

$$\delta(G) = \dim_H(L_r(G)) = \dim_H(L_{ur}(G)).$$

Also, in this survey we shall be concerned with the Patterson measure which represents a fundamental concept canonically associated to every Kleinian group G. In his pioneering work [50] Patterson laid the foundation for a comprehensive study of $L(G)$ in terms of measure theory and in particular in terms of fractal dimensions. Patterson's original construction dealt with the case of Fuchsian groups, that is the case of 2-dimensional hyperbolic space, and was then generalised by Sullivan in [73] to the general Kleinian group case (cf. [48, 75]). We now briefly recall this construction. For some sequence of positive numbers ϵ_k tending to zero (and given that G is of $\delta(G)$-divergence type[4], meaning that the Poincaré series $\mathcal{P}_s(0,0)$ diverges for $s = \delta(G)$ (otherwise, a slowly varying function has to be introduced which then forces this divergence (cf. [50]))), μ is the weak limit of the sequence of measures

$$\mu_k := \left(\mathcal{P}_{\delta(G)+\epsilon_k}(0,0)\right)^{-1} \sum_{g \in G} \exp\left(-(\delta(G)+\epsilon_k)d(0,g(0))\right) \mathbf{1}_{g(0)},$$

where $\mathbf{1}_x$ refers to the Dirac measure at $x \in \mathbb{D}^{N+1}$. Note that in the geometrically finite case it is known that μ does not depend on the choice of the sequence (ϵ_k), and hence that μ is unique (cf. [73]). However, this is certainly not the case in general.

One of the important geometric properties of the Patterson measure is that it transforms nicely under elements of G. This property is referred to as $\delta(G)$-conformality, which means that for arbitrary Borel sets $E \subset S^n$ and for every $g \in G$ we have

$$\mu(g^{-1}E) = \int_E \left(\frac{1 - \|g(0)\|^2}{\|\xi - g(0)\|^2}\right)^{\delta(G)} d\mu(\xi).$$

[4]Note, a geometrically finite Kleinian group G is always of $\delta(G)$-divergence type (cf. [73]).

Clearly, here the integrant is just the the norm of the conformal derivative or equally the Poisson kernel raised to the power δ.

Remarks.

The easiest examples of Kleinian groups with non-empty ordinary set are classical Schottky groups, that are subgroups of index 2 of groups generated by reflections at pairwise disjoint circles in $\mathbb{C}$. Further classical examples are for instance quasi-Fuchsian groups and certain subgroups of the Picard group, which in particular include a group with limit set the Apollonian packing. Interesting more advanced examples are for instance hyperbolic manifolds fibred over the circle, or normal subgroups of geometrically finite Kleinian groups. For an extensive list of examples we refer to [38, 40, 45]. For general discussions of Kleinian groups we refer to [1, 2, 5, 34, 38, 40, 43, 45, 48, 77].

Finally, we mention that there are various natural generalisations of hyperbolic manifolds. For instance, for rank 1 manifolds and higher rank symmetric spaces (see e.g. [3, 37, 18, 19, 33, 82]), and in particular for Hadamard manifolds (e.g. [29]) and for groups which act discontinuously on complex hyperbolic space (e.g. [47]), most of the fractal analysis in this survey continues to hold, although so far this has been written up only partially. We also mention that convex cocompact Kleinian groups are the cradle of 'hyperbolic groups in the sense of Gromov', and that finitely generated Kleinian groups with parabolic elements (which are *not* Gromov-hyperbolic) have motivated the concept of 'relative hyperbolicity' ([28, 16, 17]).

Throughout we shall use the notation $a \asymp b$ for two positive reals a, b to indicate that a/b is uniformly bounded away from zero and infinity. We write $a \ll b$ if a/b uniformly bounded away from infinity.

2. Geometrically finite hyperbolic manifolds

Recall that G is called geometrically finite if the action of G on $\mathbb{D}$ admits a fundamental polyhedron with finitely many sides. The following theorem goes back to Beardon and Maskit [6] (see also [7, 11]) and gives a characterisation of geometrical finiteness in terms of $L(G)$.

Theorem 2 (Theorem of Beardon and Maskit).
If G is a geometrically finite Kleinian group, then $L(G) = L_r(G) \cup L_p(G)$ (where $L_p(G)$ might be empty, in which case G is convex cocompact).

Thurston was presumably the first to realise the dynamical significance of the concept geometrical finiteness. He observed the following ([77]).

Theorem 3 (Thurston's observation).
A Kleinian group G and its associated hyperbolic $(n+1)$-manifold $\mathcal{G}$ are geometrically finite if and only if a neighbourhood of the convex hull of the geodesic core $C(\mathcal{G})$ has finite hyperbolic volume.

2.1 Diophantine analysis of the Patterson measure

In this section we discuss some of the results obtained in the study of the essential support of the Patterson measure. For this we introduce the following notation. For $\xi \in L(G)$ and $t > 0$, let ξ_t denote the unique point on the ray between 0 and ξ with hyperbolic distance t from 0. Let $b(\xi_t)$ denote the intersection of $\mathbb{S}$ with the $(n+1)$-ball whose boundary is orthogonal to $\mathbb{S}$, which contains ξ, and which intersects s_ξ orthogonally at ξ_t. Hence, $b(\xi_t)$ is an n-ball in $\mathbb{S}$ with radius comparable to e^{-t}. Also, define $k(\xi_t)$ to be equal to $k(p)$ if the projection of ξ_t onto $\mathcal{G}$ is contained in the cusp region of $\mathcal{G}$ associated with the parabolic point p; otherwise we let $k(\xi_t)$ be equal to $\delta(G)$. Furthermore, define $\Delta(\xi_t) := d(\xi_t, G(0))$, and let k_{min} and k_{max} denote the minimal and maximal occurring rank for the parabolic elements in G.

The following result provides a key observation in the investigations of the coarse geometry of the Patterson measure μ. We remark that this measure formula has recently been extended to complex hyperbolic manifolds [47] and to Hadamard manifolds [60].

Theorem 4 (Global measure formula). ([75, 72])
For all $\xi \in L(G)$ and $t > 0$, we have

$$\mu(b(\xi_t)) \asymp \exp\left(-t\delta(G) - \Delta(\xi_t)(\delta(G) - k(\xi_t))\right).$$

Immediate implications are that the Patterson measure is a doubling measure and that the limit set of a geometrically finite Kleinian group is uniformly perfect.

In order to derive further information on the fractal nature of the limit set of a geometrically finite Kleinian group, it is vital to give good approximations of the essential support of the Patterson measure. The following three theorems shed some light on the essential support from different perspectives. For the first theorem recall that an element of $L(G)$ is called a Myrberg limit point if the projection of s_ξ onto $\mathcal{G}$ has the property that it approximates every finite part of every geodesic in $C(\mathcal{G})$ with arbitrary accuracy infinitely many times. The theorem was obtained by Tukia in [79] and independently by the author in [67].

Theorem 5 (Generalised Myrberg Theorem).
For μ-almost every $\xi \in L(G)$ we have that ξ is a Myrberg limit point.

We remark that, based on Sullivan's results in [73], it was shown in [67] that for an arbitrary Kleinian groups G the set of Myrberg limit points is of full μ-measure if and only if the geodesic flow on $\mathcal{G}$ is ergodic with respect to the Liouville-Patterson measure.

For the remainder of this section we shall now assume that G has parabolic elements. We remark that for the cases in which there are no parabolic elements, analogous results can be obtained and the proofs are far less involved.

The following theorem represents a generalisation of a classical theorem of Khintchine in metrical Diophantine approximations ([36]). An important point here is that the theorem, if combined with the measure formula, gives some useful insight into the fluctuation of μ, as will be demonstrated in the sequel.

Theorem 6 (Generalised Khintchine Theorem). ([72])
For μ-almost all $\xi \in L(G)$, we have

$$\limsup_{t\to\infty}(\Delta(\xi_t)/\log t) = 1/\left(2\delta(G) - k_{max}\right) \quad .$$

An immediate implication is that for μ-almost all $\xi \in L(G)$ we have that $\lim_{t\to\infty}\Delta(\xi_t)/t = 0$ (Sullivan's ergodic law [73]).

Also, we have the following description in terms of asymptotic frequencies with which recurrent geodesics on $\mathcal{G}$ enter the cusp regions. For this let $\mathcal{C}_p(\tau)$ refer to the region inside the cuspidal region of $\mathcal{G}$ associated with the parabolic fixed point p, which has hyperbolic distance τ to the projection of 0 onto $\mathcal{G}$. Let $N_{p,\tau}(\xi, t)$ denote the number of geodesic connected components of the intersection of $\mathcal{C}_p(\tau)$ with the projection onto $\mathcal{G}$ of the geodesic segment between 0 and ξ_t. Generalising a result by Nakada [46] for imaginary quadratic fields which is closely related to the Doeblin-Lenstra conjecture proven in [10], the following result was obtained in [64].

Theorem 7 (Generalised Nakada Theorem).
For every parabolic fixed point p and for all $\tau > 0$ sufficiently large, we have for μ-almost all $\xi \in L(G)$,

$$\lim_{t\to\infty}(N_{p,\tau}(\xi, t)/t) \asymp \exp(-\tau(2\delta(G) - k(p))).$$

Finally, we state the following immediate consequence of a combination of the measure formula and the generalised Khintchine theorem. In here $\mathcal{I}_\mu$ refers to the *information dimension*, $\mathcal{R}_\mu(q)$ to the *q-th generalised Renyi dimension*, and $\mathcal{L}_\mu(q+1)$ to the *q-th logarithmic index* (see [21] for the definitions).

Theorem 8. [68]
For $q \neq 0$, we have

$$\mathcal{I}_\mu = \mathcal{R}_\mu(q) = \delta(G) \text{ and } \mathcal{L}_\mu(q+1) = q\delta(G).$$

*Hence in particular, μ is a $\delta(G)$-*regular measure, *which means that for μ-almost all $\xi \in L(G)$ we have*

$$\liminf_{t\to\infty}(\log \mu(b(\xi_t)))/(-t) = \limsup_{t\to\infty}(\log \mu(b(\xi_t)))/(-t) = \delta(G).$$

2.2 Coarse-structure fractal analysis

By a result of Beardon [4] we have that if G has parabolic elements, then $\delta(G) > k_{max}/2$. On the other hand if G does not have parabolic elements of rank n, then an immediate consequence of the theorem of Beardon and Maskit is that every geodesic in $C(\mathcal{G})$ is contained in some fixed hyperbolic neighbourhood of the boundary of $C(\mathcal{G})$. Using this observation, one easily verifies that $L(G)$ is a porous set and hence, by a standard result in fractal geometry, the box-counting dimension $\dim_B(L(G))$ is strictly less than n. If there are parabolic elements of rank n then, although then $L(G)$ is not porous, it was shown by Sullivan [73] and Tukia [78] that we still have $\dim_H(L(G)) < n$. More generally, we have the following theorem which shows that for a geometrically finite group Hausdorff-, packing- ($\dim_P$), and box-counting dimension all agree and are equal to $\delta(G)$. Clearly, an immediate consequence of the theorem is that $\dim_B(L(G)) < n$. The theorem was obtained in [71] and later extended in [8] to analytically finite groups (that are groups which have the 'Ahlfors property', meaning that $\Omega(G)/G$ is a finite union of cofinite Riemann surfaces).

Theorem 9 (Box-counting dimension Theorem).
We have that

$$\dim_H(L(G)) = \dim_P(L(G)) = \dim_B(L(G)) = \delta(G).$$

Let us look closer at the behaviour of the s-dimensional Hausdorff measure $\mathcal{H}_s$ and packing measure $\mathcal{P}_s$ of the limit set, and also how these relate to the Patterson measure μ. The following table gives the complete picture for geometrically finite Kleinian groups acting on hyperbolic 3-space. The table is an immediate consequence of the measure formula and the generalised Khintchine theorem, using the well-known 'generalised mass distribution principle' ([23, 44]). We remark that similar tables can be produced for geometrically finite Kleinian groups in higher dimensions (cf. [68, 75]).

CUSPS vs. $\delta = \delta(G)$	$0 < \delta(G) < 1$	$\delta(G) = 1$	$1 < \delta(G) < 2$
no cusps	$\mu \asymp \mathcal{H}_\delta \asymp \mathcal{P}_\delta$	$\mu \asymp \mathcal{H}_1 \asymp \mathcal{P}_1$	$\mu \asymp \mathcal{H}_\delta \asymp \mathcal{P}_\delta$
$k_{max} = 1$	$\mu \asymp \mathcal{P}_\delta, \mathcal{H}_\delta = 0$	$\mu \asymp \mathcal{H}_1 \asymp \mathcal{P}_1$	$\mu \asymp \mathcal{H}_\delta, \mathcal{P}_\delta = \infty$
$k_{min} = 2$	$\nexists$	$\nexists$	$\mu \asymp \mathcal{P}_\delta, \mathcal{H}_\delta = 0$
$k_{min} = 1, k_{max} = 2$	$\nexists$	$\nexists$	$\mathcal{H}_\delta = 0, \mathcal{P}_\delta = \infty$

Of special interest is the case described in the final row of the table, where $1 < \delta(G) < 2$ and where there are parabolic elements of rank 1 as well as of rank 2. Here the Patterson measure does not have an immediate geometric interpretation, and it is still an open problem if in this case the Patterson measure can be described in terms of generalised geometric measures. We expect that it should be helpful to employ techniques developed by Makarov [39] to answer similar questions for harmonic measures. However, by combining the measure formula and the generalised Khintchine theorem, we at least can obtain the following approximations to the geometric nature of the Patterson measure in this case. Namely, for the Hausdorff measure $\mathcal{H}_{\phi_\theta}$ with respect to the gauge function ϕ_θ, given by $\phi_\theta(r) := r^{\delta(G)} \exp\left(\left(\frac{1}{2}+\theta\right)\frac{2-\delta(G)}{\delta(G)-1}\log\log\frac{1}{r}\right)$, we find that ([68])

$$\mathcal{H}_{\phi_{-\theta}}(L(G)) \ll 1 \ll \mathcal{H}_{\phi_\theta}(L(G)) \text{ for all } \theta > 0.$$

We remark that by the same means, similar estimates can be given for gauge functions in the context of the generalised packing measure of the limit set. Finally, let us remark that our analysis strongly supports the following.

Conjecture.
If G has parabolic elements of rank 1 *and* 2*, then* μ *and* $\mathcal{H}_\phi$ *are comparable, for* ϕ *given by*

$$\phi(r) = r^{\delta(G)} \exp\left(\frac{2-\delta(G)}{2(\delta(G)-1)}\left(\log\log\frac{1}{r} + \log\log\log\log\frac{1}{r}\right)\right).$$

2.3 Fine-structure fractal analysis

In this section we give some results derived from generalisations of methods in metrical Diophantine approximations, the theory of large deviations, ergodic theory and multifractal analysis. In particular these results contribute to a finer fractal analysis of limit sets of geometrically finite Kleinian groups, and hence of the geodesic dynamics on geometrically finite hyperbolic manifolds. Throughout this section we shall assume that G is a geometrically finite Kleinian group with parabolic elements.

Coarse multifractal analysis.

Coarse multifractal analysis studies global irregularities of the distribution of $\mu(b(\xi_t))$ for large values of t. In order to perform such investigations in the context of Kleinian groups, it is helpful first to consider σ-Jarník limit sets $L^\sigma(G)$. These sets represent a canonical generalisation of the sets of well-approximable irrational numbers (see [31]).
For $0 < \sigma < 1$, let

$$L^\sigma(G) := \{\xi \in L(G) : \limsup_{t\to\infty} (\Delta(\xi_t)/t) \geq \sigma\}.$$

The following theorem was obtained by Hill and Velani in [30], and independently by the author in [65, 69]. For ease of exposition, we have here restricted the statement once more to the case of Kleinian groups acting on hyperbolic 3-space.

Theorem 10 (Generalised Jarník-Besicovitch Theorem).

(1) For $\delta(G) \leq k_{max}$, we have $\dim_H(L^\sigma(G)) = \delta(G)(1-\sigma)$.

(2) For $\delta(G) > k_{max}$, we have

$$\dim_H(L^\sigma(G)) = \begin{cases} \delta(G)(1-\sigma) & \text{for } \delta(G)(1-\sigma) \leq 1 \\ \delta(G)\left(1 - \frac{\sigma}{1+\sigma}\frac{2\delta(G)-1}{\delta(G)}\right) & \text{for } \delta(G)(1-\sigma) \geq 1. \end{cases}$$

Hence, for $\delta(G) \leq k_{max}$ the theorem shows that $\dim_H(L^\sigma(G))$ is a linear function in σ. Whereas in the second case, where $\delta(G) > 1$ and where G necessarily has parabolic elements of rank 1, this function is partially non-linear. In particular, in this case there exists a unique point σ^* at which the derivative of this function is not continuous (note, for Kleinian groups in higher dimension there can be more than one such point). The significance of σ^* is that $\dim_H(L^{\sigma^*}(G)) = 1$.

Refinements of the above results on Jarník limit sets then give rise to the following coarse multifractal analysis of the Patterson measure μ. Consider the following level sets

$$k_\mu(\theta) := \dim_H\{\xi \in L(G) : \liminf_{r\to 0} \log \mu(b(\xi_t))/(-t) \leq \theta\},$$

$$l_\mu(\theta) := \dim_H\{\xi \in L(G) : \limsup_{r\to 0} \log \mu(b(\xi_t))/(-t) \geq \theta\},$$

and their associated multifractal spectra

$$\{k_\mu(\theta) : \theta \in (2\delta(G) - k_{max}, \delta(G))\} \text{ the } \textit{upper liminf-spectrum},$$

$$\{l_\mu(\theta) : \theta \in (\delta(G), 2\delta(G) - k_{min})\} \text{ the } \textit{lower limsup-spectrum.}$$

As shown in [69], for $\mathcal{H}_{\delta(G)}(L(G)) \neq 0$ the upper liminf-spectrum is trivial, whereas for $\mathcal{P}_{\delta(G)}(L(G)) \neq \infty$ the lower limsup-spectrum turns out to be trivial. More interestingly, for the remaining cases the following results were obtained in [69].

Theorem 11 (Coarse multifractal spectra).

(1) $\mathcal{H}_{\delta(G)}(L(G)) = 0$ *if and only if we have, for all* θ *in the domain of the upper liminf-spectrum,*

$$k_\mu(\theta) = \delta(G) - (\theta - \delta(G)) \frac{\delta(G)}{\delta(G) - k_{max}}.$$

(2) $\mathcal{P}_{\delta(G)}(L(G)) = \infty$ *if and only if we have, for all* θ *in the domain of the lower limsup-spectrum,*

$$l_\mu(\theta) = \begin{cases} \delta(G) - (\theta - \delta(G)) \frac{2\delta(G)-1}{\theta-1} & \textit{for} \quad \theta \leq \theta^* \\ \delta(G) - (\theta - \delta(G)) \frac{\delta(G)}{\delta(G)-1} & \textit{for} \quad \theta \geq \theta^*, \end{cases}$$

where we have set $\theta^* := (2\delta(G) - 1) - \frac{\delta(G)-1}{\delta(G)}$.

The theorem shows that if the packing measure of $L(G)$ is infinite, then the lower limsup-spectrum is partially non-linear, and the linear and non-linear parts intersect precisely at θ^*. Again, the significance of θ^* is that $l_\mu(\theta^*) = 1$. Furthermore, the derivative of the lower limsup-spectrum is not continuous at θ^*. Therefore, θ^* can be interpreted as a point at which a *coarse multifractal phase transition* occurs. Note that our analysis in particular shows that such a phase transition occurs if and only if $\mathcal{P}_{\delta(G)}(L(G)) = \infty$.

Fine multifractal analysis.

Fine multifractal analysis studies fractal entities for level sets of certain real-valued functions, where the level sets are defined by means of strict limit processes. We remark that fine multifractal analysis imbeds into the ergodic-theoretical discipline of thermodynamical formalism. (For elementary introductions to the relevant dimension theory we refer to [24, 56]).

In this section we consider the special case in which G is an essentially free Kleinian group. This means that G is basically a free geometrically finite Kleinian group, except that there might be group-relations arising from stabilisers of parabolic fixed points of G. In order to describe the results for this type of Kleinian group, we recall the following lexicographical coding of $L(G)$. Let $F \subset \mathbb{D}$ be a fundamental polyhedron of G. Then the hyperbolic ray s_ξ from the origin to some arbitrary $\xi \in L(G)$ passes in succession through fundamental domains $F, g_{\xi,1}(F), g_{\xi,2}(F), g_{\xi,3}(F), \ldots$, where the $g_{\xi,n}^{-1} g_{\xi,n+1}$ are elements of the set of generators of G. Here we assume for ease of exposition that s_ξ passes exclusively through n-dimensional and hence through no lower-dimensional faces of F. For $\alpha \in \mathbb{R}$, we then define the level sets

$$\mathcal{F}_\alpha(G) := \left\{ \xi \in L(G) : \lim_{n\to\infty} d(0, g_{\xi,n}(0))/n = \alpha \right\}.$$

In the investigations of the Hausdorff dimensions of these level sets the following pressure function P turns out to be crucial. In here $|g|$ refers to the word length of g in G.

$$P : \mathbb{R} \to \mathbb{R} \text{ given by } P(\alpha) := \lim_{n\to\infty} \frac{1}{n} \log \sum_{\substack{g \in G \\ |g|=n}} \exp(-\alpha d(0, g(0))).$$

The following result was obtained in [35].

Theorem 12 (Fine multifractal spectra).
There exists $\alpha_-, \alpha_+ \in \mathbb{R}$ such that $\mathcal{F}_\alpha(G) \neq \emptyset$ if and only if $\alpha \in [\alpha_-, \alpha_+]$, and such that for each $\alpha \in (\alpha_-, \alpha_+)$, we have

$$\dim_H(\mathcal{F}_\alpha(G)) = \alpha^{-1} \inf_{\beta \in \mathbb{R}} \{\alpha\beta + P(\beta)\}. \qquad (*)$$

(1) For G convex cocompact we have $(*)$ holds for all $\alpha \in [\alpha_-, \alpha_+]$, and P is real analytic everywhere. Furthermore, on (α_-, α_+) we have for $(P')^{-1}$, the inverse of the derivative of P,

$$\dim_H(\mathcal{F}_\alpha(G)) = (P')^{-1}(\alpha) - P(\alpha)/\alpha.$$

(2) If G has parabolic elements, then P is real analytic on $(-\infty, \delta(G))$ and equal to 0 otherwise. Additionally, the following holds.

(i) If $\delta(G) \leq (k_{max}+1)/2$, then P is differentiable everywhere.

(ii) If $\delta(G) > (k_{max}+1)/2$, then there exists a fine multifractal phase transition, *meaning that the right derivative of P at $\delta(G)$ vanishes, whereas the left derivative of P at $\delta(G)$ is strictly negative.*

We remark that the two types of phase transitions which we discussed, namely the coarse multifractal and the fine multifractal phase transition, seem to be two completely unrelated phenomena which can be detected within the limit set of a Kleinian group. As our analysis clearly shows, a Kleinian group permits either both, or none, or exactly one of these two types, and each of these possibilities can actually occur. Further note that the coarse multifractal analysis is of pure geometric nature, whereas the fine multifractal analysis results from mixing the algebraic appearance of G (here in form of the use of the word length) with its geometric realisation.

3. Finitely generated hyperbolic manifolds

In this section we give a brief report on some of the results obtained in the fractal analysis of limit sets of finitely generated, geometrically infinite Kleinian groups G acting on hyperbolic 3-space. Such geometrically infinite groups were first shown to exist over 30 years ago by Greenberg in [27]. The first explicit examples were constructed by Jørgensen in [32]. Subsequently, these groups have attracted a great deal of attention from various different points of view. One of the most important conjectures in this area, which is still undecided but which over the years sparked off a vast amount of fruitful and stimulating research, is the following. Here λ_2 refers to the 2-dimensional Lebesgue measure in $\mathbb{S}^2$.

The Ahlfors Conjecture.
For every finitely generated Kleinian group G we have

$$\lambda_2(L(G)) = 0.$$

For geometrically finite groups this conjecture clearly holds, since then, as we already saw, we have that $\dim_B(L(G)) < 2$ which in particular implies that $\lambda_2(L(G)) = 0$. In the geometrically infinite case the following result was obtained in [8].

Theorem 13 (Theorem of Bishop and Jones (II)).
For every finitely generated, geometrically infinite Kleinian group G we have

$$\dim_H(L(G)) = 2.$$

We remark that in the light of the Ahlfors conjecture it seems worthwhile to comment on the strategy of the proof of this result. Namely, the essential part of the proof of Bishop and Jones is to show that

$$\delta(G) < 2 \Longrightarrow \lambda_2(L(G)) > 0.$$

(Clearly, this gives the theorem, since then either $\delta(G) = 2$ which implies that $\dim_H(L_{ur}(G)) = 2$ and hence $\dim_H(L(G)) = 2$, or $\delta(G) < 2$ which gives $\lambda_2(L(G)) > 0$ and hence in particular $\dim_H(L(G)) = 2$). In other words, the proof shows that a finitely generated, geometrically infinite Kleinian group G with $\dim_H(L_r(G)) < 2$ is necessarily a counterexample for the Ahlfors conjecture. Of course it is a problem to show that a group of this type does exist. For instance, it is known that such a counterexample cannot be a topologically tame Kleinian group of bounded type ([15, 14, 74]), which means that the associated hyperbolic manifold has injectivity radius uniformly bounded from below and is homeomorphic to the interior of a compact 3-manifold with boundary. In fact for the Patterson measure μ of such tame groups we have the following result. (Note that in this case the Patterson measure is not necessarily unique.)

Theorem 14 (Theorem of Bishop and Jones (III)). ([9])
Let G be a geometrically infinite, topologically tame Kleinian group of bounded type, and let $\mathcal{H}_\varphi$ refer to the Hausdorff measure with respect to the gauge function φ given by

$$\varphi(r) = r^2 \exp\left(\frac{1}{2}\left(\log\log\frac{1}{r} + \log\log\log\log\frac{1}{r}\right)\right).$$

Then the following holds:

$$\mu \asymp \mathcal{H}_\varphi \text{ and } \mu(L_r(G)) = 0.$$

4. Infinitely generated hyperbolic manifolds

Finally, we give a brief report on some of the results derived in the fractal analysis of geodesic dynamics on infinitely generated hyperbolic manifolds. It is evidently clear that the class of all infinitely generated groups as such is by far too large. Hence, one of the problems is to elaborate some structure inside this huge class of groups which makes it feasible to study these groups systematically.

As a first subclass we mention groups which can be exhausted by an increasing chain of subgroups. More precisely, here we consider a Kleinian group G such that $G = \bigcup_k G_k$, for some sequence of Kleinian subgroups $G_1 \subset G_2 \subset \cdots \subset G_k \subset \cdots$. In this situation we clearly have that $\bigcup_k L_{ur}(G_k) \subset L_{ur}(G)$. On the other hand, if $\xi \in L_{ur}(G)$ then there exists an infinite path c_ξ in the Cayley graph of G such that s_ξ is fully contained in some fixed hyperbolic neighbourhood of c_ξ, and such that the hyperbolic lengths of the geodesic segments of c_ξ are uniformly bounded from above. Therefore $L_{ur}(G) = \bigcup_k L_{ur}(G_k)$, and hence it follows, by monotonicity of Hausdorff dimension (see e.g. [23]),

$$\dim_H(L_{ur}(G)) = \dim_H(\bigcup_k L_{ur}(G_k)) = \sup_k \dim_H(L_{ur}(G_k)).$$

Combining this observation with the theorem of Bishop and Jones (I), we then immediately obtain the following result. We remark that this result was conjectured by Patterson in [51, 53, 54] and subsequently proven by Sullivan in [73], using the Patterson measure.

Theorem 15.
Let G be a Kleinian group such that $G = \bigcup_k G_k$, for some sequence of Kleinian subgroups $G_1 \subset G_2 \subset \cdots \subset G_k \subset \cdots$. We then have

$$\delta(G) = \sup_k \delta(G_k).$$

We remark that by similar means one shows that the Hausdorff dimension of $L_{ur}(G)$ is lower semi-continuous with respect to algebraic convergence (see [41]).

Another interesting class of infinitely generated Kleinian groups is provided by intermediate coverings of geometrically finite hyperbolic manifolds. Here one considers normal subgroups N of a geometrically finite Kleinian group G. As one easily verifies, in this situation we have that $L(G) = L(N)$. Hence, this class of Kleinian groups gives the opportunity to study geodesic cores which are both, the geodesic core of an infinitely generated manifold (associated with N) as well as the geodesic core of a geometrically finite manifold (associated with G).

The following result was obtained in [25].

Theorem 16.
For a normal subgroup N of a geometrically finite Kleinian group G we have

$$\dim_H(L_{ur}(N)) \geq \dim_H(L_{ur}(G))/2.$$

Recall that for a $\delta(G)$-divergence type group G the geodesic flow on $\mathcal{G}$ is ergodic with respect to the Liouville-Patterson measure. As mentioned before, geometrically finite groups are always of $\delta(G)$-divergence type, and hence the following result of Rees in particular shows under which circumstances ergodicity of the geodesic flow is preserved if one passes to an intermediate covering of a geometrically finite hyperbolic 2-manifold (see also the discussion in [55, 62]).

Theorem 17 (Theorem of Rees). ([58, 59])
Let N be a normal subgroup of a geometrically finite Fuchsian group G such that G/N is isomorphic to $\mathbb{Z}^m$, for $m \in \mathbb{N}$. Then $\delta(N) = \delta(G)$, and

(i) if G has no parabolic elements, then N is of $\delta(N)$-divergence type if and only if $m = 1$ or 2;

(ii) *if G has parabolic elements, then N is of $\delta(N)$-divergence type if and only if $m = 1$.*

This theorem seems to suggest that polynomial growth of the factor group might be a necessary and sufficient condition for having equality of the two exponents of convergence. The following theorem shows that the right condition is in fact 'amenabilty' (see [12, 83, 81]), at least for certain convex cocompact Kleinian groups. Here the condition $\delta(G) > n/2$ occurs since the proof of Brooks in [13] uses spectral theory of the Laplacian, and it seems extremely likely that this condition can be removed (see [80]).

Theorem 18 (Theorem of Brooks). ([13])
Let N be a normal subgroup of a convex cocompact Kleinian group G such that $\delta(G) > n/2$. Then $\delta(N) = \delta(G)$ if and only if G/N is amenable.

Yet another interesting class of geometrically infinite Kleinian groups is the following. A Kleinian group G is called a discrepancy group if and only if $\dim_H(L_{ur}(G)) < \dim_H(L(G))$. This type of groups seems to have been studied for the first time by Patterson in [53] (see also [57]). Note that by the previous theorem we immediately have that every normal subgroup with non-amenable factor in a convex cocompact Kleinian group (with Poincaré exponent greater than $n/2$) is a discrepancy group. In [25] we gave further examples of discrepancy groups, and started with a finer fractal analysis of the limit sets of these groups by considering the following set of τ-weakly recurrent limit points

$$L_r^{(\tau)}(G) := \{\xi : s_\xi \cap b\,(g(0), \tau\, d(0, g(0))) \neq \emptyset \text{ for infinitely many } g \in G\}.$$

Theorem 19. ([25])
For a discrepancy group G we have for all $0 < \tau < \frac{\dim_H(L(G)) - \delta(G)}{\dim_H(L(G))}$,

$$\delta(G) \le \dim_H(L_r^{(\tau)}(G)) < \dim_H(L(G)),$$

and hence in particular,

$$\dim_H(L(G) \setminus L_r^{(\tau)}(G)) = \dim_H(L(G)).$$

Finally, we mention a class of Kleinian groups which we have called geometrically tight. This type of groups has recently been studied by Matsuzaki in [42]. A Kleinian group G is called geometrically tight if there exists $\rho > 0$ and a countable exceptional set $E \subset L(G)$ (where E is either empty, or represents endings of $\mathcal{G}$ which are either 'cusp-like' or contained in a fixed neighourhood of some geodesic in $\mathcal{G}$) such that for every $\xi \in L(G) \setminus E$, the projection of s_ξ onto $\mathcal{G}$ 'returns infinitely many times' to the hyperbolic ρ-neighbourhood of the boundary of $C(\mathcal{G})$. Clearly, if the set E is empty then every geodesic in $C(\mathcal{G})$ is contained in the ρ-neighbourhood of the boundary of $C(\mathcal{G})$. In this special situation one can then argue as for geometrically finite groups without parabolic points of rank n, which gives that $L(G)$ is a porous set, and hence $\dim_B(L(G)) < n$. For E non-empty, a straight-forward adaptation of Tukia's techniques in [78] leads to the following. Here it would be interesting to show that Hausdorff dimension can be replaced in general by box-counting dimension, which can obviously be done for $E = \emptyset$.

Theorem 20 (Theorem of Matsuzaki). ([42])
For a geometrically tight Kleinian group G we have

$$\dim_H(L(G)) < n.$$

Bibliography

[1] W. Abikoff, 'Kleinian groups - geometrically finite and geometrically perverse', *Contemp. Math.* **74** (1988) 1–50.

[2] L.V. Ahlfors, *Möbius transformations in several dimensions*, University of Minnesota, Lecture Notes, 1981.

[3] P. Albuquerque, 'Patterson-Sullivan measures in higher rank symmetric spaces', *C.R. Acad. Sci. Paris Sér I. Math.* **324** (1997) 427–432.

[4] A.F. Beardon, 'Inequalities for certain Fuchsian groups', *Acta Math.* **127** (1971) 221–258.

[5] A.F. Beardon, *The geometry of discrete groups*, Springer Verlag, New York, 1983.

[6] A.F. Beardon, B. Maskit, 'Limit points of Kleinian groups and finite sided fundamental polyhedra', *Acta Math.* **132** (1974) 1–12.

[7] C.J. Bishop, 'On a theorem of Beardon and Maskit', *Ann. Acad. Sci. Fenn.* **21** (1996) 383–388.

[8] C.J. Bishop, P.W. Jones, 'Hausdorff dimension and Kleinian groups', *Acta Math.* **56** (1997) 1–39.

[9] C.J. Bishop, P.W. Jones, 'The law of iterated logarithm for Kleinian groups', *Contemp. Math.* **211** (1997) 17–50.

[10] W. Bosma, H. Jager, F. Wiedijk, 'Some metrical observations on the approximation by continued fractions', *Indag. Math.* **45** (1983) 281–299.

[11] B.H. Bowditch, 'Geometrical finiteness for hyperbolic groups', *J. Functional Analysis* **113** (1993) 245–317.

[12] R. Brooks, 'The fundamental group and the spectrum of the Laplacian', *Comm. Math. Helv.* **56** (1981) 581–598.

[13] R. Brooks, 'The bottom of the spectrum of a Riemannian covering', *J. Reine Angew. Math.* **357** (1985) 101–114.

[14] M. Burger, R.D. Canary, 'A lower bound on λ_0 for geometrically finite hyperbolic n-manifolds', *J. Reine Angew. Math.* **454** (1994) 37-57.

[15] R.D. Canary, 'On the Laplacian and the geometry of hyperbolic three-manifolds', *J. Diff. Geom.* **36** (1992) 349-367.

[16] M. Coornaert, 'Sur les groupes proprement discontinus d'isométries des espaces hyperboliques au sens de Gromov', *Dissertation, Univ. Louis Pasteur*, Strasbourg, 1990.

[17] M. Coornaert, 'Measures de Patterson-Sullivan sur le bord d'un espace hyperboliques au sens de Gromov', *Pacific J. Math.* **159** (1993) 241–270.

[18] K. Corlette, 'Hausdorff dimensions of limit sets I', *Invent. Math.* **102** (1990) 521–542.

[19] K. Corlette, A. Iozzi, 'Limit sets of discrete groups of isometries of exotic hyperbolic spaces', *Trans. Amer. Math. Soc.* **351**, 4 (1999) 1507–1530.

[20] H.S.M. Coxeter, *Introduction into geometry*, Wiley Classic Library, J. Wiley, (sec. ed.) 1989.

[21] D.D. Cutler, 'A review of the theory and estimation of fractal dimension', in *Nonlinear Time Series and Chaos*, Vol. I: Dimension Estimation and Models, ed. H. Tong, World Scientific, Singapore, 1993.

[22] S.G. Dani, 'Bounded orbits of flows on homogeneous spaces', *Comment. Math. Helv.* **61**, no. 4 (1986) page 636–660.

[23] K.J. Falconer, *Fractal geometry, mathematical foundations and applications*, J. Wiley, 1990.

[24] K.J. Falconer, *Techniques in fractal geometry*, J. Wiley, 1997.

[25] K.H. Falk, B.O. Stratmann, 'Remarks on Hausdorff dimensions for transient limit sets of Kleinian groups', *Math. Gottingensis* **05** (2003) 1–16 (preprint).

[26] J.L. Fernández, M.V. Melián, 'Bounded geodesics of Riemann surfaces and hyperbolic manifolds', *Trans. Amer. Math. Soc.* **347**, no. 9 (1995) 3533–3549.

[27] L. Greenberg, 'Fundamental polyhedra for Kleinian groups', *Ann. of Maths.* **84** (1966) 433–441.

[28] M. Gromov, 'Hyperbolic groups', in *Essays in group theory*, Math. Sci. Res. Inst. Publ. **8**, eds. S.M. Gersten, Springer (1987) 75-263.

[29] S. Hersonsky, F. Paulin, 'Hausdorff dimension of Diophantine geodesics in negatively curved manifolds', *J. Reine Angew. Math.* **539** (2001) 29–43.

[30] R. Hill, S.L. Velani, 'The Jarník-Besicovitch theorem for geometrically finite Kleinian groups', *Proc. London Math. Soc.* (3) **77** (1998) 524 - 550.

[31] V. Jarník, 'Zur metrischen Theorie der Diophantischen Approximationen', *Prace Math.-fiz* **36**, 2. Heft (1928).

[32] T. Jørgensen, 'Compact 3-manifolds of constant negative curvature fibering over the circle', *Ann. of Maths.* **106** (1977) 61–72.

[33] V. Kaimanovich, 'Invariant measures and measures at infinity', *Amm. Inst. H. Poincaré Phys. Théor.* **53** (1990) 361–393.

[34] M. Kapovich, *Hyperbolic manifolds and discrete group: lectures on Thurstons hyperbolization*, University of Utah Lecture Notes, 1999.

[35] M. Kesseböhmer, B. O. Stratmann, 'A multifractal formalism for growth rates and applications to geometrically finite Kleinian groups', *Math. Gottingensis* **07** (2001) 1–29 (preprint); *Ergod. Th. & Dynam. Sys.* **24** (2004) 141–170 (2003).

[36] A. Y. Khintchine, *Continued fractions*, Univ. of Chicago Press, Chicago and London, 1964.

[37] G. Knieper, 'Closed geodesics and the uniqueness of maximal measure for rank 1 geodesic flows', *Proc. Sympos. Pure Math*, vol. **69**, Amer. Math. Soc., Providence, RI, (2001) 573–590.

[38] S.L. Krushkal, B.N. Apanasov, N.A. Gusevskii, *Kleinian groups and uniformization in examples and problems*, Transl. Math. Monographs **62** AMS, 1986.

[39] N.G. Makarov, 'Probability methods in the theory of conformal mappings', *Leningrad Math. Jour.* **1** (1990) 1–56.

[40] B. Maskit, '*Kleinian groups*, Grundlehren d. Math. Wiss. **287**, Springer-Verlag, Berlin, 1988.

[41] K. Matsuzaki, 'Convergence of the Hausdorff dimension of the limit sets of Kleinian groups', *Contemp. Math.* **256** (1998) 243–254,

[42] K. Matsuzaki, 'Hausdorff dimension of limit sets of infinitely generated Kleinian groups', *Math. Proc. Camb. Phil. Soc* **128** (2000) 123–139.

[43] K. Matsuzaki & M. Taniguchi, *Hyperbolic manifolds and Kleinian groups*, Oxford Math. Monographs, 1998.

[44] P. Mattila, *Geometry of sets and measures in Euclidean spaces. Fractals and rectifiability*, Cambridge University Press, 1995.

[45] D. Mumford, C. Series, D. Wright, *Indra's pearls. The vision of Felix Klein*, Cambridge University Press, New York, 2002.

[46] H. Nakada, 'On metrical theory of diophantine approximation over imaginary quadratic fields', *Acta Arith.* **51** (1988) 131–154.

[47] F. Newberger, 'On the Patterson-Sullivan measure for geometrically finite groups acting on complex hyperbolic space or quaternionic hyperbolic space', preprint (2003).

[48] P.J. Nicholls, *The ergodic theory of discrete groups*, Cambridge University Press, 1989.

[49] J.R. Parker, B.O. Stratmann, 'Kleinian groups with singly cusped parabolic fixed points', *Kodai J. Math.* **24** (2001) 169–206.

[50] S.J. Patterson, 'The limit set of a Fuchsian group', *Acta Math.* **136** (1976) 241–273.

[51] S.J. Patterson, 'The exponent of convergence of Poincaré series', *Monatsh. f. Math.* **82** (1976) 297–315.

[52] S.J. Patterson, 'Diophantine approximation in Fuchsian groups', *Phil. Trans. Roy. Soc. Lond.* **282** (1976) page 527–563.

[53] S.J. Patterson, 'Some examples of Fuchsian groups', *Proc. London Math. Soc.* **(3) 39** (1979) 276–298.

[54] S.J. Patterson, 'Further remarks on the exponent of convergence of Poincaré series', *Tôhoku Math. Journ.* **35** (1983) 357–373.

[55] S.J. Patterson, 'Lectures on measures on limit sets of Kleinian groups' in *Analytical and geometric aspects of hyperbolic space*, editor D.B.A. Epstein, Cambridge University Press, 1987.

[56] Y.B. Pesin, *Dimension theory in dynamical systems. Contemporary views and applications*, Chicago Lect. in Math., Univ. of Chic. Press, Chicago, 1997.

[57] C. Pommerenke, 'On the Green's function of Fuchsian groups', *Ann. Acad. Sci. Fenn. A1 Math.* **2** (1976) 409–427.

[58] M. Rees, 'Checking ergodicity of some geodesic flows with infinite Gibbs measure', *Ergod. Th. & Dynam. Syst.* **1** (1981) 107–133.

[59] M. Rees, 'Divergence type of some subgroups of finitely generated Fuchsian groups', *Ergod. Th. & Dynam. Syst.* **1** (1981) 209–221.

[60] B. Schapira, 'Lemme de l'ombre et non divergence des horospéres d'une variété géométriquement finie', *Ann. Inst. Fourier* **54** (2004) 939–989.

[61] W.M. Schmidt, 'On badly approximable numbers and certain games", *Transactions of the AMS* **123** (1966) 178–199.

[62] B.O. Stratmann, 'Ergodentheoretische Untersuchungen Kleinscher Gruppen', *Master thesis, Univ. Göttingen* (1987) 1–150.

[63] B.O. Stratmann, 'Diophantine approximation in Kleinian groups', *Math. Proc. Camb. Phil. Soc.* **116** (1994) 57–78.

[64] B.O. Stratmann, 'A note on counting cuspidal excursions', *Annal. Acad. Sci. Fenn.* **20** (1995) 359–372.

[65] B.O. Stratmann, 'Fractal dimensions for Jarník limit sets; the semi-classical approach', *Ark. för Mat.* **33** (1995) 385–403.

[66] B.O. Stratmann, 'The Hausdorff dimension of bounded geodesics on geometrically finite manifolds', *Ergod. Th. & Dynam. Sys.* **17** (1997) 227–246.

[67] B.O. Stratmann, 'A remark on Myrberg initial data for Kleinian groups', *Geometriae Dedicata* **65** (1997) 257-0266.

[68] B.O. Stratmann, 'Multiple fractal aspects of conformal measures; a survey', in *Workshop on Fractals and Dynamics'*, eds. M. Denker, S.-M. Heinemann, B.O. Stratmann, *Math. Gottingensis* **05** (1997) 65–71.

[69] B.O. Stratmann, 'Weak singularity spectra of the Patterson measure for geometrically finite Kleinian groups with parabolic elements', *Michigan Math. Jour.* **46** (1999) 573–587.

[70] B.O. Stratmann, 'The exponent of convergence of Kleinian groups; on a theorem of Bishop and Jones', *Math. Gottingensis* **01** (2001) 1–16 (preprint); *Fractal geometry and stochastics III. Proceedings of the 3rd conference, Friedrichroda, Germany, March 17-22, 2003. Basel: Birkhuser. Progress in Probability* **57**, 93–107 (2004)

[71] B.O. Stratmann, M. Urbański, 'The box-counting dimension for geometrically finite Kleinian groups', *Fund. Matem.* **149** (1996) 83–93.

[72] B.O. Stratmann, S. Velani, 'The Patterson measure for geometrically finite groups with parabolic elements, new and old', *Proc. Lond. Math. Soc.* (3) **71** (1995) 197–220.

[73] D. Sullivan, 'The density at infinity of a discrete group of hyperbolic motions', *IHES Publ. Math.* **50** (1979) 171–202.

[74] D. Sullivan, 'Growth of positive harmonic functions and Kleinian group limit sets of zero planar measure', in *Geometry Symposium (Utrecht 1980)*, Lecture Notes in Math., Vol. **894**, Springer, Berlin (1981) 127–144.

[75] D. Sullivan, 'Entropy, Hausdorff measures old and new, and limit sets of geometrically finite Kleinian groups', *Acta Math.* **153** (1984) 259–277.

[76] D. Sullivan, 'On the ergodic theory at infinity of an arbitrary discrete group of hyperbolic motions', *Riemann surfaces and related topics*, Ann. Math. Studies **97** (1981) 465–496.

[77] W.P. Thurston, *The geometry and topology of 3-manifolds*, lecture notes, Princeton, 1979.

[78] P. Tukia, 'The Hausdorff dimension of the limit set of a geometrically finite Kleinian group', *Acta Math.* **152** (1984) 127–140.

[79] P. Tukia, 'The Poincaré series and the conformal measure of conical and Myrberg limit points', *J. Anal. Math.* **62** (1994) 241–259.

[80] N.T. Varopoulos, 'Finitely generated Fuchsian groups', *J. Reine Angew. Math.* **375/376** (1987) 394–405.

[81] S. Wagon, *The Banach-Tarski paradox*, Cambridge University Press, 1986.

[82] C. Yue, 'Mostow rigidity for rank 1 discrete subgroups with ergodic Bowen-Margulis measure', *Invent. Math.* **125** (1996) 75–102.

[83] R.J. Zimmer, '*Ergodic theory and semisimple groups*, Monographs in Math. **81**, Birkhäuser Verlag, Basel, 1984.

A VOLUME FORMULA FOR GENERALISED HYPERBOLIC TETRAHEDRA*

Akira Ushijima
Department of Mathematics,
Faculty of Science,
Kanazawa University,
920-1192 JAPAN†
ushijima@kenroku.kanazawa-u.ac.jp‡

Abstract A generalised hyperbolic tetrahedron is a polyhedron (possibly non-compact) with finite volume in hyperbolic space, obtained from a tetrahedron by the polar truncation at the vertices lying outside the space. In this paper it is proved that a volume formula for ordinary hyperbolic tetrahedra devised by J. Murakami and M. Yano can be applied to such generalised tetrahedra. There are two key tools for the proof; one is the so-called Schläfli's differential formula for hyperbolic polyhedra, and the other is a necessary and sufficient condition for given numbers to be the dihedral angles of a generalised hyperbolic simplex with respect to their dihedral angles.

Keywords: Hyperbolic tetrahedron, Gram matrix, volume formula

2000 Mathematics Subject Classifications: Primary: 52A38; secondary: 51M09.

1. Introduction

Obtaining volume formulae for polyhedra is one of the basic and important problems in geometry. In hyperbolic space this problem has been attacked from the beginning. An *orthoscheme* is a simplex which generalises a right triangle. Since any hyperbolic polyhedron can be decomposed into a finite number of orthoschemes, the first step in solving

*This research was partially supported by the Inamori Foundation.
†Current address: Mathematics Institute, University of Warwick, Coventry CV4 7AL, UK
‡Current e-mail address: ushijima@maths.warwick.ac.uk

the problem is to obtain a volume formula for orthoschemes. In the three-dimensional case, N. I. Lobachevsky [Lo] found in 1839 a volume formula for orthoschemes. Thus the next step is to find a formula for ordinary tetrahedra.

Although the lengths of the six edges of a tetrahedron determine its size and shape, in hyperbolic space it is known that the six dihedral angles also play the same role; indeed, the variables of Lobachevsky's formula mentioned above are dihedral angles. So, following this direction, we obtain in this paper a volume formula for *generalised* hyperbolic tetrahedra in terms of the six dihedral angles (the meaning of the term "generalised" will be explained later). Here we note that, by the fact mentioned in the previous paragraph, we can construct such a formula from that of orthoschemes. But in this case we have to calculate the dihedral angles of orthoschemes from those of tetrahedra. It is emphasized that what we want to obtain is a formula which can be calculated directly from the dihedral angles of a given generalised tetrahedron.

The first answer to this problem was (at least as far as the author is aware) given by W.-Y. Hsiang in 1988. In [Hs] the formula was presented through an integral expression. On the other hand, many volume formulae for hyperbolic polyhedra are presented via the Lobachevsky function or the dilogarithm function (see, for example, [Ve]). The first such presentation was given by Y. Cho and H. Kim in 1999 (see [CK]). Their formula was derived from that of not orthoschemes but ideal tetrahedra, and has the following particular property: due to the method of proof it is not symmetric with respect to the dihedral angles. Thus the next problem is to obtain a symmetrical expression. Such a formula was obtained by J. Murakami and M. Yano in about 2001. They derived their formula in [MY] from the quantum $6j$-symbol, but their proof consists in showing the equivalence between their expression and that of Cho-Kim.

In hyperbolic space it is possible to consider that vertices of a polyhedron lie "outside the space and its sphere at infinity". For each such vertex there is a canonical way to cut off a neighborhood containing the vertex. This operation is called a *truncation* at the vertex, which will be defined more precisely in Section 3. We thus obtain convex polyhedra (possibly non-compact) with finite volume. Such polyhedra appear, for example, as fundamental polyhedra for hyperbolic Coxeter groups, or as building blocks of three-dimensional hyperbolic manifolds with totally geodesic boundary. Since volumes of hyperbolic manifolds are topological invariants, it is meaningful, also for low-dimensional topology, to obtain a volume formula for such polyhedra.

R. Kellerhals showed in 1989 that the formula for orthoschemes can be applied, without any modification, to ones "mildly" truncated at *principal vertices* (see Appendix for the definition and [Ke] for the formula). This result inspires the idea that Murakami-Yano's formula may also be applied to *generalised hyperbolic tetrahedra*, that is, tetrahedra in which each vertex is either finite, ideal or truncated (see Definition 3.1 for precise definition). This is what we will prove in this paper (see Theorem 1.1). Here we note that, although the edges emanating from a vertex at infinity have infinite length, the dihedral angles may be finite. This is another reason why we take not the edge lengths but the dihedral angles as the variables of our volume formula.

The key tool for the proof is the so-called Schläfli's differential formula, a simple description for the volume differential of a polyhedron as a function of the dihedral angles and the lengths of the edges. Since the volume formula is a function of the dihedral angles, to apply Schläfli's differential formula we have to translate the dihedral angles of a generalised tetrahedron to the lengths of its edges. This requirement yields a necessary and sufficient condition on a set of positive numbers to be the dihedral angles of a generalised simplex (see Theorem 3.2).

Volume formula for generalised hyperbolic tetrahedra Let $T = T(A, B, C, D, E, F)$ be a generalised tetrahedron in the three-dimensional hyperbolic space $\mathbb{H}^3$ whose dihedral angles are A, B, C, D, E, F. Here the configuration of the dihedral angles are as follows (see also Figure 1): three edges corresponding to A, B and C arise from a vertex, and the angle D (resp. E, F) is put on the edge opposite to that of A (resp. B, C). Let G be the *Gram matrix* of T defined as follows:

$$G := \begin{pmatrix} 1 & -\cos A & -\cos B & -\cos F \\ -\cos A & 1 & -\cos C & -\cos E \\ -\cos B & -\cos C & 1 & -\cos D \\ -\cos F & -\cos E & -\cos D & 1 \end{pmatrix}.$$

Let $a := \exp(\sqrt{-1}A), b := \exp(\sqrt{-1}B), \dots, f := \exp(\sqrt{-1}F)$, and let $U(z, T)$ be the complex-valued function defined as

$$\begin{aligned} U(z,T) \;:=\; & \frac{1}{2}\{\mathrm{Li}_2(z) + \mathrm{Li}_2(abdez) + \mathrm{Li}_2(acdfz) + \mathrm{Li}_2(bcefz) \\ & - \mathrm{Li}_2(-abcz) - \mathrm{Li}_2(-aefz) - \mathrm{Li}_2(-bdfz) - \mathrm{Li}_2(-cdez)\}, \end{aligned}$$

where $\mathrm{Li}_2(z)$ is the dilogarithm function defined by the analytic continuation of the integral

$$\mathrm{Li}_2(x) := -\int_0^x \frac{\log(1-t)}{t}dt \quad \text{for a positive real number } x.$$

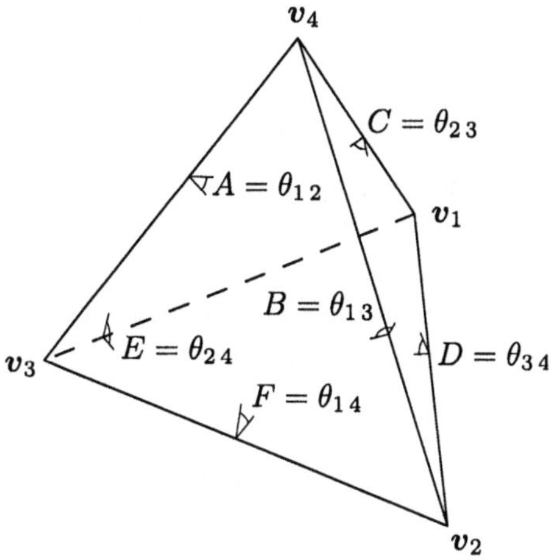

Figure 1. The dihedral angles of T

We denote by z_1 and z_2 the two complex numbers defined as follows:

$$z_1 := -2\frac{\sin A \sin D + \sin B \sin E + \sin C \sin F - \sqrt{\det G}}{ad + be + cf + abf + ace + bcd + def + abcdef},$$
$$z_2 := -2\frac{\sin A \sin D + \sin B \sin E + \sin C \sin F + \sqrt{\det G}}{ad + be + cf + abf + ace + bcd + def + abcdef}.$$

Theorem 1.1. *The volume* Vol(T) *of a generalised tetrahedron T is given as*

$$\mathrm{Vol}(T) = \frac{1}{2}\Im\left(U(z_1, T) - U(z_2, T)\right), \tag{1.1}$$

where $\Im$ means the imaginary part.

Acknowledgments

The author would like to give his thanks to Professor Jun Murakami and Masakazu Yano for useful discussions about their results. The author also would like to give his sincere gratitude to the referee for his careful reading and advices.

2. Preliminaries

In this section we review several well-known facts about hyperbolic geometry. See, for example, [Us] for more precise explanation and the proofs of the propositions.

The $(n+1)$-*dimensional Lorentzian space* $\mathbb{E}^{1,n}$ is the real vector space $\mathbb{R}^{n+1}$ of dimension $n+1$ with the *Lorentzian inner product*

$\langle \mathbf{x}, \mathbf{y} \rangle := -x_0y_0 + x_1y_1 + \cdots + x_ny_n$, where $\mathbf{x} = (x_0, x_1, \ldots, x_n)$ and $\mathbf{y} = (y_0, y_1, \ldots, y_n)$. Let $H_T := \{\mathbf{x} \in \mathbb{E}^{1,n} \mid \langle \mathbf{x}, \mathbf{x} \rangle = -1\}$ be the (standard) hyperboloid of two sheets, and let $H_T^+ := \{\mathbf{x} \in \mathbb{E}^{1,n} \mid \langle \mathbf{x}, \mathbf{x} \rangle = -1 \text{ and } x_0 > 0\}$ be its upper sheet. The restriction of the quadratic form induced by $\langle \cdot, \cdot \rangle$ on $\mathbb{E}^{1,n}$ to the tangent spaces of H_T^+ is positive definite and gives a Riemannian metric on H_T^+. The space obtained from H_T^+ equipped with this metric is called the *hyperboloid model* of the n-dimensional hyperbolic space, and we denote it by $\mathbb{H}^n$. The hyperbolic distance, say d, between two points $\mathbf{x}$ and $\mathbf{y}$ can be measured by the formula

$$\langle \mathbf{x}, \mathbf{y} \rangle = -\cosh d\,. \tag{2.1}$$

Let $L := \{\mathbf{x} \in \mathbb{E}^{1,n} \mid \langle \mathbf{x}, \mathbf{x} \rangle = 0\}$ be the (standard) cone and let $L^+ := \{\mathbf{x} \in \mathbb{E}^{1,n} \mid \langle \mathbf{x}, \mathbf{x} \rangle = 0 \text{ and } x_0 > 0\}$ be its upper half. Then a ray in L^+ started from the origin $\mathbf{o}$ corresponds to a point in the ideal boundary of $\mathbb{H}^n$. The set of such rays forms the *sphere at infinity*, and we denote it by $\mathbf{S}_\infty^{n-1}$. Then each ray in L^+ becomes a point at infinity of $\mathbb{H}^n$.

Let us denote by $\mathcal{P}$ the radial projection from $\mathbb{E}^{1,n} - \{\mathbf{x} \in \mathbb{E}^{1,n} \mid x_0 = 0\}$ to the affine hyperplane $\mathbb{P}_1^n := \{\mathbf{x} \in \mathbb{E}^{1,n} \mid x_0 = 1\}$ along the ray from the origin $\mathbf{o}$. The projection $\mathcal{P}$ induces a homeomorphism on $\mathbb{H}^n$ to the n-dimensional open unit ball $\mathbf{B}^n$ in $\mathbb{P}_1^n$ centered at the origin $(1, 0, 0, \ldots, 0)$ of $\mathbb{P}_1^n$, which is called the *projective model* of $\mathbb{H}^n$. The hyperplane $\mathbb{P}_1^n$ contains not only $\mathbf{B}^n$ and its set theoretic boundary $\partial \mathbf{B}^n$ in $\mathbb{P}_1^n$, which is canonically identified with $\mathbf{S}_\infty^{n-1}$, but also the outside of the compactified projective model $\overline{\mathbf{B}^n} := \mathbf{B}^n \sqcup \partial \mathbf{B}^n \approx \mathbb{H}^n \sqcup \mathbf{S}_\infty^{n-1}$. So $\mathcal{P}$ can be naturally extended to the mapping from $\mathbb{E}^{1,n} - \{\mathbf{o}\}$ to the n-dimensional real projective space $\mathbb{P}^n := \mathbb{P}_1^n \sqcup \mathbb{P}_\infty^n$, where $\mathbb{P}_\infty^n$ is the set of lines in the affine hyperplane $\{\mathbf{x} \in \mathbb{E}^{1,n} \mid x_0 = 0\}$ through $\mathbf{o}$. We denote by $\operatorname{Ext} \overline{\mathbf{B}^n}$ the exterior of $\overline{\mathbf{B}^n}$ in $\mathbb{P}^n$.

The (standard) *hyperboloid of one sheet* H_S is defined as $H_S := \{\mathbf{x} \in \mathbb{E}^{1,n} \mid \langle \mathbf{x}, \mathbf{x} \rangle = 1\}$. For an arbitrary point $\mathbf{u}$ in H_S, we define a half-space $R_\mathbf{u}$ and a hyperplane $P_\mathbf{u}$ in $\mathbb{E}^{1,n}$ as follows:

$$R_\mathbf{u} := \{\mathbf{x} \in \mathbb{E}^{1,n} \mid \langle \mathbf{x}, \mathbf{u} \rangle \leq 0\}\,,$$

$$P_\mathbf{u} := \{\mathbf{x} \in \mathbb{E}^{1,n} \mid \langle \mathbf{x}, \mathbf{u} \rangle = 0\} = \partial R_\mathbf{u}\,.$$

We denote by $\Gamma_\mathbf{u}$ (resp. $\Pi_\mathbf{u}$) the intersection of $R_\mathbf{u}$ (resp. $P_\mathbf{u}$) and $\mathbf{B}^n$. Then $\Pi_\mathbf{u}$ is a geodesic hyperplane in $\mathbb{H}^n$, and the correspondence between the points in H_S and the half-spaces $\Gamma_\mathbf{u}$ in $\mathbb{H}^n$ is bijective. We call $\mathbf{u}$ a *normal vector* to $P_\mathbf{u}$ (or $\Pi_\mathbf{u}$). The following two propositions are on relationships between the Lorentzian inner product and geometric objects.

Proposition 2.1. *Let* $\mathbf{x}$ *and* $\mathbf{y}$ *be two arbitrary points in* H_S, *and we assume that, as vectors, they are not parallel. Then the following hold:*

(1) *Two geodesic hyperplanes* $\Pi_{\mathbf{x}}$ *and* $\Pi_{\mathbf{y}}$ *intersect if and only if* $|\langle \mathbf{x}, \mathbf{y} \rangle| < 1$. *In this case the (hyperbolic) angle* θ *between them measured in* $\Gamma_{\mathbf{x}}$ *and* $\Gamma_{\mathbf{y}}$ *is obtained from the formula*

$$\langle \mathbf{x}, \mathbf{y} \rangle = -\cos\theta\,. \tag{2.2}$$

(2) *Two geodesic hyperplanes* $\Pi_{\mathbf{x}}$ *and* $\Pi_{\mathbf{y}}$ *do not intersect in* $\overline{\mathbf{B}^n}$, *i.e., they intersect in* $\operatorname{Ext}\overline{\mathbf{B}^n}$, *if and only if* $|\langle \mathbf{x}, \mathbf{y} \rangle| > 1$. *In this case the (hyperbolic) distance* d *between* $\Pi_{\mathbf{x}}$ *and* $\Pi_{\mathbf{y}}$ *is obtained from*

$$|\langle \mathbf{x}, \mathbf{y} \rangle| = \cosh d\,, \tag{2.3}$$

and they are said to be ultraparallel.

(3) *Two geodesic hyperplanes* $\Pi_{\mathbf{x}}$ *and* $\Pi_{\mathbf{y}}$ *intersect not in* $\mathbf{B}^n$ *but in* $\partial\mathbf{B}^n$ *if and only if* $|\langle \mathbf{x}, \mathbf{y} \rangle| = 1$. *In this case the angle and the distance between* $\Pi_{\mathbf{x}}$ *and* $\Pi_{\mathbf{y}}$ *is* 0, *and they are said to be* parallel.

Proposition 2.2. *Let* $\mathbf{x}$ *be a point in* $\mathbf{B}^n$ *and let* $\Pi_{\mathbf{y}}$ *be a geodesic hyperplane whose normal vector is* $\mathbf{y} \in H_S$ *with* $\langle \mathbf{x}, \mathbf{y} \rangle < 0$. *Then the distance* d *between* $\mathbf{x}$ *and* $\Pi_{\mathbf{y}}$ *is obtained from the formula*

$$\langle \mathbf{x}, \mathbf{y} \rangle = -\sinh d\,. \tag{2.4}$$

Let $\mathbf{v}$ be a point in $\operatorname{Ext}\overline{\mathbf{B}^n}$. Then $\mathcal{P}^{-1}(\mathbf{v}) \cap H_S$ consists of two points and, independent of the choice of $\tilde{\mathbf{v}} \in \mathcal{P}^{-1}(\mathbf{v}) \cap H_S$, we can define the same hyperplane $\Pi_{\tilde{\mathbf{v}}}$. We call $\Pi_{\tilde{\mathbf{v}}}$ the *polar geodesic hyperplane to* $\mathbf{v}$ and $\mathbf{v}$ the *pole* of $\Pi_{\tilde{\mathbf{v}}}$. Then the following proposition holds:

Proposition 2.3. *Let* $\mathbf{v}$ *be a point in* $\operatorname{Ext}\overline{\mathbf{B}^n}$.

(1) *Any hyperplane through* $\mathbf{v}$ *with intersecting* $\mathbf{B}^n$ *is perpendicular to* $\Pi_{\tilde{\mathbf{v}}}$ *in* $\mathbb{H}^n$.

(2) *Let* $\mathbf{u}$ *be a limit point of* $\Pi_{\tilde{\mathbf{v}}}$, *i.e.,* $\mathbf{u} \in P_{\mathbf{v}} \cap \partial\mathbf{B}^n$. *Then the line through* $\mathbf{u}$ *and* $\mathbf{v}$ *is tangent to* $\partial\mathbf{B}^n$.

3. A characterization theorem for the dihedral angles of a generalised hyperbolic simplex

Unlike spherical geometry or Euclidean geometry, in hyperbolic geometry we can consider not only points at infinity but also points "beyond infinity". This situation can be easily seen in the projective ball model, and it extends the idea of simplices. Such a simplex is called a *generalised simplex.* We start this section with its precise definition. In this section we assume $n \geq 3$.

Definition 3.1. Let $\triangle$ be a simplex in $\mathbb{P}_1^n$ containing the origin of $\mathbb{P}_1^n$ (remark; since any polyhedron in $\overline{\mathbf{B}^n}$ can be moved to the one containing the origin by the action of an orientation-preserving isomorphism of $\mathbb{H}^n$, this assumption does not lose generality in hyperbolic geometry). Suppose that all interiors of the ridges (i.e., $(n-2)$-dimensional faces) intersect $\overline{\mathbf{B}^n}$.

1. Let $\mathbf{v}$ be a vertex of $\triangle$ in $\operatorname{Ext}\overline{\mathbf{B}^n}$. The *(polar) truncation* of $\triangle$ at $\mathbf{v}$ is an operation to omit the pyramid of apex $\mathbf{v}$ whose base is an $(n-1)$-dimensional simplex $\Pi_{\tilde{\mathbf{v}}} \cap \triangle$, and then to cap the open end of $\triangle$ with $\Pi_{\tilde{\mathbf{v}}} \cap \triangle$ (see Figure 2).

2. The *truncated simplex*, say $\triangle'$, is a polyhedron in $\overline{\mathbf{B}^n}$ obtained from $\triangle$ by the truncation at all vertices in $\operatorname{Ext}\overline{\mathbf{B}^n}$.

3. A *generalised simplex* in $\mathbb{H}^n$ is a polyhedron which is either a simplex in the ordinary sense or a truncated simplex described above.

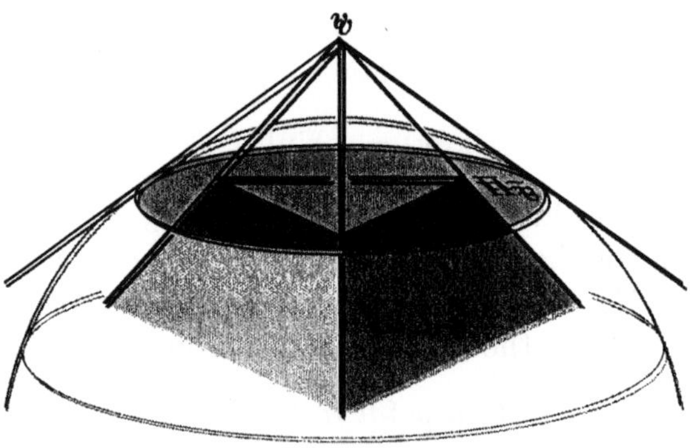

Figure 2. The truncation of $\triangle$ at $\mathbf{v}$

In this paper we will regard vertices of $\triangle$ as those of $\triangle'$. This means, for example, a vertex of $\triangle$ in $\operatorname{Ext}\overline{\mathbf{B}^n}$ is no longer that of $\triangle'$ in the ordinary sense, since it is not an element of $\triangle'$, but we will call it a "vertex of $\triangle'$". Conversely, the vertices (in the ordinary sense) arisen by the truncation will not be mentioned as the "vertices of $\triangle'$" in this paper. Under this usage, we call a vertex $\mathbf{v}$ of $\triangle'$ *finite* (resp. *ideal, ultraideal*) when $\mathbf{v} \in \mathbf{B}^n$ (resp. $\partial\mathbf{B}^n, \operatorname{Ext}\overline{\mathbf{B}^n}$).

Let σ^n be an n-dimensional generalised simplex in the projective model, and let $\{\mathbf{v}_i\}_{i\in I}$, where $I := \{1, 2, \dots, n+1\}$, be its vertex set. Then we regard the lift of the vertices to $\mathbb{E}^{1,n}$ as follows: if a vertex $\mathbf{v}$ is finite, then the lift is uniquely determined by $\mathcal{P}^{-1}(\mathbf{v}) \cap H_T^+$. If a vertex is ultraideal, then there are two choices of the lift, and we choose the one defining the half-space containing σ^n. If a vertex is ideal, then we do not need at this point to determine the exact lift in L^+. The *i-th facet* of σ^n is defined to be the $(n-1)$-dimensional face of σ^n opposite to $\mathbf{v}_i$.

Let G be a matrix of order $n+1$. We prepare several notations on matrices which will be used later.

1. We denote by G_{ij} the submatrix of order n obtained from G by deleting its i-th row and j-th column.

2. We denote a cofactor of G by $c_{ij} := (-1)^{i+j} \det G_{ij}$.

3. Suppose G is a real symmetric matrix. Then we denote by $\operatorname{sgn} G$ the signature of G, i.e., if G has a positive and b negative eigenvalues, then $\operatorname{sgn} G = (a, b)$.

The following theorem tells us a necessary and sufficient condition for given positive numbers to be the dihedral angles of a generalised simplex. Calculations given in the proof will be used in the proof of the main theorem.

Theorem 3.2. *Suppose that the following set of positive numbers is given:*

$$\{\theta_{ij} \in [0, \pi] \mid i, j \in I, \theta_{ij} = \theta_{ji}, \theta_{ij} = \pi \text{ iff } i = j\}.$$

Then the following two conditions are equivalent.

(1) *There exists a generalised hyperbolic simplex in $\mathbb{H}^n$ whose dihedral angle between i-th facet and j-th facet is θ_{ij}.*

(2) *The real symmetric matrix $G := (-\cos\theta_{ij})$ of order $n+1$ satisfies the following two conditions:*

   (a) $\operatorname{sgn} G = (n, 1)$,

   (b) $c_{ij} > 0$ *for any $i, j \in I$ with $i \neq j$.*

Proof. First we assume that (2) holds. By (a) there exists an orthogonal matrix, say U, of order $n+1$ such that the conjugate of G by U is as follows:

$$G = {}^tU \begin{pmatrix} -1 & 0 & 0 & \cdots & 0 \\ 0 & 1 & 0 & \cdots & 0 \\ 0 & 0 & 1 & \cdots & 0 \\ \vdots & \vdots & \vdots & \ddots & \vdots \\ 0 & 0 & 0 & \cdots & 1 \end{pmatrix} U.$$

Let $U = (\mathbf{u}_1, \mathbf{u}_2, \ldots, \mathbf{u}_{n+1})$. Since (a) holds, $\det U \neq 0$. So $\mathbf{u}_1, \mathbf{u}_2, \ldots, \mathbf{u}_{n+1}$ are linearly independent, and thus they are a basis of $\mathbb{E}^{n,1}$. Now we define a vector

$$\mathbf{w}_i := \sum_{k=1}^{n+1} c_{i\,k}\, \mathbf{u}_k.$$

It is easy to see that, for any $i, j \in I$, the equation $\langle \mathbf{w}_i, \mathbf{u}_j \rangle = \delta_{i\,j} \det G$ holds. Here $\delta_{i\,j}$ means the Kronecker delta. Using this relation, we obtain $\langle \mathbf{w}_i, \mathbf{w}_j \rangle = c_{i\,j} \det G$. Now we define

$$\mathbf{v}_i := \begin{cases} \dfrac{\mathbf{w}_i}{\sqrt{|c_{i\,i} \det G|}} & \text{if } c_{i\,i} \neq 0, \\ \mathbf{w}_i & \text{if } c_{i\,i} = 0, \end{cases} \tag{3.1}$$

and we show that $\{\mathbf{v}_i\}_{i \in I}$ is the set of vertices of the generalised simplex in question.

Here we calculate $\langle \mathbf{v}_i, \mathbf{v}_j \rangle$;

$$\langle \mathbf{v}_i, \mathbf{v}_j \rangle = \begin{cases} \dfrac{-c_{i\,j}}{\sqrt{|c_{i\,i} c_{j\,j}|}} & \text{if } c_{i\,i} c_{j\,j} \neq 0, \\ -c_{i\,j} \sqrt{\left| \dfrac{\det G}{c_{i\,i}} \right|} & \text{if } c_{i\,i} \neq 0 \text{ and } c_{j\,j} = 0, \\ c_{i\,j} \det G & \text{if } c_{i\,i} = c_{j\,j} = 0. \end{cases} \tag{3.2}$$

When $c_{i\,i} > 0$, $\langle \mathbf{v}_i, \mathbf{v}_i \rangle = -1$, i.e., $\mathbf{v}_i \in H_T$, and when $c_{i\,i} < 0$, $\langle \mathbf{v}_i, \mathbf{v}_i \rangle = 1$, i.e., $\mathbf{v}_i \in H_S$. Suppose $c_{i\,i} = 0$. Then, by (3.2), $\langle \mathbf{v}_i, \mathbf{v}_i \rangle = 0$. Furthermore we have $\mathbf{v}_i = \mathbf{w}_i \neq \mathbf{o}$ since $\mathbf{u}_1, \mathbf{u}_2, \ldots, \mathbf{u}_{n+1}$ are linearly independent and $c_{i\,j} \neq 0$ as $i \neq j$. So, in this case, $\mathbf{v}_i \in L - \{\mathbf{o}\}$.

Previous calculations also show that $\langle \mathbf{v}_i, \mathbf{v}_j \rangle > 0$ when $i \neq j$. So all $\mathbf{v}_i$'s contained in $H_T \sqcup (L - \{\mathbf{o}\})$ are in the upper or lower half-space of $\mathbb{R}^{n+1}$. Thus, exchanging all $\mathbf{v}_i$'s with $-\mathbf{v}_i$'s if necessary, we can assume that $\mathbf{v}_i \in H_T^+ \sqcup L^+ \sqcup H_S$.

By the construction of $\mathbf{w}_i$, the vectors $\mathbf{w}_1, \mathbf{w}_2, \ldots, \mathbf{w}_{n+1}$ are linearly independent, and so are $\mathbf{v}_1, \mathbf{v}_2, \ldots, \mathbf{v}_{n+1}$. Thus the convex hull of $\{\mathbf{v}_1, \mathbf{v}_2, \ldots, \mathbf{v}_{n+1}\}$ is an n-simplex in $\mathbb{E}^{n+1}$. Let σ be the intersection of the hull and $\bigcap_{\mathbf{v}_i \in H_S} R_{\mathbf{v}_i}$. Then $P(\sigma)$ is the simplex in question and we obtain (1).

Secondly we assume (1). Let σ be a generalised n-simplex, and let $\{\mathbf{v}_i\}_{i\in I}$ be its vertex set. We denote by $\mathbf{u}_i$ the outward unit normal vector to the facet opposite to $\mathbf{v}_i$. Then, by the definition of the generalised simplex, $\mathbf{u}_i \in H_S$. Since $-\cos\theta_{i\,j} = \langle \mathbf{u}_i, \mathbf{u}_j\rangle$, we have $\operatorname{sgn} G = (n, 1)$, which is the condition (a).

As before, we define a vector $\mathbf{w}_i$ as

$$\mathbf{w}_i := \sum_{k=1}^{n+1} c_{i\,k}\, \mathbf{u}_k.$$

By the definition of $\mathbf{u}_j$, $\langle \mathbf{v}_i, \mathbf{u}_j\rangle = 0$ when $i \neq j$. So $\mathbf{v}_i \in \bigcap_{i\neq j\in I} \mathbf{u}_j^{\perp}$. On the other hand, $\mathbf{w}_i \in \bigcap_{i\neq j\in I} \mathbf{u}_j^{\perp}$ since $\langle \mathbf{w}_i, \mathbf{u}_j\rangle = \delta_{i\,j} \det G$. The dimension of $\bigcap_{i\neq j\in I} \mathbf{u}_j^{\perp}$ is 1, which implies that $\mathbf{w}_i$ is parallel to $\mathbf{v}_i$. Especially $\mathbf{w}_i = d_i \mathbf{v}_i$ for some $d_i > 0$. Thus we have

$$\langle \mathbf{w}_i, \mathbf{w}_j\rangle = d_i d_j \langle \mathbf{v}_i, \mathbf{v}_j\rangle = c_{i\,j} \det G.$$

So $c_{i\,j} = d_i d_j \langle \mathbf{v}_i, \mathbf{v}_j\rangle / \det G$. By the definition of generalised simplices together with the lifts of the vertices, $\langle \mathbf{v}_i, \mathbf{v}_j\rangle < 0$ for any $i \neq j \in I$. Thus we have the condition (b), and the proof is complete. □

Remarks.

1. *The strategy of the previous proof is the same as that of THEOREM in [Lu], where it is proved a necessary and sufficient condition for a given set of positive real numbers to be the dihedral angles of a hyperbolic simplex in the ordinary sense. Moreover such conditions for spherical and Euclidean simplices are also presented in [Lu].*

2. *We call the matrix G that appeared in the previous theorem the* Gram matrix *of an n-dimensional generalised simplex σ. Here it should be noted that this definition is slightly different from the ordinary one (see, for example, [Vi]), since we do not deal with the normal vectors to the faces obtained by truncation.*

3. *By the definition (3.1), we can characterize vertices from the Gram matrix; if $c_{i\,i} > 0$ (resp. $c_{i\,i} = 0, c_{i\,i} < 0$), then the vertex opposite to the i-th facet is finite (resp. ideal, ultraideal).*

4. Schläfli's differential formula and the maximal volume amongst the generalised hyperbolic tetrahedra

Now we go on to the main purpose of this paper, namely a proof of a volume formula for generalised hyperbolic tetrahedra. So, from now on, we only consider the three-dimensional case. The key tool for our proof is the so-called "Schläfli's differential formula", which gives us a relationship between volume formulae and their dihedral angles. In this section we present the formula and also give its other usage, which is a determination of the maximal volume amongst the generalised hyperbolic tetrahedra.

In the three-dimensional case, Schläfli's differential formula can be written as follows (see [Ke, §2]).

Theorem 4.1. *Let T be a generalised tetrahedron in $\mathbb{H}^3$ with dihedral angles θ_i and corresponding edges of length l_i for $i = 1, 2, \ldots, 6$. We denote by* Vol *the volume function on the set of the generalised tetrahedra. Then the following equation holds:*

$$\mathrm{d}\,\mathrm{Vol}(T) = -\frac{1}{2}\sum_{i=1}^{6} l_i \,\mathrm{d}\theta_i.$$

U. Haagerup and H. Munkholm proved in [HM] that, in hyperbolic space of dimension greater than 1, a simplex is of maximal volume if and only if it is ideal and regular[1]. Once we extend the class of tetrahedra to the generalised ones, by using this formula, we can determine generalised tetrahedra having the maximal volume.

Theorem 4.2. *The maximal volume amongst the generalised hyperbolic tetrahedra is attained by the ideal right-angled octahedron, and the volume is $8\Lambda(\pi/4)$, where $\Lambda(x)$ is the function, called the* Lobachevsky function, *defined as*

$$\Lambda(x) := -\int_0^x \log|2\sin t|\,\mathrm{d}t.$$

Proof. Since any ordinary tetrahedron is contained in an ideal hyperbolic tetrahedron, the maximal volume is attained by generalised tetrahedra whose vertices are outside hyperbolic space. Furthermore Schläfli's differential formula says that the smaller the length of an internal edge is,

[1] More generally, it is proved by N. Peyerimhoff that the regular simplex is of maximal volume amongst all simplices in a closed geodesic ball in $\mathbb{H}^n$; see [Pe, THEOREM 1].

the larger the volume of a generalised tetrahedron is. So the maximal volume is attained by a truncated tetrahedron with all internal edges having length 0 and, by the definition of the truncation, all dihedral angles are then equal to $\pi/2$. Namely the right-angled ideal octahedron is a unique generalised tetrahedron with maximal volume. Its actual volume will be calculated in the proof of the main theorem. □

5. Proof of a volume formula for generalised hyperbolic tetrahedra

Let $\mathcal{V}(T) := \Im\left(U(z_1,T) - U(z_2,T)\right)/2$ be the right-hand side of the formula (1.1). Then, by Theorem 4.1, it is necessary to check that the partial derivative of $\mathcal{V}(T)$ with respect to each dihedral angle is equal to $-1/2$ times the length of the edge corresponding to it. Since the formula is symmetric under permutations of the dihedral angles, from now on we focus on a dihedral angle, say A.

We first calculate the length of the edge corresponding to A. Let $\mathbf{v}_4$ be the vertex which is the common endpoint of three edges corresponding to A, B and C, and let $\mathbf{v}_1$ (resp. $\mathbf{v}_2$, $\mathbf{v}_3$) be the other endpoint of the edge corresponding to C (resp. B, A). It should be noted that we also regard $\mathbf{v}_i$ as the vector in $\mathbb{E}^{1,3}$ corresponding the vertex. Recall that G is the Gram matrix of T defined as follows:

$$G := \begin{pmatrix} 1 & -\cos A & -\cos B & -\cos F \\ -\cos A & 1 & -\cos C & -\cos E \\ -\cos B & -\cos C & 1 & -\cos D \\ -\cos F & -\cos E & -\cos D & 1 \end{pmatrix}.$$

We here note that the following identity[2] holds:

$$c_{34}^2 - c_{33}\,c_{44} = (-\det G)\sin^2 A. \tag{5.1}$$

We first consider the case where the endpoints of A are both finite or ultraideal vertices. In this case, since $c_{33}\,c_{44} > 0$, using (2.1), (2.3) and (3.2), we have

$$-\cosh l = \langle \mathbf{v}_3, \mathbf{v}_4\rangle = -\frac{c_{34}}{\sqrt{c_{33}\,c_{44}}},$$

where l is the length of the edge joining $\mathbf{v}_3$ and $\mathbf{v}_4$. Solve this equation by $\exp l$ and we have

$$\exp l = \frac{c_{34} + \sqrt{c_{34}^2 - c_{33}\,c_{44}}}{\sqrt{c_{33}\,c_{44}}}.$$

[2]Professors Derevnin and Mednykh thankfully mentioned to the author that a more general result was already known as Jacobi's theorem (see [Pr, 2.5.1. THEOREM (Jacobi)]).

Thus, using (5.1), we have

$$\exp 2l = \frac{2c_{34}^2 - c_{33}\,c_{44} + 2c_{34}\sqrt{-\det G}\sin A}{c_{33}\,c_{44}}. \tag{5.2}$$

We secondly consider the case where the endpoints of A are a finite vertex and a ultraideal vertex. In this case $c_{33}\,c_{44} < 0$. So, using (2.4), we have

$$-\sinh l = \langle \mathbf{v}_3, \mathbf{v}_4 \rangle = -\frac{c_{34}}{\sqrt{-c_{33}\,c_{44}}}.$$

Solve this equation by $\exp l$ and, as (5.2), we have

$$\exp l = \frac{c_{34} + \sqrt{c_{34}^2 - c_{33}\,c_{44}}}{\sqrt{-c_{33}\,c_{44}}}$$

$$\iff \exp 2l = \frac{2c_{34}^2 - c_{33}\,c_{44} + 2c_{34}\sqrt{-\det G}\sin A}{-c_{33}\,c_{44}}. \tag{5.3}$$

Next we calculate the partial derivative of $\mathcal{V}(T)$ with respect to A. We first calculate $\partial U(z_i, T)/\partial A$;

$$\begin{aligned} \frac{\partial U(z_i,T)}{\partial A} &= \frac{\partial U(z_i,T)}{\partial a}\frac{\mathrm{d}a}{\mathrm{d}A} \\ &= \frac{\sqrt{-1}}{2}\log\frac{\varphi(z_i)}{\psi(z_i)} + \frac{\partial U(z_i,T)}{z_i}\frac{\partial z_i}{\partial a}\frac{\mathrm{d}a}{\mathrm{d}A}, \end{aligned} \tag{5.4}$$

where $\varphi(z) := (1 + abcz)\,(1 + aefz)$ and $\psi(z) := (1 - abdez)\,(1 - acdfz)$. It is easily checked that the second term of the right-hand side is real, so we obtain

$$\Im\frac{\partial U(z_i,T)}{\partial A} = \frac{1}{2}\log\left|\frac{\varphi(z_i)}{\psi(z_i)}\right|,$$

and thus we have

$$\frac{\partial \mathcal{V}(T)}{\partial A} = \frac{1}{4}\log\left|\frac{\varphi(z_1)\psi(z_2)}{\varphi(z_2)\psi(z_1)}\right|.$$

By Theorem 4.1, all we have to show is that

$$\frac{1}{4}\log\left|\frac{\varphi(z_1)\psi(z_2)}{\varphi(z_2)\psi(z_1)}\right| = -\frac{l}{2} \iff \left|\frac{\varphi(z_2)\psi(z_1)}{\varphi(z_1)\psi(z_2)}\right| = \exp 2l.$$

Substitute z_1 and z_2 for $\varphi(z_2)\psi(z_1)/\left(\varphi(z_1)\psi(z_2)\right)$ and we have

$$\begin{aligned} \frac{\varphi(z_2)\psi(z_1)}{\varphi(z_1)\psi(z_2)} &= \frac{c_{34} + \sqrt{-\det G}\sin A}{c_{34} - \sqrt{-\det G}\sin A} \\ &= \frac{2c_{34}^2 - c_{33}c_{44} + 2c_{34}\sqrt{-\det G}\sin A}{c_{33}c_{44}}. \end{aligned}$$

Thus, as we saw in (5.2) and (5.3), the absolute value of the above is equal to $\exp 2l$. Furthermore $\mathrm{Vol}(T)$ and $\mathcal{V}(T)$ are analytic functions whose variables are $A, B, \ldots, F$. This means that $\mathcal{V}(T)$ is a volume formula for generalised tetrahedra up to some constant. Now we check that $\mathcal{V}(T)$ is the formula indeed.

Consider the regular tetrahedron in $\mathbb{P}^3$ circumscribing $\overline{\mathbf{B}^3}$. Then the polyhedron obtained by the truncation at all vertices is the regular ideal octahedron with all dihedral angles being $\pi/2$, which we denote by T. Its volume is calculated as follows: since T is symmetric, it can be carved like sections of an orange. Then we obtain four isometric ideal tetrahedra, and its dihedral angles are $\pi/2$, $\pi/4$ and $\pi/4$. Using Milnor's volume formula for ideal tetrahedra (see [Mi]) we have

$$\begin{aligned}\mathrm{Vol}(T) &= 4\left(\Lambda(\frac{\pi}{2}) + \Lambda(\frac{\pi}{4}) + \Lambda(\frac{\pi}{4})\right)\\ &= 8\Lambda(\frac{\pi}{4})\,(\approx 3.66386)\,.\end{aligned}$$

Here we note that the Lobachevsky function $\Lambda(x)$ is a periodic odd function with period π and $\Lambda(\pi/2) = 0$. Furthermore there is the following relationship with the dilogarithm function (see, for example, [Ki, §1.1.4])

$$\Im\,\mathrm{Li}_2(\exp\sqrt{-1}x) = 2\Lambda(\frac{x}{2}) \quad \text{for any } x \in \mathbb{R}. \tag{5.5}$$

On the other hand, T is regarded as the truncated regular tetrahedron whose dihedral angle is 0. So we have

$$(z_1, z_2) = (-\sqrt{-1}, \sqrt{-1}).$$

Moreover in this case

$$U(z,T) = 2\left(\mathrm{Li}_2(z) - \mathrm{Li}_2(-z)\right).$$

So, using (5.5), we have

$$\begin{aligned}\mathcal{V}(T) &= \Im\left(\mathrm{Li}_2(z_2) - \mathrm{Li}_2(-z_2) - \mathrm{Li}_2(z_1) + \mathrm{Li}_2(z_1)\right)\\ &= 2\Im\left(\mathrm{Li}_2(\sqrt{-1}) - \mathrm{Li}_2(-\sqrt{-1})\right)\\ &= 4\left(\Lambda(\frac{\pi}{4}) - \Lambda(\frac{-\pi}{4})\right)\\ &= 8\Lambda(\frac{\pi}{4})\\ &= \mathrm{Vol}(T).\end{aligned}$$

Thus we have proved Theorem 1.1. □

Appendix: Murakami-Yano's volume formula

Murakami-Yano's volume formula comes from the quantum $6j$-symbol, which is a source of invariants for three-dimensional manifolds. To obtain the formula they applied R. Kashaev's method for the computation of the quantum $6j$-symbol. He discovered in [Ka] a relationship between certain quantum invariants and the volumes of several hyperbolic knots in the three-dimensional sphere. A technical key tool for his results is the so-called *saddle point approximation*, which is an asymptotic approximation of some integral by saddle points of an integrand. The values z_1 and z_2 in this paper correspond to saddle points of $U(z,T)$. Roughly speaking, z_1 and z_2 arise from the solutions of $\mathrm{d}U(z,T)/\mathrm{d}z = 0$.

They also proved in [MY] that, by taking an appropriate branch of $U(z,T)$, $\Re\left(U(z_1,T) - U(z_2,T)\right)/2 = 0$ if T is a hyperbolic tetrahedron, and $\left(U(z_1,T) - U(z_2,T)\right)/2 = \mathrm{Vol}(T)$ if T is a spherical tetrahedron.

Appendix: Graphs on the volume and the edge length of a regular tetrahedron

The following graphs show the relations between the dihedral angle and the volume or the edge length of a regular tetrahedron. The dihedral angle varies from 0 to $\arccos(1/3) \approx 1.230959$. As we saw in Theorem 4.2, the angle 0 corresponds to the regular ideal octahedron. The angle $\arccos(1/3)$ corresponds to the Euclidean regular tetrahedron, which can be regarded as the infinitesimally small hyperbolic regular tetrahedron. The dihedral angle of the regular ideal tetrahedron is $\pi/3$.

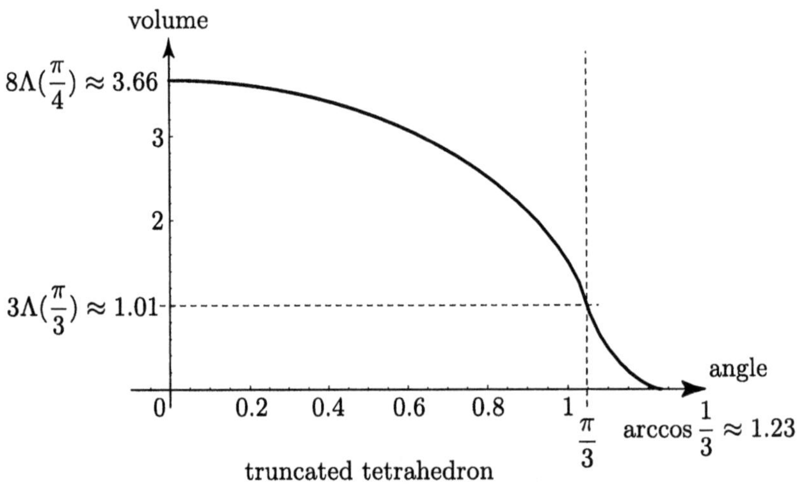

Figure B.1. Volume of a regular tetrahedron

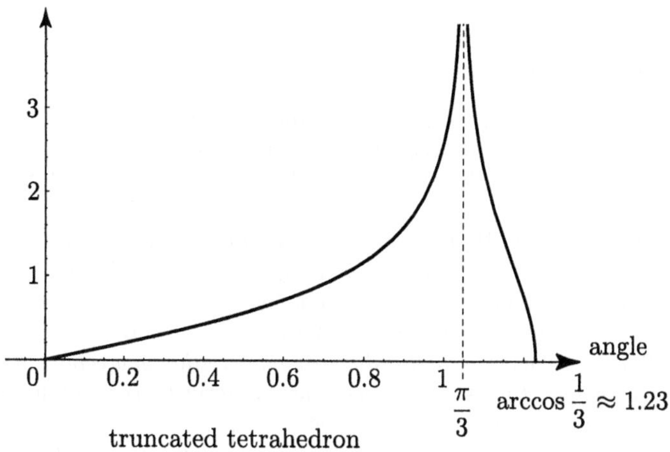

Figure B.2. Edge length of a regular tetrahedron

Appendix: Further generalisation and open problems; when the truncation is not "mild"

In this paper it is assumed that the truncation of a tetrahedron is "mild", namely the faces obtained by the truncations do not intersect each other in $\mathbb{H}^3$. What happens if we do not assume this condition, i.e., if the truncation is "intense"?

A three-dimensional *orthoscheme* is a tetrahedron with dihedral angles $B = E = F = \pi/2$ (see [Ke] for details; another basic reference is [BH][3]). Then the edge spanned by $\mathbf{v}_1$ and $\mathbf{v}_2$ is perpendicular to the face spanned by $\mathbf{v}_2, \mathbf{v}_3, \mathbf{v}_4$, and also the face spanned by $\mathbf{v}_1, \mathbf{v}_2, \mathbf{v}_3$ is perpendicular to the edge spanned by $\mathbf{v}_3$ and $\mathbf{v}_4$, which is the meaning of the definition in own Introduction. Two vertices $\mathbf{v}_1$ and $\mathbf{v}_4$ are called *principal vertices.*

For an orthoscheme, if two faces obtained by truncations at principal vertices intersect each other in hyperbolic space, then the orthoscheme becomes a cube with all faces being Lambert quadrilaterals (i.e., quadrilaterals with one acute and three right angles). Such a cube is called a *Lambert cube.* In [Ke] she also constructed a volume formula for Lambert cubes. The formula is closely related to the one for orthoschemes, so the following problem naturally arises:

Problem 1. *Construct a volume formula for intensely truncated tetrahedra from our volume formula.*

A possible first step to solve this problem is to see the relationship between Kellerhals formula for mildly truncated orthoschemes and our formula. Her formula is presented via the Lobachevsky function, and we can easily rewrite our formula with respect to it (see [MY, (1.19)]). Though ours theoretically contains hers, there seems no direct relationship between them. So the following problem also arises:

[3]Professor Molnár kindly told the author about this book.

Problem 2. *For orthoschemes, find a relationship between our volume formula and Kellerhals's formula.*

Bibliography

[BH] Johannes Böhm and Eike Hertel, *Polyedergeometrie in n-dimensionalen Räumen konstanter Krümmung* , Lehrbücher und Monographien aus dem Gebiete der Exakten Wissenschaften. Mathematische Reihe **70**, Birkhäuser Verlag, 1981.

[CK] Yunhi Cho and Hyuk Kim, *On the Volume Formula for Hyperbolic Tetrahedra*, Discrete & Computational Geometry **22** (1999), 347–366.

[HM] Uffe Haagerup and Hans J. Munkholm, *Simplices of maximal volume in hyperbolic n-space*, Acta Mathematica **147** (1981), 1–11.

[Hs] Wu-Yi Hsiang, *On infinitesimal symmetrization and volume formula for spherical or hyperbolic tetrahedrons*, The Quarterly Journal of Mathematics, Oxford, Second Series **39** (1988), 463–468.

[Ka] R. M. Kashaev, *The Hyperbolic Volume of Knots from the Quantum Dilogarithm*, Letters in Mathematical Physics **39** (1997), 269–275.

[Ke] Ruth Kellerhals, *On the volume of hyperbolic polyhedra*, Mathematische Annalen **285** (1989), 541–569.

[Ki] Anatol N. Kirillov, *Dilogarithm identities*, Progress of Theoretical Physics. Supplement **118** (1995), 61–142.

[Lo] N. I. Lobatschefskij, *Imaginäre Geometrie und ihre Anwendung auf einige Integrale*, translated into German by H. Liebmann (Leipzig, 1904).

[Lu] Feng Luo, *On a Problem of Fenchel*, Geometriae Dedicata **64** (1997), 277–282.

[Mi] John Milnor, *Hyperbolic geometry: the first 150 years*, Bulletin (New Series) of the American Mathematical Society **6** (1982), 9–24.

[MY] Jun Murakami and Masakazu Yano, *On the volume of a hyperbolic tetrahedron*, available at
`http://www.f.waseda.jp/murakami/papers/tetrahedronrev4.pdf`.

[Pe] Norbert Peyerimhoff, *Simplices of maximal volume or minimal total edge length in hyperbolic space*, Journal of the London Mathematical Society (2) **66** (2002), 753–768.

[Pr] V. V. Prasolov, *Problems and Theorems in Linear Algebra*, Translations of Mathematical Monographs **134**, American Mathematical Society, 1994.

[Us] Akira Ushijima, *The Tilt Formula for generalised Simplices in Hyperbolic Space*, Discrete & Computational Geometry **28** (2002), 19–27.

[Ve] Andrei Vesnin, *On Volumes of Some Hyperbolic 3-manifolds*, Lecture Notes Series **30**, Seoul National University, 1996.

[Vi] E. B. Vinberg (Ed.), *Geometry II*, Encyclopaedia of Mathematical Sciences **29**, Springer-Verlag, 1993.

IV

TILINGS, ORBIFOLDS AND MANIFOLDS, VISUALIZATION

THE GEOMETRY OF HYPERBOLIC MANIFOLDS OF DIMENSION AT LEAST 4

John G. Ratcliffe
Department of Mathematics, Vanderbilt University, Nashville, TN 37240
ratclifj@math.vanderbilt.edu

Abstract In this paper, we will discuss some new developments in the theory of hyperbolic manifolds of dimension at least 4. The main concern of the paper is the problem of understanding the distribution of volumes of hyperbolic manifolds in dimensions 4 and 5.

Keywords: hyperbolic manifolds, 3-manifolds, 4-manifolds, 5-manifolds, volume, arithmetic manifolds, Coxeter groups, Coxeter simplices

1. Introduction

A hyperbolic n-manifold is a Riemannian manifold isometric to the orbit space H^n/Γ of a torsion-free discrete group Γ of isometries of hyperbolic n-space H^n. The most important invariant of a hyperbolic n-manifold is its volume. Besides H^n only manifolds of finite volume will be considered in this paper.

In the early nineteen seventies, H.-C. Wang [13] proved that for each real number $V > 0$ and dimension $n \geq 4$, there are only finitely many hyperbolic n-manifolds with volume at most V. Wang's theorem is not true for dimension $n = 2$ or 3. If $n = 2$, there is a continuum of different closed hyperbolic surfaces of the same area. When $n = 3$, there can be a countably infinite number of hyperbolic 3-manifolds of bounded volume.

Recently, M. Burger, T. Gelander, A. Lubotzky, and S. Mozes [1] proved that if $N(V)$ is the number of hyperbolic n-manifolds, $n \geq 4$, with volume at most V, then $N(V)$ grows superexponentially with V. More specifically, there are positive constants a and b, depending on n, such that

$$V^{aV} \leq N(V) \leq V^{bV}.$$

If M is a hyperbolic n-manifold of finite volume, with n even, then the Gauss-Bonnet theorem says that

$$\mathrm{Vol}(M) = (-1)^{n/2}(2\pi)^{n/2}\chi(M)/(n-1)!!.$$

Thus the volume of M is a constant multiple of the Euler characteristic of M. For example, if $n = 4$, then

$$\mathrm{Vol}(M) = \tfrac{4}{3}\pi^2\chi(M).$$

2. Hyperbolic 4-Manifolds of Small Volume

In 1985, M. Davis [3] gave the first explicit geometric construction of a closed hyperbolic 4-manifold by gluing together the opposite sides of a regular hyperbolic 120-cell P with dihedral angle 72°. The 120-cell P has 120 sides each of which is a regular dodecahedron. The 120-cell P is hard to visualize, but it is helpful to realize that P has a central cross-section that is a regular decagon, see Figure 1. The opposite sides of the 120-cell P are paired by a hyperbolic translation whose axis passes through the centers of the sides. Figure 1 shows the axis of a hyperbolic translation that glues the top and bottom sides of P. It is worth noting that the side-pairing of P restricts to a side-pairing of the regular decagon in Figure 1 that glues up to a closed orientable hyperbolic surface of genus 2.

Note that

$$\chi(M) = 26,$$

and so the volume of the Davis manifold is fairly small. The Davis 4-manifold M is very beautiful. Its symmetry group has order 28,800. The Davis manifold M is orientable and a spin manifold. For the geometry of the Davis manifold, see Ratcliffe-Tschantz [9].

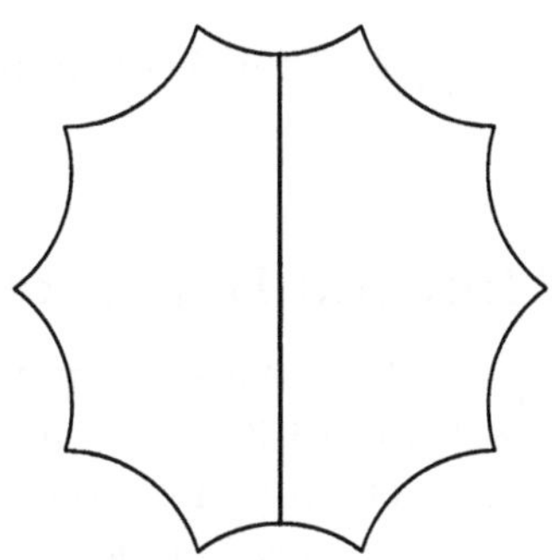

Figure 1. A 2-dimensional central cross-section of a regular 120-cell

In 1993, Tschantz constructed a nonorientable hyperbolic 4-manifold N by gluing together two right-angled regular hyperbolic 120-cells along their sides. Now

$$\chi(N) = 17,$$

and so the Tschantz 4-manifold N has smaller volume than the Davis 4-manifold. The Tschantz 4-manifold N is the hyperbolic 4-manifold of smallest known Euler characteristic and volume.

Question: Is there a closed hyperbolic 4-manifold M with $\chi(M) = 1$? Such a manifold would have minimum volume among all closed hyperbolic 4-manifolds. One possible way of constructing M would be to glue together pairs of sides of a regular hyperbolic 120-cell P with dihedral angle 120°, since P is part of regular tessellation of H^4 and

$$\mathrm{Vol}(P) = \tfrac{4}{3}\pi^2.$$

Note that there are $119!! \cdot 120^{60}$ possible choices for a side-pairing!

The next theorem was conjectured by J. Ratcliffe and S. Tschantz in 1993 after spending two months of computer time unsuccessfully trying to construct a closed orientable hyperbolic 4-manifold of Euler characteristic 1 by gluing together the sides of a regular hyperbolic 120-cell with dihedral angle 120°. The conjecture was proved by J. Ratcliffe and M. Davis in 1995.

Theorem 1. *Every closed orientable hyperbolic manifold has an even Euler characteristic.*

Proof. Every closed orientable n-manifold has an even Euler characteristic when n is not a multiple of 4, so we may assume that n is a multiple of 4. In 1955, Chern [2] proved that all the Pontrjagin numbers of a closed orientable hyperbolic n-manifold M are zero; therefore, the signature $\sigma(M)$ of M is zero by the signature theorem, and so

$$\chi(M) \equiv \beta_{n/2}(M) \equiv \sigma(M) \equiv 0 \bmod 2.$$

□

3. Volumes of Integral Hyperbolic Manifolds

In her 1992 paper [4], R. Kellerhals computed the volume of the hyperbolic Coxeter orthoscheme $(3,3,3,4,3)$ to be $7\zeta(3)/46080$. The appearance of the Riemann zeta function spurred J. Ratcliffe and S. Tschantz to try to understand the arithmetic significance of the volume of $(3,3,3,4,3)$. S. Tschantz first observed that the hyperbolic Coxeter simplex Δ^5 whose Coxeter graph is

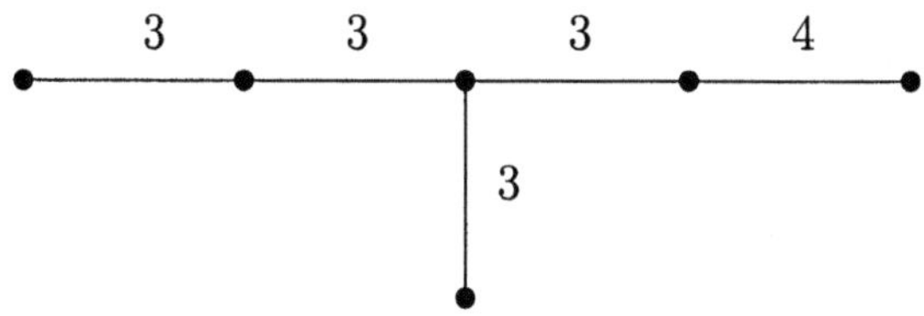

can be subdivided into three copies of the orthoscheme $(3,3,3,4,3)$. The hyperbolic Coxeter simplex Δ^5 has a nice arithmetic interpretation which we now explain.

A real $(n+1)\times(n+1)$ matrix A is said to be Lorentzian if A preserves the inner product

$$x \circ y = x_1y_1 + x_2y_2 + \cdots + x_ny_n - x_{n+1}y_{n+1}.$$

The hyperboloid model of hyperbolic n-space is the metric space

$$H^n = \{x \in \mathbb{R}^{n+1} : x \circ x = -1 \text{ and } x_{n+1} > 0\}$$

with metric d defined by

$$\cosh d(x,y) = -x \circ y.$$

A Lorentzian $(n+1)\times(n+1)$ matrix A is said to be positive if A maps H^n to H^n. The isometries of H^n correspond to the positive Lorentzian $(n+1)\times(n+1)$ matrices.

Let Γ^n be the group of all positive Lorentzian $(n+1)\times(n+1)$ matrices with integer entries. Then Γ^n acts as a discrete group of isometries of H^n and the hyperbolic orbifold $M^n = H^n/\Gamma^n$ has finite volume for all $n > 1$.

The orbifold M^n has a nice geometric description for $n \leq 9$. E. Vinberg [12] proved that the orbifold

$$M^n = H^n/\Gamma^n$$

is a hyperbolic Coxeter n-simplex for $n \leq 9$ (in particular $M^5 = \Delta^5$). Hence, the group Γ^n is a Coxeter group for $n \leq 9$.

In 1995, J. Ratcliffe and S. Tschantz [7] computed the volume of M^n, for all n, in terms of powers of π, the Riemann zeta function

$$\zeta(s) = 1 + \frac{1}{2^s} + \frac{1}{3^s} + \frac{1}{4^s} + \cdots$$

and the Dirichlet L-function

$$L(s) = 1 - \frac{1}{3^s} + \frac{1}{5^s} - \frac{1}{7^s} + \cdots.$$

Table 1. The Coxeter diagram of the Coxeter group Γ^n and the volume of the orbifold M^n for $n = 2, \ldots, 9$.

M^2		$\frac{\pi}{4}$
M^3	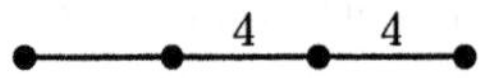	$\frac{L(2)}{12}$
M^4		$\frac{\pi^2}{1440}$
M^5	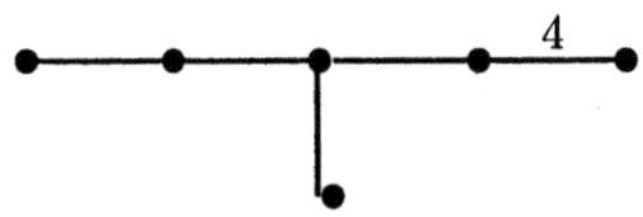	$\frac{7\zeta(3)}{15360}$
M^6		$\frac{\pi^3}{777600}$
M^7	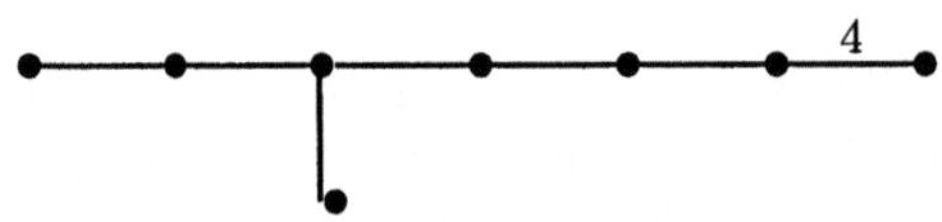	$\frac{L(4)}{362880}$
M^8	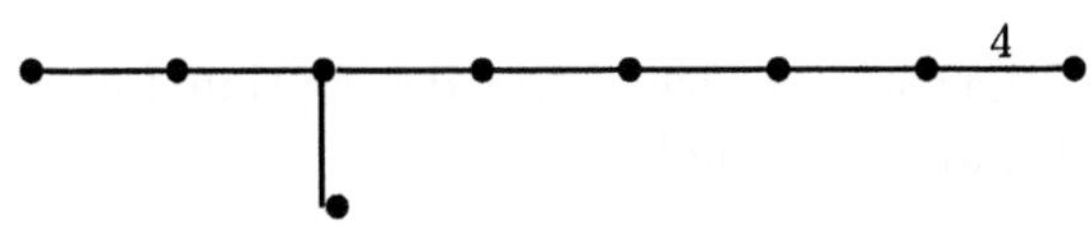	$\frac{17\pi^4}{9144576000}$
M^9	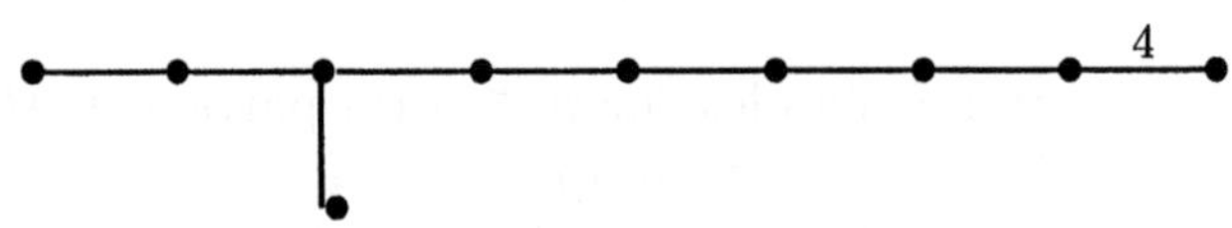	$\frac{527\zeta(5)}{44590694400}$

Table 2. The values of $\chi(M_p^4)$ and $\mathrm{Vol}(M_p^5)$ for the first five odd primes

$\chi(M_3^4)$	=	54,	$\mathrm{Vol}(M_3^5)$	=	$22113\,\zeta(3)/2$,
$\chi(M_5^4)$	=	9750,	$\mathrm{Vol}(M_5^5)$	=	$26446875\,\zeta(3)$,
$\chi(M_7^4)$	=	144060,	$\mathrm{Vol}(M_7^5)$	=	$2112387795\,\zeta(3)$,
$\chi(M_{11}^4)$	=	26793030,	$\mathrm{Vol}(M_{11}^5)$	=	$7545654056325\,\zeta(3)/2$,
$\chi(M_{13}^4)$	=	142747878,	$\mathrm{Vol}(M_{13}^5)$	=	$46355018040513\,\zeta(3)$.

For an integer $k > 1$, the congruence k subgroup of Γ^n is the group Γ_k^n of all matrices in Γ^n that are congruent to the identity matrix modulo k. The group Γ_k^n is torsion-free if and only if $k > 2$. Define

$$M_k^n = H^n/\Gamma_k^n.$$

Then M_k^n is a hyperbolic n-orbifold for $k = 2$, and a hyperbolic n-manifold for $k > 2$. The group Γ_k^n is a subgroup of finite index of Γ^n. Hence, we have

$$\mathrm{Vol}(M_k^n) = [\Gamma^n : \Gamma_k^n]\,\mathrm{Vol}(M^n).$$

Let p be an odd prime number, let $n > 1$, and let $\mathbb{Z}_p = \mathbb{Z}/p\mathbb{Z}$. J. Ratcliffe and S. Tschantz [7] showed that

$$[\Gamma^n : \Gamma_p^n] = \begin{cases} \frac{1}{2}|\mathrm{O}(n,1;\mathbb{Z}_p)|, & \text{if } p \equiv 1, 7 \bmod 8; \\ |\mathrm{O}(n,1;\mathbb{Z}_p)|, & \text{if } p \equiv 3, 5 \bmod 8. \end{cases}$$

The order of the group $\mathrm{O}(n,1;\mathbb{Z}_p)$ is well known and so we have explicit formulas for the volumes of M_p^n for all odd prime numbers p. The values of $\chi(M_p^4)$ and $\mathrm{Vol}(M_p^5)$ for the first five odd primes are given in Table 2. The manifolds M_p^n are large but beautiful, since they have large symmetry groups.

4. Congruence 2, Hyperbolic Manifolds

Let $e_1, \ldots, e_n$ be the standard basis of $\mathbb{R}^n$, and let T^n be the n-simplex with an actual vertex at the origin O and ideal vertices $e_1, \ldots, e_n$ in the conformal ball model B^n of hyperbolic n-space. We call T^n a corner n-simplex and the origin O the corner vertex of T^n. The n-simplex has n sides incident with the origin, which we call corner sides of T^n, and one side which is a regular ideal $(n-1)$-simplex. The dihedral angles between adjacent corner sides of T^n are $\pi/2$.

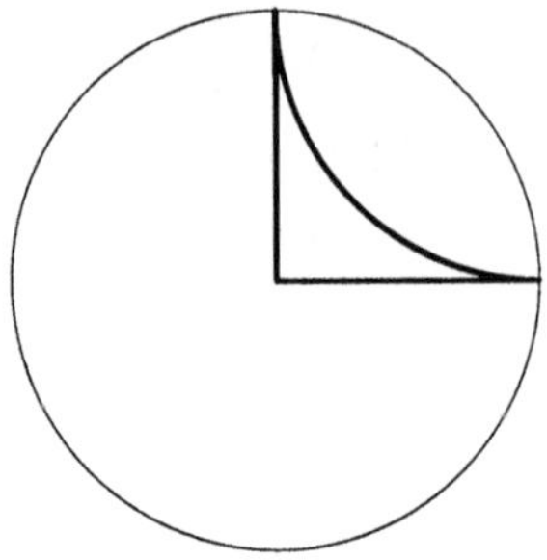

Figure 2. A corner triangle

The congruence 2 subgroup Γ_2^n has only 2-torsion. The maximum order of a finite subgroup of Γ_2^n is 2^n which is the order of the subgroup

$$K^n = \{\mathrm{diag}(\pm 1, \ldots, \pm 1, 1)\}.$$

The group K^n is generated by the reflections in the corner sides of the corner n-simplex T^n.

Let Γ be a torsion-free subgroup of Γ_2^n. The finite group K^n acts freely on the left cosets of Γ in Γ_2^n by right multiplication,

$$\Gamma A \mapsto \Gamma AD,$$

and so $|K^n| = 2^n$ divides $[\Gamma_2^n : \Gamma]$.

We are interested in finding torsion-free subgroups of Γ_2^n of minimal index 2^n for dimensions $n = 4, 5$, but first let us consider the lower dimensional cases $n = 2, 3$.

The group Γ^2 is a reflection group with respect to a 45°–0° right triangle Δ^2 in H^2. The corner triangle T^2 is subdivided into two copies of Δ^2. See Figures 2 and 3.

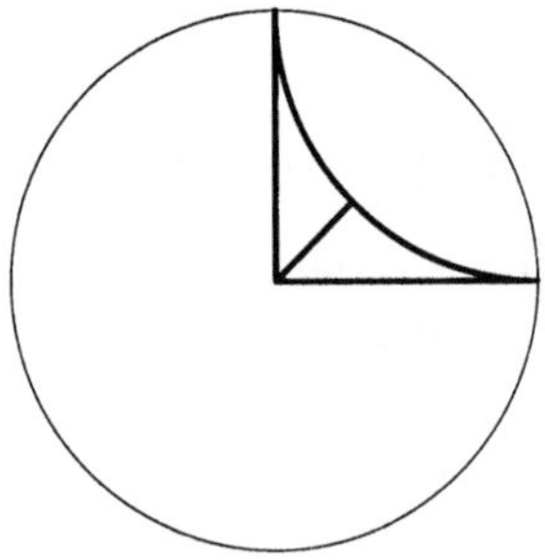

Figure 3. A bisected corner triangle

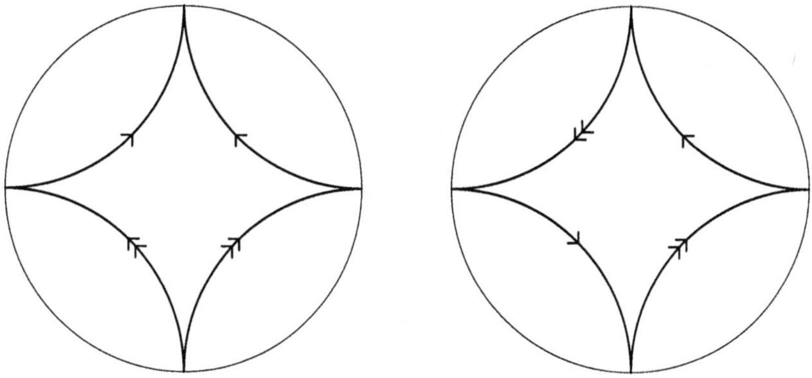

Figure 4. The 3-punctured sphere and the 2-punctured projective plane

The index of Γ_2^2 in Γ^2 is 2. The group Γ_2^2 is a reflection group with respect to the corner triangle $P^2 = T^2$. The Lorentzian matrices that represent the reflections in the sides of P^2 are

$$\begin{pmatrix} -1 & 0 & 0 \\ 0 & 1 & 0 \\ 0 & 0 & 1 \end{pmatrix}, \begin{pmatrix} 1 & 0 & 0 \\ 0 & -1 & 0 \\ 0 & 0 & 1 \end{pmatrix}, \begin{pmatrix} -1 & -2 & 2 \\ -2 & -1 & 2 \\ -2 & -2 & 3 \end{pmatrix}.$$

Let Γ be a torsion-free subgroup of Γ_2^2 of minimal index 4. Then the set $Q^2 = K^2P^2$ is a fundamental polygon for Γ, since K^2 is a set of coset representatives for Γ in Γ_2^2. The set Q^2 is the ideal square with vertices $\pm e_1$ and $\pm e_2$.

The group Γ determines a side-pairing of the sides of Q^2 such that the sides of Q^2 are paired by elements of K^2. There are two such side-pairings for Q^2. One yields the 3-punctured sphere and the other a 2-punctured projective plane. See Figure 4.

We next consider the case $n = 3$. The group Γ^3 is a reflection group with respect to a noncompact 3-simplex Δ^3 in H^3 whose Coxeter diagram is

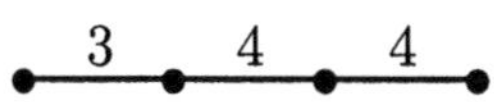

The corner 3-simplex T^3 is subdivided into six copies of Δ^3. See Figure 5. The index of Γ_2^3 in Γ^3 is 12. The group Γ_2^3 is a reflection group with respect to the right-angled polyhedron P^3 which is the union of T^3 and the reflected image of T^3 along its ideal side. The polyhedron P^3 has six sides each of which is congruent to the right-triangle P^2. See Figure 6.

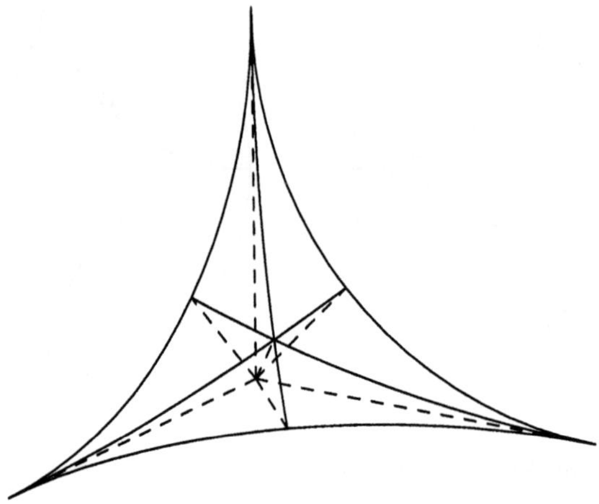

Figure 5. The subdivision of T^3 into six copies of Δ^3

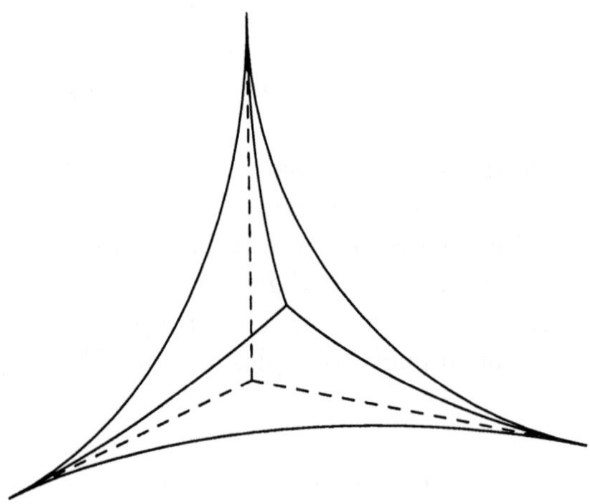

Figure 6. The polyhedron P^3

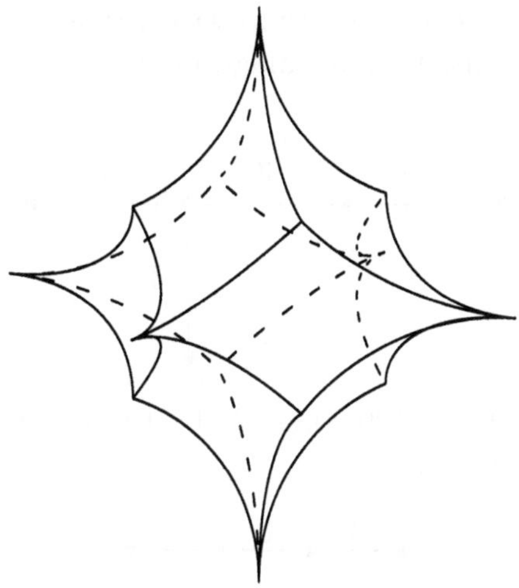

Figure 7. The rhombic dodecahedron Q^3

Let Γ be a torsion-free subgroup of Γ_2^3 of minimal index 8. The set $Q^3 = K^3P^3$ is a fundamental polyhedron for Γ. The polyhedron Q^3 is a rhombic dodecahedron with 14 vertices, 8 actual, and 6 ideal. See Figure 7. All the dihedral angles of Q^3 are $\pi/2$. The sides of Q^3 are paired by elements of K^3. One obtains 13 different hyperbolic 3-manifolds by gluing together the sides of Q^3 by elements of K^3. These manifolds are open with three or four cusps. Only three of these manifolds are orientable. The three orientable manifolds are homeomorphic to the complement of a link in the 3-sphere. In particular, we obtain the complement of the Borromean rings in this way. See Figure 8.

Figure 8. The Borromean rings

5. Congruence 2, Hyperbolic 4-Manifolds

The group Γ^4 is a reflection group with respect to a noncompact 4-simplex Δ^4 in H^4 whose Coxeter diagram is

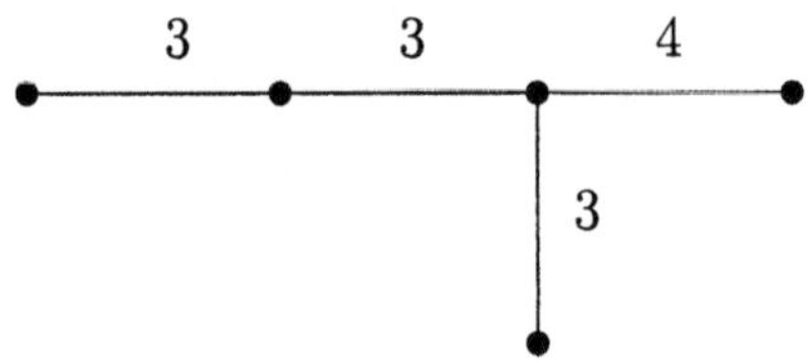

The index of Γ_2^4 in Γ^4 is 120. Let Σ^4 be the Coxeter subgroup of Γ^4 whose Coxeter diagram is

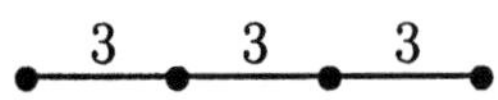

Then Σ^4 has order 120. The group Γ_2^4 is a reflection group with respect to the right-angled polytope

$$P^4 = \Sigma^4 \Delta^4,$$

since Σ^4 is a set of coset representatives for Γ_2^4 in Γ^4.

The ideal vertices of P^4 are the vertices of a regular ideal 4-simplex S^4 in B^4. The polytope P^4 is obtained from S^4 by gluing onto each side of S^4 a corner 4-simplex whose corner vertex becomes an actual vertex of P^4. One of the attached corner simplices is T^4. The polytope P^4 has 10 sides each of which is congruent to P^3. See Figure 9 for a 2-dimensional analogue of P^4.

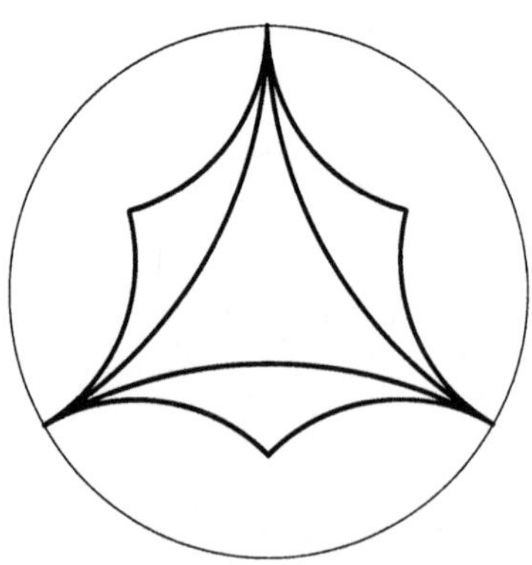

Figure 9. Three corner triangles attached to the sides of an ideal triangle

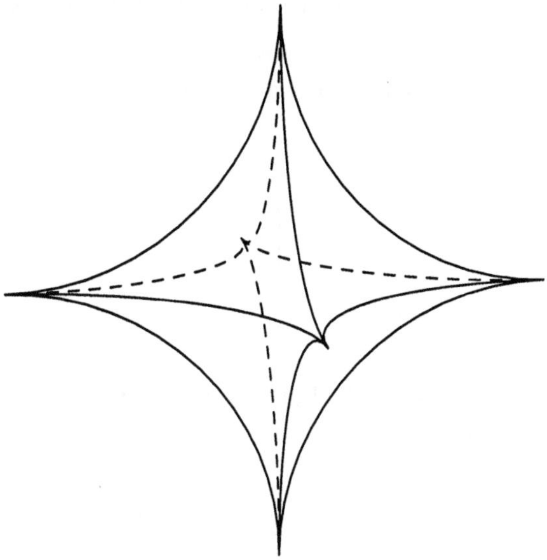

Figure 10. An ideal regular octahedron

Let Γ be a torsion-free subgroup of Γ_2^4 of minimal index 16. The set $Q^4 = K^4P^4$ is a fundamental polytope for Γ. The polytope Q^4 is a regular ideal 24-cell with vertices $\pm e_1, \ldots, \pm e_4$ and

$$(\pm 1/2, \pm 1/2, \pm 1/2, \pm 1/2).$$

The polytope Q^4 has 24 sides each of which is a regular ideal octahedron. See Figure 10. The dihedral angle of Q^4 is $\pi/2$.

The sides of Q^4 are paired by elements of K^4. In 1993, J. Ratcliffe and S. Tschantz [8] showed that one obtains 1171 different hyperbolic 4-manifolds by gluing together the sides of Q^4 by elements of K^4. These manifolds are open with five or six cusps. Only 22 of these manifolds are orientable.

All these manifolds have Euler characteristic 1, and so are minimum volume hyperbolic 4-manifolds. Now 15 of the orientable manifolds have a positive first Betti number, and so these manifolds have n-fold covers for all positive integers n. Thus we have the following theorem.

Theorem 2. *There are orientable open complete hyperbolic 4-manifolds of finite volume whose Euler characteristic is any given positive integer. Therefore, the volume spectrum of hyperbolic 4-manifolds is the set of all positive integral multiples of* $4\pi^2/3$.

6. Congruence 2, Hyperbolic 5-Manifolds

The group Γ^5 is a reflection group with respect to a noncompact 5-simplex Δ^5 in H^5 whose Coxeter diagram is

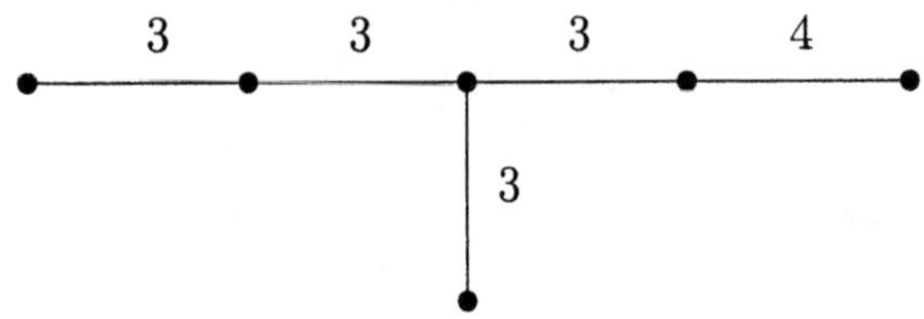

The index of Γ^5_2 in Γ^5 is 1920. Let Σ^5 be the Coxeter subgroup of Γ^5 whose Coxeter diagram is

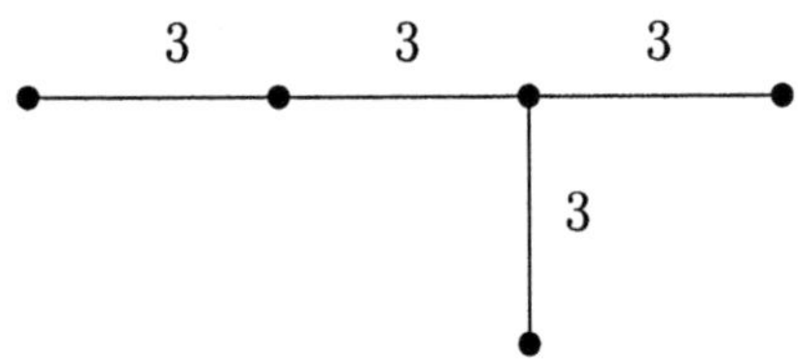

Then Σ^5 has order 1920. The group Γ^5_2 is a reflection group with respect to the right-angled polytope

$$P^5 = \Sigma^5\Delta^5.$$

The polytope P^5 has 16 actual vertices and 10 ideal vertices. The 10 ideal vertices of P^5 are the vertices of a regular ideal 5-dimensional cross polytope R^5. The cross polytope R^5 has 32 sides each of which is a regular ideal 4-simplex.

The polytope P^5 is obtained from the cross polytope R^5 by gluing onto each of 16 mutually nonadjacent sides of R^5 a corner 5-simplex with one of the attached corner 5-simplices being T^5. The corner vertices of the attached corner 5-simplices become the actual vertices of P^5. The polytope P^5 has 16 sides each of which is congruent to P^4.

Let Γ be a torsion-free subgroup of Γ^5_2 of minimal index 32. The set $Q^5 = K^5P^5$ is a fundamental polytope for Γ. All the dihedral angles of Q^5 are $\pi/2$. The polytope Q^5 has 160 actual vertices and 90 ideal vertices. Ten of the ideal vertices, $\pm e_1, \ldots, \pm e_5$, are the vertices of a regular ideal cross polytope O^5.

The polytope Q^5 has 72 sides of two different types. The polytope Q^5 has 32 small sides congruent to P^4 with one small side centered in the middle of each of the 32 orthants of B^5.

The polytope Q^5 has 40 large sides made up of eight copies of P^4 centered about a common infinite edge which we call the axis of the large side. The axes of the large sides of Q^5 are the 40 edges of the cross polytope O^5. The polytope Q^5 has the same symmetry group as the regular cross polytope O^5.

The sides of Q^5 are paired by elements of K^5. In 2002, J. Ratcliffe and S. Tschantz [10] showed that one obtains 3607 different orientable hyperbolic 5-manifolds by gluing together the sides of Q^5 by elements of K^5. These manifolds are open with 10 or 12 cusps. Only 26 of these manifolds have 12 cusps. These manifolds are the first examples of explicitly described hyperbolic 5-manifolds of finite volume.

All these hyperbolic 5-manifolds have volume equal to $28\,\zeta(3)$ which is known to be an irrational number [11]. This stands in contrast to dimension 3, where the irrationality of volumes of hyperbolic 3-manifolds is unknown.

7. A Small Volume Hyperbolic 5-Manifold

We now describe a hyperbolic 5-manifold, of volume $7\zeta(3)/4$, that is obtained as a quotient space of one of our hyperbolic 5-manifolds. The most symmetric 10-cusped hyperbolic 5-manifold M, obtained in the previous section, has a symmetry group of order 1536. The group of symmetries of M has a subgroup of 16 orientation preserving symmetries that acts freely on M. See [10] for details.

Let N be the quotient manifold under the action of this group of order 16. Then N is an orientable hyperbolic 5-manifold with two cusps. The manifold N has $H_1(N) = \mathbb{Z}\oplus\mathbb{Z}_4$, $H_2(N) = \mathbb{Z}_4^2$, $H_3(N) = \mathbb{Z}$, $H_4(N) = \mathbb{Z}$. The first homology groups of the cusps are $\mathbb{Z} \oplus \mathbb{Z}_4^2$ and $\mathbb{Z} \oplus \mathbb{Z}_4$.

The polytope Q^5 is subdivided into 32 copies of the polytope P^5 by the five coordinate hyperplanes $x_i = 0$ for $i = 1, \ldots 5$, and so the manifold M can be subdivided into 32 copies of P^5. The subgroup of symmetries of order 16 of M whose orbit space is N acts freely on the 32 copies of P^5 subdividing M, and so N can be described by gluing together two copies of P^5.

Let ρ be the reflection in the second coordinate hyperplane $x_2 = 0$. Then the polytopes P^5 and $\rho(P^5)$ are adjacent along a common side. We shall describe a side-pairing for the two polytopes P^5 and $\rho(P^5)$ that glues up to the hyperbolic 5-manifold N. Table 3 lists the Lorentz normal vectors of the sides of P^5 and $\rho(P^5)$.

We order the sides of P^5 and $\rho(P^5)$ by the ordering in Table 3. The ith side of $\{P^5, \rho(P^5)\}$ is paired to the jth side of $\{P^5, \rho(P^5)\}$ for $(i, j) = (1, 13)$, $(2, 18)$, $(3, 8)$, $(4, 6)$, $(5, 20)$, $(7, 22)$, $(9, 16)$, $(10, 19)$,

Table 3. The Lorentz normal vectors of the 32 sides of P^5 and $\rho(P^5)$

$$\begin{array}{llll}
s_1 = (-1,\ 0,\ 0,\ 0,\ 0,\ 0) & & s_{17} = (-1,\ 0,\ 0,\ 0,\ 0,\ 0) \\
s_2 = (\ 0,-1,\ 0,\ 0,\ 0,\ 0) & & s_{18} = (\ 0,\ 1,\ 0,\ 0,\ 0,\ 0) \\
s_3 = (\ 0,\ 0,-1,\ 0,\ 0,\ 0) & & s_{19} = (\ 0,\ 0,-1,\ 0,\ 0,\ 0) \\
s_4 = (\ 0,\ 0,\ 0,-1,\ 0,\ 0) & & s_{20} = (\ 0,\ 0,\ 0,-1,\ 0,\ 0) \\
s_5 = (\ 0,\ 0,\ 0,\ 0,-1,\ 0) & & s_{21} = (\ 0,\ 0,\ 0,\ 0,-1,\ 0) \\
s_6 = (\ 0,\ 0,\ 0,\ 1,\ 1,\ 1) & & s_{22} = (\ 0,\ 0,\ 0,\ 1,\ 1,\ 1) \\
s_7 = (\ 0,\ 0,\ 1,\ 0,\ 1,\ 1) & & s_{23} = (\ 0,\ 0,\ 1,\ 0,\ 1,\ 1) \\
s_8 = (\ 0,\ 0,\ 1,\ 1,\ 0,\ 1) & & s_{24} = (\ 0,\ 0,\ 1,\ 1,\ 0,\ 1) \\
s_9 = (\ 0,\ 1,\ 0,\ 0,\ 1,\ 1) & & s_{25} = (\ 0,-1,\ 0,\ 0,\ 1,\ 1) \\
s_{10} = (\ 0,\ 1,\ 0,\ 1,\ 0,\ 1) & & s_{26} = (\ 0,-1,\ 0,\ 1,\ 0,\ 1) \\
s_{11} = (\ 0,\ 1,\ 1,\ 0,\ 0,\ 1) & & s_{27} = (\ 0,-1,\ 1,\ 0,\ 0,\ 1) \\
s_{12} = (\ 1,\ 0,\ 0,\ 0,\ 1,\ 1) & & s_{28} = (\ 1,\ 0,\ 0,\ 0,\ 1,\ 1) \\
s_{13} = (\ 1,\ 0,\ 0,\ 1,\ 0,\ 1) & & s_{29} = (\ 1,\ 0,\ 0,\ 1,\ 0,\ 1) \\
s_{14} = (\ 1,\ 0,\ 1,\ 0,\ 0,\ 1) & & s_{30} = (\ 1,\ 0,\ 1,\ 0,\ 0,\ 1) \\
s_{15} = (\ 1,\ 1,\ 0,\ 0,\ 0,\ 1) & & s_{31} = (\ 1,-1,\ 0,\ 0,\ 0,\ 1) \\
s_{16} = (\ 1,\ 1,\ 1,\ 1,\ 1,\ 2) & & s_{32} = (\ 1,-1,\ 1,\ 1,\ 1,\ 2)
\end{array}$$

(11, 25), (12, 29), (14, 24), (15, 31), (17, 30), (21, 23), (26, 32), (27, 28). The Lorentzian matrices that pair the ith side to the jth side are given below. The second matrix is the identity matrix because side 2 of P^5 is equal to side 18 of $\rho(P^5)$.

$$\begin{pmatrix} 1 & 0 & -1 & 0 & 0 & 1 \\ 0 & 0 & 0 & 1 & 0 & 0 \\ 0 & 0 & -1 & 0 & -1 & 1 \\ 1 & 0 & 0 & 0 & -1 & 1 \\ 0 & 1 & 0 & 0 & 0 & 0 \\ 1 & 0 & -1 & 0 & -1 & 2 \end{pmatrix}, \begin{pmatrix} 1 & 0 & 0 & 0 & 0 & 0 \\ 0 & 1 & 0 & 0 & 0 & 0 \\ 0 & 0 & 1 & 0 & 0 & 0 \\ 0 & 0 & 0 & 1 & 0 & 0 \\ 0 & 0 & 0 & 0 & 1 & 0 \\ 0 & 0 & 0 & 0 & 0 & 1 \end{pmatrix},$$

$$\begin{pmatrix} 0 & 1 & 0 & 0 & 0 & 0 \\ 1 & 0 & 0 & 0 & 0 & 0 \\ 0 & 0 & 1 & -1 & 0 & 1 \\ 0 & 0 & 1 & 0 & -1 & 1 \\ 0 & 0 & 0 & -1 & -1 & 1 \\ 0 & 0 & 1 & -1 & -1 & 2 \end{pmatrix}, \begin{pmatrix} -1 & 0 & 0 & 0 & -1 & 1 \\ 0 & 0 & 1 & 0 & 0 & 0 \\ 0 & 1 & 0 & 0 & 0 & 0 \\ 0 & 0 & 0 & 1 & -1 & 1 \\ -1 & 0 & 0 & 1 & 0 & 1 \\ -1 & 0 & 0 & 1 & -1 & 2 \end{pmatrix},$$

$$\begin{pmatrix} 0 & -1 & -1 & 0 & 0 & 1 \\ 0 & 0 & 1 & 1 & 0 & -1 \\ 1 & 0 & 0 & 0 & 0 & 0 \\ 0 & 0 & 0 & 0 & -1 & 0 \\ 0 & -1 & 0 & -1 & 0 & 1 \\ 0 & -1 & -1 & -1 & 0 & 2 \end{pmatrix}, \begin{pmatrix} 0 & -1 & -1 & 0 & 0 & 1 \\ 0 & 0 & 1 & 1 & 0 & -1 \\ 1 & 0 & 0 & 0 & 0 & 0 \\ 0 & 0 & -2 & 0 & -1 & 2 \\ 0 & -1 & -2 & -1 & -2 & 3 \\ 0 & -1 & -3 & -1 & -2 & 4 \end{pmatrix},$$

$$\begin{pmatrix} -1 & -2 & 0 & -1 & -2 & 3 \\ 0 & -2 & 0 & 0 & -1 & 2 \\ -1 & -2 & -1 & 0 & -2 & 3 \\ 0 & -1 & 0 & 0 & -2 & 2 \\ 0 & -2 & -1 & -1 & -2 & 3 \\ -1 & -4 & -1 & -1 & -4 & 6 \end{pmatrix}, \begin{pmatrix} 0 & 0 & 0 & 0 & 1 & 0 \\ 1 & 0 & 0 & 1 & 0 & -1 \\ 0 & 1 & 0 & 1 & 0 & -1 \\ 0 & 0 & 1 & 0 & 0 & 0 \\ -1 & -1 & 0 & 0 & 0 & 1 \\ -1 & -1 & 0 & -1 & 0 & 2 \end{pmatrix},$$

$$\begin{pmatrix} 0 & -1 & 0 & -1 & 0 & 1 \\ 0 & 2 & 2 & 1 & 1 & -3 \\ 0 & -1 & 0 & 0 & -1 & 1 \\ 1 & 0 & 0 & 0 & 0 & 0 \\ 0 & -2 & -1 & 0 & 0 & 2 \\ 0 & -3 & -2 & -1 & -1 & 4 \end{pmatrix}, \begin{pmatrix} -2 & -1 & 0 & -1 & -2 & 3 \\ 0 & 0 & 0 & 1 & 1 & -1 \\ 0 & -1 & 0 & 0 & -1 & 1 \\ -1 & 0 & 0 & 0 & -2 & 2 \\ 0 & 0 & 1 & 0 & 0 & 0 \\ -2 & -1 & 0 & -1 & -3 & 4 \end{pmatrix},$$

$$\begin{pmatrix} 0 & 0 & 0 & 0 & 1 & 0 \\ 1 & 0 & 0 & 1 & 0 & -1 \\ -2 & -1 & -2 & -1 & 0 & 3 \\ -2 & 0 & -1 & 0 & 0 & 2 \\ -1 & -1 & 0 & 0 & 0 & 1 \\ -3 & -1 & -2 & -1 & 0 & 4 \end{pmatrix}, \begin{pmatrix} -1 & -2 & 0 & 0 & 0 & 2 \\ 2 & 1 & 0 & 0 & 0 & -2 \\ 0 & 0 & 1 & 0 & 0 & 0 \\ 0 & 0 & 0 & 1 & 0 & 0 \\ 0 & 0 & 0 & 0 & 1 & 0 \\ -2 & -2 & 0 & 0 & 0 & 3 \end{pmatrix},$$

$$\begin{pmatrix} 1 & 0 & 0 & -1 & 0 & 1 \\ 0 & 0 & 0 & 0 & -1 & 0 \\ 1 & 0 & -1 & 0 & 0 & 1 \\ 0 & -1 & 0 & 0 & 0 & 0 \\ 0 & 0 & -1 & -1 & 0 & 1 \\ 1 & 0 & -1 & -1 & 0 & 2 \end{pmatrix}, \begin{pmatrix} 0 & 0 & -1 & -1 & 0 & 1 \\ 0 & -1 & 1 & 0 & 0 & -1 \\ -1 & 1 & -1 & -1 & 1 & 2 \\ -1 & 0 & -1 & 0 & 0 & 1 \\ 0 & 0 & -1 & 0 & 1 & 1 \\ -1 & 1 & -2 & -1 & 1 & 3 \end{pmatrix},$$

$$\begin{pmatrix} -1 & 2 & -1 & -2 & 0 & 3 \\ 0 & -2 & 0 & 1 & 0 & -2 \\ 0 & 2 & -1 & -2 & -1 & 3 \\ -1 & 2 & 0 & -2 & -1 & 3 \\ 0 & 1 & 0 & -2 & 0 & 2 \\ -1 & 4 & -1 & -4 & -1 & 6 \end{pmatrix}, \begin{pmatrix} 0 & 1 & -2 & 0 & 0 & 2 \\ -1 & 0 & 0 & 0 & 0 & 0 \\ 0 & 0 & -1 & -1 & 0 & 1 \\ 0 & 0 & -1 & 0 & -1 & 1 \\ 0 & 2 & -2 & -1 & -1 & 3 \\ 0 & 2 & -3 & -1 & -1 & 4 \end{pmatrix}.$$

The set $P = P^5 \cup \rho(P^5)$ is a right-angled convex polytope with 20 sides. The above side-pairing of P determines a facet-pairing of P in the sense of §11.1 of [6] so that P is an inexact fundamental polytope for the discrete group Γ generated by the above 16 matrices. The hyperbolic manifold N is isometric to the space form H^5/Γ.

All the entries of the generators of Γ are integers, and so Γ is a torsion-free subgroup of Γ^5. The volume of N is $7\zeta(3)/4$. The volume of the fundamental domain Δ^5 of Γ^5 is $7\zeta(3)/15360$, and so the index of Γ in Γ^5 is 3840. The spherical Coxeter group $(3,3,3,4)$ has order 3840, and so Γ is a torsion-free subgroup of Γ^5 of minimal index. Thus N is a minimal volume, integral, hyperbolic 5-manifold.

R. Kellerhals [5] has proved, by a horoball packing argument, that if M is an m-cusped hyperbolic 5-manifold, then

$$\mathrm{vol}(M)/m \;>\; 0.3922.$$

Now N has two cusps and

$$\mathrm{vol}(N)/2 \;=\; 1.0518\ldots.$$

This suggests that N has very small volume, and perhaps N is a minimal volume open hyperbolic 5-manifold.

Acknowledgements

The author wishes to thank the organizers of the János Bolyai conference for putting on such a splendid conference and making these proceedings possible. The author also wishes to express his sadness in the passing away of H.S.M. Coxeter who attended the author's talk and was a source of inspiration for many of the ideas in the author's work.

Bibliography

[1] M. Burger, T. Gelander, A. Lubotzky, and S. Mozes, Counting hyperbolic manifolds, *Geom. Funct. Anal.* **12** (2002), 1161-1173.

[2] S.-S. Chern, On the curvature and characteristic classes of a Riemannian manifold, *Abh. Math. Sem. Univ. Hamburg*, **20** (1955), 117-126.

[3] M. W. Davis, A hyperbolic 4-manifold, *Proc. Amer. Math. Soc.* **93** (1985), 325-328.

[4] R. Kellerhals, On the volumes of hyperbolic 5-orthoschemes and the Trilogarithm, *Comment. Math. Helvetici*, **67** (1992), 648-663.

[5] R. Kellerhals, Volumes of cusped hyperbolic manifolds, *Topology*, **37** (1998), 719-734.

[6] J. G. Ratcliffe, *Foundations of Hyperbolic Manifolds*, Graduate Texts in Math., vol. **149**, Springer-Verlag, Berlin, Heidelberg, and New York, 1994.

[7] J. G. Ratcliffe and S. T. Tschantz, Volumes of integral congruence hyperbolic manifolds, *J. Reine Angew. Math.* **488** (1997), 55-78.

[8] J. G. Ratcliffe and S. T. Tschantz, The volume spectrum of hyperbolic 4-manifolds, *Experimental Math.* **9** (2000), 101-125.

[9] J. G. Ratcliffe and S. T. Tschantz, On the Davis hyperbolic 4-manifold, *Topology Appl.* **111** (2001), 327-342.

[10] J. G. Ratcliffe and S. T. Tschantz, Integral congruence two hyperbolic 5-manifolds, preprint, arXiv:math.GT/0308125 v1, 13 Aug 2003.

[11] A. van der Poorten, A proof that Euler missed ... Apéry's proof of the irrationality of $\zeta(3)$, *Math. Intelligencer*, **1** (1979), 195-203.

[12] E. B. Vinberg, Discrete groups generated by reflections in Lobacevskii spaces, *Math. USSR-Sbornik*, **1** (1967), 429-444.

[13] H.-C. Wang, Topics on totally discontinuous groups, In: *Symmetric Spaces*, W. M. Boothby and G. L. Weiss, eds., Pure Appl. Math., Vol. 8, Marcel Dekker, New York (1972), 459-487.

REAL-TIME ANIMATION IN HYPERBOLIC, SPHERICAL, AND PRODUCT GEOMETRIES

Jeffrey R. Weeks
15 Farmer Street, Canton, NY 13617, USA

Abstract Standard 3D graphics algorithms for flat space extend naturally to hyperbolic and spherical spaces, allowing real-time interactive graphics to run in hardware on ordinary desktop computers. Small modifications extend the algorithms to the anisotropic product spaces $H^2 \times E$ and $S^2 \times E$.

Keywords: Hyperbolic geometry, visualization, interactive computer graphics, product spaces.

1. Introduction

For roughly 2000 years people tried unsuccessfully to prove that Euclid's fifth postulate followed from his other postulates and axioms. By 1823 János Bolyai made the conceptual leap of negating the fifth postulate and taking seriously the resulting geometry, namely hyperbolic geometry. Bolyai's intuitive image of hyperbolic space was surely very good, but it wasn't until 1868 that Eugenio Beltrami constructed several models of hyperbolic geometry including those that later became known as the Klein model, the Poincaré disk model, and the upper half space model. These models showed that hyperbolic geometry is consistent if Euclidean geometry is, and provided a technical foundation for further research. Nevertheless, to visualize hyperbolic geometry it's fun to start with a physical model and feel with your hands what hyperbolic geometry is really like.

Activity #1. Construct a physical model of the hyperbolic plane by assembling paper or plastic triangles so that exactly seven triangles meet at each vertex. If you start at a point and work outward, tiling a disk of radius 1, then radius 2, then radius 3, etc., you will experience the exponential growth of area in hyperbolic geometry. Construct the analogous model of a sphere by assembling triangles so that five meet at each vertex, and construct the Euclidean plane with triangles meeting six per vertex. (This construction is originally due to Bill Thurston. Detailed instructions and a template for the triangles appear in Chapter 10 of [1].)

In the tiling of the Euclidean plane you constructed in Activity #1, the unit circle is approximated as a hexagon, with perimeter 6. In the tiling of the hyperbolic plane, by contrast, the unit circle appears as a heptagon, with perimeter 7, while in the spherical tiling it is a pentagon with perimeter 5. More importantly, as the radius r increases ($r = 1, 2, 3, \dots$), the perimeter p of the "combinatorial circle" grows linearly in the Euclidean case ($p = 6, 12, 18, \dots$), grows exponentially in the hyperbolic case ($p = 7, 21, 56, \dots$), but collapses to zero in the spherical case ($p = 5, 5, 0$). On a smooth hyperbolic plane, Euclidean plane, or sphere, it is straightforward to prove that the circumference of a circle grows as $2\pi \sinh r$, $2\pi r$, or $2\pi \sin r$, respectively.

If a 2-dimensional person living in the Euclidean plane views an object exactly one unit away, the object subtends some angle in her field of view. If she views the same object, from the same distance, in the hyperbolic plane, the object occupies a smaller fraction of the unit circle, not because the object is smaller (it isn't) but because the circumference of the unit circle is larger; therefore the object subtends a narrower angle in her field of view and appears smaller to her. Conversely, if she views the same object from the same distance on the sphere, it occupies a larger fraction of the unit circle, subtends a wider angle in her field of view, and looks bigger to her. The same principle applies in 3-dimensional spaces: distant objects look smaller in hyperbolic geometry and bigger in spherical geometry.

Activity #2. Download the Curved Spaces software from `www.geometrygames.org`. Start the program *Curved Spaces Complete D3D.exe* and read the brief **Instructions** from the **Help** menu to learn to fly the spaceship. Choose **Open** from the **File** menu, go to the folder *Sample Spaces/basic*, and explore the various regular and semi-regular tilings of hyperbolic, Euclidean, and spherical space. Figures 1, 2, and 3 show screenshots of typical tilings. The program itself lets you fly freely, with an option for stereoscopic 3D if you have red-blue glasses. (Note: If you have trouble starting the program on an older computer, you may use the OpenGL version *Curved Spaces Complete GL.exe* instead, or download a free copy of the DirectX graphics library required by the D3D version, as explained in the file *Instructions-Complete.txt*.)

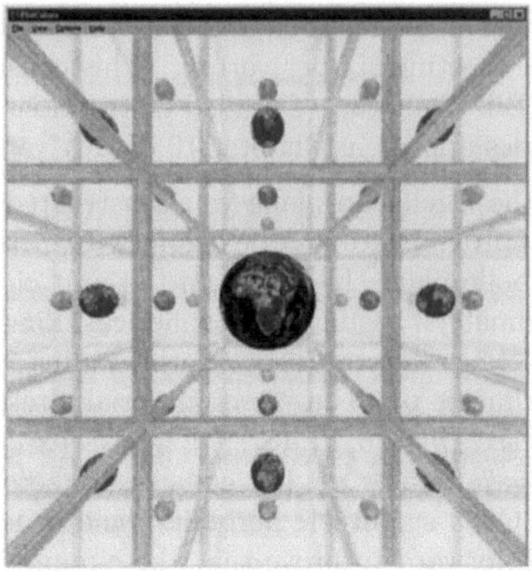

Figure 1. A tiling of Euclidean space by cubes. The Curved Spaces software also lets you explore more exotic Euclidean tilings using rhombic dodecahedra, truncated octahedra, and other shapes.

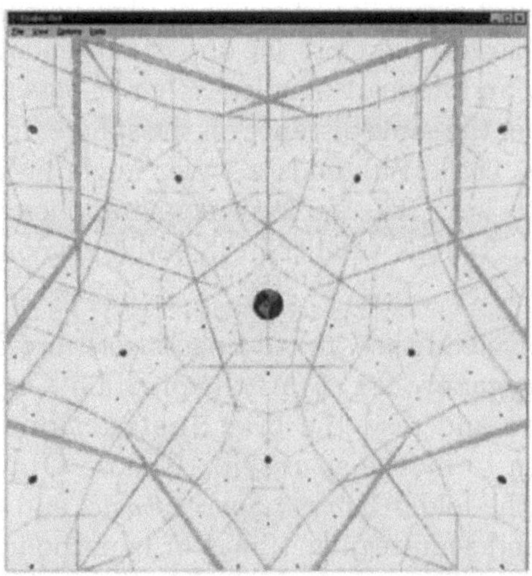

Figure 2. A tiling of hyperbolic space by dodecahedra. If you start with any face (such as the pentagon at the center of the figure) and extend it to all sides, you'll see a copy of the hyperbolic plane H^2 sitting inside H^3. Each such copy of H^2 approaches a limit circle on the sphere at infinity. Look carefully and you will see that the face planes meet at right angles. This is the tiling used in the video *Not Knot* [2]

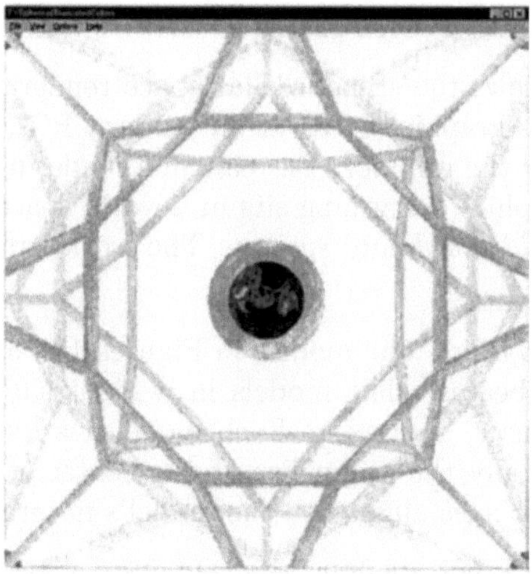

Figure 3. A tiling of the 3-sphere by truncated cubes. Even though they are really the same size, the nearer planet appears smaller than the more distant planet it eclipses, because the more distant planet subtends a larger angle in our field of view. This is a direct consequence of the fact that the circumference of a circle in the 3-sphere decreases when its radius exceeds $\pi/2$.

2. Overview

Interactive real-time hyperbolic and spherical animations (illustrated in Figures 2 and 3) run in hardware on the 3D graphics card of any modern PC. They do not require a custom-built graphics card, because the same circuitry that does flat-space graphics works equally well for curved spaces. The article "Real-time rendering in curved spaces" [3] provides an elementary yet complete explanation of hyperbolic and spherical rendering and the beautiful geometry supporting it. Here we summarize the main ideas in Sections 3, 4 and 5.

The rendering algorithm used in the isotropic spaces H^3, E^3 and S^3 generalizes naturally to the anisotropic product spaces $H^2 \times E$ and $S^2 \times E$. Even though the algorithm differs from the isotropic case enough that it cannot run entirely in hardware, it is still simple and beautiful. Section 6 describes this new algorithm (which has not been published elsewhere) and tells where to obtain the software that implements it.

Section 7 describes ongoing work to develop and implement a rendering algorithm for the remaining homogeneous 3-dimensional geometries, namely the twisted geometries and solv geometry.

3. Flat Space Rendering

This section reviews the standard flat space rendering algorithm, as implemented in the graphics cards of all modern PCs. For color illustrations, see [3]; for the full story on real-time rendering, see [4].

In computer graphics everything sits in a world space, parameterized by an (x, y, z) *world coordinate system.* The world space contains the following:

- *Models (or objects).* The models in Figures 1, 2 and 3 are planets and wooden beams. The models in a flight simulator might be airplanes, houses, and trees. Each model has its own coordinate system: typically the origin lies at the center of the model, and the coordinate axes align with the model's natural axes.

- *Camera.* The camera has its own coordinate system. By convention the camera always looks down its own z-axis, with its x-axis pointing right and its y-axis pointing up[1].

- *View Frustum.* When you look out through the viewfinder of a camera the volume of space you see is a pyramid. In computer graphics this pyramid is usually clipped by what are known as the near clipping plane and the far clipping plane (Figure 4). That is, the computer doesn't draw anything nearer than the near clipping plane nor anything farther than the far clipping plane. So the volume of space you see is not a full pyramid, but only a frustum of a pyramid. The view frustum is parameterized by *projection coordinates*, which run from -1 to $+1$ in both the x- and y-directions and from 0 to 1 in the z-direction, as shown in Figure 4.

We have now defined four coordinate systems: world coordinates, model coordinates, camera coordinates, and projection coordinates. For clarity of thought, I'd like to separate these four coordinate systems into their own spaces (Figure 5).

This separation into four distinct spaces makes it easier to keep track of the various transformations mapping one coordinate system to another. Specifically:

- The *model transformation* maps a model into world space. A typical scene has many models, each with its own model transformation.

[1] In the DirectX graphics library the camera looks down the positive z-axis while in the OpenGL graphics library it looks down the negative z-axis. Henceforth we won't worry about such technical details. Programmers may consult [3] or [4] for library-specific details.

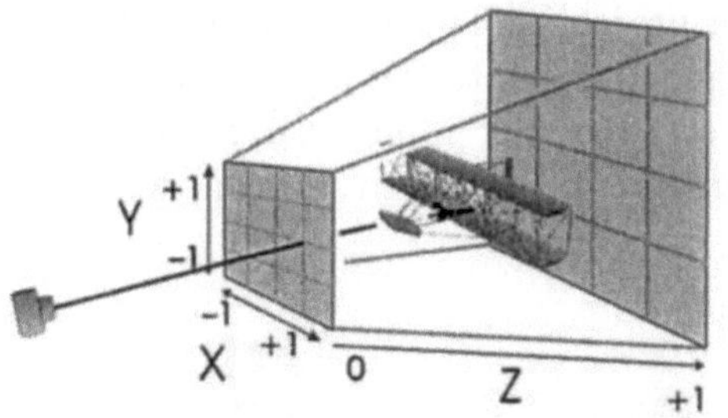

Figure 4. The volume of space seen by the camera is a pyramid. The near and far clipping planes truncate the pyramid to a frustum. Projection coordinates parameterize the frustum in a standard way.

$$\underset{\text{model space}}{R^3} \xrightarrow[]{\text{model transformation}} \underset{\text{world space}}{R^3} \xrightarrow[]{\text{view transformation}} \underset{\text{camera space}}{R^3} \xrightarrow[]{\text{projection transformation}} \underset{\text{projection space}}{R^3}$$

Figure 5. To render an image of an object, for example an airplane, your computer's graphics card maps the airplane from model space to world space, thence to camera space, and finally to projection space (cf. Figure 6).

- The *camera transformation* maps the camera into world space. Its inverse is called the *view transformation*, which maps the world into camera space. It's more intuitive to think of placing the camera into world space, but for computer graphics it's more useful to think of the inverse operation. That is, you image the camera sitting still in its own space, and moving the whole world to sit in front of the camera.

- The *projection transformation* maps the view frustum (Figure 4) to a standard cube (Figure 6).

Let us pause to consider what it means to make a perspectively correct 2-dimensional image of a 3-dimensional scene. Imagine you hold up a pane of glass in front of your favorite scene and – keeping both the glass and your eye perfectly still – you paint each point on the glass according to the color of the object you see beyond it. The painted image provides a perfect record of the scene, in the sense that anybody who looks at it will see the same objects in the same directions in which you saw them in real life.

Computer graphics uses a similar system: to render an image of a 3-dimensional object, we run a line from each point on the object to the camera, and where the line crosses the near clipping plane we color a pixel (Figure 4). In projection space it's even easier: each point on the object projects orthogonally onto the plane of the picture (Figure 6). In other words, given an object at point (x, y, z) in projection space, (x, y) tells where to draw that object on the screen, while z tells how deep it is into the scene.

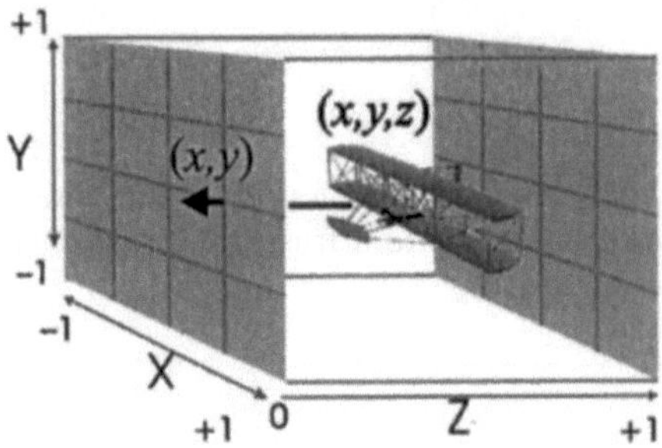

Figure 6. In projection space the object's (x, y) coordinates tell its position on the rendered image, while its z coordinate tells the depth into the scene.

The model, view, and projection transformations are realized using matrices. Rotation matrices are easily constructed using sines and cosines. Translation is more difficult, because no 3×3 matrix can possibly take the origin $(0, 0, 0)$ anywhere but to itself. The standard solution in computer graphics is to add a fourth coordinate, writing (x, y, z) as $(x, y, z, 1)$. As long as the fourth coordinate is 1, the fourth column of the matrix determines the translation:

$$\begin{pmatrix} x+a \\ y+b \\ z+c \\ 1 \end{pmatrix} = \begin{pmatrix} 1 & 0 & 0 & a \\ 0 & 1 & 0 & b \\ 0 & 0 & 1 & c \\ 0 & 0 & 0 & 1 \end{pmatrix} \begin{pmatrix} x \\ y \\ z \\ 1 \end{pmatrix}$$

To realize an arbitrary rigid motion of Euclidean 3-space, replace the 3×3 identity submatrix with any desired rotation:

$$\begin{pmatrix} x' \\ y' \\ z' \\ 1 \end{pmatrix} = \begin{pmatrix} * & * & * & a \\ * & * & * & b \\ * & * & * & c \\ 0 & 0 & 0 & 1 \end{pmatrix} \begin{pmatrix} x \\ y \\ z \\ 1 \end{pmatrix}$$

Geometrically, appending the fourth coordinate with a value of $w = 1$ realizes the model, world, and camera spaces as slices at height $w = 1$ in 4-dimensional Euclidean space (Figure 7).

The model and view transformations are now linear, represented as 4×4 matrices, and your graphics card is hard-wired to apply these

$$\underset{\text{model space}}{R^4} \xrightarrow{\text{model transformation}} \underset{\text{world space}}{R^4} \xrightarrow{\text{view transformation}} \underset{\text{camera space}}{R^4} \xrightarrow{\text{projection transformation}} \underset{\text{projection space}}{R^4}$$

Figure 7. Writing (x, y, z) as $(x, y, z, 1)$ realizes the model, world, camera and projection spaces as 3-dimensional slices of R^4 at height $w = 1$, and the model and view transformation become linear, represented by 4×4 matrices.

matrices. However, when we map camera space into projection space, something different is going on. We still use a 4×4 matrix, with the pleasingly simple form

$$\begin{pmatrix} 1 & 0 & 0 & 0 \\ 0 & 1 & 0 & 0 \\ 0 & 0 & 1 & -N \\ 0 & 0 & 1 & 0 \end{pmatrix} \begin{pmatrix} x \\ y \\ z \\ 1 \end{pmatrix} = \begin{pmatrix} x \\ y \\ z - N \\ z \end{pmatrix}$$

where N is the distance from the camera to the near clipping plane. The only complication is that this matrix violates our convention that the last coordinate be 1. Fortunately the cure is simple: divide through by the last coordinate to push everything back to height 1:

$$\begin{pmatrix} 1 & 0 & 0 & 0 \\ 0 & 1 & 0 & 0 \\ 0 & 0 & 1 & -N \\ 0 & 0 & 1 & 0 \end{pmatrix} \begin{pmatrix} x \\ y \\ z \\ 1 \end{pmatrix} = \begin{pmatrix} x \\ y \\ z - N \\ z \end{pmatrix} \to \begin{pmatrix} x/z \\ y/z \\ 1 - N/z \\ 1 \end{pmatrix}.$$

Geometrically the matrix maps the view frustum in camera space (Figure 4) to a sloping hyperplane in projection space, and then dividing through by the last coordinate projects it onto the standard cube at height $w = 1$ (Figure 6). For further details and illustrations, please see [3].

In summary, when you ask your computer to draw a picture of an object, it

(1) applies the model, view, and projection matrices, and then

(2) divides through by the fourth coordinate,

giving you an image of the object in perfect perspective. Your 3D graphics card implements the whole procedure in hardware, and can therefore draw detailed scenes very, very fast.

4. Spherical Rendering

Our goal is to understand computer graphics not only in flat space (Figure 1), but also in hyperbolic space (Figure 2) and in the 3-sphere

(Figure 3). Luckily hyperbolic and spherical rendering are no harder than flat space rendering. In fact, they're almost exactly the same! Let's begin with spherical rendering.

The spherical rendering algorithm (Figure 8) is almost identical to the flat space algorithm (Figure 7), except that the model, world, and camera spaces are now 3-spheres rather than flat slices at height $w = 1$. The model and view transformations are given by orthogonal 4×4 matrices, elements of the orthogonal group $O(4)$.

$$\underset{\text{model space}}{S^3} \xrightarrow[]{\text{model transformation}} \underset{\text{world space}}{S^3} \xrightarrow[]{\text{view transformation}} \underset{\text{camera space}}{S^3} \xrightarrow[]{\text{projection transformation}} \underset{\text{projection space}}{R^3}$$

Figure 8. For spherical rendering, the model, world and camera spaces are 3-spheres, and the model and view transformations are rotations in $O(4)$.

The projection transformation merits careful thought. How do we create a perspectively correct image in a 3-sphere? Assume the object of interest (for example the airplane of Figure 4) lies in front of the camera in camera space. A naive – but correct! – approach would be to project the object radially from the 3-sphere onto the tangent space. The tangent space is the flat Euclidean 3-space touching the 3-sphere exactly where the camera sits. We can then apply the standard flat space rendering algorithm (Section 3), because a camera in the tangent space sees the same objects in the same directions that the original camera sees on the 3-sphere itself, and so the resulting image is guaranteed to be correct. Algebraically the algorithm looks like this:

$$\begin{pmatrix} x \\ y \\ z \\ w \end{pmatrix} \xrightarrow{\substack{\text{project} \\ \text{to tangent} \\ \text{space}}} \begin{pmatrix} x/w \\ y/w \\ z/w \\ 1 \end{pmatrix} \xrightarrow{\substack{\text{apply} \\ \text{projection} \\ \text{matrix}}} \begin{pmatrix} x/w \\ y/w \\ z/w - N \\ z/w \end{pmatrix} \xrightarrow{\substack{\text{devide by} \\ \text{fourth} \\ \text{coordinate}}} \begin{pmatrix} x/z \\ y/z \\ 1 - Nw/z \\ 1 \end{pmatrix}$$

What if we're feeling lazy and skip the first step? What if we start with a point (x, y, z, w) on the 3-sphere and immediately apply the flat space algorithm? As the following computation shows, we get the same result!

$$\begin{pmatrix} x \\ y \\ z \\ w \end{pmatrix} \xrightarrow{\substack{\text{apply} \\ \text{projection} \\ \text{matrix}}} \begin{pmatrix} x \\ y \\ z - Nw \\ z \end{pmatrix} \xrightarrow{\substack{\text{devide by} \\ \text{fourth} \\ \text{coordinate}}} \begin{pmatrix} x/z \\ y/z \\ 1 - Nw/z \\ 1 \end{pmatrix}$$

Indeed we must get the same result, because (x, y, z, w) and $(x/w, y/w, z/w, 1)$ are scalar multiples of each other, thus their images after applying the projection matrix must also be scalar multiples of each other, and normalizing the last coordinate to 1 then makes them equal.

This is good news! The preceding paragraph proves we can naively apply the same old flat space projection algorithm, and we'll get a perspectively correct picture in a 3-sphere!

Technical note: The usual flat space projection matrix works in the 3-sphere, but defines a view frustum restricted to the nearest quarter of the space. A minor adjustment extends it to include the entire hemisphere in front of the camera, and an additional adjustment includes the back hemisphere as well. Please see [3] for details.

In summary, the algorithm for rendering perspectively correct images in the 3-sphere is identical to the flat space algorithm, namely

(1) apply the model, view, and projection matrices, and then

(2) divide through by the fourth coordinate,

except that the 4×4 matrices are now orthogonal matrices rather than the flat space matrices of Section 3. Fortunately the circuitry on your 3D graphics card works equally well for all types of matrices, and thus renders spherical images in hardware just as quickly and accurately as it does flat space images. The fact that the graphics card works perfectly in situations beyond the flat space graphics envisioned by the engineers who created it is a real tribute to the beauty and elegance of mathematics, as well as the good design on the part of the engineers.

5. Hyperbolic Rendering

Hyperbolic rendering works just like flat space rendering (Section 3) and spherical rendering (Section 4). The key is to work in the Minkowski space model of hyperbolic space. Just as a 3-sphere may be defined as $S^3 = \{v \in R^4 \text{ such that } |v|^2 = 1\}$, where length is given by the Euclidean metric $|v|^2 = x^2 + y^2 + z^2 + w^2$, hyperbolic 3-space may be defined as $H^3 = \{v \in R^4 \text{ such that } |v|^2 = -1\}$, where length is given by the Lorentz metric $|v|^2 = x^2 + y^2 + z^2 - w^2$. This definition of H^3 yields the equation $x^2 + y^2 + z^2 - w^2 = -1$, which corresponds to a hyperboloid of two sheets. By convention we ignore the lower sheet $(w < 0)$ and work only with the upper sheet $(w > 0)$. The hyperboloid looks positively curved to our Euclidean eyes, but has constant negative curvature when measured using the Lorentz metric.

Readers unfamiliar with the Lorentz metric may be interested to know that the Lorentz metric is Mother Nature's definition of length, not in space but in spacetime. That is, just as the Euclidean squared length $x^2 + y^2 + z^2$ is a physically meaningful, coordinate independent quantity in flat space, the Lorentz squared length $x^2 + y^2 + z^2 - t^2$ is a physically meaningful, coordinate independent quantity in flat spacetime [5] and easily resolves the paradoxes of special relativity.

Just as rigid motions of the 3-sphere are those that preserve Euclidean distances, the rigid motions of the hyperbolic plane are those that preserve Lorentz distances. The former comprise the orthogonal group $O(4)$, while the latter comprise the Lorentz group $O(3,1)$. One may construct an arbitrary matrix in $O(4)$ using sines and cosines, while matrices in $O(3,1)$ typically require hyperbolic sines and cosines as well[2].

Activity #3. If you want to visualize Lorentz transformations, download the program *KaleidoTile* from `www.geometrygames.org` and select one of the hyperbolic tilings, say (2,3,7). KaleidoTile will put you inside the hyperboloid, looking down towards the origin. As you drag the tiling with the mouse, you will see the Lorentz transformations. To see the concavity of the hyperboloid in true perspective, choose **Stereoscopic 3D** from *KaleidoTile's* **Help** menu and follow the instructions.

The hyperbolic rendering algorithm (Figure 9) is almost identical to the flat space algorithm (Figure 7) and the spherical algorithm (Figure 8), except that the model, world, and camera spaces are now hyperboloids in Minkowski space. The model and view transformations are given by 4×4 Lorentz matrices, elements of the group $O(3,1)$. Just as in the spherical case, the projection transformation can be realized by projecting onto the tangent space and then applying the flat space algorithm or, more simply, applying the flat space algorithm directly, without projecting onto the tangent space.

In summary, the algorithm for rendering perspectively correct images in hyperbolic 3-space is identical to the flat space algorithm, namely

(1) apply the model, view, and projection matrices, and then

(2) divide through by the fourth coordinate,

[2] Just as the ordinary sine and cosine are defined to be the coordinates of the point you reach by going a distance d along a circle $S^1 = \{(x,y) \in R^2 \text{ such that } x^2 + y^2 = 1\}$, the hyperbolic sine and cosine are defined to be the coordinates of the point you reach by going a distance d along a hyperbolic line $H^1 = \{(x,y) \in R^2 \text{ such that } x^2 - y^2 = -1\}$. Just as the distance d along the circle is measured using the Euclidean metric, the distance d along the hyperbolic line is measured using the Lorentz metric.

$$\underset{\text{model space}}{H^3} \xrightarrow[]{\text{model transformation}} \underset{\text{world space}}{H^3} \xrightarrow[]{\text{view transformation}} \underset{\text{camera space}}{H^3} \xrightarrow[]{\text{projection transformation}} \underset{\text{projection space}}{R^3}$$

Figure 9. For hyperbolic rendering, the model, world and camera spaces are copies of H^3, and the model and view transformations are elements of $O(3,1)$.

except that the 4×4 matrices are now Lorentz matrices rather than the flat space matrices of Section 3 or the orthogonal matrices of Section 4. Fortunately the circuitry on your 3D graphics card works well with Lorentz matrices and renders hyperbolic images quickly and accurately.

6. Product Space Rendering

A cylinder is a stack of circles (Figure 10). More formally, it is the product $S^1 \times E$ of a circle S^1 and a line E (see Chapter 6 of [1] for an elementary introduction to products). Similarly, a stack of surfaces M^2 makes a 3-dimensional product space $M^2 \times E$ (Chapter 18 of [1]). In the present article the surface M^2 may be the hyperbolic plane H^2, the Euclidean plane E^2, or the sphere S^2. The product $E^2 \times E$ gives Euclidean 3-space E^3, but we include it here for sake of completeness.

Activity #4. Download the *Product Spaces* program from the Chapter 6 listing at `www.geometrygames.org/SoS`, read the **Navigation** instructions on its **Help** menu to learn to fly, and explore the forests in $H^2 \times E$, $E^2 \times E$, and $S^2 \times E$. Figures 11-14 show sample grayscale images of each space – the program itself lets you fly in real time in color.

The forest in Figure 11 shows the effects of exponential growth in hyperbolic geometry. Even at a modest distance the number of trees becomes very large. The trees look thin because their cross sections lie in a horizontal hyperbolic slice of $H^2 \times E$ and subtend a narrow angle in our field of view (cf. Section 1), while their heights lie in a vertical Euclidean slice of $H^2 \times E$ and subtend the same angle they would in E^3.

You might be surprised to see that the forest floor looks flat in Figure 11. Even though the forest floor is intrinsically a hyperbolic plane H^2, it is extrinsically flat ("totally geodesic") in $H^2 \times E$, just as a cross section of the cylinder $S^1 \times E$ in Figure 10 is intrinsically a circle S^1 yet lies straight ("geodesic") in the ambient $S^1 \times E$, bending to neither one side nor the other.

The forest floor looks flat near ground level in $S^2 \times E$ as well (Figure 13), but now the trees look fat because their cross sections lie in a hor-

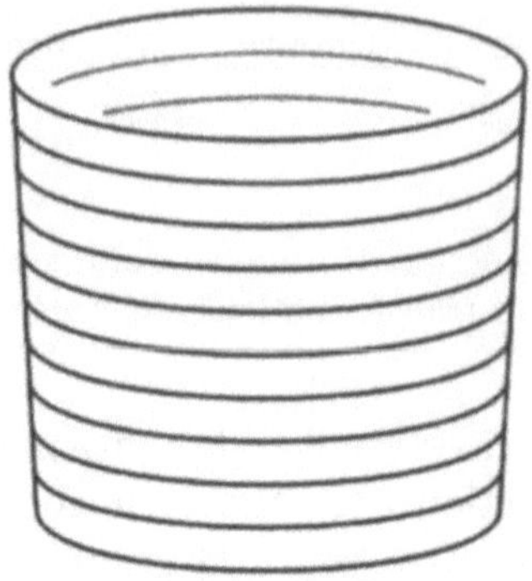

Figure 10. A cylinder is a product $S^1 \times E$, visualized here as a stack of circles. If you replace each circle S^1 with a curved surface M^2, you'll get the product space $M^2 \times E$.

Figure 11. A forest in $H^2 \times E$. The number of trees grows exponentially with the distance.

izontal spherical slice S^2 of $S^2 \times E$ while their heights lie in a vertical Euclidean slice $E \times E$. If we view the forest from above the treetops (Figure 14), then even though $S^2 \times E$ is simply connected we see repeating images of each tree: our line of sight can spiral several times around $S^2 \times E$ before reaching a given tree, just as an observer's line of sight could spiral several times around the cylinder in Figure 10 before

Figure 12. A forest in $E^2 \times E$. The number of trees grows as the square of the distance.

Figure 13. A forest in $S^2 \times E$, viewed near ground level. The whole forest contains only six trees.

Figure 14. The same forest in $S^2 \times E$, viewed from above the treetops. We see infinitely many images of each tree, repeating off into the fog.

arriving at a given object. If there were no fog we would see infinitely many images of each tree in $S^2 \times E$. However, when we are close to the ground (Figure 13) the repeating images are less noticeable because the nearest image of a given tree trunk exactly eclipses all the rest.

You might be wondering why gaps appear in the forest floor in Figures 13 and 14. If we stand at the north pole of S^2 and look a distance π in any direction we see the south pole. In other words, the south pole of S^2 gets stretched out to become a circle of radius π on the forest floor in Figures 13 and 14. Stretching a single point to become a circle is difficult in computer graphics, so the program simply omits a small neighborhood of the south pole. Similarly, if you look a distance 2π in any direction you see all the way around the sphere, back to the north pole, so the second gap is where the north pole gets stretched out to become a circle of radius 2π on the forest floor. The third gap is the south pole again, this time at a radius of 3π, and so on to infinity.

The algorithm for rendering images in the product geometries $H^2 \times E$, $E^2 \times E$ and $S^2 \times E$ is similar in spirit to the algorithm for the isotropic geometries H^3, E^3, and S^3 (Figures 9, 7, and 8, respectively). The algorithm of Figure 15 defines *rigid mode*, in which both observer and objects conform exactly to the anisotropic geometry of $M^2 \times E$. That is, the observer and the objects may rotate about the preferred vertical axis, but may not rotate about other axes because that would (very slightly) distort their internal geometry. In the more realistic *flexible mode* (Figure 16) we assume both observer and objects have enough internal flexibility to rotate freely about any axis, even though such motions would slightly deform their internal structure. Flexible mode is surely more realistic than rigid mode, because even a steel bar has flexibility between its atoms. In flexible mode the model (resp. the camera) lives in E^3 and maps into the model space $M^2 \times E$ (resp. the camera space $M^2 \times E$) via the exponential map, perhaps after an arbitrary rotation $E^3 \to E^3$ about its own center.

$$\underset{\text{model space}}{M^2 \times E} \xrightarrow[\ ]{\substack{\text{model}\\ \text{transformation}}} \underset{\text{world space}}{M^2 \times E} \xrightarrow[\ ]{\substack{\text{view}\\ \text{transformation}}} \underset{\text{camera space}}{M^2 \times E} \xrightarrow[\ ]{\substack{\text{projection}\\ \text{transformation}}} \underset{\text{proj. space}}{R^3}$$

Figure 15. In rigid mode both model and camera respect the anisotropic geometry $M^2 \times E$.

$$\underset{\text{model space}}{E^3} \xrightarrow{exp} \underset{\text{model space}}{M^2 \times E} \xrightarrow{model} \underset{\text{world space}}{M^2 \times E} \xrightarrow{view} \underset{\text{camera space}}{M^2 \times E} \xrightarrow{exp^{-1}} \underset{\text{camera space}}{E^3} \xrightarrow{proj} \underset{\text{proj. space}}{R^3}$$

Figure 16. In flexible mode the model begins in a Euclidean model space E^3, rotates about its own center, and then goes via the exponential map into the anisotropic model space $M^2 \times E$. Similarly, the inverse of the exponential map takes the contents of the anisotropic camera space $M^2 \times E$ to the Euclidean camera space E^3, where they rotate about the origin and then go on to projection space.

A point in $M^2 \times E$ may be represented as (x, y, z, h), where (x, y, z) gives the coordinates in the 2-dimensional factor M^2 and h gives the height in the linear factor E. A mapping $M^2 \times E \to M^2 \times E$ may be represented

(1) as a 3×3 matrix to rotate the M^2 factor and an increment Δh to the height h, or

(2) as a 5×5 matrix

$$\begin{pmatrix} * & * & * & 0 & 0 \\ * & * & * & 0 & 0 \\ * & * & * & 0 & 0 \\ 0 & 0 & 0 & 1 & \Delta h \\ 0 & 0 & 0 & 0 & 1 \end{pmatrix} \begin{pmatrix} x \\ y \\ z \\ h \\ 1 \end{pmatrix}$$

in which the upper-left 3×3 block rotates the M^2 factor while the lower-right 2×2 block translates the E factor, or

(3) as a 4×4 matrix which handles the increment Δh multiplicatively

$$\begin{pmatrix} * & * & * & 0 \\ * & * & * & 0 \\ * & * & * & 0 \\ 0 & 0 & 0 & \exp \Delta h \end{pmatrix} \begin{pmatrix} x \\ y \\ z \\ \exp h \end{pmatrix}$$

as suggested by Daryl Cooper.

The projection from the curved camera space $M^2 \times E$ onto the flat camera space E^3 (Figure 15) is given by the inverse exponential map

$$(x, y, z, h) \to \left(\frac{x \text{ acosh } z}{\sqrt{x^2 + y^2}}, \frac{y \text{ acosh } z}{\sqrt{x^2 + y^2}}, h \right) \quad \text{in } H^2 \times E \,,$$

$$(x, y, z, h) \to (x, y, h) \quad \text{in } E^2 \times E \,,$$

$$(x, y, z, h) \to \left(\frac{x \text{ acos } z}{\sqrt{x^2 + y^2}}, \frac{y \text{ acos } z}{\sqrt{x^2 + y^2}}, h \right) \quad \text{in } S^2 \times E \,.$$

One must resist the temptation to use simple radial projection from the M^2 factor onto its tangent plane because it would spoil the aspect ratio between the M^2 factor and the E factor. Note that the multiple images that arise geometrically in $S^2 \times E$ as a result of our line of sight spiraling several times around the space (Figure 13), arise computationally as a result of the multi-valued acos() function. The mapping from the flat model space E^3 onto the curved model space $M^2 \times E$ is less sensitive than the camera space mapping, because a model typically occupies only a small volume of space and a simple approximation to the true exponential map may be used without danger.

In summary, real-time rendering in a product space $M^2 \times E$ is reasonably straightforward, but nevertheless differs sufficiently from the isotropic algorithm that it cannot run fully in hardware on a standard graphics card.

7. Current Work: Twisted Space Rendering

William P. Thurston's famous Geometrization Conjecture [6] classifies homogeneous 3-dimensional geometries as follows:

Table 1. Thurston's eight homogeneous 3-dimensional geometries.

H^3	E^3	S^3
$H^2 \times E$	$E^2 \times E$ (E^3)	$S^2 \times E$
twisted $H^2 \times E$ (SL_2R)	twisted $E^2 \times E$ (nil geometry)	twisted $S^2 \times E$ (S^3)
solv geometry		

For an elementary introduction to these geometries, please see [1], from which the above terminology is taken.

Sections 3, 4, and 5 of the present paper explained rendering in the isotropic geometries E^3, S^3, and H^3. Section 6 explained rendering in the product geometries $H^2 \times E$, $E^2 \times E$, and $S^2 \times E$. The author is currently developing an algorithm for rendering in the three twisted geometries. The first challenge was to find a convenient matrix representation for isometries from each twisted space to itself. Luckily Emil Molnár's work [7] provides a ready solution. The second challenge is to explicitly determine the exponential map. While simple in theory, this is tricky in practice because geodesics appear as helices relative to the fibers. Furthermore, the helical nature of geodesics means that the twisted geometries exhibit multiple imaging, even in the cases of twisted $H^2 \times E$ and twisted $E^2 \times E$, which are diffeomorphic to E^3. Nevertheless, these difficulties do not seem insurmountable, and the author hopes to develop software for real-time animation in these spaces and to publish the results as a future article.

If all goes well solv geometry will yield to the same approach. Solv geometry is the most unusual of the homogeneous 3-dimensional geometries, yet it is easy to see from first principles that a vertical slice is a hyperbolic plane. János Bolyai would be pleased.

Bibliography

[1] Jeff Weeks, *The Shape of Space*, 2nd edition, Marcel Dekker, 2002.

[2] Charlie Gunn and Delle Maxwell (directors), *Not Knot*, A K Peters, 1991.

[3] Jeff Weeks, "*Real-time rendering in curved spaces*", IEEE Computer Graphics & Applications **22** No. 6 (Nov-Dec 2002) 90-99.

[4] Thomas Möller and Eric Haines, *Real-Time Rendering*, 2nd edition, A K Peters, 2002.

[5] Edwin F. Taylor and John Archibald Wheeler, *Spacetime Physics*, W.H. Freeman and Co., 1963.

[6] William P. Thurston, "*Three dimensional manifolds, Kleinian groups and hyperbolic geometry*", Bulletin (New Series) of the American Mathematical Society **6** (1982) 357-381.

[7] Emil Molnár, "*The projective interpretation of the eight 3-dimensional homogeneous geometries*", Beiträge zur Algebra und Geometrie **38** (1997) 261-288.

ON SPONTANEOUS SURGERY ON KNOTS AND LINKS*

A.D. Mednykh
Sobolev Institute of Mathematics, Novosibirsk, 630090, Russia
mednykh@math.nsc.ru

V.S. Petrov
Sobolev Institute of Mathematics, Novosibirsk, 630090, Russia
petrov@math.nsc.ru

Abstract Geometrical properties of cone-manifolds obtained by orbifold and spontaneous Dehn surgeries of knots and links are investigated. Explicit hyperbolic volume formulae for the Figure-eight knot, Whitehead link, and Borromean rings link cone-manifolds are obtained.

1. Introduction

In 1975 R. Riley ([16]) discovered the existence of complete hyperbolic structure on some knots and links complements in the 3-sphere. Later W. P. Thurston showed that a complement of a simple knot (except torical and spherical) admits a hyperbolic structure. It follows from this result ([17]) that almost all Dehn surgeries on a hyperbolic knot complement produce a hyperbolic manifold.

A number of works in the last 20 years has been devoted to precise descriptions of manifolds, orbifolds, and cone-manifolds obtained in this manner (see for example [2, 3, 4, 6, 11, 18]).

This paper is a review of recent results pertaining to the geometrical properties of cone-manifolds obtained by spontaneous Dehn surgeries on knots and links. This work is part of a talk given by the first named author on the "János Bolyai Conference on Hyperbolic Geometry" held in Budapest on 8-12 July, 2002. Mostly, it contains a survey of results

*Supported by the Russian Foundation for Basic Research (grant 03-01-00104).

obtained by the authors and their collaborators, but some new results are also given.

We remind the reader of some basic definitions:

Definition 1.1. *A 3-dimensional hyperbolic cone-manifold is a Riemannian 3-dimensional manifold of constant negative sectional curvature with cone-type singularity along simple closed geodesics. To each component of a singular set we associate a real number $n \geq 1$ such that the cone-angle around the component is $\alpha = 2\pi/n$. The concept of the hyperbolic cone-manifold generalizes the hyperbolic manifold which appears in the partial case when all cone-angles are 2π. The hyperbolic cone-manifold is also a generalization of the hyperbolic 3-orbifold which arises when all associated numbers n are integers. Euclidean and spherical cone–manifolds are defined similarly.*

We identify the group of orientation preserving isometries of $\mathbb{H}^3$ with the group $PSL(2, \mathbb{C})$ consisting of linear fractional transformations

$$A : z \in \mathbb{C} \to \frac{az + b}{cz + d}.$$

By the canonical procedure the linear transformation A can be uniquely extended to the isometry of $\mathbb{H}^3$. We prefer to deal with the matrix

$$\widetilde{A} = \begin{pmatrix} a & b \\ c & d \end{pmatrix} \in SL(2, \mathbb{C})$$

rather than the element $A \in PSL(2, \mathbb{C})$. The matrix $\widetilde{A}$ is uniquely determined by the element A up to a sign. If there is no confusion we shall use the same letter A for both A and $\widetilde{A}$.

Let C be a hyperbolic cone–manifold with singular set $\Sigma = \Sigma_1 \cup \Sigma_2 \cup \cdots \cup \Sigma_k$ being a link consisting of components $\Sigma_j = \Sigma_{\alpha_j}$, $j = 1, 2, \ldots, k$ with cone-angles $\alpha_1, \ldots, \alpha_k$ respectively. Then C defines a nonsingular but incomplete hyperbolic manifold $N = C - \Sigma$. Denote by Φ the fundamental group of the manifold N.

The hyperbolic structure of N defines, up to conjugation in $PSL(2, \mathbb{C})$, a holonomy homomorphism

$$\hat{h} : \Phi \to PSL(2, \mathbb{C}).$$

It is shown in [19] that the monodromy homomorphism of an orientable cone-orbifold can be lifted to $SL(2, \mathbb{C})$ if all cone angles are less than π. Denote by $h : \Phi \to SL(2, \mathbb{C})$ this lifting homomorphism. Choose an orientation on the link $\Sigma = \Sigma_1 \cup \Sigma_2 \cup \cdots \cup \Sigma_k$ and fix a meridian-longitude pair $\{m_j, l_j\}$ for each component $\Sigma_j = \Sigma_{\alpha_j}$. Then the matrices

$M_j = h(m_j)$ and $L_j = h(l_j)$ satisfy the following properties:

$$M_jL_j = L_jM_j,\ j = 1, 2, \ldots, k.$$

Definition 1.2. *Cone-manifold C is said to be obtained by orbifold Dehn surgery with cone angle $\alpha_j = \frac{2\pi}{m}$ (or with a slope $\frac{m}{0}$) on the component Σ_j if $tr(M_j) = 2\cos(\frac{\alpha_j}{2})$.*

Definition 1.3. *Cone-manifold C is said to be obtained by spontaneous Dehn surgery with cone angle $\alpha_j = \frac{2\pi}{m}$ (or with a slope $\frac{0}{m}$) on the component Σ_j if $tr(L_j) = 2\cos(\frac{\alpha_j}{2})$.*

See [3] for details.

Definition 1.4. *A complex length γ_j of the singular component Σ_j of the cone-manifold C is defined as displacement of the isometry L_j of $\mathbb{H}^3$, where $L_j = h(l_j)$ is represented by the longitude l_j of Σ_j.*

Immediately from the definition we get [1, p.46]

$$2\cosh\gamma_j = \mathrm{tr}(L_j^2).$$

We note that the meridian-longitude pair $[m_j, l_j]$ is uniquely determined up to a common conjugating element of the group Φ. Hence the complex length $\gamma_j = l_j + i\phi_j$ is uniquely determined up to sign and $(mod\, 2\pi i)$ by the above definition. Since $\mathrm{tr}({L_j}^2) = \mathrm{tr}^2(L_j) - 2$ we have also $\mathrm{tr}^2(L_j) = 4\cosh^2(\frac{\gamma_j}{2})$

The main tool for volume calculation is the following Schläfli formula [4].

Theorem 1.5. *Suppose that C_t is a smooth 1-parameter family of (curvature K) cone–manifold structures on an n-manifold, with singular locus Σ of a fixed topological type. Then the derivative of volume of C_t satisfies*

$$(n-1)KdV(C_t) = \sum_{\sigma} V_{n-2}(\sigma)d\theta(\sigma)$$

where the sum is over all components σ of the singular locus Σ and $\theta(\sigma)$ is the cone angle along σ.

In the present paper we will deal with three-dimensional cone-manifolds of negative constant curvature $K = -1$. The Schläfli formula in this case reduces to

$$dV = -\frac{1}{2}\sum_{i} l_{\alpha_i} d\alpha_i,$$

where the sum is taken over all components of the singular set Σ with lengths l_{α_i} and cone angles α_i.

2. Figure-eight knot

2.1 Orbifold surgery on the figure-eight knot

Denote by E the compliment to the figure-eight knot in a 3-sphere (see Figure 1).

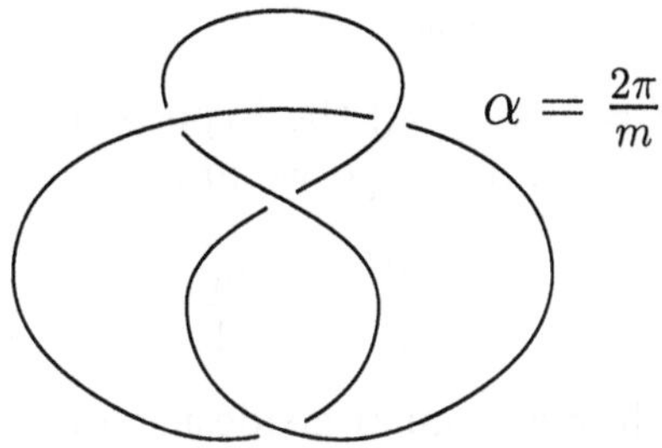

Figure 1. The figure-eight cone-manifold $E(\frac{m}{0})$.

The following theorem was obtained by Mednykh and Rasskazov in [9].

Theorem 2.1. *Let $E(\frac{m}{0})$ be a cone-manifold obtained by orbifold surgery on the figure-eight knot with cone angle $\alpha = \frac{2\pi}{m}$. Then $E(\frac{m}{0})$ is hyperbolic for $0 \leq \alpha < \frac{2\pi}{3}$, Euclidean for $\alpha = \frac{2\pi}{3}$, and spherical for $\frac{2\pi}{3} < \alpha < \frac{4\pi}{3}$. The hyperbolic volume of the cone-manifold is given by the formula:*

$$\mathrm{Vol}(E(\frac{m}{0})) = \int_{\alpha}^{\frac{2\pi}{3}} \operatorname{arcosh}(1 + \cos t - \cos 2t)\, dt\,.$$

2.2 Spontaneous surgery on the figure-eight knot

Recall that the first example of a complete hyperbolic 3-manifold of finite volume was constructed by Gieseking in 1912. This manifold can be obtained by identification of faces of regular ideal tetrahedra by orientation reversing isometries of $\mathbb{H}^3$ (see [12] for details).

Spontaneous Dehn surgery on a Gieseking manifold was considered in [13] by E. Molnár, I. Prok, and J. Szirmai with obvious and easily recoverable mistakes, noticed also by the second named author. In the improvements to the paper [14], sent to the reviewer J. Böhm (see Zbl pre01604732), it was proven that $G(\frac{0}{m})$ is hyperbolic if $0 \leq \alpha < 2\pi$, $\alpha = \frac{2\pi}{m}$ and the fundamental polyhedron was constructed in $\mathbb{H}^3$. Also the hyperbolic volume was obtained as a sum of three Lobachevsky functions. We give a more simple hyperbolic volume formula in the following:

Theorem 2.2. *Let $G(\frac{0}{m})$ be a hyperbolic cone-manifold obtained by spontaneous surgery on the Gieseking manifold with cone angle $\alpha = \frac{2\pi}{m}$. Then the volume of $G(\frac{0}{m})$ is given by the formula:*

$$\mathrm{Vol}(G(\frac{0}{m})) = \frac{1}{2}\int_{\pi/m}^{\pi} \operatorname{arcosh}(\frac{1+\sqrt{17-8\cos x}}{4})\, dx\,.$$

Proof. Denote by $V = \mathrm{Vol}(G(\alpha))$ the hyperbolic volume of $G(\alpha)$. Then by virtue of the Schläfli formula [4] we have

$$\frac{\partial V}{\partial \alpha} = -\frac{l_\alpha}{2} \tag{1}$$

where l_α is the length of a singular geodesic corresponding to cone angle α. Moreover, by [3] we note that

$$V \to 0 \text{ as } \alpha \to 2\pi\,. \tag{2}$$

We set

$$\begin{aligned}\tilde{V} &= \frac{1}{2}\int_{\pi/m}^{\pi} \operatorname{arcosh}(\frac{1+\sqrt{17-8\cos x}}{4})\, dx \\ &= \frac{1}{4}\int_{\alpha}^{2\pi} \operatorname{arcosh}(\frac{1+\sqrt{17-8\cos \frac{x}{2}}}{4})\, dx\end{aligned}$$

and show that $\tilde{V}$ satisfies conditions (1) and (2). Then $\tilde{V} = V$ and the theorem is proven.

To verify (1) we note that l_α can be found from the following equation (see [13] and [14] for a geometric basis):

$$l_\alpha = \log|z-1|\,,$$

where $(z-1)$ is derived from the equation ([14])

$$\frac{z}{(z-1)^2} = -e^{\frac{i\alpha}{2}}\,.$$

Hence l_α is represented by the expression

$$l_\alpha = \log|e^{\operatorname{arcosh}((-\frac{1}{2})e^{\frac{-i\alpha}{4}})}|\,.$$

Keeping in mind that $|e^{\zeta}| = e^{\Re(\zeta)}$ we get after simplifying

$$l_\alpha = \frac{1}{2}\operatorname{arcosh}(\frac{1+\sqrt{17-8\cos\frac{\alpha}{2}}}{4})\,,$$

hence $\dfrac{\partial \tilde{V}}{\partial \alpha} = -\dfrac{l_\alpha}{2}$.

The boundary condition

$$\tilde{V} = \frac{1}{4}\int_{\alpha}^{2\pi} \operatorname{arcosh}\left(\frac{1+\sqrt{17-8\cos\frac{x}{2}}}{4}\right) dx \to 0$$

as $\alpha \to 2\pi$ follows from the convergence of the integral. □

We recall that a double sheeted covering of the Gieseking manifold is a complement to the figure-eight knot (see [12]). Hence $E(\frac{0}{m})$ obtained by spontaneous surgery on the figure-eight knot is a double sheeted covering over the cone-manifold $G(\frac{0}{m})$ obtained by spontaneous surgery on the Gieseking manifold with cone angle $\alpha = \frac{2\pi}{m}$.

Hilden, Lozano, and Montesinos-Amilibia have shown (see [3]) that the cone-manifold $E(\frac{0}{m})$ is hyperbolic for $0 \leq \alpha < 2\pi$, $\alpha = \frac{2\pi}{m}$. Some complicated formula for the hyperbolic volume was also obtained. We found a very simple version of this formula as a consequence of Theorem 2.2.

Theorem 2.3. *Let $E(\frac{0}{m})$ be a hyperbolic cone-manifold obtained by spontaneous surgery on the figure-eight knot manifold with cone angle $\alpha = \frac{2\pi}{m}$. Then the volume of $E(\frac{0}{m})$ is given by the formula:*

$$\operatorname{Vol}(E(\frac{0}{m})) = \int_{\pi/m}^{\pi} \operatorname{arcosh}\left(\frac{1+\sqrt{17-8\cos x}}{4}\right) dx.$$

3. Whitehead link cone-manifold

3.1 Orbifold surgery on the Whitehead link

We denote by W the Whitehead link shown on Figure 2. Recall ([17]) that $\mathbb{S}^3 \setminus W$ is a hyperbolic manifold. Denote by $h : \Phi = \pi_1(\mathbb{S}^3 \setminus W) \to SL(2, \mathbb{C})$ the lifting of its holonomy homomorphism.

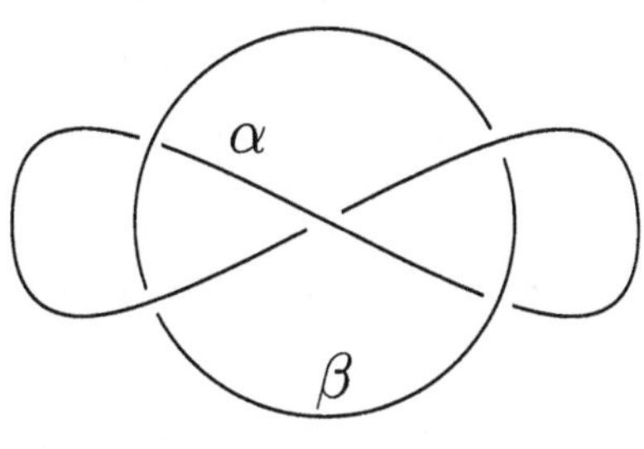

Figure 2. The Whitehead link cone-manifold $W(\alpha, \beta)$.

By slight modification of arguments from [8] we obtain the following two propositions:

Proposition 3.1. *Up to conjugation in $SL(2,\mathbb{C})$ the matrices $M_\alpha = h(m_\alpha)$ and $M_\beta = h(m_\beta)$ can be represented in the following form:*

$$M_\alpha = \begin{pmatrix} \cos\frac{\hat{\alpha}}{2} & ie^{\frac{\rho}{2}}\sin\frac{\hat{\alpha}}{2} \\ ie^{-\frac{\rho}{2}}\sin\frac{\hat{\alpha}}{2} & \cos\frac{\hat{\alpha}}{2} \end{pmatrix},$$

$$M_\beta = \begin{pmatrix} \cos\frac{\hat{\beta}}{2} & ie^{\frac{\rho}{2}}\sin\frac{\hat{\beta}}{2} \\ ie^{-\frac{\rho}{2}}\sin\frac{\hat{\beta}}{2} & \cos\frac{\hat{\beta}}{2} \end{pmatrix},$$

where $\hat{\alpha}$ and $\hat{\beta}$ satisfy relations $\mathrm{tr}(M_\alpha) = 2\cos\frac{\hat{\alpha}}{2}$, $\mathrm{tr}(M_\beta) = 2\cos\frac{\hat{\beta}}{2}$, and ρ is a complex distance between axes of M_α and M_β. Moreover, $u = \cosh(\rho)$ is a complex root of equation

$$u^3 - \hat{A}\hat{B}u^2 + \frac{\hat{A}^2\hat{B}^2 + \hat{A}^2 + \hat{B}^2 - 1}{2}u + \hat{A}\hat{B} = 0\,,$$

where $\hat{A} = \cot\frac{\hat{\alpha}}{2}$ and $\hat{B} = \cot\frac{\hat{\beta}}{2}$.

Setting $z = \dfrac{\hat{A}\hat{B}}{u}$ and multiplying the obtained polynomial equation by $(z+1)$ we have (see also [10])

$$2(z^2 + \hat{A}^2)(z^2 + \hat{B}^2) = (1 + \hat{A}^2)(1 + \hat{B}^2)(z^2 - z^3)\,. \tag{3}$$

Proposition 3.2. *Let $W(S,T) = STS^{-1}T^{-1}ST^{-1}S^{-1}T$, M_α and M_β be the same matrices as in Proposition 3.1, $L_\alpha = W(M_\alpha, M_\beta)$, $L_\beta = W(M_\beta, M_\alpha)$, and the condition $M_\alpha L_\alpha = L_\alpha M_\alpha$ be satisfied. Then*

$$\frac{\coth\frac{i\hat{\alpha}}{2}}{\coth\frac{\hat{l}_\alpha}{4}} = \frac{\coth\frac{i\hat{\beta}}{2}}{\coth\frac{\hat{l}_\beta}{4}} = z\,,$$

where z is a root of Equation (3), $\Im(z) > 0$ and $\hat{l}_\alpha$ and $\hat{l}_\beta$ can be derived from $2\cosh(\hat{l}_\alpha) = \mathrm{tr}({L_\alpha}^2)$ and $2\cosh(\hat{l}_\beta) = \mathrm{tr}({L_\beta}^2)$.

The following result was obtained in [7] and [10].

Theorem 3.3. *Let $W(\frac{m}{0}, \frac{n}{0})$ be a hyperbolic cone-manifold obtained by orbifold surgeries on the components of the Whitehead link with cone angles $\alpha = \frac{2\pi}{m}$ and $\beta = \frac{2\pi}{n}$. Then*

$$\mathrm{Vol}(W(\frac{m}{0}, \frac{n}{0})) = i\int_{\zeta_1}^{\zeta_2} \log\left[\frac{2(\zeta^2 + A^2)(\zeta^2 + B^2)}{(1 + A^2)(1 + B^2)(\zeta^2 - \zeta^3)}\right]\frac{d\zeta}{\zeta^2 - 1}$$

where $A = \cot\frac{\alpha}{2}$, $B = \cot\frac{\beta}{2}$, $\zeta_1 = \overline{z}$, $\zeta_2 = z$, $\Im(z) > 0$ *and* z *is a root of the cubic equation*

$$z^3 + \frac{1}{2}(A^2B^2 + A^2 + B^2 - 1)z^2 - A^2B^2z + A^2B^2 = 0.$$

3.2 Spontaneous surgery on the Whitehead link

Proposition 3.4. *Let* $W(\frac{0}{m}, \frac{0}{n})$ *be a hyperbolic cone-manifold obtained by spontaneous surgery on the components of the Whitehead link with cone angles equal to* $\frac{2\pi}{m}$ *and* $\frac{2\pi}{n}$ *respectively. Denote by* l_α *and* l_β *the complex lengths of singular geodesics of* $W(\frac{0}{m}, \frac{0}{n})$ *with cone angles* $\alpha = \frac{2\pi}{m}$ *and* $\beta = \frac{2\pi}{n}$ *respectively. Then*

$$\frac{\coth\frac{l_\alpha}{2}}{\coth\frac{i\alpha}{4}} = \frac{\coth\frac{l_\beta}{2}}{\coth\frac{i\beta}{4}} = z\,,$$

where $\Im(z) > 0$, *and* z *is a root of the equation*

$$2z^2(1 + \hat{A}^2)(1 + \hat{B}^2) = (z^2 + \hat{A}^2)(z^2 + \hat{B}^2)(1 - z)\,,$$

$\hat{A} = \tan\frac{\alpha}{4}$, $\hat{B} = \tan\frac{\beta}{4}$.

Proof. The result follows from Proposition 3.2 for $l_\alpha = i\hat{\alpha}$, $\hat{l}_\alpha = i\alpha$, $l_\beta = i\hat{\beta}$, and $\hat{l}_\beta = i\beta$. □

Theorem 3.5. *Let* $W(\frac{0}{m}, \frac{0}{n})$ *be a hyperbolic cone-manifold obtained by spontaneous surgery on the components of the Whitehead link with cone angles* $\alpha = \frac{2\pi}{m}$ *and* $\beta = \frac{2\pi}{n}$. *Then*

$$\mathrm{Vol}(W(\frac{0}{m}, \frac{0}{n})) = i\int_{\zeta_1}^{\zeta_2} \log\left[\frac{1+\hat{A}^2}{\zeta^2+\hat{A}^2}\frac{1+\hat{B}^2}{\zeta^2+\hat{B}^2}\frac{2\zeta^2}{1-\zeta}\right]\frac{d\zeta}{\zeta^2-1}\,,$$

where $\hat{A} = \tan\frac{\alpha}{4}$, $\hat{B} = \tan\frac{\beta}{4}$, $\zeta_1 = \overline{z}$, $\zeta_2 = z$, $\Im(z) > 0$, *and*

$$2z^2(1 + \hat{A}^2)(1 + \hat{B}^2) = (z^2 + \hat{A}^2)(z^2 + \hat{B}^2)(1 - z)\,. \qquad (4)$$

Proof. Denote by $V = \mathrm{Vol}(W(\alpha, \beta))$ the hyperbolic volume of $W(\alpha, \beta)$. Then by virtue of the Schläfli formula [4] we have

$$\frac{\partial V}{\partial \alpha} = -\frac{\Re(l_\alpha)}{2}, \quad \frac{\partial V}{\partial \beta} = -\frac{\Re(l_\beta)}{2}, \qquad (5)$$

where l_α and l_β are complex lengths of singular geodesics corresponding to cone angles α and β respectively. Moreover, by [10] we note that

$$V \to Vol(W(0,0)) \text{ as } \alpha \to 0 \text{ and } \beta \to 0, \tag{6}$$

where

$$\mathrm{Vol}(W(0,0)) = i\int_{1-i}^{1+i} \log\left[\frac{2}{z^2-z^3}\right]\frac{dz}{z^2-1} = 3.66386\ldots$$

is the hyperbolic volume of the Whitehead link complement $W(0,0) = S^3 \setminus W$.

We set $\tilde{V} = i\int_{\zeta_1}^{\zeta_2} \log\left[\frac{1+\hat{A}^2}{\zeta^2+\hat{A}^2}\frac{1+\hat{B}^2}{\zeta^2+\hat{B}^2}\frac{2\zeta^2}{1-\zeta}\right]\frac{d\zeta}{\zeta^2-1}$ and show that $\tilde{V}$ satisfies conditions (5) and (6). Then $\tilde{V} = V$ and the theorem is proven.

To verify (5) we introduce the function

$$F(\zeta,\hat{A},\hat{B}) = \frac{i}{\zeta^2-1}\log\left[\frac{1+\hat{A}^2}{\zeta^2+\hat{A}^2}\frac{1+\hat{B}^2}{\zeta^2+\hat{B}^2}\frac{2\zeta^2}{1-\zeta}\right].$$

Then by the Leibniz formula we get

$$\frac{\partial\tilde{V}}{\partial\alpha} = F(\zeta_2,\hat{A},\hat{B})\frac{\partial\zeta_2}{\partial\alpha} - F(\zeta_1,\hat{A},\hat{B})\frac{\partial\zeta_1}{\partial\alpha} + \int_{\zeta_1}^{\zeta_2}\frac{\partial F(\zeta,\hat{A},\hat{B})}{\partial\hat{A}}\frac{\partial\hat{A}}{\partial\alpha}d\zeta. \tag{7}$$

We note that $F(\zeta_1,\hat{A},\hat{B}) = F(\zeta_2,\hat{A},\hat{B}) = 0$ if ζ_1, ζ_2, $\hat{A}$, and $\hat{B}$ are as stated in the theorem. Moreover, since $\alpha = 4\arctan\hat{A}$ we have $\frac{\partial\hat{A}}{\partial\alpha} = \frac{1+\hat{A}^2}{4}$ and

$$\frac{\partial F(\zeta,\hat{A},\hat{B})}{\partial\hat{A}}\frac{\partial\hat{A}}{\partial\alpha} = \frac{i\hat{A}}{2(\zeta^2+\hat{A}^2)}.$$

Hence, by Proposition 3.4 we obtain from Equation (7)

$$\frac{\partial\tilde{V}}{\partial\alpha} = \frac{i}{2}\int_{\zeta_1}^{\zeta_2}\frac{\hat{A}\,d\zeta}{\zeta^2+\hat{A}^2} = \frac{i}{2}\arctan\frac{\zeta_2}{\hat{A}} - \frac{i}{2}\arctan\frac{\zeta_1}{\hat{A}} = -\frac{\Re(l_\alpha)}{2}.$$

The equation $\frac{\partial\tilde{V}}{\partial\beta} = -\frac{\Re(l_\beta)}{2}$ can be obtained in the same way.

Given $\Im(z) > 0$ we have $z \to 1+i$, as $\alpha \to 0$ and $\beta \to 0$. Then the boundary condition (6) for the function $\tilde{V}$ follows from the integral formula. □

Remark 3.6. *There is exactly one root of Equation 4 such that $\Im(z) > 0$. Indeed, Equation 4 is equivalent to*

$$(z+1)(z^4 - 2z^3 + (\hat{A}^2 + \hat{B}^2 + 2)z^2 + 2\hat{A}^2\hat{B}^2 z - \hat{A}^2\hat{B}^2) = 0.$$

Let

$$P(z) = z^4 - 2z^3 + (\hat{A}^2 + \hat{B}^2 + 2)z^2 + 2\hat{A}^2\hat{B}^2 z - \hat{A}^2\hat{B}^2,$$

then by Mathematica we have

$$\begin{aligned} R = \text{Resultant}\left[P'(z)/2, P(z), z\right] &= -\hat{A}^2(1+\hat{A}^2)\hat{B}^2(1+\hat{B}^2)(8+12\hat{A}^2 \\ &+ 6\hat{A}^4 + \hat{A}^6 + 12\hat{B}^2 + 39\hat{A}^2\hat{B}^2 + 6\hat{B}^4 + 30\hat{A}^2\hat{B}^4 + 27\hat{A}^4\hat{B}^4 + \hat{B}^6) \\ &= -\hat{A}^2\hat{B}^2(1+\hat{A}^2)(1+\hat{B}^2)Q, \end{aligned}$$

where $Q \geq 8$.

Since $R < 0$ for all nonzero $\hat{A}$ and $\hat{B}$ one can easily deduce that equation $P(z) = 0$ always has a pair of real and a pair of complex conjugate roots for all $\hat{A}$ and $\hat{B}$.

Proposition 3.7. *Let $W(\frac{0}{m}, \frac{n}{0})$ be a hyperbolic cone-manifold obtained by a spontaneous surgery on the first component of the Whitehead link and an orbifold surgery on the second component with cone angles equal to $\frac{2\pi}{m}$ and $\frac{2\pi}{n}$ respectively. Denote by l_α and l_β complex lengths of singular geodesics of $W(\frac{0}{m}, \frac{n}{0})$ with cone angles $\alpha = \frac{2\pi}{m}$ and $\beta = \frac{2\pi}{n}$ respectively. Then*

$$\frac{\coth\frac{l_\alpha}{2}}{\coth\frac{i\alpha}{4}} = \frac{\coth\frac{i\beta}{2}}{\coth\frac{-l_\beta}{4}} = z,$$

where $\Im(z) > 0$ and z is a root of the equation

$$2(1+A^2)(z^2+B) = (1+A^2z^2)(1+B^2)(1-z), \tag{8}$$

$A = \cot\frac{\alpha}{4}$ and $B = \cot\frac{\beta}{2}$.

Proof. The result follows from Proposition 3.2 for $l_\alpha = i\hat{\alpha}$, $\hat{l}_\alpha = i\alpha$, $\hat{\beta} = \beta$, and $\hat{l}_\beta = -l_\beta$. □

Following the plot of the proof of Theorem 3.5 and applying Proposition 3.7 we obtain the following:

Theorem 3.8. *Let $W(\frac{0}{m}, \frac{n}{0})$ be a cone-manifold obtained by spontaneous surgery on the first component and orbifold surgery on the second component of the Whitehead link with cone angles $\alpha = \frac{2\pi}{m}$ and $\beta = \frac{2\pi}{n}$ respectively. Then*

$$\mathrm{Vol}(W(\frac{0}{m}, \frac{n}{0})) = i\int_{\zeta_1}^{\zeta_2} \log\left[2\frac{1+A^2}{1+A^2\zeta^2}\frac{\zeta^2+B^2}{1+B^2}\frac{1}{1-\zeta}\right]\frac{d\zeta}{\zeta^2-1},$$

where $A = \cot\frac{\alpha}{4}$, $B = \cot\frac{\beta}{2}$, $\zeta_1 = \overline{z}$, $\zeta_2 = z$, $\Im(z) > 0$, *and*

$$2(1+A^2)(z^2+B) = (1+A^2z^2)(1+B^2)(1-z)\,.$$

4. Boromean rings cone-manifold

4.1 Orbifold surgery on the Boromean rings

In this subsection we study the geometrical properties of cone-manifolds $B(\alpha,\beta,\gamma)$ obtained by orbifold Dehn surgery on three components of the Borromean rings with cone angles α, β, and γ (see Figure 3).

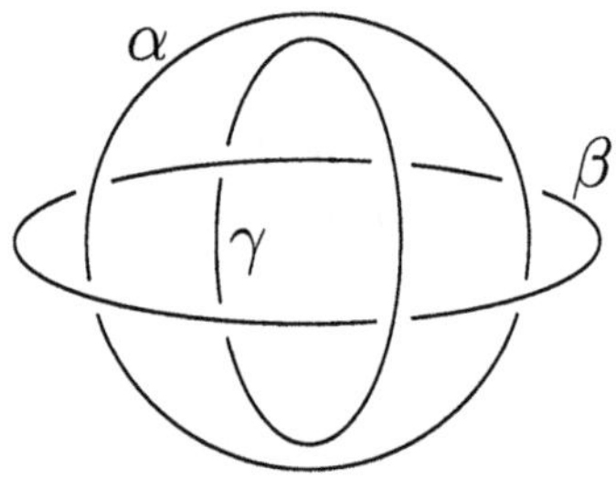

Figure 3. The Borromean cone–manifold $B(\alpha,\beta,\gamma)$.

The following result was essentially obtained by R. Kellerhals [5] (see [7] for details of the proof):

Theorem 4.1. *Let* $B(\frac{k}{0},\frac{l}{0},\frac{m}{0})$ *be a cone-manifold obtained by orbifold surgery on the components of the Borromean rings with cone angles* $\alpha = \frac{2\pi}{k}$, $\beta = \frac{2\pi}{l}$ *and* $\gamma = \frac{2\pi}{m}$. *Then* $B(\frac{k}{0},\frac{l}{0},\frac{m}{0})$ *is hyperbolic for* $0 < \alpha,\beta,\gamma < \pi$ *and its volume is given by the formula:*

$$\mathrm{Vol}(B(\frac{k}{0},\frac{l}{0},\frac{m}{0})) = 2\int_T^{\infty} \log\left[\frac{(t^2-A^2)(t^2-B^2)(t^2-C^2)}{(1+A^2)(1+B^2)(1+C^2)t^2}\right]\frac{dt}{t^2+1}\,,$$

where T *is a positive root of the equation*

$$T^4 - (A^2+B^2+C^2+1)T^2 - A^2B^2C^2 = 0\,,$$

$A = \tan\frac{\alpha}{2}$, $B = \tan\frac{\beta}{2}$, *and* $C = \tan\frac{\gamma}{2}$.

4.2 Spontaneous surgery on the Boromean rings

The following three results were proved by M. Pashkevich in [15]:

Theorem 4.2. *Let* $B(\frac{0}{k},\frac{l}{0},\frac{m}{0})$ *be a hyperbolic cone-manifold obtained by a spontaneous surgery with cone angle* $\alpha = \frac{2\pi}{k}$ *on one component of* *the Borromean rings and an orbifold surgery with cone angles* $\beta = \frac{2\pi}{l}$, $\gamma = \frac{2\pi}{m}$ *on the other two components. Then*

$$\mathrm{Vol}(B(\frac{0}{k}, \frac{l}{0}, \frac{m}{0})) = 2\int_T^{\infty} \log\left|\frac{(1+A^2)(t^2-B^2)(t^2-C^2)}{(1-t^2A^2)(1+B^2)(1+C^2)}\right|\frac{dt}{t^2+1},$$

where T *is a positive root of the equation*

$$(1+A^2)T^2 - (1+B^2+C^2-A^2B^2C^2) = 0,$$

$A = \tan\frac{\pi}{2k}$, $B = \tan\frac{\pi}{l}$, $C = \tan\frac{\pi}{m}$.

Theorem 4.3. *Let* $B(\frac{0}{k}, \frac{0}{l}, \frac{m}{0})$ *be a hyperbolic cone-manifold obtained by spontaneous surgery with cone angles* $\alpha = \frac{2\pi}{k}$, $\beta = \frac{2\pi}{l}$ *on two components of the Borromean rings and an orbifold surgery with cone angle* $\gamma = \frac{2\pi}{m}$ *on the third component. Then*

$$\mathrm{Vol}(B(\frac{0}{k}, \frac{0}{l}, \frac{m}{0})) = -2\int_0^{T} \log\left|\frac{(1+A^2)(1+B^2)(t^2-C^2)t^2}{(1-t^2A^2)(1-t^2B^2)(1+C^2)}\right|\frac{dt}{t^2+1}$$

where T *is a positive root of the equation*

$$(1+A^2+B^2-A^2B^2C^2)T^2 - (1+C^2) = 0,$$

$A = \tan\frac{\pi}{2k}$, $B = \tan\frac{\pi}{2l}$, $C = \tan\frac{\pi}{m}$.

Theorem 4.4. *Let* $B(\frac{0}{k}, \frac{0}{l}, \frac{0}{m})$ *be a hyperbolic cone-manifold obtained by spontaneous surgery with cone angles* $\alpha = \frac{2\pi}{k}$, $\beta = \frac{2\pi}{l}$, $\gamma = \frac{2\pi}{m}$ *on three components of the Borromean rings. Then*

$$\mathrm{Vol}(B(\frac{0}{k}, \frac{0}{l}, \frac{0}{m})) = -2\int_0^{T} \log\left|\frac{(1+A^2)(1+B^2)(1+C^2)t^4}{(1-t^2A^2)(1-t^2B^2)(1-t^2C^2)}\right|\frac{dt}{t^2+1}$$

where T *is a positive root of the equation*

$$A^2B^2C^2T^4 + (1+A^2+B^2+C^2)T^2 - 1 = 0,$$

$A = \tan\frac{\pi}{2k}$, $B = \tan\frac{\pi}{2l}$, $C = \tan\frac{\pi}{2m}$.

Bibliography

[1] Fenchel W. *Elementary geometry in hyperbolic space.* De Gruyter, Berlin, 1989.

[2] Helling H., Kim A. C., Mennicke J. L. *A geometric study of Fibonacci groups.* Journal of Lie Theory, Vol. 8 (1998), 1-23.

[3] Hilden H. M., Lozano M. T., Montesinos-Amilibia J. M. *On a remarkable polyhedron geometrizing the figure eight knot cone manifolds.* J. Math. Sci. Univ. Tokyo, Vol. 2, 1995, 501–561.

[4] Hodgson C. D. *Schläfli revisited: Variation of volume in constant curvature spaces.* Preprint.

[5] Kellerhals R. *On the volume of hyperbolic polyhedra.* Math. Ann. 285, 541-569 (1989).

[6] Kojima S. *Deformation of hyperbolic 3-cone-manifolds.* J. Differential Geometry, Vol. 49 (1998), 469-516.

[7] Mednykh A. D. *On hyperbolic and spherical volumes for knot and link cone manifolds.* Kleinian Groups and Hyperbolic 3-Manifolds, Lond. Math. Soc. Lec. Notes 299, 1-19, Y. Komori, V. Markovic & C. Series (Eds)/ Cambridge Univ. Press, 2003

[8] Mednykh A. *On the the Remarkable Properties of the Hyperbolic Whitehead Link Cone–Manifolds.* Knots in Hellas '98 (C.McA.Gordon, V.F.R.Jones, L.H.Kauffman, S.Lambropoulou, J.H.Przytycki Eds.), World Scientific, 2000, pp. 290–305.

[9] Mednykh A., Rasskazov A. *Volumes and degeneration of cone-structures on the figure-eight knot.* preprint, 2002, available in http://cis.paisley.ac.uk/research/reports/index.html

[10] Mednykh A.D., Vesnin A.Yu. *On the Volume of Hyperbolic Whitehead Link Cone-Manifolds.* SCIENTIA, Series A: Mathematical Sciencies, Vol. 8 (2002), 1-11, Universidad Tecnica Federico Santa Maria, Valparaiso, Chile

[11] Mednykh A.D., Vesnin A.Yu. *Covering properties of small volume 3-dinemsional hyperbolic manifolds.* Knot theory and its ramifications, 1998, Vol.7, No.3, 381-392.

[12] Milnor J. *Hyperbolic geometry: the first 150 years.* 1982, Bull. A.M.S. 6, 9-24

[13] Molnár E., Prok I., Szirmai J. *The Gieseking manifold and its surgery orbifolds.* Novi Sad J. Math., Vol.29, No. 3, 1999, 187-197.

[14] Molnár E., Prok I., Szirmai J. *Classification of hyperbolic manifolds and related orbifolds with charts up to two ideal simplices.* Karáné, G. (ed.) et al., Topics in algebra, analysis and geometry. Proceedings of the Gyula Strommer national memorial conference, Balatonfüred, Hungary, May 1-5, 1999. Budapest: BPR Kiadó. 293-315 (2000).

[15] Pashkevich M. *Spontaneous surgery on the Borromean rings.* Siberian Math. J., Vol. 44, 4, (2003), 821–836.

[16] Riley R. *An elliptical path from parabolic representations to hyperbolic structure.* Topology of Low-Dimension manifolds, LNM, 722, Springer-Verlag, 1979, 99-133.

[17] Thurston W.P. *The geometry and topology of 3-manifolds.* Princeton University Mathematics Department. Lecture notes, 1992.

[18] Weeks J. *Computer program SnapPea and tables of volumes and isometries of knots, links, and manifolds.* available by ftp from *geom.umn.edu.*

[19] Qing Zhou *The Moduli Space of Hyperbolic Cone Structures.* J. Differential Geometry, vol. 51 (1999), 517-550.

CLASSIFICATION OF TILE-TRANSITIVE 3-SIMPLEX TILINGS AND THEIR REALIZATIONS IN HOMOGENEOUS SPACES

E. Molnár — I. Prok — J. Szirmai
Budapest University of Technology and Economics
Department of Geometry
H-1521 Budapest, Hungary
geom@math.bme.hu

Abstract This is a survey on the tilings $(\mathcal{T}, \Gamma)$ in the title where the vertex stabilizers in Γ are finite spherical $\mathbf{S}^2$ or infinite Euclidean $\mathbf{E}^2$ (cocompact) plane groups. The results are collected in figures and tables and illustrated by an infinite family series Family 30 in Section 4. The obtained orbifolds, maybe after splitting procedure, are realized in seven homogeneous Riemannian 3-spaces by means of projective metrics.

Keywords: Simplex tilings, homogeneous Riemann geometries, D-symbols

1. Introduction

In our papers [MP91, MPSz97], on the base of [M92] and [DHM93], we *combinatorially classified* the above *tilings* $(\mathcal{T}, \Gamma)$, with a group Γ *acting transitively on the simplices* of $\mathcal{T}$, into 32 *maximal family series* (for short, families) by a maximal group $\Gamma^* = \operatorname{Aut} \mathcal{T}$. Each tiling $\mathcal{T}$ has local *neighbour parameters* to the simplex *edge classes* under Γ, showing how many simplices in $\mathcal{T}$ surround any edge in the Γ-class considered. Thus we get also the *rotational orders* to the edge classes, determining the group Γ up to a *presentation (up to a homeotopy class).* The principle of *equivariance classification* above can be formulated by introducing *barycentric subdivision of* $(\mathcal{T}, \Gamma)$ and D-*symbols* (Delone-Delaney-Dress-symbols, see e.g. [DHM93]) and by their D-*morphisms.*

By D-*diagrams* we have 64 *fundamental* and 35 other series (where a simplex has a non-trivial stabilizer in Γ), ordered uniquely into the above 32 families. These are collected in our figures and Table 5. The neighbour parameters, thus the rotational orders of Γ, tell us whether a maximal tiling, representing a family, can be *realized* in a *homogeneous Riemannian space* from among the *Thurston geometries* [S83, Th97]:

$$\mathbf{E}^3,\ \mathbf{S}^3,\ \mathbf{H}^3,\ \mathbf{S}^2\times\mathbf{R},\ \mathbf{H}^2\times\mathbf{R},\ \widetilde{\mathbf{SL}_2\mathbf{R}},\ \mathbf{Nil},\ (\mathbf{Sol}\text{ will not occur!}).$$

Or, we have to solve some problems of metric realization of various kinds.

One is that *we allow Euclidean* $\mathbf{E}^2$ *plane groups* besides the spherical ($\mathbf{S}^2$) groups *for vertex stabilizers* (Table 1). The former ones lead to *ideal vertices* (at infinity), e.g. in $\mathbf{E}^3$, $\mathbf{H}^2\times\mathbf{R}$, $\widetilde{\mathbf{SL}_2\mathbf{R}}$ and $\mathbf{H}^3$, respectively. Then certain *surgery effects* may lead to various metrics and to a deformed simplex (called a *degenerate simplex tile,* see Family 8) e.g. in $\mathbf{H}^2\times\mathbf{R}$ and $\widetilde{\mathbf{SL}_2\mathbf{R}}$ as well.

Moreover, *bad spherical* 2*-orbifolds* ($\mathbf{p}$, $\mathbf{pq}$, $*\mathbf{p}$, $*\mathbf{pq}$; $2 \leq p < q$ with CONWAY's *denotation* by Table 1 [ZVC80, LM90]) can occur for a vertex stabilizer which excludes the metric realizability (Family 14).

Another phenomenon can occur with a spherical (this will not be our case here) or Euclidean surface subgroup in Γ leaving invariant a 2-surface within a simplex tile of $(\mathcal{T},\Gamma)$. Then we get a *non-geometric splitting orbifold* (as in the geometrization conjecture of W. P. THURSTON [S83, Th97]). Namely, by *splitting* (and occasional null measure modification) along the neighbourhood of such surfaces we get orbifolds which can be realized metrically, though the pieces may be in different spaces above (e.g. in Family 16).

It can turn out that the orbifold, induced in $(\mathcal{T},\Gamma)$ by Γ-orbits, has a spherical ($\mathbf{S}^3$) metric, but we get a degenerate *simplex tile,* namely we get a *lens tile* (as in Family 12) instead of a usual simplex.

The projective metric models of the above eight Thurston geometries [M97] can be adopted to the simplex tile as projective coordinate simplex. Therefore *we can compute all essential data of our simplex tile in each case,* sometimes with free parameters, e.g. in $\mathbf{E}^3$, $\mathbf{S}^2\times\mathbf{R}$, $\mathbf{H}^2\times\mathbf{R}$.

Our figures and tables summarize our complete classification, from which the Reader obtains all our tile-transitive simplex tilings for each of the above 3*-geometries* (Table 4). E.g. for the hyperbolic space $\mathbf{H}^3$ of J. BOLYAI and N. I. LOBACHEVSKII we get 92 maximal tile-transitive simplex tilings and 245 non-maximal ones.

Thus, we generalized, extended and completed the particular results of M. J. M. HILL, D. M. Y. SOMMERVILLE, L. BAUMGARTNER, M. GOLDBERG, moreover, the similarity and congruence classes of fundamental simplices of finite volume in $\mathbf{E}^3$ and $\mathbf{H}^3$, respectively, by I. K. ZHUK [Zh83]. The fundamental $\mathbf{E}^3$-tilings with compact simplices have already been classified in [MP88], completely into 26 tilings under 20 crystallographic space groups.

After some preliminaries we shall illustrate our method by interpreting the famous tiling family (Family 30) $K_{mn} \sim \Gamma_{39}$ of I. K. ZHUK. He started these investigations (as E. MOLNÁR learned of N. P. DOLBILIN) by initiative of B. N. DELONE, by enumerating the 64 fundamental simplex face pairings by hand. In a general program of E. MOLNÁR [M92], I. PROK's implementation reproduced these $64 = 2^6$ pairings by computer. Furthermore, I. PROK enumerated the $4096 = 2^{12}$ fundamental 4-simplex facet pairings, disproving the conjecture of I. K. ZHUK ($1024 = 2^{10}$). I. PROK [P99] proved a $\approx d^{\frac{1}{2}d^2}$ asymptotics for the different (up to simplex automorphisms) facet pairings of a d-simplex.

Our realization method by projective metrics was started in [M92, M97]. J. SZIRMAI elaborated computer programs for this e.g. in [Sz96, MPSz97, MPSz98].

The projective interpretation of the eight Thurston geometries has been summarized in Table 3 by [M97]. We have been working on more complete algorithmization of our method for more general polyhedron tilings, related to D-symbols as well [M96]. This survey hints at some facts from [MP91] and [MPSz97] and makes more precise some data in our tables.

2. Tile transitive simplex tiling and its D-symbol

Let $(\mathcal{T}, \Gamma)$ denote a *face-to-face tiling* $\mathcal{T}$ by topological simplices T_1, T_2, ... as a usual incidence structure of 0-, 1-, ..., d = 3-dimensional constituents in a simply connected topological 3-space $(\mathcal{X}^3, G)$ with a group $\Gamma < G$ that acts on $\mathcal{T}$ by topological mappings of $\mathcal{X}^3$, preserving the incidences.

Let Γ be *tile-transitive* on $\mathcal{T}$, i.e. for any two 3-tiles $T_1, T_2 \in \mathcal{T}$ there exists (at least one) $\gamma \in \Gamma$ such that

$$\gamma : \mathcal{T} \to \mathcal{T}, \quad T_1 \mapsto T_2 := T_1^{\gamma}.$$

Two tilings $(\mathcal{T}, \Gamma)$ and $(\mathcal{T}', \Gamma')$ are called *combinatorially* (topologically) *equivariant,* and lie in the same *equivariance class,* iff there is a bijective incidence preserving (topological) mapping

$$\varphi : \mathcal{T} \to \mathcal{T}', \quad T \mapsto T' := T^{\varphi} \quad \text{such that} \quad \Gamma' = \varphi^{-1}\Gamma\varphi.$$

More visually, we have the commutative diagram

$$\begin{array}{ccc} \mathcal{T} \ni T_1 & \stackrel{\varphi}{\longmapsto} & T_1' \in \mathcal{T}' \\ {\scriptstyle \Gamma\ni\gamma}\big\downarrow & & \big\downarrow{\scriptstyle \gamma'\in\Gamma'} \\ \mathcal{T} \ni T_2 & \stackrel{\varphi}{\longmapsto} & T_2' \in \mathcal{T}'. \end{array}$$

Table 1. Spherical and Euclidean plane groups by their signatures

Macbeath signature with parameters	Conditions	Pln	Hermann – – Mauguin	Sch*	Coxeter – – Moser	Conway
$(+, 0; [q, q]; \{\ \})$	$q \geq 1$	$\mathbf{S}^2$	**q**	C_q	$[Q]^+$	**qq**
$(+, 0; [\ ]; \{(q, q)\})$ $q \geq 1$	$q \geq 1$ $q = 1$ $q = 2$	$\mathbf{S}^2$	**qm** **m** $= \bar{\mathbf{2}}$ **mm2**	C_{qv} C_{1v} C_{2v}	$[q]$ $[1]$ $[2]$	***qq** **q** ≥ 1
$(+, 0; [2, 2, q]; \{\ \})$ $q \geq 2$	$q \geq 2$ q even	$\mathbf{S}^2$	**q2** **q22**	D_q	$[2, q]^+$	**22q** **q** ≥ 2
$(+, 0; [\ ]; \{(2, 2, q)\})$ $q \geq 2$	q even $q = q'$ odd $q = 2$	$\mathbf{S}^2$	**q/mmm** **(2q')m2** **mmm**	D_{qh}	$[2, q]$	***22q** **q** ≥ 2
$(+, 0; [q]; \{(1)\})$ $q \geq 2$	q even q odd	$\mathbf{S}^2$	**q/m** $(\overline{\mathbf{2q}})$	C_{qh}	$[2, q^+]$	**q*** **q** ≥ 2
$(+, 0; [2]; \{(q)\})$ $q \geq 2$	q odd $q = q'$ even	$\mathbf{S}^2$	$\bar{\mathbf{q}}$**m** $(\overline{\mathbf{2q'}})$**2m**	D_{qd}	$[2^+, 2q]$	**2*q** **q** ≥ 2
$(-, 1; [q]; \{\ \})$ $q \geq 1$	q odd q' even	$\mathbf{S}^2$	$\bar{\mathbf{q}}$ $(\overline{\mathbf{2q'}})$	S_{2q}	$[2^+, 2q^+]$	**q⊗** **q** ≥ 1
$(+, 1; [\ ]; \{\ \})$ $(+, 0; [2, 2, 2, 2]; \{\ \})$ $(+, 0; [\ ]; \{(1), (1)\})$ $(-, 2; [\ ]; \{\ \})$ $(-, 1; [\ ]; \{(1)\})$ $(+, 0; [\ ]; \{(2, 2, 2, 2)\})$ $(+, 0; [2, 2]; \{(1)\})$ $(-, 1; [2, 2]; \{\ \})$ $(+, 0; [2]; \{(2, 2)\})$		$\mathbf{E}^2$	**1. p1** **2. p2** **3. pm** **4. pg** **5. cm** **6. pmm** **7. pmg** **8. pgg** **9. cmm**		Hermann– –Mauguin	**O** **2222** **** **⊗⊗** ***⊗** ***2222** **22*** **22⊗** **2*22**
$(+, 0; [p, q, r]; \{\ \})$ (p, q, r) $2 \leq p \leq q \leq r \leq 6$	(2, 3, 3) (2, 3, 4) (2, 3, 5) (2, 3, 6) (2, 4, 4) (3, 3, 3)	$\mathbf{S}^2$ $\mathbf{S}^2$ $\mathbf{S}^2$ $\mathbf{E}^2$ $\mathbf{E}^2$ $\mathbf{E}^2$	**23** **432** **532** **16. p6** **10. p4** **13. p3**	T O I	$[3, 3]^+$ $[3, 4]^+$ $[3, 5]^+$ $[3, 6]^+$ $[4, 4]^+$ Δ^+	**233** **234** **235** **236** **244** **333**
$(+, 0; [\ ]; \{(p, q, r)\})$ (p, q, r) $2 \leq p \leq q \leq r \leq 6$	(2, 3, 3) (2, 3, 4) (2, 3, 5) (2, 3, 6) (2, 4, 4) (3, 3, 3)	$\mathbf{S}^2$ $\mathbf{S}^2$ $\mathbf{S}^2$ $\mathbf{E}^2$ $\mathbf{E}^2$ $\mathbf{E}^2$	$\bar{\mathbf{4}}$**3m** **m**$\bar{\mathbf{3}}$**m** **5**$\bar{\mathbf{3}}$**m** **17. p6m** **11. p4m** **14. p3m1**	T_d O_h I_h	$[3, 3]$ $[3, 4]$ $[3, 5]$ $[3, 6]$ $[4, 4]$ Δ	***233** ***234** ***235** ***236** ***244** ***333**
$(+, 0; [p]; \{(q)\})$ $(p; q)$ $3 \leq p,\ 2 \leq q$	(3; 2) (3; 3) (4; 2)	$\mathbf{S}^2$ $\mathbf{E}^2$ $\mathbf{E}^2$	**m**$\bar{\mathbf{3}}$ **15. p31m** **12. p4g**	T_h	$[3^+, 4]$ $[3^+, 6]$ $[4^+, 4]$	**3*2** **3*3** **4*2**

*Sch = Schoenflies

$(\mathcal{T},\Gamma)$ is a *maximal tiling* iff $\Gamma = \operatorname{Aut}\mathcal{T}$ is a (maximal) *group of automorphisms* of $\mathcal{T}$. A maximal tiling $(\mathcal{T}^* = \mathcal{T},\ \Gamma^* = \operatorname{Aut}\mathcal{T})$ represents a *family of tilings* $(\mathcal{T},\Gamma)$ *such that there is a bijection* (topological mapping) above

$$\varphi : \mathcal{T} \to \mathcal{T}^*, \quad \text{with} \quad \varphi^{-1}\Gamma\varphi \le \Gamma^* = \operatorname{Aut}\mathcal{T}.$$

Such a tiling $(\mathcal{T},\Gamma)$ above is also called a *symmetry breaking* of $(\mathcal{T}^*,\Gamma^*)$. That means, *a family consists of the* (non-equivariant) *symmetry breakings of its maximal representative tiling.*

The principle of classification into equivariance classes can be concisely formulated by D-*symbols* (initiated by A. W. M. DRESS [D87, DHM93, H93, M96] on an observation of M. S. DELANEY; B. N. DELONE also introduced analogous adjacency symbols in less concise form).

2.1. We take a combinatorial *barycentric subdivision* $\mathcal{C}$ of our tiling $(\mathcal{T},\Gamma)$ which is *invariant under the group* Γ. Any barycentric simplex $C \in \mathcal{C}$ has a 3-vertex as a formal 3-midpoint of a tile; then a 2-vertex as a 2-midpoint of a tile face; then a 1-vertex, a midpoint of an edge of the former face; then a 0-vertex of the former edge. This induces also an i-face $(i = 0,\,1,\,2,\,3 = d)$ of C, opposite to its i-vertex. Moreover, natural *involutive adjacency bijections* (operations)

$$\sigma_i : \mathcal{C} \to \mathcal{C}, \quad C \mapsto \sigma_i C, \quad i = 0,1,2,3 = d$$

are assumed to be introduced with the Γ-invariance property

$$(\sigma_i C)^\gamma = \sigma_i(C^\gamma) =: \sigma_i C^\gamma, \quad \gamma \in \Gamma, \quad \sigma_i \in \Sigma_I.$$

Here the free Coxeter group

$$\Sigma_I := (\sigma_0,\sigma_1,\sigma_2,\sigma_3 \,-\, \sigma_0^2 = \sigma_1^2 = \sigma_2^2 = \sigma_3^2 = 1), \quad I = \{0,1,2,3 = d\}$$

acts (on the left) on the barycentric tiling $\mathcal{C}$ commuting with the (right) action of Γ, as our convention by

$$(\sigma_j\sigma_i)C^{(\gamma_1\gamma_2)} := \sigma_j(\sigma_i C^{\gamma_1\gamma_2}) = (\sigma_j\sigma_i C^{\gamma_1})^{\gamma_2}$$

shows.

2.2. A Γ-*orbit* of any $C \in \mathcal{C}$ will be defined as an element

$$D := C^\Gamma := \{C^\gamma : \gamma \in \Gamma\} \quad \text{of a set} \quad \mathcal{D} := \mathcal{C}/\Gamma,$$

called a D-*set,* whose *adjacency operations* are induced by

$$\sigma_i : \mathcal{D} \to \mathcal{D}, \quad D \mapsto \sigma_i D = \{\sigma_i C^\gamma : \gamma \in \Gamma, C \in D\}, \quad i = 0,1,2,3;$$

i.e. Σ_I acts on the D-set $\mathcal{D} =: (\Sigma_I, \mathcal{D})$ as well.

Now the tile-transitivity of $(\mathcal{T}, \Gamma)$ implies that $(\Sigma_I, \mathcal{D})$ to $(\mathcal{T}, \Gamma)$ is a connected *finite* D-*diagram* (graph) with $4 = (d+1)$-marked (or coloured) branches (edges, loops are allowed as well)

$$\sigma_0 : \cdots\cdots, \quad \sigma_1 : - - -, \quad \sigma_2 : ——, \quad \sigma_3 : \sim\sim\sim, \quad (d = 3)$$

illustrated in our figures. Any $D = C^\Gamma$ will be a node (vertex) of our diagram. The possible cardinalities are $|\mathcal{D}| := 24\ (= (d+1)!)$, 12, 8, 6, 4, 3, 2, 1 in our 3-simplex case, according to the automorphism subgroups of a simplex of order 1, 2, 3, 4, 6, 8, 12, 24, respectively, as in our figures and tables.

2.3. A fundamental domain $\mathbf{F}_\Gamma$ for Γ will be glued from $|\mathcal{D}| =: n$ barycentric simplices of $\mathcal{C}$,

$$\mathbf{F}_\Gamma = \bigcup_{j=1}^{n} C_j,$$

where each C_j represents its orbit D_j whose set amounts $(\Sigma_I, \mathcal{D})$.

Starting with any $C =: C_1 \in \mathcal{C}$, there are well-defined numberings by Σ_I (see in [M96]) for gluing a simply connected $\mathbf{F}_\Gamma$ as a simplex or its part (see figures for families 1–12).

2.4. Such an $\mathbf{F}_\Gamma$ will represent *a good orbifold* $\mathcal{T}/\Gamma$, and it will provide a presentation for Γ by the generalized Poincaré algorithm not detailed here [M92, M96] (see at our family Table 5). To this we fix a *symmetric matrix function with natural* ($\mathbb{N}$) *entries*

$$M : (\Sigma_I, \mathcal{D}) \to \mathbb{N}_{I \times I}, \quad D \mapsto M(D) := m_{ij}(D), \quad i, j \in I$$

as follows.

(i) First we introduce another matrix function

$$R : D \mapsto r_{ij}(D) := \min\{r \in \mathbb{N} : (\sigma_j \sigma_i)^r(D) = D\}$$

($r = 0$ is excluded) for describing the base manifold of the orbifold we are introducing.

(ii) Then the matrix V of rotation orders for Γ (valences, branching numbers) will be introduced with natural ($\mathbb{N}$) entries

$$V : D \mapsto v_{ij}(D) \quad \text{such that} \quad m_{ij}(D) := v_{ij}(D) r_{ij}(D)$$

will be our D-*matrix function* M with

$$m_{ij}(D) = \min\{m \in \mathbb{N} : (\sigma_j \sigma_i)^m(C) = C, \quad C \in D\}.$$

(iii) To $\mathcal{T}/\Gamma$ by $(\Sigma_I, \mathcal{D}, M)$ we get the properties, later on as requirements (or *axioms*) for D-*matrix function* M and for D-symbol $(\Sigma_I, \mathcal{D}, M)$, in general:

(1) $m_{ii}(D) = 1$ for each $D \in \mathcal{D}$ and $i \in I = \{0, 1, 2, 3\}$;
(2) $m_{ij}(D) = m_{ij}(\sigma_i D) = m_{ji}(D) \geq 2$; $i \neq j \in I$; $D \in \mathcal{D}$;
(3) $m_{ij}(D) = 2$ if $|j - i| \geq 2$; $D \in \mathcal{D}$;
(4) $(\sigma_j \sigma_i)^{m_{ij}(D)}(D) = D$, $D \in \mathcal{D}$, $i, j \in I$.

In this survey we exclude ∞ (or 0) from the entries of M and so of V. Then (4) would not give any relation for Γ.

Furthermore, for a simplex tiling $(\mathcal{T}, \Gamma) \rightsquigarrow (\Sigma_I, \mathcal{D}, M)$,

$$m_{01}(D) = m_{12}(D) = 3, \quad D \in \mathcal{D}$$

always hold, because of triangle faces and trihedral corners of a simplex.

The matrix entries $m_{23}(D) \geq 3$ *are required for the number of surrounding simplices in* $(\mathcal{T}, \Gamma)$ *round the simplex edge at any barycentric simplex* $C \in D$ *in its orbit.* The entry

$$v_{23}(D) = m_{23}(D)/r_{ij}(D)$$

provides just the rotational order of the Γ-stabilizer of this simplex edge.

2.5. *Thus, in our figures and tables the groups* Γ *are given (each up to a presentation, i.e. up to a homeotopy* [ZVC80]) *as follows.*

(1) We indicate the simplex stabilizer in Γ on the left-hand side of the letter Γ with its order, and a lower index on its right-hand side for numbering the cases. E.g. we have three cases (in Table 5) for stabilizer $\bar{4} = \mathbf{2}\otimes$ of order 4 (by Table 1–2), and 64 fundamental cases for the trivial stabilizer. The simplex stabilizer, denoted by Γ^3, is described by the subsymbol $(\Sigma_I^3, \mathcal{D}, M^3)$, defined by the σ_0, σ_1, σ_2 operations and the restricted matrix function

$$M^3 : D \to m_{rs}(D), \quad r, s \in \{0, 1, 2\}.$$

(2) In parentheses we give the m_{23} matrix entries for the Γ-classes of edges in a form $r_{23}v_{23}$. The entries r_{23} will be determined finally by the side pairings σ_3 of the 'free faces' of the simplex. Thus $D \mapsto$

Table 2. Spatial Coxeter–Lanner simplices with proper or ideal vertices

Coxeter diagram with parameters	Conditions	Spc	Ideal vertex if any, its stabilizer	Self-symmetry group with generators. Extension in $\mathbf{E}^3$
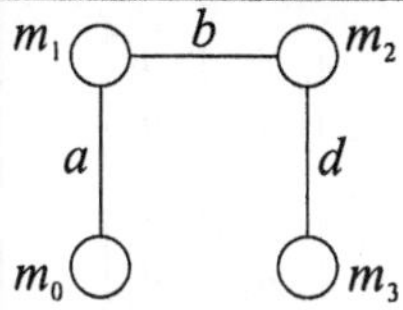	(2, 3, 4)	$\mathbf{S}^3$		
(a, b, d)	(2, 3, 5)	$\mathbf{S}^3$		
$2 \leq a;\ 3 \leq b \leq d \leq 6$	(2, 3, 6)	$\mathbf{E}^3$	A_0 **(p6m)**	**1**
$a < d$	(3, 3, 4)	$\mathbf{S}^3$		
if $a = d$ then **2** is the subgroup of self-symmetries	(3, 3, 5)	$\mathbf{S}^3$		$\begin{bmatrix} A_0 A_1 A_2 A_3 \\ A_0 A_1 A_2 A_3 \end{bmatrix}$
	(3, 3, 6)	$\mathbf{H}^3$	A_0 **(p6m)**	
	(3, 4, 4)	$\mathbf{H}^3$	A_0 **(p4m)**	
	(4, 3, 5)	$\mathbf{H}^3$		**Family 13**
	(4, 3, 6)	$\mathbf{H}^3$	A_0 **(p6m)**	
	(5, 3, 6)	$\mathbf{H}^3$	A_0 **(p6m)**	
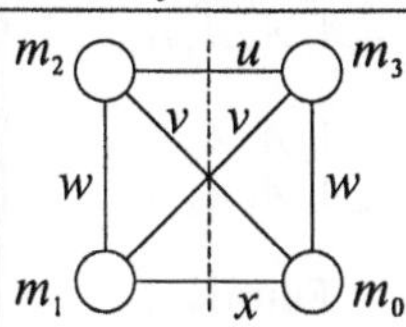	(2; 2, 3; 3)	$\mathbf{S}^3$		
$(u;\ v,\ w;\ x)$	(2; 2, 3; 4)	$\mathbf{S}^3$		
$2 \leq v < w \leq 6$	(2; 2, 3; 5)	$\mathbf{H}^3$		
$2 \leq u < x \leq 6$	(2; 2, 3; 6)	$\mathbf{H}^3$	A_2, A_3 **(p6m)**	
if $v = w$ then **mm2**	(2; 2, 4; 3)	$\mathbf{E}^3$	**221. Pm$\bar{3}$m < 229. Im$\bar{3}$m**	
if $u = x$ then **222**	(2; 2, 4; 4)	$\mathbf{H}^3$	A_2, A_3 **(p4m)**	
	(2; 2, 5; 3)	$\mathbf{H}^3$		
	(2; 2, 6; 3)	$\mathbf{H}^3$	A_2, A_3 **(p6m)**	**2**
	(3; 2, 3; 4)	$\mathbf{H}^3$		
	(3; 2, 3; 5)	$\mathbf{H}^3$		$\begin{bmatrix} A_0 A_1 A_2 A_3 \\ A_1 A_0 A_3 A_2 \end{bmatrix}$
	(3; 2, 3; 6)	$\mathbf{H}^3$	A_2, A_3 **(p6m)**	
	(3; 2, 4; 4)	$\mathbf{H}^3$	A_2, A_3 **(p4m)**	
	(4; 2, 3; 5)	$\mathbf{H}^3$		**Family 9**
	(4; 2, 3; 6)	$\mathbf{H}^3$	A_2, A_3 **(p6m)**	
	(5; 2, 3; 6)	$\mathbf{H}^3$	A_2, A_3 **(p6m)**	
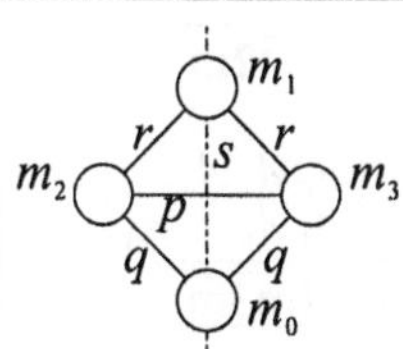	(2; 2, 3; 2)	$\mathbf{S}^3$		
$(p;\ q,\ r;\ s)$	(2; 2, 3; 4)	$\mathbf{E}^3$	**225. Fm$\bar{3}$m < 221. Pm$\bar{3}$m**	
$2 \leq p \leq 3;\ 2 \leq s \leq 6$	(2; 2, 3; 5)	$\mathbf{H}^3$		
$2 \leq q < r \leq 4$	(2; 2, 3; 6)	$\mathbf{H}^3$	A_2, A_3 **(p6m)**	
if $q = r$ then **mm2**	(2; 2, 4; 2)	$\mathbf{E}^3$	A_0 **(p4m)**	**m**
	(2; 2, 4; 3)	$\mathbf{H}^3$	A_0 **(p4m)**	
	(2; 3, 4; 2)	$\mathbf{H}^3$	A_0 **(p4m)**	$\begin{bmatrix} A_0 A_1 A_2 A_3 \\ A_0 A_1 A_3 A_2 \end{bmatrix}$
	(3; 2, 3; 3)	$\mathbf{H}^3$	A_0 **(p3m1)**	
	(3; 2, 3; 4)	$\mathbf{H}^3$	A_0 **(p3m1)**	
	(3; 2, 3; 5)	$\mathbf{H}^3$	A_0 **(p3m1)**	**Family 6**
	(3; 2, 3; 6)	$\mathbf{H}^3$	A_2, A_3 **(p6m)** / A_0 **(p3m1)**	

Table 2 (continued)
Spatial Coxeter–Lanner simplices with proper or ideal vertices

Coxeter diagram with parameters	Conditions	Spc	Ideal vertex if any, its stabilizer	Self-symmetry group with generators. Extension in $\mathbf{E}^3$
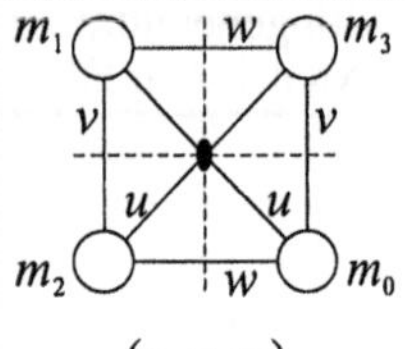 (u, v, w) $2 \le u < v < w \le 6$ if e.g. $v = w$ then **$\bar{4}$2m**	(2, 3, 4) (2, 3, 5) (2, 3, 6)	$\mathbf{H}^3$ $\mathbf{H}^3$ $\mathbf{H}^3$	every A_i **(p6m)**	**222** $\begin{bmatrix} A_0 & A_1 & A_2 & A_3 \\ A_1 & A_0 & A_3 & A_2 \end{bmatrix}$ $\begin{bmatrix} A_0 & A_1 & A_2 & A_3 \\ A_2 & A_3 & A_0 & A_1 \end{bmatrix}$ **Family 5**
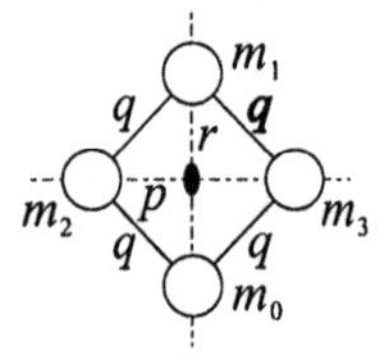 $(p; q; r)$ $p < r$ if $p = r$ then **$\bar{4}$2m**	$(p; 2; r)$ (2; 3; 3)	$\mathbf{S}^3$ $\mathbf{H}^3$	A_2, A_3 **(p3m1)**	**mm2** $\begin{bmatrix} A_0 & A_1 & A_2 & A_3 \\ A_1 & A_0 & A_2 & A_3 \end{bmatrix}$ $\begin{bmatrix} A_0 & A_1 & A_2 & A_3 \\ A_0 & A_1 & A_3 & A_2 \end{bmatrix}$ **Family 4**
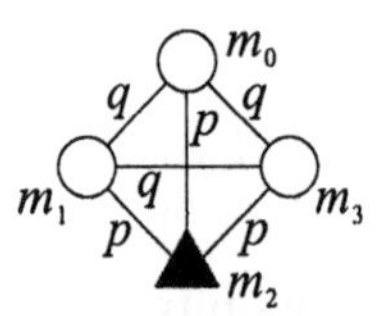 $(p; q)$ $p \ne q$ if $p = q$ then **$\bar{4}$3m**	(2; 3) (3; 2) (4; 2)	$\mathbf{E}^3$ $\mathbf{S}^3$ $\mathbf{H}^3$	A_2 **(p3m1)** A_0, A_1, A_3 **(p4m)**	**3m** $\begin{bmatrix} A_0 & A_1 & A_2 & A_3 \\ A_1 & A_0 & A_2 & A_3 \end{bmatrix}$ $\begin{bmatrix} A_0 & A_1 & A_2 & A_3 \\ A_0 & A_3 & A_2 & A_1 \end{bmatrix}$ **Family 3**
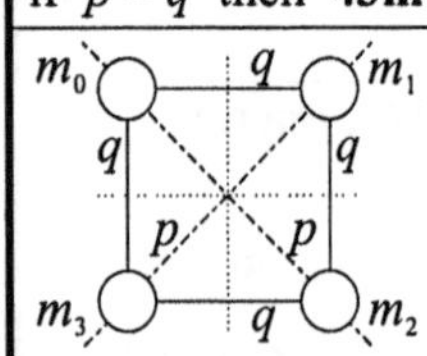 $(p; q)$ $p \ne q$ if $p = q$ then **$\bar{4}$3m**	$(p; 2)$ (2; 3) (2; 4)	$\mathbf{S}^3$ $\mathbf{E}^3$ $\mathbf{H}^3$	every A_i **(p4m)**	**$\bar{4}$2m** $\begin{bmatrix} A_0 & A_1 & A_2 & A_3 \\ A_1 & A_0 & A_2 & A_3 \end{bmatrix}$ **216. F$\bar{4}$3m < 229. Im$\bar{3}$m** $\begin{bmatrix} A_0 & A_1 & A_2 & A_3 \\ A_2 & A_3 & A_0 & A_1 \end{bmatrix}$ **Family 2**
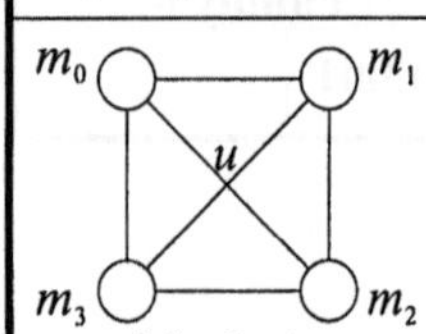 (u) $2 \le u$	(2) (3)	$\mathbf{S}^3$ $\mathbf{H}^3$	every A_i **(p3m1)**	**$\bar{4}$3m** **Family 1**

$v_{23}(D)$ $(= a, b, \ldots, u, v, \ldots)$ has been fixed by a Γ, or varies if we search for a convenient Γ and a desired simplex tiling $(\mathcal{T}, \Gamma)$ from an appropriate D-symbol $(\Sigma_I, \mathcal{D}, M)$ to be constructed.

(3) Any partial subsymbols of $(\Sigma_I, \mathcal{D}, M)$ to an edge (1-face) as a connected component, i.e. $({}^c\Sigma^1, {}^c\mathcal{D}^1, {}^c\mathcal{M}^1)$ is defined by the σ_0, σ_2, σ_3 operations and the restricted matrix function ${}^c\mathcal{M}^1$ for a component ${}^c\mathcal{D}^1 \subset \mathcal{D}$. This characterizes the partial 2-dimensional tiling $({}^c\mathcal{T}^1, {}^c\Gamma^1)$ surrounding an edge midpoint by corner domains, bounded by 0-, 2-, 3-marked side faces of the corresponding barycentric simplices. *Any fundamental domain ${}^c\mathbf{F}^1$, glued together from the representatives of ${}^c\mathcal{D}^1$ in ${}^c\mathcal{C}^1$, has to describe a good spherical $\mathbf{S}^2$ orbifold.* These have been classified by the well-known finite spherical groups (Table 1).

(4) Any partial subsymbols $({}^c\Sigma^0, {}^c\mathcal{D}^0, {}^c\mathcal{M}^0)$ of $(\Sigma_I, \mathcal{D}, M)$, defined by the σ_1, σ_2, σ_3 operations and its restricted matrix function ${}^c\mathcal{M}^0$ for a well-defined component ${}^c\mathcal{D}^0 \subset \mathcal{D}$, describes a vertex Γ-class by a partial 2-tiling $({}^c\mathcal{T}^0, {}^c\Gamma^0)$, surrounding the corresponding vertex (and its Γ-images), by '123-corner' domains. Again, a fundamental domain ${}^c\mathbf{F}^0$ can be glued from the representatives of ${}^c\mathcal{D}^0$. *This ${}^c\mathbf{F}^0$ has to be either a good spherical $\mathbf{S}^2$ orbifold or a Euclidean $\mathbf{E}^2$ orbifold.* The latter ones are also well-known by the 17 plane crystallographic groups (Table 1). Such a domain ${}^c\mathbf{F}^0$ is glued together from Γ-image corner domains of the simplex tile.

Again, there is an algorithm, implemented on computer for deciding these necessary conditions (see e.g. [LM90]). Our figures and tables contain these data.

3. Maximal tilings and families by D-symbols

To a later guarantee of the metric existence of a tiling $(\mathcal{T}, \Gamma)$ from a well-formed D-symbol, we formulate the concept of *the family* and the maximal tiling $(\mathcal{T}, \Gamma^* = \operatorname{Aut} \mathcal{T})$ from Section 2 by means of D-symbols (due to A. W. M. Dress, see e.g. [D87]).

Theorem 3.1. *Two tilings $(\mathcal{T}, \Gamma)$ and $(\mathcal{T}', \Gamma')$ are (combinatorially) equivariant iff their* D*-symbols $(\Sigma_I, \mathcal{D}, M)$ and $(\Sigma'_{I'}, \mathcal{D}', M')$ are* D*-isomorphic. That means $\Sigma_I \cong \Sigma'_{I'}$ are isomorphic free Coxeter groups (identified), moreover, there is a bijection*

$$\psi : \mathcal{D} \to \mathcal{D}', \quad D \mapsto D' := D^\psi \quad \text{with} \quad \sigma'_i(D^\psi) = (\sigma_i D)^\psi \quad \text{and}$$
$$m'_{ij}(D^\psi) = m_{ij}(D) \quad \text{for any} \quad D \in \mathcal{D}, \quad i, j \in I = \{0, 1, \ldots, d\}.$$

We speak about D-*morphism* of D-symbols iff $\Sigma_I \cong \Sigma'_{I'}$ and ψ above is surjective but it preserves the two above assumptions.

Theorem 3.2. $(\mathcal{T}, \Gamma)$ *is a symmetry breaking of* $(\mathcal{T}', \Gamma')$ *iff there is a* D-*morphism* ψ *from* $(\Sigma_I, \mathcal{D}, M)$ *onto* $(\Sigma'_{I'}, \mathcal{D}', M')$.

Theorem 3.3. $(\mathcal{T}, \Gamma^* = \mathrm{Aut}\, \mathcal{T})$ *is a maximal tiling iff its* D-*symbol does not allow any proper* D-*morphism onto a smaller* D-*symbol.*

Such a maximal tiling with minimal D-*symbol represents a family of tilings, the set of its symmetry breakings or* D-*morphic preimages of its* D-*symbols.* Of course, we restrict ourselves in forming a family. Namely, we require the prescribed tiling property for the family members, e.g. simplex-transitivity now. In our figures some places, e.g. at Families 2–12, we have illustrated minimal D-diagrams and their D-morphic preimages (Families 2, 3, 5).

We have indicated only the tiling groups. E.g.

$$ {}^{\bar{4}2m}_{\quad 8}\Gamma(\bar{u}; 2\bar{v}) > {}^{\bar{4}}_{4}\Gamma_1(2u; 2v) $$

expresses that the second group (tiling, orbifold), by its fundamental domain in the picture, is a symmetry breaking (invariant of index 2) of the maximal group ${}^{\bar{4}2m}_{\quad 8}\Gamma(\bar{u}; 2\bar{v})$, if $\bar{u} = 2u$, $2\bar{v} = 2v$ for the ▬ and ═ edges, respectively (as entries of the matrix functions).

Sometimes the *parameters involve (non-invariant) symmetry breaking,* e.g. ${}^{\bar{4}2m}_{\quad 8}\Gamma(u; 2v)$ as representative for Family (series) 2 is maximal iff $u \neq 2v$, else ${}^{\bar{4}3m}_{\quad 24}\Gamma(\bar{u} = u = 2v)$ is the supergroup (of index 3) in Family (series) 1. Thus ${}^{\bar{4}2m}_{\quad 8}\Gamma(u; 2v)$ for $(u; v) = (\mathbf{6}; \mathbf{3})$ has the supergroup (realizable in $\mathbf{H}^3$) ${}^{\bar{4}3m}_{\quad 24}\Gamma(\bar{u} = u = 2v = 6)$ only with 1 node in its D-symbol. Then the subgroup (of index 3) *is not invariant.* We indicated this 'boldface' rotational orders in our Table 5 of families.

We define the metric realization of our simplex tiling $(\mathcal{T}, \Gamma)$ *as* $(\mathbf{T}, \mathbf{\Gamma})$, *in a homogeneous Riemannian space* $(\mathbf{X}^3, G = \mathrm{Isom}\, \mathbf{X}^3)$ *from among the eight Thurston geometries, iff there is a topological equivariance bijection* $\varphi : \mathcal{X} \to \mathbf{X}^3$ *mapping the tiling* $\mathcal{T}$ *onto* $\mathbf{T}$ *(a topological simplex tiling in* $\mathbf{X}^3$*) such that* $\mathbf{\Gamma} = \varphi^{-1}\Gamma\varphi$, *i.e.* φ *carries the action of* Γ *onto the isometric action of* $\mathbf{\Gamma}$.

It is important to emphasize that our $(\mathcal{T}, \Gamma)$ can always be endowed with a simply connected piece-wise linear (PL) topology (through the barycentric simplices and the Σ_I action by 'local reflection' [M96]).

So we build up our metric embedding through simplicial mapping of the fundamental domain $\mathbf{F}_\Gamma \subset \mathcal{X}^3$ onto a metric fundamental domain $\mathbf{F}_\Gamma$ in $\mathbf{X}^3$ by its projective model (Table 3). Now a simplex tile will be a projective coordinate simplex in a projective sphere $\mathcal{PS}^3$. A side pairing

Table 3. The eight homogeneous geometries

Space **X**	Signature of polarity $\Pi(.)$ or scalar product $\langle\ ,\ \rangle$ in V_4	Domain of proper points of **X** in $\mathcal{PS}^3$ ($\mathbf{V}^4(\mathbf{R})$, V_4)	The group G = Isom **X** as a special transformation group of $\mathcal{PS}^3$
$\mathbf{S}^3$	$(++++)$	$\mathcal{PS}^3$	Coll. $\mathcal{PS}^3$ preserving $\Pi(.)$
$\mathbf{H}^3$	$(-+++)$	$\{(\mathbf{x})\in\mathcal{P}^3 : \langle\mathbf{x},\mathbf{x}\rangle < 0\}$	Coll. $\mathcal{P}^3$ preserving $\Pi(.)$
$\widetilde{\mathbf{SL_2R}}$	$(--++)$ with skew line fibering	Universal covering of $\mathcal{H} := \{[\mathrm{x}] \in\mathcal{PS}^3 : \langle\mathbf{x},\mathbf{x}\rangle < 0\}$ by fibering transformations	Coll. $\mathcal{PS}^3$ preserving $\Pi(.)$ and fibres
$\mathbf{E}^3$	$(0+++)$	$\mathcal{A}^3 = \mathcal{P}^3\setminus\{\omega^\infty\}$ where $\omega^\infty := (\boldsymbol{b}^0)$, $\boldsymbol{b}^0_\cdot = \mathbf{0}$	Coll. $\mathcal{P}^3$ preserving $\Pi(.)$ generated by plane reflections
$\mathbf{S}^2\times\mathbf{R}$	$(0+++)$ with O-line bundle fibering	$\mathcal{A}^3\setminus\{O\}$ O is a fixed origin	G is generated by plane reflections and sphere inversions, leaving invariant the O-concentric 2-spheres of $\Pi(.)$
$\mathbf{H}^2\times\mathbf{R}$	$(0-++)$ with O-line bundle fibering	$\mathcal{C}^+ = \{X\in\mathcal{A}^3 : \langle\overrightarrow{OX},\overrightarrow{OX}\rangle < 0$, half cone$\}$ by fibering	G is generated by plane reflections and hyperboloid inversions, leaving invariant the O concentric half-hyperboloids in the half cone $\mathcal{C}^+$ by $\Pi(.)$
Sol	$(0-++)$ with parallel plane fibering	$\mathcal{A}^3$ with parallel plane fibering	Coll. $\mathcal{P}^3$ preserving $\Pi(.)$ and the parallel plane fibering
Nil	$(0\ 0\ \pm 0\ +)$ with parallel line bundle fibering	$\mathcal{A}^3$ with a distinguished parallel plane pencil along each fibre line	Coll. of $\mathcal{P}^3$ (•) preserving $\Pi(.)$ and the line bundle with the plane pencil along each fibre line

(•) more precisely: these are conjugate to quadratic mappings by a fixed transform $x' = x,\ y' = y,\ z' = z - \frac{1}{2}xy$

of the simplex tile will be described by matrices, as at Zhuk tilings in the next section, or generated by quadratic inversions [M97] in addition, as for $\mathbf{S}^2\times\mathbf{R}$, $\mathbf{H}^2\times\mathbf{R}$, not discussed more in this survey.

A crucial point of our method is that the metric realization of a maximal tiling will imply the metric realization of its family. And vice-versa, if a maximal tiling, by its orbifold, can not be metrically realized, then neither can any member of its family in most cases, however, not in all cases. We do not want to go into details with respect to some delicate problems, as possible but not existing parallel realizations, e.g. in $\mathbf{H}^2\times\mathbf{R}$ and $\widetilde{\mathbf{SL}_2\mathbf{R}}$, where the maximal group can only be realized in $\mathbf{H}^2\times\mathbf{R}$ (see Γ_{16} in Family 8). Similar phenomena occur in $\mathbf{E}^3$ and **Nil** iff an ideal vertex appears and Γ consists only of orientation preserving mappings (see also Γ_{16} in Family 8 and Table 4 with the '+' cases). *Thus we have positively and constructively decided* THURSTON*'s orbifold conjecture for orbifold $\mathcal{T}/\Gamma$, derived from our tile-transitive simplex tilings.* At the same time we can disprove his theorem in [Th97], Theorem 4.7.10.

As we have seen in [MPSz97], it is enough to realize some essential maximal families as Table 5 for Family 30 and Family 11 will show in the next section.

4. Zhuk's tilings in spaces $\mathbf{S}^3$, $\mathbf{H}^3$, Nil and $\widetilde{\mathbf{SL}_2\mathbf{R}}$

We start with a 3-simplex $A_0A_1A_2A_3$ as a combinatorial incidence structure of its vertices $A_i(\mathbf{a}_i)$, $i \in 0, 1, 2, 3$; edges A_iA_j, $i < j$; faces $b^i(\boldsymbol{b}^i) = A_jA_kA_l$, $\{i, j, k, l\} = I$; and the solid itself. Parallelly we have just introduced the basis vectors $\{\mathbf{a}_i\}$ for a *real vector space* $\mathbf{V}^4(\mathbb{R})$ and the dual basis $\{\boldsymbol{b}^i\}$ for the *dual space* $\boldsymbol{V}_4$ of linear forms with $\mathbf{a}_i\boldsymbol{b}^j = \delta_i^j$ (the Kronecker symbol, Einstein-Schouten index conventions will also be used). By

$$\mathbf{x} \overset{+}{\sim} \mathbf{y} \Leftrightarrow \mathbf{y} = c\mathbf{x}; \quad \mathbf{x}, \mathbf{y} \in \mathbf{V}^4 \setminus \{\mathbf{0}\}, \quad c \in \mathbb{R}^+ \quad \text{(positive real numbers)}$$
$$\Leftrightarrow (\mathbf{x}) = (\mathbf{y})$$

we introduce the rays of $\mathbf{V}^4$ as *points of the projective* 3-*sphere* $\mathcal{PS}(\mathbf{V}^4)$, *and by*

$$\boldsymbol{u} \overset{+}{\sim} \boldsymbol{v} \Leftrightarrow \boldsymbol{v} = \tfrac{1}{c}\boldsymbol{u}; \quad \boldsymbol{u}, \boldsymbol{v} \in \boldsymbol{V}_4 \setminus \{\boldsymbol{0}\}, \quad c \in \mathbb{R}^+,$$
$$\Leftrightarrow (\boldsymbol{u}) = (\boldsymbol{v})$$

the *oriented (positive) planes* of $\mathcal{PS}(\boldsymbol{V}_4)$ will be defined. Then

$$\mathbf{x}\boldsymbol{u} = \mathbf{y}\boldsymbol{v} \lesseqqgtr 0$$

means: either $(\mathbf{x})$ lies in the negative halfspace of $(\boldsymbol{u})$ $(<)$; or $(\mathbf{x})$ is incident to the plane $(\boldsymbol{u})$ $(=)$; or $(\mathbf{x})$ lies in the positive halfspace of $(\boldsymbol{u})$ $(>)$; respectively.

$\mathcal{PS}^3(\mathbb{R})$ is defined as a *projective* 3-*sphere* by the above incidence structure of (non-zero) subspaces of $\mathbf{V}^4$ or dually of $\boldsymbol{V}_4$. The *projective* 3-*space* $\mathcal{P}^3(\mathbb{R})$ is clearly defined by unifying opposite rays as a projective point $(\mathbf{x})$ and by a form-class $(\boldsymbol{u})$ as a plane *up to non-zero real factors.*

We prefer $\mathcal{PS}^3(\mathbb{R})$ not to exclude the spherical space $\mathbf{S}^3$ from a unified treatment.

Let the tiling $\mathcal{T}$ and the group $\Gamma = \Gamma_{39}(a;2b;6c)$ be introduced by the face pairings in Table 5, at Family 30 and in the corresponding picture of our figures, as we repeat here for convenience of the Reader (since the structure is the same throughout this paper).

$$\mathbf{r}_0 : A_1A_2A_3 \to A_1A_3A_2; \quad \mathbf{r}_1 : A_2A_3A_0 \to A_2A_0A_3 \tag{4.1}$$

are defined as involutive 'halfturns' of the faces, denoted also by r_0 and r_1, respectively, in our picture.

$$\mathbf{r} : A_0A_1A_3 =: r^{-1} \to r := A_0A_1A_2 \tag{4.2}$$

is a 'combinatorial rotation'. The pairings map the vertices, edges and faces as introduced, preserving the incidences. The solid of $A_0A_1A_2A_3$ will be mapped onto the neighbouring simplices along the corresponding faces and further, we can form the simplex images under the product of generators.

In the basis – dual basis pair $\{\mathbf{a}_i\}$ – $\{\boldsymbol{b}^j\}$ we have the matrix presentation e.g. for the mapping $\mathbf{r}$ (and its inverse $\mathbf{r}^{-1}$, too)

$$\mathbf{r}\big((r_i^j),\,(R_j^k)\big), \quad \text{with} \quad r_i^jR_j^k = \delta_i^k; \quad \mathbf{a}_i \mapsto r_i^j\mathbf{a}_j, \quad \text{i.e.} \tag{4.3}$$

$$(r_i^j) : \begin{pmatrix}\mathbf{a}_0\\ \mathbf{a}_1\\ \mathbf{a}_2\\ \mathbf{a}_3\end{pmatrix} \mapsto \begin{pmatrix}1&0&0&0\\0&1&0&0\\s&t&A&-1\\0&0&1&0\end{pmatrix}\begin{pmatrix}\mathbf{a}_0\\ \mathbf{a}_1\\ \mathbf{a}_2\\ \mathbf{a}_3\end{pmatrix} \quad \text{with} \begin{cases} A = 2\cos\frac{2\pi}{a},\\ 3 \le a \in \mathbb{N},\\ \text{i.e. } -1 \le A < 2,\end{cases}$$

$$\text{and} \quad (R_j^k) : \begin{pmatrix}\boldsymbol{b}^0 & \boldsymbol{b}^1 & \boldsymbol{b}^2 & \boldsymbol{b}^3\end{pmatrix} \mapsto \begin{pmatrix}\boldsymbol{b}^0 & \boldsymbol{b}^1 & \boldsymbol{b}^2 & \boldsymbol{b}^3\end{pmatrix}\begin{pmatrix}1&0&0&0\\0&1&0&0\\0&0&0&1\\s&t&-1&A\end{pmatrix},$$

(by row-column multiplication). Here the parameter a, i.e. $A = 2\cos\frac{2\pi}{a}$, is fixed for $\Gamma := \Gamma_{39}(a;2b;6c)$ as a rotational order. The relation

$$\mathbf{r}^a = 1 \tag{4.4}$$

holds, indeed. The line of A_0A_1 is pointwise fixed. The parameters s and t will be fixed later by the other relations and some projective freedom. In such a way we attempt to define for our group

$$\Gamma := \Gamma_{39}(a;2b;6c) = \big(\mathbf{r}_0, \mathbf{r}_1, \mathbf{r} - 1 = \mathbf{r}_0^2 = \mathbf{r}_1^2 = \mathbf{r}^a = (\mathbf{r}\mathbf{r}_0)^b$$
$$= (\mathbf{r}_1\mathbf{r}\mathbf{r}_1\mathbf{r}_0\mathbf{r}_1\mathbf{r}^{-1})^c\big), \quad 3 \le a \in \mathbb{N}, \quad 2 \le b \in \mathbb{N}, \quad 1 \le c \in \mathbb{N} \tag{4.5}$$

generated in (4.1–2), a true projective representation, i.e. the relations in (4.5) will be fulfilled, and hold only the group theoretical consequences of these defining relations. *This will be the extension of the so-called Poincaré algorithm to the projective sphere* $\mathcal{PS}^3$ *in the sense of* [M92].

We further continue describing our example. The halfturns

$$\mathbf{r}_0: \begin{pmatrix}\mathbf{a}_0\\ \mathbf{a}_1\\ \mathbf{a}_2\\ \mathbf{a}_3\end{pmatrix} \mapsto \begin{pmatrix}-1 & x & y & y\\ 0 & 1 & 0 & 0\\ 0 & 0 & 0 & 1\\ 0 & 0 & 1 & 0\end{pmatrix}\begin{pmatrix}\mathbf{a}_0\\ \mathbf{a}_1\\ \mathbf{a}_2\\ \mathbf{a}_3\end{pmatrix}, \quad \mathbf{r}_1: \begin{pmatrix}\mathbf{a}_0\\ \mathbf{a}_1\\ \mathbf{a}_2\\ \mathbf{a}_3\end{pmatrix} \mapsto \begin{pmatrix}0 & 0 & 0 & 1\\ v & -1 & u & v\\ 0 & 0 & 0 & 1\\ 1 & 0 & 0 & 0\end{pmatrix}\begin{pmatrix}\mathbf{a}_0\\ \mathbf{a}_1\\ \mathbf{a}_2\\ \mathbf{a}_3\end{pmatrix} \tag{4.6}$$

indeed describe involutive point transformations for (4.1). The mapping $\mathbf{r}_0\mathbf{r}$ has to be a rotation of order $2 \leq b \in \mathbb{N}$ as a new parameter, which fixes A_1A_2 pointwise. Thus we obtain, say in the dual basis $\{\boldsymbol{b}^j\}$,

$$\mathbf{r}_0\mathbf{r}\colon \begin{pmatrix}\boldsymbol{b}^0 & \boldsymbol{b}^1 & \boldsymbol{b}^2 & \boldsymbol{b}^3\end{pmatrix} \mapsto \begin{pmatrix}\boldsymbol{b}^0 & \boldsymbol{b}^1 & \boldsymbol{b}^2 & \boldsymbol{b}^3\end{pmatrix}\begin{pmatrix}-1 & x & y & y\\ 0 & 1 & 0 & 0\\ 0 & 0 & 1 & 0\\ -s & sx+t & sy+A & sy-1\end{pmatrix}, \tag{4.7}$$

with $B := 2\cos\frac{2\pi}{b} = sy - 2$, for $2 \leq b$; and $b = 2$ iff $s = y = 0$.

Take the special case $c = 1$, else outer vertices occur (not detailed in our Table 5). The equation $\mathbf{r}_1\mathbf{r}_0\mathbf{r}_1 = \mathbf{r}^{-1}\mathbf{r}_1\mathbf{r}$, equivalent then to the 5th relation in (4.5), yields the matrix equation by $\{\boldsymbol{b}^j\}$ as follows:

$$\begin{pmatrix}0 & 0 & 1 & 0\\ v+us & -1+ut & uA+v & -u\\ 1 & 0 & 0 & 0\\ tv+A+(tu-1)s & (tu-2)t & (tu-1)A+s+tv & 1-tu\end{pmatrix}$$
$$= \begin{pmatrix}0 & 0 & 1 & 0\\ (vx-1)v+vy+u & 1-vx & (vx-1)u+vy+v & v(vx-2)\\ 1 & 0 & 0 & 0\\ xv+y & -x & xu+y & -1+vx\end{pmatrix}. \tag{4.8}$$

a) First, we consider the cases (in (4.7))

$$2 < b, \quad B + 2 = sy > 0. \tag{4.9}$$

Then we obtain a compatibility criterion for tu,

$$B + 2 = tu(A + tu - 1)^2,$$
$$\text{with} \quad A = 2\cos\tfrac{2\pi}{a} \geq -1, \quad B = 2\cos\tfrac{2\pi}{b} \geq -1 \tag{4.10}$$

and the solutions, with some projective freedom of basis change

$$t = 1 = v, \quad x = 2 - u, \quad s = A + u - 1, \quad y = u(A + u - 1). \qquad (4.11)$$

ZHUK [Zh83] derived equivalent conditions by absolute trigonometry, only for the spherical space $\mathbf{S}^3$ and the hyperbolic space $\mathbf{H}^3$.

To specify the representation above, we look for a polarity $(_*) : \boldsymbol{V}_4 \to \mathbf{V}^4$, $\boldsymbol{b}^i \mapsto b^{ij}\mathbf{a}_j$, *invariant under* $\mathbf{r}$, $\mathbf{r}_0$, $\mathbf{r}_1$ *above. The plane → point correspondence, expressing the orthogonality, will be equivalently given by the scalar product* $\langle\ ,\ \rangle$ *and the quadratic form* $\xi_i b^{ij} \xi_j$ *over the dual space* $\boldsymbol{V}_4$. We get an 'eigenvector–eigenvalue' problem for the symmetric matrix (b^{ij}) above, and a result by computation (see also [MPSz98])

$$\begin{aligned}\xi_i b^{ij} \xi_j := {} & \xi_0\xi_0 - \xi_0(2-u)\xi_1 - \xi_0 u(A+u-1)\xi_2 - \xi_0 u(A+u-1)\xi_3 \\ & +\xi_1\xi_1 - \xi_1 u\xi_2 - \xi_1 u\xi_3 + \xi_2 u\xi_2 - \xi_2 Au\xi_3 + \xi_3 u\xi_3 \qquad (4.12) \\ = {} & \Big[\xi_0 - \frac{1}{2}(2-u)\xi_1 - \frac{1}{2}u(A+u-1)\xi_2 - \frac{1}{2}u(A+u-1)\xi_3\Big]^2 \\ & +\frac{1}{4}(4-u)u\Big[\xi_1 - \frac{(2-u)(A+u-1)+2}{4-u}\xi_2 - \frac{(2-u)(A+u-1)+2}{4-u}\xi_3\Big]^2 \\ & +\frac{u(2-A-u)(A+2)}{4-u}\Big[\xi_2 - \frac{2A+u}{2(2-A-u)}\xi_3\Big]^2 + \frac{u(A+2)(4-4A-3u)}{4(2-A-u)}\xi_3\xi_3.\end{aligned}$$

Example 4.1. $3 = a = b$, i.e. $A = B = -1$ and the solution of $1 = u(u-2)^2$ by (4.10) leads uniquely(!) to

$$u = c_{33} = \frac{3+\sqrt{5}}{2} = \Big(\frac{1+\sqrt{5}}{2}\Big)^2, \quad x = -\frac{\sqrt{5}-1}{2} = 2\cos\frac{3\pi}{5},$$
$$s = \frac{\sqrt{5}-1}{2} = 2\cos\frac{2\pi}{5}, \quad y = \frac{1+\sqrt{5}}{2} = 2\cos\frac{\pi}{5},$$

and to a quadratic form of signature $(+ + + +)$ in (4.12) Thus (by our Table 3), we get the spherical projective metric space $\mathbf{S}^3$, and our simplex has the face angles by (4.12).

$$\beta^{01} = \frac{3\pi}{5}, \quad \beta^{02} = \frac{\pi}{3} = \beta^{03}, \quad \beta^{12} = \frac{\pi}{5} = \beta^{13}, \quad \beta^{23} = \frac{2\pi}{3}.$$

$$\text{E.g.} \quad \cos\beta^{01} = -\frac{b^{01}}{\sqrt{b^{00}b^{11}}} = \frac{2-c_{33}}{2\sqrt{1\cdot 1}} = \frac{1-\sqrt{5}}{4} \quad \Rightarrow \quad \beta^{01} = \frac{3\pi}{5}$$

serves the face angle at the simplex edge A_2A_3.

We can check that our representation is true, indeed. The face angle $\beta^{23} = \frac{2\pi}{3}$ at A_0A_1 provides $a = 3$ surrounding simplices in $\mathcal{T}$, $\beta^{02} + \beta^{03} = \frac{2\pi}{3}$ at A_1A_3 and A_1A_2 provides $2b = 6$ simplices in $\mathcal{T}$, and $b = 3$

is just the rotational order of the stabilizers of these edges in Γ. The angle sum at the six Γ-equivalent halfedges in A_0A_2, A_2A_3 and A_3A_0 amounts to $2(\beta^{13}+\beta^{01}+\beta^{12})=2\pi$. Thus the six image simplices exactly surround these edges in $\mathcal{T}$, the halfedge stabilizer is trivial here as the 5th relation of (4.5) dictates by $c=1$.

We could compute any metric data of our simplex $A_0A_1A_2A_3$, e.g. for the distances $d(A_jA_k)$ we have

$$\cos\big[d(A_jA_k)\big]=\frac{a_{ij}}{\sqrt{a_{jj}a_{kk}}}\quad\text{by}\quad (a_{jk}):=(b^{ij})^{-1}\quad\text{from (4.12).}$$

Examples 4.2 – 4.8. The following parameter pairs analogously lead to signature $(-\,+\,+\,+)$ in (4.12). These interesting Zhuk tilings are metrically realized in the hyperbolic space $\mathbf{H}^3$ by (4.10–11), uniquely.

2. $(a;b)=(4;3)$, i.e. $(A;B)=(0;-1)$; $1=u(u-1)^2$, $u=c_{43}=1.754877667$;

3. $(a;b)=(3;4)$, i.e. $(A;B)=(-1;0)$; $2=u(u-2)^2$, $u=c_{34}=2.839286755$;

4. $(a;b)=(5;3)$, $u=c_{53}=1.269497961$;

5. $(a;b)=(3;5)$; $u=c_{35}=2.943151259$;

6. $(a;b)=(\mathbf{6};\mathbf{3})$, leads to a **non-maximal** $\mathbf{H}^3$-tiling with regular simplex of face angle $\frac{\pi}{3}$, i.e. all vertices are ideal (at the absolute).

7. $(a;b)=(3;6)$, $\beta^{01}=\beta^{23}=\frac{2\pi}{3}$, $\beta^{02}=\beta^{03}=\beta^{12}=\beta^{13}=\frac{\pi}{6}$,

8. $(a;b)=(4;4)$, $\beta^{01}=\beta^{23}=\frac{\pi}{2}$, $\beta^{02}=\beta^{03}=\beta^{12}=\beta^{13}=\frac{\pi}{4}$,

provide maximal $\mathbf{H}^3$-tilings with simplices of ideal vertices.

b) Second, let us consider the cases (not discussed by ZHUK in general)

$$b=2,\quad 0=B+2=s=y;\quad 3\le a,\quad -1\le A=2\cos\tfrac{2\pi}{a}<2;$$

and the solution of (4.8) is as follows

$$t=1=v,\qquad u=1-A,\qquad x=1+A. \tag{4.13}$$

It turns out that the invariant polarity, or equivalently, the invariant quadratic form over $\boldsymbol{V}_4$ can be obtained from (4.12) just by substituting $u=1-A$. The coefficients in the diagonal form of (4.12) will then be

$$1;\quad \frac{1}{4}(3+A)(1-A);\quad \frac{(1-A)(A+2)}{3+A};\quad \frac{1}{4}(A-1)^2(A+2). \tag{4.14}$$

(i) For $3 \le a < 6$, i.e. $-1 \le A < 1$ the signature by (4.14) is $(+ + + +)$ and we shall get $\mathbf{S}^3$-tilings with interesting Hopf–Seifert fiberings.

(ii) For $a = 6$, i.e. $A = 1$ the signature is $(0\ 0\ 0\ +)$, and by Table 3 we shall get $\mathcal{T}$ in the **Nil** space with $\Gamma_{39}(a = 6;\ 2b = 4;\ c = 1) =: \Gamma$ and Seifert fibering $*\mathbf{6 3_1 2_1}$ over the Euclidean base orbifold $*\mathbf{632}$.

(iii) For $6 < a \in \mathbb{N}$, i.e. $-1 < A < 2$, the signature by (4.14) is $(- - + +)$, and we shall obtain simplex tilings in the space $\widetilde{\mathbf{SL_2R}}$ by Table 3, again by Seifert fiberings.

In general, for parameters $(a; b; c) = (a; 2; 1)$, $3 \le a$ we have

$$\begin{aligned}\Gamma &:= \Gamma_{39}(a;\ 2b = 4;\ 6c = 6)\\ &= \left(\mathbf{r}_0, \mathbf{r}_1, \mathbf{r} - 1 = \mathbf{r}_0^2 = \mathbf{r}_1^2 = (\mathbf{r}\mathbf{r}_0)^2 = \mathbf{r}_1\mathbf{r}\mathbf{r}_1\mathbf{r}_0\mathbf{r}_1\mathbf{r}^{-1}\right), \quad (4.15)\end{aligned}$$

as the presentation, specialized from (4.5). As a consequence, the transformation

$$\mathbf{t} := (\mathbf{r}_1\mathbf{r}_0)^3 = (\mathbf{r}_1\mathbf{r}\mathbf{r}_0)^2 \quad (4.16)$$

then generates an invariant cyclic subgroup $\langle \mathbf{t} \rangle$ in Γ and also in its invariant (of index 2) subgroup $\Gamma_i := \langle \mathbf{r},\ \mathbf{s} := \mathbf{r}_0\mathbf{r}_1,\ \mathbf{h} = \mathbf{r}_1\mathbf{r}\mathbf{r}_0 \rangle \lhd \Gamma$.

The factor group $\Gamma_i / \langle \mathbf{t} \rangle$ has a presentation

$$\Gamma_i / \langle \mathbf{t} \rangle = (\bar{\mathbf{r}}, \bar{\mathbf{s}}, \bar{\mathbf{h}} - 1 = \bar{\mathbf{r}}^a = \bar{\mathbf{s}}^3 = \bar{\mathbf{h}}^2 = \bar{\mathbf{r}}\bar{\mathbf{s}}\bar{\mathbf{h}}) \quad (4.17)$$

as a plane group of (orbifold) signature

$$\left(+,\ 0;\ [2, 3, a];\ \{\ \}\right) = \mathbf{23a}, \quad \text{(Table 1)}. \quad (4.18)$$

This is a 2-sphere with three cone points (rotation centres) of orders 2, 3, a. Then $\Gamma_i = \mathbf{2_1 3_1 a}$ is a logical denotation (as we see at $\Gamma_{55}(u;\ 10v)$ in Family 11). The starting group $\Gamma := \Gamma_{39}(a;\ 2b = 4;\ 6c = 6)$ factorized by $\langle \mathbf{t} \rangle$ will be

$$\Gamma / \langle \mathbf{t} \rangle = *\mathbf{23a}, \quad \text{a reflection group.} \quad (4.18')$$

In this sense we apply $\Gamma = *\mathbf{2_1 3_1 a}$, as denotation. If $7 \le a$ then the base 2-orbifold is obviously hyperbolic ($\mathbf{H}^2$), but Γ will be an $\widetilde{\mathbf{SL_2R}}$ space group, as we shall see.

The skew transformation $\mathbf{t}$ by (4.16), acting on points, is described by the matrix

$$\mathbf{t}: \begin{pmatrix} -1 & A+1 & 0 & 0 \\ (A-2)(A+1) & -A^3+3A+1 & 0 & 0 \\ -(A+1) & A(A+1) & A & -A^2+1 \\ -(A+1) & A+1 & A^2-1 & -A(A^2-2) \end{pmatrix} \quad (4.19)$$

with conjugate complex eigenvalues of multiplicity 2:

$$\lambda = \frac{1}{2}\big[A(3-A^2) \pm i(A^2-1)\sqrt{4-A^2}\big] =: \cos\psi \pm i\sin\psi.$$

Now $\cos\psi := \cos\frac{2\pi}{a}\big(3 - 4\cos^2\frac{2\pi}{a}\big) = -\cos\frac{6\pi}{a} = \cos\big(\pi - \frac{6\pi}{a}\big)$ and

$$\psi = \left|\pi\frac{a-6}{a}\right| \tag{4.20}$$

can be taken.

Examples 4.9 – 4.11. For $a = 3,\ 4,\ 5$ we have spherical $\mathbf{S}^3$ simplices, obtained from (4.12) with $u = 1 - A = 1 - 2\cos\frac{2\pi}{a}$ as follows:

$$\beta^{23} = \frac{2\pi}{a}, \quad \beta^{02} = \beta^{03} = \frac{\pi}{2}, \quad \cos\beta^{01} = \frac{1}{2} + \cos\frac{2\pi}{a},$$

$$\cos\beta^{12} = \cos\beta^{13} = \frac{1}{2}\sqrt{1-\cos\frac{2\pi}{a}}, \quad \text{i.e. for } a = 3,\ 4,\ 5 \text{ we get}$$

$$\beta^{01} = \frac{\pi}{2}, \frac{\pi}{3}, \frac{\pi}{5} \quad \text{and} \quad \beta^{12} = \beta^{13} = \frac{\pi}{4}, \frac{\pi}{3}, \frac{2\pi}{5}; \quad \text{respectively.}$$

The skew transformation $\mathbf{t}$, to cyclic fibre transformations $\langle\mathbf{t}\rangle$, has ψ as corresponding angle parameter:

$$a = 3: \quad \psi = \pi, \quad \text{i.e.} \quad \mathbf{t} = -\mathbf{id}, \quad \text{the order} \quad |\Gamma| = 2\cdot 24 = 48;$$

$$a = 4: \quad \psi = \frac{\pi}{2}, \quad |\Gamma| = 4\cdot 48 = 192;$$

$$a = 5: \quad \psi = \frac{\pi}{5}, \quad |\Gamma| = 10\cdot 120 = 1200; \quad \text{all by (4.18}'\text{).}$$

Example 4.12. If $a = 6$, i.e. $A = 1$, our polarity by (4.12) is of signature $(0\ 0\ 0\ +)$ (Table 3). Our **Nil** orbifold to Γ fibers over the $\mathbf{E}^2$ 2-orbifold to $*\mathbf{236} = \mathbf{p6m}$ (Table 1). Moreover, our generators $\mathbf{r}_0$, $\mathbf{r}_1$ and $\mathbf{r}$ have the common invariant plane $\boldsymbol{b}^0\cdot 1 + \boldsymbol{b}^1\cdot 1 + \boldsymbol{b}^2\cdot 1 + \boldsymbol{b}^3\cdot 1 =: \boldsymbol{b}^{0'}$, chosen as ideal plane $\omega^\infty = \boldsymbol{b}^{0'}$, and they have the common fixed point $-\mathbf{a}_0 + \mathbf{a}_1 =: \mathbf{a}_{1'}$ the common pole of our polarity as a new ideal point $A_{1'}^\infty$ for our new coordinate simplex. We define the basis change

$$\begin{pmatrix}\boldsymbol{b}^{0'} & \boldsymbol{b}^{1'} & \boldsymbol{b}^{2'} & \boldsymbol{b}^{3'}\end{pmatrix} = \begin{pmatrix}\boldsymbol{b}^{0} & \boldsymbol{b}^{1} & \boldsymbol{b}^{2} & \boldsymbol{b}^{3}\end{pmatrix}\begin{pmatrix}1&0&0&0\\1&1&0&0\\1&0&1&0\\1&0&0&1\end{pmatrix}, \quad \text{thus}$$

$$\begin{pmatrix}\mathbf{a}_{0'}\\ \mathbf{a}_{1'}\\ \mathbf{a}_{2'}\\ \mathbf{a}_{3'}\end{pmatrix} = \begin{pmatrix}1&0&0&0\\-1&1&0&0\\-1&0&1&0\\-1&0&0&1\end{pmatrix}\begin{pmatrix}\mathbf{a}_{0}\\ \mathbf{a}_{1}\\ \mathbf{a}_{2}\\ \mathbf{a}_{3}\end{pmatrix}.$$

Then we can express our generators and the fibre transformation **t** for the points of our affine **Nil** space model

$$\mathcal{A}^3 := \{(\mathbf{x}) \in \mathcal{PS}^3 : \mathbf{x}\boldsymbol{b}^{0'} =: x^{0'} = 1\} =: \mathbf{Nil},$$

$$\mathbf{r}_0 : \begin{pmatrix} \mathbf{a}_{0'} \\ \mathbf{a}_{1'} \\ \mathbf{a}_{2'} \\ \mathbf{a}_{3'} \end{pmatrix} \mapsto \begin{pmatrix} 1 & 2 & 0 & 0 \\ 0 & -1 & 0 & 0 \\ 0 & -2 & 0 & 0 \\ 0 & -2 & 1 & 0 \end{pmatrix} \begin{pmatrix} \mathbf{a}_{0'} \\ \mathbf{a}_{1'} \\ \mathbf{a}_{2'} \\ \mathbf{a}_{3'} \end{pmatrix}, \quad \mathbf{r}_1 : \begin{pmatrix} 1 & 0 & 0 & 1 \\ 0 & -1 & 0 & 0 \\ 0 & 0 & 1 & -1 \\ 0 & 0 & 0 & -1 \end{pmatrix},$$

$$\mathbf{r} : (R^{k'}_{j'}) = \begin{pmatrix} 1 & 0 & 0 & 0 \\ 0 & 1 & 0 & 0 \\ 0 & 1 & 1 & -1 \\ 0 & 0 & 1 & 0 \end{pmatrix}, \quad \mathbf{t} : \begin{pmatrix} 1 & 2 & 0 & 0 \\ 0 & 1 & 0 & 0 \\ 0 & 0 & 1 & 0 \\ 0 & 0 & 0 & 1 \end{pmatrix}.$$

The fibre transformation **t** is just a translation in **Nil** by the vector $2 \cdot \mathbf{a}_{1'}$. Now we can see that **r** is a 6-fold rotation about $A_0A_1 = A_{0'}A_{1'}$. Furthermore, $\mathbf{r}_1\mathbf{r}_0$ is a 3_1-screw motion about the axis $\left(1', x^{1'}, \frac{1}{3}', \frac{1}{3}'\right)$ with $\frac{2}{3}'\mathbf{a}_1$ translation, and $\mathbf{r}_1\mathbf{r}\mathbf{r}_0$ is a 2_1-screw motion about the axis $\left(1', x^{1'}, 0', \frac{1}{2}'\right)$ with $1'\mathbf{a}_{1'}$ translation.

Examples 4.13 – ∞. If $a > 6$, i.e. $A > 1$, we get an $\widetilde{\mathbf{SL_2R}}$-orbifold which fibers over the corresponding $\mathbf{H}^2$-orbifold (4.18′). Our invariant polarity by (4.12) is of signature $(--++)$ from (4.14). Now the fibering skew transformation **t** by (4.19) has the eigenvalues by (4.20). We can perform the above computations on the projective model of $\widetilde{\mathbf{SL_2R}}$, i.e. the covering space $\tilde{\mathcal{H}}$ of the one-parted (one-sheet) hyperboloid solid

$$\mathcal{H} = \{(\mathbf{x}) \in \mathcal{PS}^3 : \langle \mathbf{x}, \mathbf{x} \rangle := x^i a_{ij} x^j > 0\} \ \text{ in } \mathcal{PS}^3, \tag{4.21}$$

where (a_{ij}) is the inverse of (b^{ij}) by (4.12), and the action of the cyclic group **t** is extended as universal covering transformations of $\mathcal{H}$. We know from Table 3 (and [M97]) that we obtain the universal covering space $\tilde{\mathcal{H}}$, fibred by $\mathbb{R}$-lines over the hyperbolic plane $\mathbf{H}^2$. The only modification is that we have chosen '>' in (4.21), since A_0, A_1, A_2, A_3 lie in this domain now. From (4.19) we could express a new basis $\{\mathbf{a}_{i'}\}$ by which the fibering transformation in the form

$$\mathbf{t} : \begin{pmatrix} \mathbf{a}_{0'} \\ \mathbf{a}_{1'} \\ \mathbf{a}_{2'} \\ \mathbf{a}_{3'} \end{pmatrix} \mapsto \begin{pmatrix} \cos\psi & \sin\psi & 0 & 0 \\ -\sin\psi & \cos\psi & 0 & 0 \\ 0 & 0 & \cos\psi & -\sin\psi \\ 0 & 0 & \sin\psi & \cos\psi \end{pmatrix} \begin{pmatrix} \mathbf{a}_{0'} \\ \mathbf{a}_{1'} \\ \mathbf{a}_{2'} \\ \mathbf{a}_{3'} \end{pmatrix}, \tag{4.22}$$

acts on $\tilde{\mathcal{H}}$. The parameter $\psi = \left|\pi\frac{a-6}{a}\right|$ by (4.20) shows (e.g. for $a = 11$) that the universal covering $\tilde{\mathcal{H}}$ is necessary, indeed. Further computations would express $\mathbf{r}_1\mathbf{r}_0 = \mathbf{s}^{-1}$ and $\mathbf{r}_1\mathbf{r}\mathbf{r}_0 = \mathbf{h}$ in the new basis $\{\mathbf{a}_{i'}\}$ related to $\mathbf{t}$ above, like that in Example 4.12.

Table 4. The metric realizations of tile transitive simplex tilings with our vertex conditions

Space $\mathbf{X}^3$	compact simplex		simplex with ideal vertex	
	maximal	non-maximal	maximal	non-maximal
$\mathbf{E}^3$	9	37	5	28 + **1**
$\mathbf{S}^3$	41; $\infty : 5$	80; $\infty : 41$	—	—
$\mathbf{H}^3$	40	54	52	191
$\mathbf{S}^2{\times}\mathbf{R}$	2	4	—	—
$\mathbf{H}^2{\times}\mathbf{R}$	$\infty : 2$	$\infty : 4$	deg. $\infty : 3$	deg. $\infty : 8$
$\widetilde{\mathbf{SL}_2\mathbf{R}}$	$\infty : 2$	$\infty : 2$	—	—
Nil	2	2	—	0 + **1**
Sol	—	—	—	—
bad orbifold	$\infty : 2$	$\infty : 4$	—	—
$\mathbf{S}^3$ **lens**	$\infty : 1$	$\infty : 2$	—	—
splitting: $\mathbf{E}^3\,\#\,\mathbf{H}^2{\times}\mathbf{R}$	—	—	3 + **2**	3
splitting: $\mathbf{E}^3\,\#\,\widetilde{\mathbf{SL}_2\mathbf{R}}$	—	—	0 + **2**	—
splitting: $\mathbf{H}^2{\times}\mathbf{R}\,\#\,\mathbf{H}^2{\times}\mathbf{R}$	—	—	1	1

The limit $a = \infty$, $A = 2$ in (4.13) would also lead to a signature $(- - + +)$, i.e. an $\widetilde{\mathbf{SL}_2\mathbf{R}}$ orbifold. Then the simplex edge A_0A_1 will be the axis at infinity, $\mathbf{r}$ will be a *horocyclic rotation,* giving no relation in (4.15). This case will appear in the splitting $\mathbf{E}^3\,\#\,\widetilde{\mathbf{SL}_2\mathbf{R}}$, at Family 24 and 31. Surprisingly, $\mathbf{E}^3\,\#\,\mathbf{H}^2{\times}\mathbf{R}$ is also possible there by the projective-inversive model of $\mathbf{H}^2{\times}\mathbf{R}$ (Table 3) where A_0A_1 degenerates to an ideal point, representing an $\mathbb{R}$-generator as axis at infinity [MSzW, MSz].

Family 11. The simplex $A_0A_1A_2A_3$ has a 2-rotation axis $A_{01}A_{23}$ here to the group $\Gamma = {}^2_2\Gamma_5(u; 5v)$ with a concave fundamental domain $\mathbf{F}_\Gamma$. But this domain can be cut and glued to the convex simplex $A_0A_{01}A_2A_3$ that is just a Zhuk simplex with $\Gamma \sim \Gamma_{39}(a = u,\ b = 2,\ c = v)$. Again, $c = v \geq 2$ would lead to an outer vertex. The Zhuk tilings for the cases $b = 2$ serve us complete descriptions for these maximal and non-maximal tilings (as well with outer vertices see [S93]).

Our projective metric models by Table 3 [M97] provide us in Table 5 the other realizations as well. For further details we refer to [MP88, MP91, MPSz97, MPSz98].

Table 5 for Families

Family 30. $\Gamma_{39}(a;2b;6c)$ $3 \le a$; $2 \le b$; $1 \le c$. Γ_{39} is maximal iff $a \neq 2b$, else ${}^{3m}_{6}\Gamma(2\bar{u} = 6c,\ \bar{v} = a = 2b)$ is a supergroup (Family 3); ZHUK's cases K_{mn} in [Zh83] $a = m$, $b = n$, $c = 1$. $\mathbf{r}_0 : A_1A_2A_3 \to A_1A_3A_2$, $\mathbf{r}_1 : A_0A_2A_3 \to A_3A_2A_0$, $\mathbf{r} : A_0A_3A_1 \to A_0A_2A_1$ — $\mathbf{r}_0^2, \mathbf{r}_1^2$; ▬ $\mathbf{r}^a$, ═ $(\mathbf{r}\mathbf{r}_0)^b$, ┅ $(\mathbf{r}_1\mathbf{r}\mathbf{r}_1\mathbf{r}_0\mathbf{r}_1\mathbf{r}^{-1})^c$. $\Gamma^0(A_1) = \mathbf{ab2}$, $\Gamma^0(A_0, A_2, A_3) = \mathbf{abc2}$. $(a;b;c)$ — $\mathbf{S}^3$: $(3;2;1)$, $(\mathbf{4};\mathbf{2};\mathbf{1})$, $(5;2;1)$, $(3;3;1)$ — **Nil**: $(6;2;1)$ $\Gamma := *\mathbf{63_12_1}$ — $\widetilde{\mathbf{SL_2R}}$: $(a;2;1)$ $7 \le a$ $\Gamma := *\mathbf{a3_12_1}$ — $\mathbf{H}^3$: $(4;3;1)$, $(3;4;1)$, $(5;3;1)$, $(3;5;1)$ / $(4;4;1)$, $(\mathbf{6};\mathbf{3};\mathbf{1})$, $(3;6;1)$ all id.v. / else outer vertex occurs.

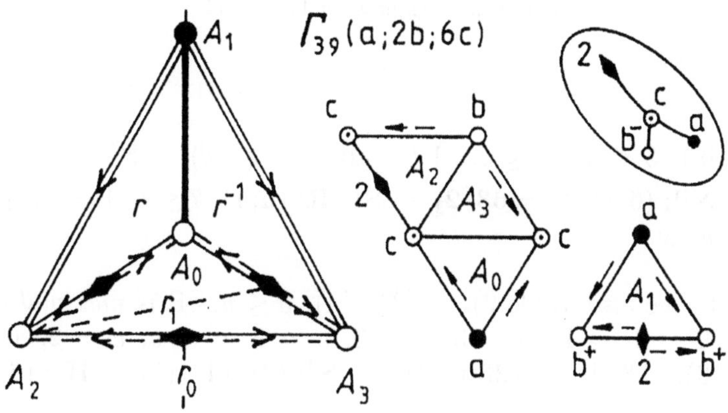

Figure 1. Figures for Family 30

Family 11. ${}^{2}_{2}\Gamma_5(u;5v)$ $3 \le u$; $1 \le v$. $\Gamma = \operatorname{Aut}\mathcal{T}$ is maximal iff $u \neq 5v$, else ${}^{\bar{4}3m}_{24}\Gamma(\bar{u} = u = 5v)$ is its supergroup (Family 1). $\Gamma \sim \Gamma_{39}(a = u;\ b = 2;\ c = v)$ (Family 30). $\Gamma^0 = \mathbf{22uv}$. $(u;v)$ — $\mathbf{S}^3$: $(3;1)$, $(4;1)$, $(\mathbf{5};\mathbf{1})$; ZHUK's cases K_{m2} in [Zh83] — **Nil**: $(6;1)$ $\Gamma = *\mathbf{63_12_1}$ — $\widetilde{\mathbf{SL_2R}}$: $(u;1)$ $7 \le u$ $\Gamma = *\mathbf{u3_12_1}$ — else $\mathbf{H}^3$: outer vertex occurs.

- $\Gamma_{14}(2u;10v)$ $2 \le u$; $1 \le v$. $\Gamma^0 = \mathbf{2222uv}$ $\mathbf{H}^2$-group. Out.v.
- $\Gamma_{36}(2u;5v)$ $2 \le u$; $1 \le v$. $\Gamma^0 = \mathbf{22uvv}$. $(u;v)$ — $\mathbf{S}^3$: $(2;1)$ — **Nil**: $(3;1)$ $\Gamma := *\mathbf{6_23_12_1}$ — $\widetilde{\mathbf{SL_2R}}$: $(u;1)$ $4 \le u$ $\Gamma = *(\mathbf{2u})_\mathbf{2}\mathbf{3_12_1}$ — $\mathbf{H}^3$: out.v.
- $\Gamma_{45}(u;10v)$ $3 \le u$; $1 \le v$. $\Gamma^0 = \mathbf{22uuv}$ $\mathbf{H}^2$-group. Out.v.

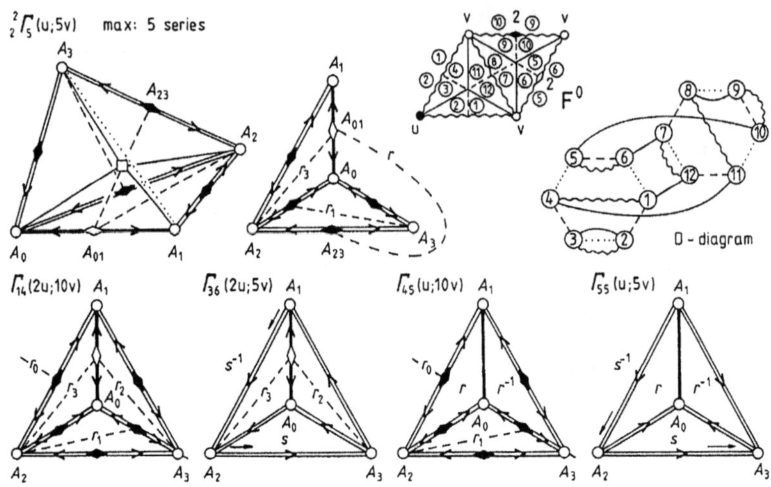

Figure 2. Figures for Family 11

• $\Gamma_{55}(u;5v)$ $3 \leq u$; $1 \leq v$. $\Gamma^0 = \mathbf{uuvv}$. $(u;v)$ — $\mathbf{S}^3$: $(3;1)$, $(4;1)$, $(\mathbf{5};\mathbf{1})$ — **Nil**: $(6;1)$ $\Gamma := \mathbf{63_1 2_1}$ — $\widetilde{\mathbf{SL_2R}}$: $(u;1)$ $7 \leq u$ $\Gamma := \mathbf{u3_1 2_1}$ — else $\mathbf{H}^3$: out.v.

Family 1 (see also Table 2). ${}^{\bar{4}3m}_{24}\Gamma(u)$ $3 \leq u$. Γ *is maximal group:* $\Gamma = \operatorname{Aut}\mathcal{T}$. $\Gamma \sim \Gamma_1(a = 3,\ b = 3,\ c = 2,\ d = u,\ e = 2,\ f = 2)$ (Family 13). $\Gamma^0(A_1) = \mathbf{*23u}$. (u) — $\mathbf{S}^3$: (3), (4), (5) — $\mathbf{H}^3$: (6) id.v. / else out.v.
• ${}^{23}_{12}\Gamma_1(2u)$ $2 \leq u$. $\Gamma \sim \Gamma_{24}(\bar{u} = u;\ \bar{v} = 3,\ \bar{w} = 1;\ \bar{x} = 3)$ (Family 6). $\Gamma^0(A_0, A_1) = \mathbf{3{*}u}$. (u) — $\mathbf{S}^3$: (2) — $\mathbf{H}^3$: (3) id.v. / out.v.
• ${}^{23}_{12}\Gamma_2(u)$ $3 \leq u$. $\Gamma \sim \Gamma_{35}(\bar{u} = u;\ \bar{v} = 2,\ \bar{w} = 3;\ \bar{x} = 3)$ (Family 6). $\Gamma^0(A_0, A_1) = \mathbf{23u}$. (u) — $\mathbf{S}^3$: (3), (4), (5) — $\mathbf{H}^3$: (6) id.v. / else out.v.
• ${}^{\bar{4}}_{4}\Gamma_3(3u)$ $1 \leq u$. $\Gamma \sim \Gamma_{42}(\bar{u} = u;\ \bar{v} = 1)$ (Family 3). $\Gamma^0(A_0, A_1, A_2) = \mathbf{2u\otimes}$. (u) — $\mathbf{S}^3$: (1) — $\mathbf{H}^3$: (2) id.v. / else out.v.
• ${}^{3}_{3}\Gamma_5(4u)$ $1 \leq u$. $\Gamma \sim \Gamma_{39}(a = 3;\ b = 2;\ c = u)$ (Family 30). $\Gamma^0(A_0, A_1, A_2) = \mathbf{223u}$. (u) — $\mathbf{S}^3$: (1) — $\mathbf{H}^3$: out.v.
• ${}^{3}_{3}\Gamma_6(8u)$ $1 \leq u$. $\Gamma \sim \Gamma_{29}(a = 3;\ b = 1;\ c = u)$ (Family 26). (u) — $\mathbf{H}^3$: out.v.
• ${}^{2}_{2}\Gamma_{11}(6u)$ $1 \leq u$. $\Gamma^0(A_0, A_1, A_3) = \mathbf{u\otimes\otimes}$. (u) — $\mathbf{H}^3$: (1) id.v. **with possible surgeries** GIESEKING (1912) / out.v.
• $\Gamma_{40}(12u)$ $1 \leq u$. $\mathbf{H}^3$: out.v. • $\Gamma_{41}(12u)$ $1 \leq u$. $\mathbf{H}^3$: out.v.
• $\Gamma_{48}(12u)$ $1 \leq u$. $\mathbf{H}^3$: out.v. • $\Gamma_{51}(12u)$ $1 \leq u$. $\mathbf{H}^3$: out.v.
• $\Gamma_{52}(12u)$ $1 \leq u$. $\mathbf{H}^3$: out.v. • $\Gamma_{60}(12u)$ $1 \leq u$. $\mathbf{H}^3$: out.v.

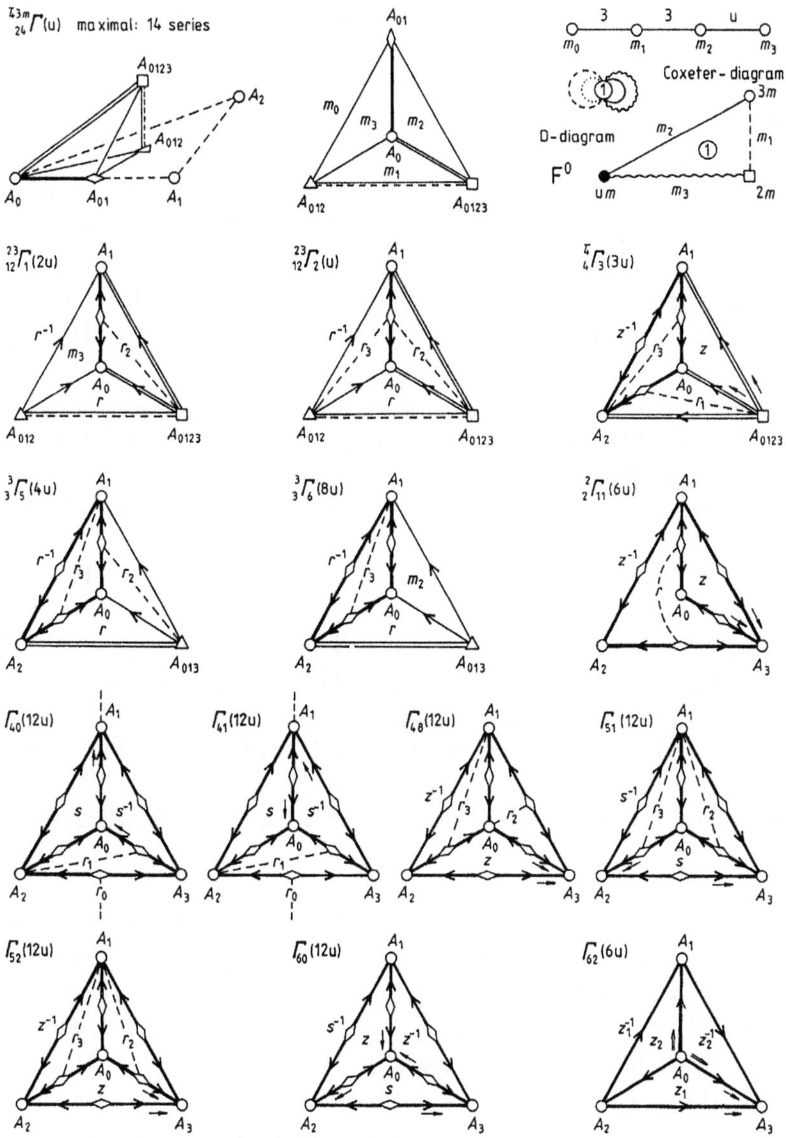

Figure 3. Figures for Family 1

- $\Gamma_{62}(6u)$ $1 \leq u$. $\Gamma^0 = \mathbf{uu}\otimes\otimes$. (u) — $\mathbf{H}^3$: (1) id.v. Gieseking manifold (1912) **with surgery orbifolds** / out.v.

Family 2 (Table 2). ${}^{\bar{4}2m}_{8}\Gamma(u; 2v)$ $3 \leq u$; $2 \leq v$. Γ is maximal iff $u \neq 2v$, else ${}^{\bar{4}3m}_{24}\Gamma(\bar{u} = u = 2v)$ is its supergroup. $\Gamma \sim \Gamma_2(a = v,\ b = 1,\ c = 2,\ d = 2,\ e = u)$ (Family 14). $\Gamma^0(A_0, A_2) = *\mathbf{uv2}$. $(u; v)$ —

$\mathbf{S}^3$: $(u;2)$, $(3;3)$, $(3;4)$ — $\mathbf{E}^3$: $(4;3)$ 229.$\mathbf{Im\bar{3}m}$ — $\mathbf{H}^3$: $(3;5)$, $(5;3)$ / $(3;6)$, $(4;4)$, $(\mathbf{6};\mathbf{3})$ id.v. / else out.v.

• ${}^{mm2}_{4}\Gamma_2(2u;2v)$ $2 \le u$; $2 \le v$. $\Gamma \sim \Gamma_{23}(\bar{u} = v;\ \bar{v} = 1,\ \bar{w} = u;\ \bar{x} = 2)$ (Family 9). $(u;v)$ — $\mathbf{S}^3$: $(u;2)$ — $\mathbf{E}^3$: $(2;3)$ 204.$\mathbf{Im\bar{3}}$ — $\mathbf{H}^3$: $(2;4)$, $(\mathbf{3};\mathbf{3})$ id.v. / out.v.

• ${}^{\bar{4}}_{4}\Gamma_1(2u;2v)$ $2 \le u$; $2 \le v$. $\Gamma \sim \Gamma_{26}(\bar{u} = u;\ \bar{v} = v;\ \bar{w} = 1)$ (Family 8). $(u;v)$ — $\mathbf{S}^3$: $(u;2)$ — $\mathbf{E}^3$: $(2;3)$ 217.$\mathbf{I\bar{4}3m}$ — $\mathbf{H}^3$: $(2;4)$, $(\mathbf{3};\mathbf{3})$ id.v. / out.v.

• ${}^{\bar{4}}_{4}\Gamma_2(u;2v)$ $3 \le u$; $2 \le v$. $\Gamma \sim \Gamma_{37}(\bar{u} = u;\ \bar{v} = v;\ \bar{w} = 1)$ (Family 8). $(u;v)$ — $\mathbf{S}^3$: $(u;2)$, $(3;3)$, $(3;4)$ — $\mathbf{E}^3$: $(4;3)$ 222.$\mathbf{Pn\bar{3}n}$ — $\mathbf{H}^3$: $(3;5)$, $(5;3)$ / $(3;6)$, $(4;4)$, $(\mathbf{6};\mathbf{3})$ id.v. / out.v.

• ${}^{222}_{4}\Gamma_2(u;2v)$ $3 \le u$; $2 \le v$. $\Gamma \sim \Gamma_{16}(\bar{u} = u;\ \bar{v} = v;\ \bar{w} = 1)$ (Family 8). $(u;v)$ — $\mathbf{S}^3$: $(u;2)$, $(3;3)$, $(3;4)$ — $\mathbf{E}^3$: $(4;3)$ 211.$\mathbf{I432}$ — $\mathbf{H}^3$: $(3;5)$, $(5;3)$ / $(3;6)$, $(4;4)$, $(\mathbf{6};\mathbf{3})$ id.v. / out.v.

• ${}^{2}_{2}\Gamma_7(2u;4v)$ $2 \le u$; $1 \le v$. $(u;v)$ — $\mathbf{S}^3$: $(u;1)$ — $\mathbf{H}^3$: $(2;2)$ id.v. / out.v.

• ${}^{2}_{2}\Gamma_8(u;4v)$ $3 \le u$; $1 \le v$. $(u;v)$ — $\mathbf{S}^3$: $(u;1)$, $(3;2)$ — $\mathbf{H}^3$: $(3;3)$, $(4;2)$ id.v. / out.v.

• ${}^{2}_{2}\Gamma_9(2u;2v)$ $2 \le u$; $2 \le v$. $(u;v)$ — $\mathbf{S}^3$: $(u;2)$ — $\mathbf{E}^3$: $(2;3)$ 197.$\mathbf{I23}$ (fundamental simplex does not exist) — $\mathbf{H}^3$: $(2;4)$, $(\mathbf{3};\mathbf{3})$ id.v. / out.v.

• $\Gamma_{30}(8u;8v)$ $1 \le u$; $1 \le v$. $\Gamma^0 = \mathbf{2{*}uv{*}v}$ $\mathbf{H}^2$-group. Out.v.

• $\Gamma_{47}(4u;4v)$ $1 \le u$; $1 \le v$. $(u;v)$ — $\mathbf{S}^3$: $(u;1)$ — $\mathbf{H}^3$: $(1;2)$ id.v. / out.v.

• $\Gamma_{58}(2u;4v)$ $2 \le u$; $1 \le v$. $(u;v)$ — $\mathbf{S}^3$: $(u;1)$ — $\mathbf{H}^3$: $(2;2)$ id.v. / out.v.

Family 3 (Table 2). ${}^{3m}_{6}\Gamma(2u;v)$ $2 \le u$; $3 \le v$. Γ is maximal iff $2u \ne v$, else ${}^{\bar{4}3m}_{24}\Gamma(\bar{u} = 2u = v)$ is its supergroup (Family 1). $\Gamma \sim \Gamma_1(a = 3,\ b = 2,\ c = 2;\ d = u,\ e = 2,\ f = v)$ (Family 13). $\Gamma^0(A_0) = \mathbf{{*}2uv}$, $\Gamma^0(A_2) = \mathbf{{*}23v}$. $(u;v)$ — $\mathbf{S}^3$: $(2;3)$, $(\mathbf{2};\mathbf{4})$, $(2;5)$, $(3;3)$, $(3;4)$, $(4;3)$, $(5;3)$ — $\mathbf{E}^3$: $(2;6)$ A_2 id.v. — $\mathbf{H}^3$: $(3;5)$ / $(4;4)$, $(6;3)$, A_0 id.v. / $(\mathbf{3};\mathbf{6})$ all id.v. / else out.v.

• ${}^{3}_{3}\Gamma_1(2u;2v)$ $2 \le u$; $2 \le v$. $\Gamma \sim \Gamma_{20}(\bar{u} = u;\ \bar{v} = v,\ \bar{w} = 1;\ \bar{x} = 3)$ (Family 6). $(u;v)$ — $\mathbf{S}^3$: $(\mathbf{2};\mathbf{2})$, $(3;2)$ — $\mathbf{E}^3$: $(2;3)$ A_2 id.v. — $\mathbf{H}^3$: $(4;2)$ A_0, A_1 id.v. / $(\mathbf{3};\mathbf{3})$ all id.v. / out.v.

• ${}^{3}_{3}\Gamma_2(4u;2v)$ $1 \le u$; $2 \le v$. $\Gamma \sim \Gamma_{24}(\bar{u} = u;\ \bar{v} = 2,\ \bar{w} = v;\ \bar{x} = 3)$ (Family 6). $(u;v)$ — $\mathbf{S}^3$: $(\mathbf{1};\mathbf{2})$ — $\mathbf{E}^3$: $(1;3)$ A_2 id.v. — $\mathbf{H}^3$: $(2;2)$ A_0, A_1 id.v. / out.v.

• ${}^{3}_{3}\Gamma_3(4u;v)$ $1 \le u$; $3 \le v$. $\Gamma \sim \Gamma_{24}(\bar{u} = u;\ \bar{v} = v,\ \bar{w} = 1;\ \bar{x} = 3)$ (Family 6). $(u;v)$ — $\mathbf{S}^3$: $(1;3)$, $(\mathbf{1};\mathbf{4})$, $(1;5)$, $(2;3)$ — $\mathbf{E}^3$: $(1;6)$ A_2 id.v. — $\mathbf{H}^3$: $(2;4)$, $(3;3)$ A_0, A_1 id.v.

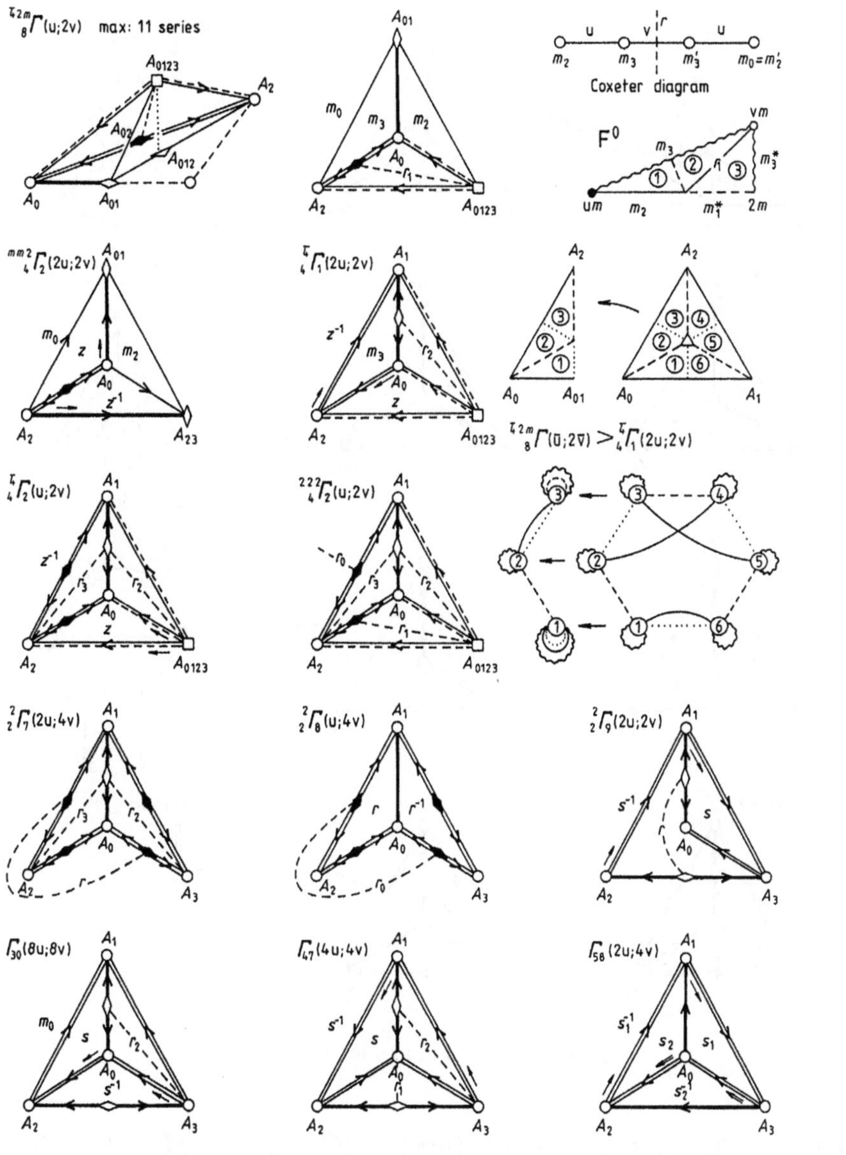

Figure 4. Figures for Family 2

• $^3_3\Gamma_4(2u;v)$ $2 \le u$; $3 \le v$. $\Gamma \sim \Gamma_{35}(\bar{u} = u;\ \bar{v} = v,\ \bar{w} = 2;\ \bar{x} = 3)$ (Family 6). $(u;v)$ — $\mathbf{S}^3$: $(2;3)$, $(\mathbf{2};\mathbf{4})$, $(2;5)$, $(3;3)$, $(3;4)$, $(4;3)$, $(5;3)$ — $\mathbf{E}^3$ **but not Nil**: $(2;6)$ A_2 id.v. — $\mathbf{H}^3$: $(3;5)$ / $(4;4)$, $(6;3)$ A_0 id.v. / $(\mathbf{3};\mathbf{6})$ all id.v. / out.v.

• $\Gamma_{33}(12u;6v)$ $1 \le u$; $1 \le v$. $\Gamma^0(A_0, A_1, A_3) = \mathbf{2{*}uv}\otimes$ $\mathbf{H}^2$-group, $\Gamma^0(A_2) = *\mathbf{v}\otimes$. $\mathbf{H}^3$: out.v.

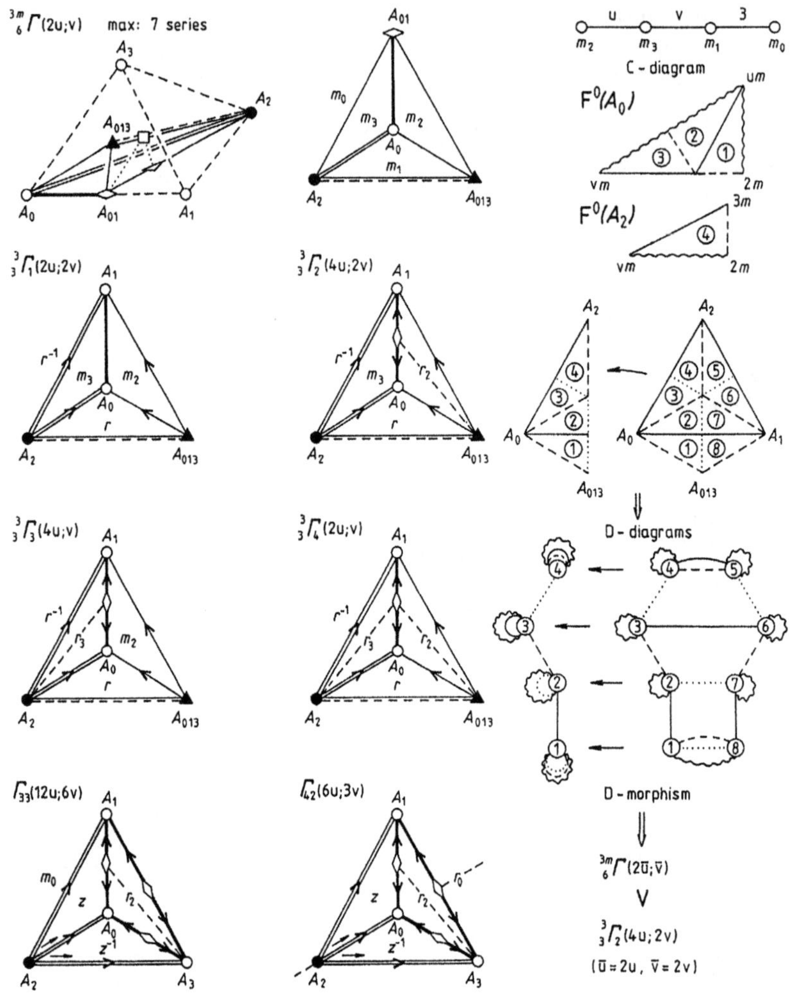

Figure 5. Figures for Family 3

• $\Gamma_{42}(6u; 3v)$ $1 \leq u$; $1 \leq v$. $(u; v)$ — $\mathbf{S}^3$: $(1; 1)$ — $\mathbf{H}^3$: $(2; 1)$ A_0, A_1, A_3 id.v. / $(\mathbf{1}; \mathbf{2})$ all id.v. / out.v. — $W_{\alpha\beta}$ ZHUK's case [Zh83].

Family 4 (Table 2). ${}^{mm2}_{4}\Gamma_1(u; 2v; w)$ $3 \leq u \leq w$, $2 \leq v$. Γ is maximal iff $u \neq w$, else ${}^{\bar{4}2m}_{8}\Gamma(\bar{u} = u = w;\ 2\bar{v} = 2v)$ is its supergroup (Family 2). $\Gamma \sim \Gamma_1(a = w,\ b = 2,\ c = 2;\ d = u,\ e = 2,\ f = v)$ (Family 13). $\Gamma^0(A_0) = \mathbf{*uv2}$, $\Gamma^0(A_2) = \mathbf{*2vw}$. $(u; v; w)$ — $\mathbf{S}^3$: $(u; 2; w)$, $(\mathbf{3}; \mathbf{3}; \mathbf{3})$, $(\mathbf{3}; \mathbf{4}; \mathbf{3})$, $(3; 3; 4)$, $(3; 3; 5)$ — $\mathbf{E}^3$: $(\mathbf{4}; \mathbf{3}; \mathbf{4})$ 221. $\mathbf{Pm\bar{3}m}$ — $\mathbf{H}^3$: $(\mathbf{3}; \mathbf{5}; \mathbf{3})$, $(\mathbf{5}; \mathbf{3}; \mathbf{5})$, $(4; 3; 5)$ / $(3; 4; 4)$, $(3; 3; 6)$ A_2 id.v. / $(\mathbf{4}; \mathbf{4}; \mathbf{4})$, $(\mathbf{3}; \mathbf{6}; \mathbf{3})$, $(\mathbf{6}; \mathbf{3}; \mathbf{6})$ all id.v. / else out.v.

• ${}^{m}_{2}\Gamma_2(u;4v;w)$ $3 \le u$; $1 \le v$; $3 \le w$. $\Gamma \sim \Gamma_{20}(\bar{u} = w;\ \bar{v} = 2,\ \bar{w} = v;\ \bar{x} = u)$ (Family 6). $(u;v;w)$ — $\mathbf{S}^3$: $(u;1;w)$, $(\mathbf{3};\mathbf{2};\mathbf{3})$ — $\mathbf{H}^3$: $(3;2;4)$, $(4;2;3)$ A_2, A_3 id.v. / $(\mathbf{4};\mathbf{2};\mathbf{4})$, $(\mathbf{3};\mathbf{3};\mathbf{3})$ all id.v. / out.v.
• ${}^{m}_{2}\Gamma_3(2u;4v;2w)$ $2 \le u$; $1 \le v$; $2 \le w$. $\Gamma \sim \Gamma_2(a = w;\ b = v;\ c = u;\ d = 2,\ e = 2)$ (Family 14). $(u;v;w)$ — $\mathbf{S}^3$: $(u;1;w)$ — $\mathbf{H}^3$: $(\mathbf{2};\mathbf{2};\mathbf{2})$ all id.v. / out.v.
• ${}^{m}_{2}\Gamma_5(u;2v;2w)$ $3 \le u$; $2 \le v$; $2 \le w$. $\Gamma \sim \Gamma_{24}(\bar{u} = w;\ \bar{v} = v,\ \bar{w} = 2;\ \bar{x} = u)$ (Family 6). $(u;v;w)$ — $\mathbf{S}^3$: $(u;2;w)$, $(3;3;2)$ — $\mathbf{E}^3$: $(\mathbf{4};\mathbf{3};\mathbf{2})$ 226. $\mathbf{Fm\bar{3}c}$ — $\mathbf{H}^3$: $(5;3;2)$ / $(3;4;2)$, $(3;3;3)$ A_2, A_3 id.v. / $(\mathbf{4};\mathbf{4};\mathbf{2})$, $(\mathbf{6};\mathbf{3};\mathbf{3})$ all id.v. / out.v.
• ${}^{2}_{2}\Gamma_3(u;4v;2w)$ $3 \le u$; $1 \le v$; $2 \le w$. $\Gamma \sim \Gamma_6(\bar{u} = u;\ \bar{v} = 1,\ \bar{w} = v;\ \bar{x} = w)$ (Family 6). $(u;v;w)$ — $\mathbf{S}^3$: $(u;1;w)$ — $\mathbf{H}^3$: $(3;2;2)$ A_2, A_3 id.v. / $(\mathbf{4};\mathbf{2};\mathbf{2})$ all id.v. / out.v.
• ${}^{2}_{2}\Gamma_4(u;2v;w)$ $3 \le u \le w$; $2 \le v$. $\Gamma \sim \Gamma_{35}(\bar{u} = u;\ \bar{v} = 2,\ \bar{w} = v;\ \bar{x} = w)$ (Family 6). $(u;v;w)$ — $\mathbf{S}^3$: $(u;2;w)$, $(\mathbf{3};\mathbf{3};\mathbf{3})$, $(\mathbf{3};\mathbf{4};\mathbf{3})$, $(3;3;4)$, $(3;3;5)$ — $\mathbf{E}^3$: $(\mathbf{4};\mathbf{3};\mathbf{4})$ 207. $\mathbf{P432}$ — $\mathbf{H}^3$: $(\mathbf{3};\mathbf{5};\mathbf{3})$, $(\mathbf{5};\mathbf{3};\mathbf{5})$, $(4;3;5)$ / $(3;4;4)$, $(3;3;6)$ A_2, A_3 id.v. / $(\mathbf{4};\mathbf{4};\mathbf{4})$, $(\mathbf{3};\mathbf{6};\mathbf{3})$, $(\mathbf{6};\mathbf{3};\mathbf{6})$ all id.v. / out.v.
• $\Gamma_{10}(2u;8v;4w)$ $2 \le u$; $1 \le v$; $1 \le w$. $(u;v;w)$ — $\mathbf{H}^3$: $(\mathbf{2};\mathbf{1};\mathbf{1})$ all id.v. / out.v.
• $\Gamma_{17}(2u;4v;2w)$ $2 \le u \le w$; $1 \le v$. $(u;v;w)$ — $\mathbf{S}^3$: $(u;1;w)$ — $\mathbf{H}^3$: $(\mathbf{2};\mathbf{2};\mathbf{2})$ all id.v. / out.v.
• $\Gamma_{28}(2u;8v;4w)$ $2 \le u$; $1 \le v$; $1 \le w$. $(u;v;w)$ — $\mathbf{H}^3$: $(\mathbf{2};\mathbf{1};\mathbf{1})$ all id.v. / out.v.
• $\Gamma_{38}(2u;4v;2w)$ $2 \le u$; $1 \le v$; $2 \le w$. $(u;v;w)$ — $\mathbf{S}^3$: $(u;1;w)$ — $\mathbf{H}^3$: $(\mathbf{2};\mathbf{2};\mathbf{2})$ all id.v. / out.v.
• $\Gamma_{54}(u;4v;w)$ $3 \le u \le w$; $1 \le v$. $(u;v;w)$ — $\mathbf{S}^3$: $(u;1;w)$, $(\mathbf{3};\mathbf{2};\mathbf{3})$ — $\mathbf{H}^3$: $(3;2;4)$, A_2, A_3 id.v. / $(\mathbf{4};\mathbf{2};\mathbf{4})$, $(\mathbf{3};\mathbf{3};\mathbf{3})$ all id.v. / out.v.
• $\Gamma_{57}(u;4v;2w)$ $3 \le u$; $1 \le v$; $2 \le w$. $(u;v;w)$ — $\mathbf{S}^3$: $(u;1;w)$ — $\mathbf{H}^3$: $(3;2;2)$, A_2, A_3 id.v. / $(\mathbf{4};\mathbf{2};\mathbf{2})$ all id.v. / out.v.

Family 5 (Table 2). ${}^{222}_{4}\Gamma_1(2u;2v,2w)$ $2 \le u \le v \le w$. Γ is maximal iff u, v, w are different, else e.g. $v = w$ ${}^{\bar{4}2m}_{8}\Gamma(\bar{u} = 2u;\ 2\bar{v} = 2v = 2w)$ is its supergroup (Family 2). $\Gamma \sim \Gamma_{11}(a = u,\ b = v,\ c = w;\ d = 1)$ (Family 20). $\Gamma^0(A_0, A_1, A_2) = *\mathbf{uvw}$. $(u;v,w)$ — $\mathbf{S}^3$: $(\mathbf{2};\mathbf{2},\mathbf{w})$ — $\mathbf{E}^3$: $(\mathbf{2};\mathbf{3},\mathbf{3})$ 224. $\mathbf{Pn\bar{3}m}$ — $\mathbf{H}^3$: $(2;3,4)$, $(2;3,5)$ / $(\mathbf{3};\mathbf{3},\mathbf{3})$, $(\mathbf{2};\mathbf{4},\mathbf{4})$, $(2;3,6)$ all id.v. / else out.v.
• ${}^{2}_{2}\Gamma_{10}(2u;2v,2w)$ $2 \le u$; $2 \le v \le w$. $(u;v,w)$ — $\mathbf{S}^3$: $(\mathbf{u};\mathbf{2},\mathbf{2})$, $(\mathbf{2};\mathbf{2},\mathbf{w})$ — $\mathbf{E}^3$: $(\mathbf{2};\mathbf{3},\mathbf{3})$ 201. $\mathbf{Pn\bar{3}}$ (no fundamental simplex), $(\mathbf{3};\mathbf{2},\mathbf{3})$ 228. $\mathbf{Fd\bar{3}c}$ (no fundamental simplex) — $\mathbf{H}^3$: $(2;3,4)$, $(3;2,4)$, $(4;2,3)$, $(2;3,5)$, $(3;2,5)$, $(5;2,3)$ / $(\mathbf{3};\mathbf{3},\mathbf{3})$, $(\mathbf{2};\mathbf{4},\mathbf{4})$, $(\mathbf{4};\mathbf{2},\mathbf{4})$, $(2;3,6)$, $(3;2,6)$, $(6;2,3)$ all id.v. / out.v.

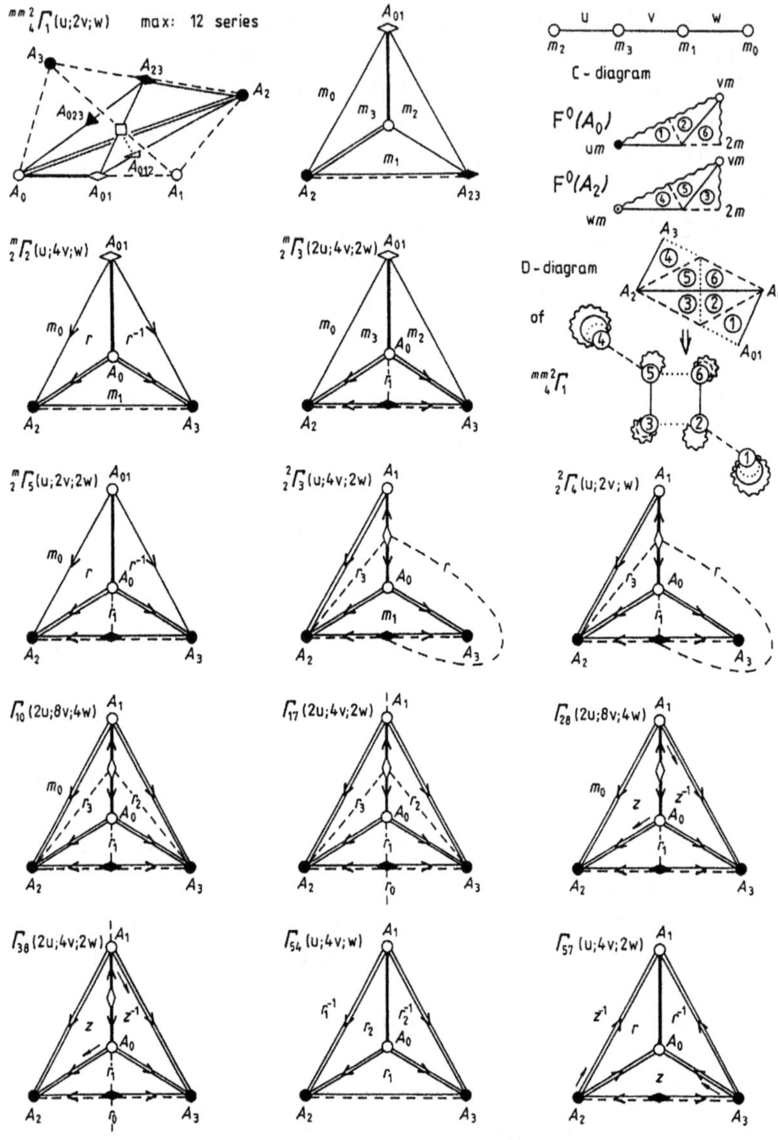

Figure 6. Figures for Family 4

Family 6 (Table 2). ${}^{m}_{2}\Gamma_1(2u;2v,2w;x)$ $2 \le u$; $2 \le v \le w$; $3 \le x$. Γ is maximal iff $2v \neq 2w$, else ${}^{mm2}_{4}\Gamma_1(\bar{u} = 2u;\ 2\bar{v} = 2v = 2w;\ \bar{w} = x)$ is its supergroup (Family 4). $\Gamma \sim \Gamma_1(a = x,\ b = w,\ c = 2;\ d = u,\ e = 2,\ f = v)$ (Family 13). $\Gamma^0(A_0) = *\mathbf{uvw}$, $\Gamma^0(A_2) = *\mathbf{2vx}$, $\Gamma^0(A_3) = *\mathbf{2wx}$. $(u;v;w;x)$ — $\mathbf{S}^3$: $(\mathbf{u};\mathbf{2},\mathbf{2};\mathbf{x})$, $(2;2,3;3)$, $(2;2,3;4)$, $(2;2,4;3)$ $(2;2,3;5)$, $(2;2,5;3)$, $(3;2,3;3)$, $(\mathbf{2};\mathbf{3},\mathbf{3};\mathbf{3})$, $(3;2,4;3)$, $(3;2,3;4)$,

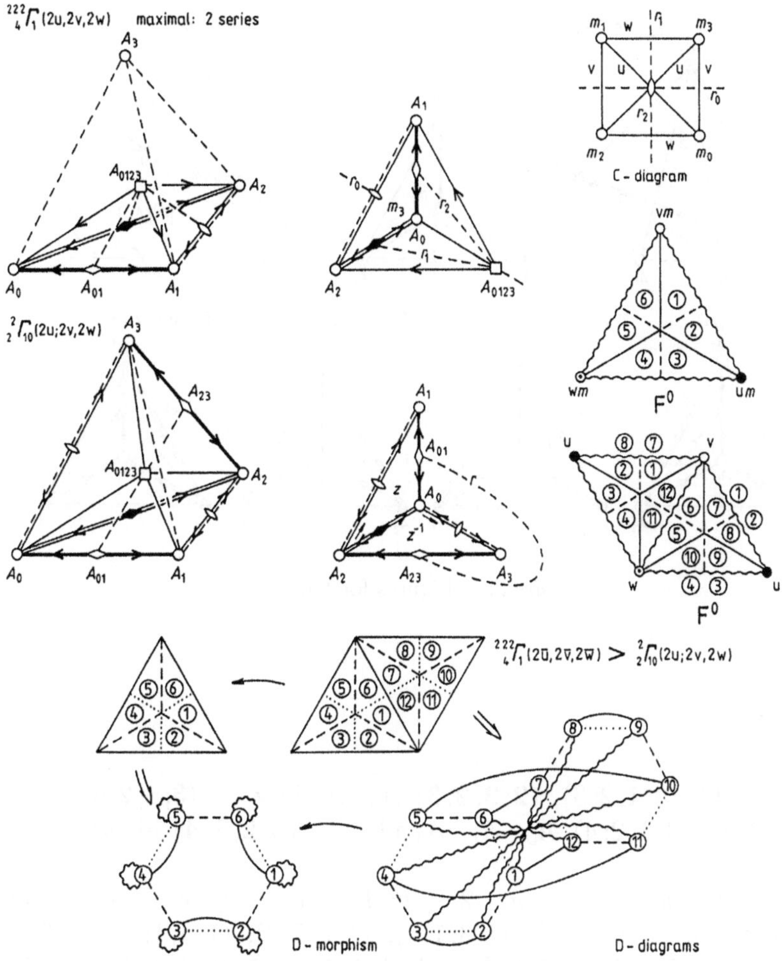

Figure 7. Figures for Family 5

(4; 2, 3; 3), (3; 2, 3; 5), (5; 2, 3; 3) — $\mathbf{E}^3$: **(2; 3, 3; 4)** (2; 3, 4; 3) 225. **Fm$\bar{3}$m**, (4; 2, 3; 4) 221. **Pm$\bar{3}$m**, / (2; 2, 4; 4), (2; 2, 3; 6), (2; 2, 6; 3) A_3 id.v. — $\mathbf{H}^3$: **(2; 3, 3; 5)**, (3; 2, 5; 3), (5; 2, 3; 5), (4; 2, 3; 5), (5; 2, 3; 4), / (6; 2, 3; 3), **(3; 3, 3; 3)**, (4; 2, 4; 3), **(2; 4, 4; 3)** A_0 id.v. / (2; 3, 6; 3), (3; 2, 6; 3), (4; 2, 4; 4), (6; 2, 3; 6) A_0, A_3 id.v. / (3; 2, 3; 6), (3; 2, 4; 4), (2; 3, 4; 4), A_3 id.v. / **(2; 3, 3; 6)** A_2, A_3 id.v. / **(2; 4, 4; 4)** all id.v. / else out.v.

• $\Gamma_6(2u; 4v, 4w; 2x)$ $\quad 2 \le u;\quad 1 \le v \le w;\quad 2 \le x.$ $\quad (u; v, w; x)$ — $\mathbf{S}^3$: **(u; 1, 1; x)** — $\mathbf{E}^3$: (2; 1, 2; 2) A_3 id.v. — $\mathbf{H}^3$: (3; 1, 2; 2) A_3 id.v. / (4; 1, 2; 2) A_0, A_1, A_3 id.v. / **(2; 2, 2; 2)** all id.v. / else out.v.

• $\Gamma_{20}(2u; 4v, 4w; x)$ $\quad 2 \le u;\quad 1 \le v \le w;\quad 3 \le x.$ $\quad (u; v, w; x)$ — $\mathbf{S}^3$: **(u; 1, 1; x)**, (2; 1, 2; 3), (3; 1, 2; 3) — $\mathbf{E}^3$: (2; 1, 2; 4), (2; 1, 3; 3) A_3

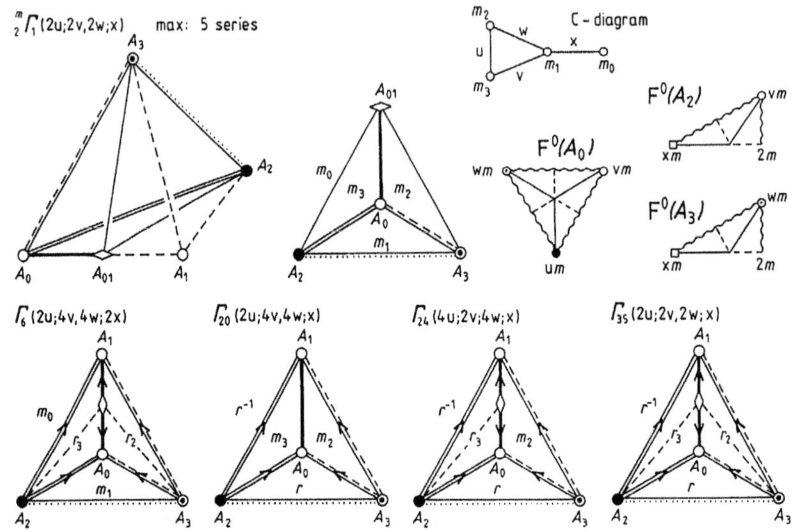

Figure 8. Figures for Family 6

id.v. — $\mathbf{H}^3$: (4; 1, 2; 3), (**2**; **2**, **2**; **3**) A_0, A_1 id.v. / (3; 1, 2; 4) A_3 id.v. / (3; 1, 3; 3), (4; 1, 2; 4) A_0, A_1, A_3 id.v. / (**2**; **2**, **2**; **4**) all id.v. / else out.v.

• $\Gamma_{24}(4u; 2v; 4w; x)$ $1 \le u$; $2 \le v$; $1 \le w$; $3 \le x$. $(u; v; w; x)$ — $\mathbf{S}^3$: (**u**; **2**; **1**; **x**), (1; 3; 1; 3), (1; 3; 1; 4), (1; 2; 2; 3), (1; 4; 1; 3), (1; 3; 1; 5), (1; 5; 1; 3), (2; 3; 1; 3) — $\mathbf{E}^3$: (1; 3; 2; 3) 202. $\mathbf{Fm\bar{3}}$, (2; 3; 1; 4) 226. $\mathbf{Fm\bar{3}c}$ / (1; 2; 2; 4), (1; 4; 1; 4), (1; 3; 1; 6), (1; 2; 3; 3), (1; 6; 1; 3) A_3 id.v. — $\mathbf{H}^3$: (2; 3; 1; 5) / (3; 3; 1; 3), (2; 2; 2; 3), (2; 4; 1; 3), (**1**; **4**; **2**; **3**) A_0, A_1 id.v. / (1; 3; 2; 4) A_3 id.v. / (1; 3; 3; 3), (2; 2; 2; 4), (2; 4; 1; 4), (3; 3; 1; 6) A_0, A_1, A_2 (A_3) id.v. / (**1**; **4**; **2**; **4**) all id.v. / else out.v.

• $\Gamma_{35}(2u; 2v, 2w; x)$ $2 \le u$; $2 \le v \le w$; $3 \le x$. $(u; v, w; x)$ — $\mathbf{S}^3$: (**u**; **2**, **2**; **x**), (2; 2, 3; 3), (2; 2, 3; 4), (2; 2, 4; 3), (2; 2, 3; 5), (2; 2, 5; 3), (3; 2, 3; 3), (**2**; **3**, **3**; **3**), (3; 2, 4; 3), (3; 2, 3; 4), (4; 2, 3; 3), (3; 2, 3; 5), (5; 2, 3; 3) — $\mathbf{E}^3$: (**2**; **3**, **3**; **4**) (2; 3, 4; 3) 209. **F432**, (4; 2, 3; 4) 207. **P432** — $\mathbf{E}^3$ **but not Nil**: (2; 2, 4; 4), (2; 2, 3; 6), (2; 2, 6; 3) A_3 id.v. — $\mathbf{H}^3$: (**2**; **3**, **3**; **5**), (3; 2, 5; 3), (5; 2, 3; 5), (4; 2, 3; 5), (5; 2, 3; 4) / (6; 2, 3; 3), (**3**; **3**, **3**; **3**), (4; 2, 4; 3), (**2**; **4**, **4**; **3**) A_0, A_1 id.v. / (3; 2, 3; 6), (3; 2, 4; 4), (2; 3, 4; 4) A_3 id.v. / (2; 3, 6; 3), (3; 2, 6; 3), (4; 2, 4; 4), (6; 2, 3; 6) A_0, A_1, A_3 id.v. / (**2**; **3**, **3**; **6**) A_2, A_3 id.v. / (**2**; **4**, **4**; **4**) all id.v. / else out.v.

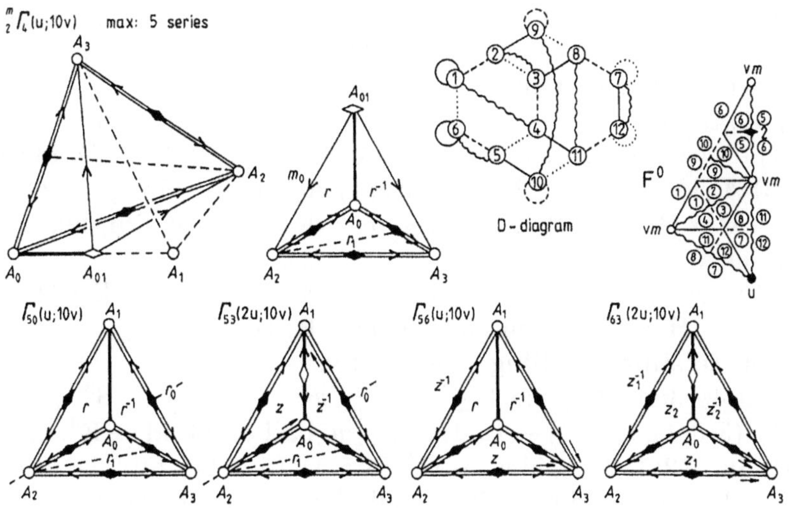

Figure 9. Figures for Family 7

Family 7. ${}^{m}_{2}\Gamma_4(u;10v)$ $3 \le u$, $1 \le v$. Γ is maximal iff $u \ne 10v$, else ${}^{\bar{4}2m}_{24}\Gamma(\bar{u} = u = 10v)$ is its supergroup (Family 1). $\Gamma \sim \Gamma_{29}(a = u$, $b = 1$, $c = v)$ (Family 26). $\Gamma^0(A_0, A_2, A_3) =$ **2u∗v** $\mathbf{H}^2$-group. Out.v.
• $\Gamma_{50}(u;10v)$ $3 \le u$; $1 \le v$. Out.v. • $\Gamma_{53}(2u;10v)$ $2 \le u$; $1 \le v$. Out.v.
• $\Gamma_{56}(u;10v)$ $3 \le u$; $1 \le v$. Out.v. • $\Gamma_{63}(2u;10v)$ $2 \le u$; $1 \le v$. Out.v.

Family 8 (Table 2). ${}^{m}_{2}\Gamma_6(2u;4v;3w)$ $2 \le u$; $1 \le v$, $1 \le w$. Γ is maximal iff $2u \ne 4v$, else ${}^{3m}_{6}\Gamma(2\bar{u} = 2u = 4v;\ \bar{v} = 3w)$ is its supergroup (Family 3). $\Gamma \sim \Gamma_2(a = v;\ b = w;\ c = 2;\ d = 2,\ e = u)$ (Family 14). $\Gamma^0(A_0, A_2) =$ **∗2vuw**, $\Gamma^0(A_3) =$ **2∗2w**. $(u;v;w)$ — $\mathbf{S}^3$: $(u;2;1)$, $(2;v;1)$, $(3;3;1)$, $(3;4;1)$ — $(u;1;1)$ $3 \le u$ **bad orbifold** — $\mathbf{E}^3$: $(4;3;1)$ 229.**Im$\bar{\mathbf{3}}$m** / $(\mathbf{2};\mathbf{1};\mathbf{2})$ A_3 id.v. — $\mathbf{H}^3$: $(3;5;1)$, $(5;3;1)$ / $(4;4;1)$, $(3;6;1)$, $(\mathbf{6};\mathbf{3};\mathbf{1})$ A_0, A_2 id.v. / $(2;2;2)$ all id.v. — $\mathbf{H}^2\times\mathbf{R}$: $(u;1;2)$ $3 \le u$ A_3 id.v. **degenerate simplex tiling** — else out.v.
• $\Gamma_9(4u;4v;6w)$ $1 \le u$; $1 \le v$; $1 \le w$. $(u;v;w)$ — $\mathbf{E}^3$: $(\mathbf{1};\mathbf{1};\mathbf{1})$ A_3 id.v. — $\mathbf{H}^3$: $(1;2;1)$ all id.v. — $\mathbf{H}^2\times\mathbf{R}$: $(u;1;1)$ $2 \le u$ A_3 id.v. **degenerate simplex tiling** — else out.v.
• $\Gamma_{16}(2u;4v;3w)$ $2 \le u$; $1 \le v$; $1 \le w$. $(u;v;w)$ — $\mathbf{S}^3$: $(u;2;1)$, $(2;v;1)$, $(3;3;1)$, $(3;4;1)$ — $(u;1;1)$ $3 \le u$ **bad orbifold** — $\mathbf{E}^3$: $(4;3;1)$ 211.**I432** — $\mathbf{E}^3$ **and Nil**: $(\mathbf{2};\mathbf{1};\mathbf{2})$ A_3 id.v. — $\mathbf{H}^3$: $(3;5;1)$, $(5;3;1)$ / $(4;4;1)$, $(3;6;1)$, $(\mathbf{6};\mathbf{3};\mathbf{1})$ A_0, A_1, A_2 id.v. / $(2;2;2)$ all id.v. — $\mathbf{H}^2\times\mathbf{R}$ (but not $\widetilde{\mathbf{SL}_2\mathbf{R}}$) $(u;1;2)$ $3 \le u$ A_3 id.v. **degenerate simplex tiling** — else out.v.

• $\Gamma_{22}(2u;4v;6w)$ $2 \le u$; $1 \le v$; $1 \le w$. $(u;v;w)$ — $\mathbf{E}^3$: $(\mathbf{2};\mathbf{1};\mathbf{1})$ A_3 id.v. — $\mathbf{H}^3$: $(2;2;1)$ all id.v. — $\mathbf{H}^2\times\mathbf{R}$: $(u;1;1)$ $3 \le u$ A_3 id.v. **degenerate simplex tiling,** — else out.v.

• $\Gamma_{26}(4u;4v;3w)$ $1 \le u$; $1 \le v$; $1 \le w$. $(u;v;w)$ — $\mathbf{S}^3$: $(u;2;1)$, $(1;v;1)$ — $(u;1;1)$ $2 \le u$ **bad orbifold** — $\mathbf{E}^3$: $(2;3;1)$ 217. $\mathbf{I\bar{4}3m}$ / $(\mathbf{1};\mathbf{1};\mathbf{2})$ A_3 id.v. — $\mathbf{H}^3$: $(2;4;1)$, $(\mathbf{3};\mathbf{3};\mathbf{1})$ A_0, A_1, A_2 id.v. / $(1;2;2)$ all id.v. — $\mathbf{H}^2\times\mathbf{R}$: $(u;1;2)$ $2 \le u$, A_3 id.v. **degenerate simplex tiling** — else out.v.

• $\Gamma_{27}(4u;4v;6w)$ $1 \le u$; $1 \le v$; $1 \le w$. $(u;v;w)$ — $\mathbf{E}^3$: $(\mathbf{1};\mathbf{1};\mathbf{1})$ A_3 id.v. — $\mathbf{H}^3$: $(1;2;1)$ all id.v. — $\mathbf{H}^2\times\mathbf{R}$: $(u;1;1)$ $2 \le u$ A_3 id.v. **degenerate simplex tiling** — else out.v.

• $\Gamma_{37}(2u;4v;3w)$ $2 \le u$; $1 \le v$; $1 \le w$. $(u;v;w)$ — $\mathbf{S}^3$: $(u;2;1)$, $(2;v;1)$, $(3;3;1)$, $(3;4;1)$ — $(u;1;1)$ $3 \le u$ **bad orbifold** — $\mathbf{E}^3$: $(4;3;1)$ 222. $\mathbf{Pn\bar{3}n}$ / $(\mathbf{2};\mathbf{1};\mathbf{2})$ A_3 id.v. — $\mathbf{H}^3$: $(3;5;1)$, $(5;3;1)$ / $(4;4;1)$, $(3;6;1)$, $(\mathbf{6};\mathbf{3};\mathbf{1})$ A_0, A_1, A_2 id.v. / $(2;2;2)$ all id.v. — $\mathbf{H}^2\times\mathbf{R}$: $(u;1;2)$ $3 \le u$ A_3 id.v. **degenerate simplex tiling** — else out.v.

Family 9 (Table 2). ${}^{2}_{2}\Gamma_1(2u;2v,2w;2x)$ $2 \le u \le x$; $2 \le v \le w$. Γ is maximal iff $2v \ne 2w$ and $2u \ne 2x$, else ${}^{mm2}_{4}\Gamma_1(\bar{u} = 2u;\ 2\bar{v} = 2v = 2w;\ \bar{w} = 2x)$ (Family 4) or ${}^{222}_{4}\Gamma_1(2\bar{u} = 2u = 2x,\ 2\bar{v} = 2v,\ 2\bar{w} = 2w)$ (Family 5) are supergroups, respectively. $\Gamma \sim \Gamma_2(a = u;\ b = 1;\ c = x;\ d = v,\ e = w)$ (Family 14). $\Gamma^0(A_0, A_1) = *\mathbf{uvw}$, $\Gamma^0(A_2, A_3) = *\mathbf{vwx}$. $(u;v,w;x)$ — $\mathbf{S}^3$: $(\mathbf{2};\mathbf{2},\mathbf{w};\mathbf{2})$, $(\mathbf{u};\mathbf{2},\mathbf{2};\mathbf{x})$, $(2;2,3;3)$, $(2;2,3;4)$ — $\mathbf{E}^3$: $(\mathbf{2};\mathbf{3},\mathbf{3};\mathbf{2})$ 215. $\mathbf{P\bar{4}3m}$, $(\mathbf{3};\mathbf{2},\mathbf{3};\mathbf{3})$ 227. $\mathbf{Fd\bar{3}m}$, $(2;2,4;3)$ 229. $\mathbf{Im\bar{3}m}$ — $\mathbf{H}^3$: $(2;2,3;5)$, $(2;2,5;3)$, $(3;2,3;4)$, $(3;2,3;5)$, $(4;2,3;5)$, $(\mathbf{2};\mathbf{3},\mathbf{4};\mathbf{2})$, $(\mathbf{3};\mathbf{2},\mathbf{4};\mathbf{3})$, $(\mathbf{4};\mathbf{2},\mathbf{3};\mathbf{4})$, $(\mathbf{2};\mathbf{3},\mathbf{5};\mathbf{2})$, $(\mathbf{3};\mathbf{2},\mathbf{5};\mathbf{3})$, $(\mathbf{5};\mathbf{2},\mathbf{3};\mathbf{5})$ / $(2;2,3;6)$, $(2;2,4;4)$, $(2;2,6;3)$, $(3;2,3;6)$, $(3;2,4;4)$, $(4;2,3;6)$, $(5;2,3;6)$, $(\mathbf{2};\mathbf{3},\mathbf{3};\mathbf{3})$ A_2, A_3 id.v. / $(\mathbf{2};\mathbf{3},\mathbf{6};\mathbf{2})$, $(\mathbf{3};\mathbf{2},\mathbf{6};\mathbf{3})$, $(\mathbf{6};\mathbf{2},\mathbf{3};\mathbf{6})$, $(\mathbf{2};\mathbf{4},\mathbf{4};\mathbf{2})$, $(\mathbf{4};\mathbf{2},\mathbf{4};\mathbf{4})$, $(\mathbf{3};\mathbf{3},\mathbf{3};\mathbf{3})$ all id.v. / else out.v.

• $\Gamma_{23}(2u;4v,4w;2x)$ $2 \le u$; $1 \le v \le w$; $2 \le x$. $(u;v,w;x)$ — $\mathbf{S}^3$: $(\mathbf{2};\mathbf{1},\mathbf{w};\mathbf{2})$, $(\mathbf{u};\mathbf{1},\mathbf{1};\mathbf{x})$ — $\mathbf{E}^3$: $(2;1,2;3)$ 217. $\mathbf{I\bar{4}3m}$, $(3;1,2;2)$ 204. $\mathbf{Im\bar{3}}$ — $\mathbf{H}^3$: $(\mathbf{3};\mathbf{1},\mathbf{2};\mathbf{3})$ / $(2;1,2;4)$, $(4;1,2;2)$, $(2;1,3;3)$, $(3;1,3;2)$, $(3;1,2;4)$, $(4;1,2;3)$ 2 id.v. / $(\mathbf{3};\mathbf{1},\mathbf{3};\mathbf{3})$, $(\mathbf{2};\mathbf{2},\mathbf{2};\mathbf{2})$, $(\mathbf{4};\mathbf{1},\mathbf{2};\mathbf{4})$ all id.v. / else out.v.

• $\Gamma_{64}(2u;2v,2w;2x)$ $2 \le u \le x$; $2 \le v \le w$. $(u;v,w;x)$ — $\mathbf{S}^3$: $(\mathbf{2};\mathbf{2},\mathbf{w};\mathbf{2})$, $(\mathbf{u};\mathbf{2},\mathbf{2};\mathbf{x})$, $(2;2,3;3)$, $(2;2,3;4)$ — $\mathbf{E}^3$: $(\mathbf{2};\mathbf{3},\mathbf{3};\mathbf{2})$ 219. $\mathbf{F\bar{4}3c}$, $(\mathbf{3};\mathbf{2},\mathbf{3};\mathbf{3})$ 203. $\mathbf{Fd\bar{3}}$, $(2;2,4;3)$ 222. $\mathbf{Pn\bar{3}n}$— $\mathbf{H}^3$: $(2;2,3;5)$, $(2;2,5;3)$, $(3;2,3;4)$, $(3;2,3;5)$, $(4;2,3;5)$, $(\mathbf{2};\mathbf{3},\mathbf{4};\mathbf{2})$, $(\mathbf{3};\mathbf{2},\mathbf{4};\mathbf{3})$, $(\mathbf{4};\mathbf{2},\mathbf{3};\mathbf{4})$, $(\mathbf{2};\mathbf{3},\mathbf{5};\mathbf{2})$, $(\mathbf{3};\mathbf{2},\mathbf{5};\mathbf{3})$, $(\mathbf{5};\mathbf{2},\mathbf{3};\mathbf{5})$, / $(2;2,3;6)$, $(2;2,4;4)$, $(2;2,6;3)$, $(3;2,3;6)$, $(3;2,4;4)$, $(4;2,3;6)$, $(5;2,3;6)$, $(\mathbf{2};\mathbf{3},\mathbf{3};\mathbf{3})$ A_2, A_3 id. v. / $(\mathbf{2};\mathbf{3},\mathbf{6};\mathbf{2})$, $(\mathbf{3};\mathbf{2},\mathbf{6};\mathbf{3})$, $(\mathbf{6};\mathbf{2},\mathbf{3};\mathbf{6})$, $(\mathbf{2};\mathbf{4},\mathbf{4};\mathbf{2})$, $(\mathbf{4};\mathbf{2},\mathbf{4};\mathbf{4})$, $(\mathbf{3};\mathbf{3},\mathbf{3};\mathbf{3})$ all id.v. / else out.v.

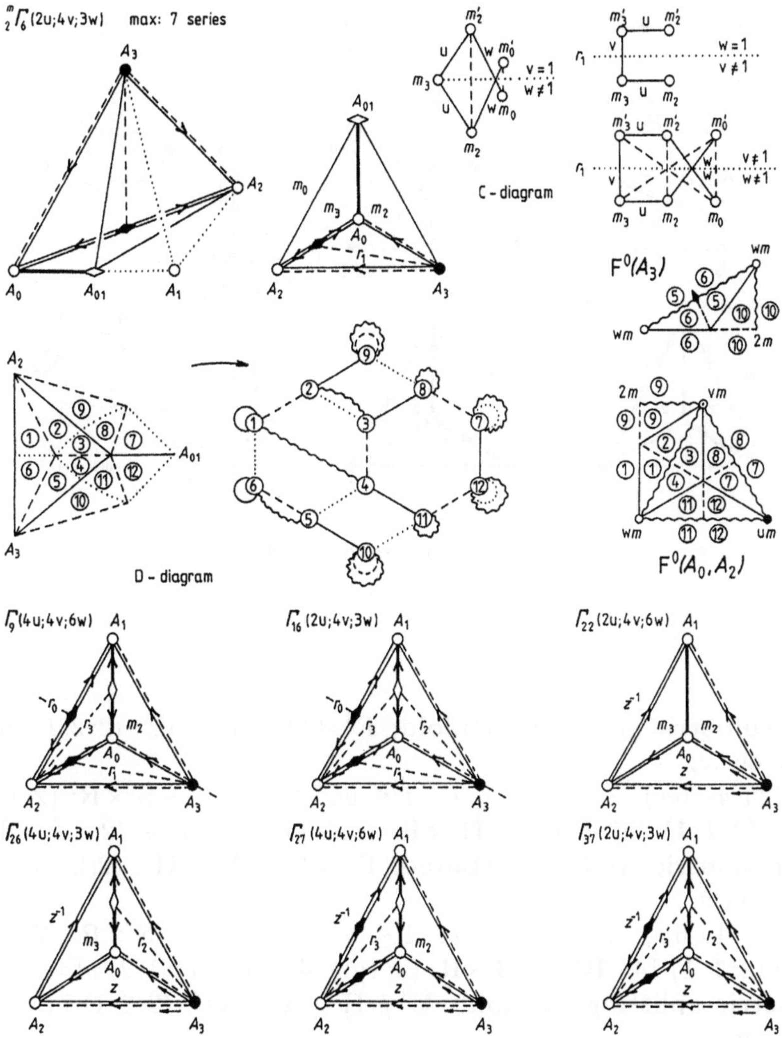

Figure 10. Figures for Family 8

Family 10. $^{2}_{2}\Gamma_2(2u;4v;6w)$ $2 \le u$; $1 \le v$; $1 \le w$. Γ is maximal iff $2u \neq 6w$, else $^{\bar{4}2m}_{8}\Gamma(\bar{u} = 4v;\ 2\bar{v} = 2u = 6w)$ is its supergroup (Family 2). $\Gamma \sim \Gamma_5(a = u;\ b = w;\ c = v;\ d = 1)$ (Family 17). $\Gamma^0 = \mathbf{2{*}uvw}$. $(u;v;w)$ — $\mathbf{S}^2{\times}\mathbf{R}$: $(2;1;1)$ — $\mathbf{E}^3$: $(\mathbf{3;1;1})$ 166. $\mathbf{R\bar{3}m}$ — $\mathbf{H}^2{\times}\mathbf{R}$: $(u;1;1)$ $4 \le u$ — $(2;1;2)$ id.v. **non-geometric splitting orbifold:** $\mathbf{E}^3 \,\#\, \mathbf{H}^2{\times}\mathbf{R}$ — $\mathbf{H}^3$: $(2;2;1)$ id.v. — else out.v.

- $\Gamma_{21}(2u;4v;6w)$ $2 \le u$; $1 \le v$; $1 \le w$. $(u;v;w)$ — $\mathbf{S}^2{\times}\mathbf{R}$: $(2;1;1)$ — $\mathbf{E}^3$: $(\mathbf{3;1;1})$ 160. $\mathbf{R3m}$ — $\mathbf{H}^2{\times}\mathbf{R}$: $(u;1;1)$ $4 \le u$ — $(2;1;2)$ id.v.

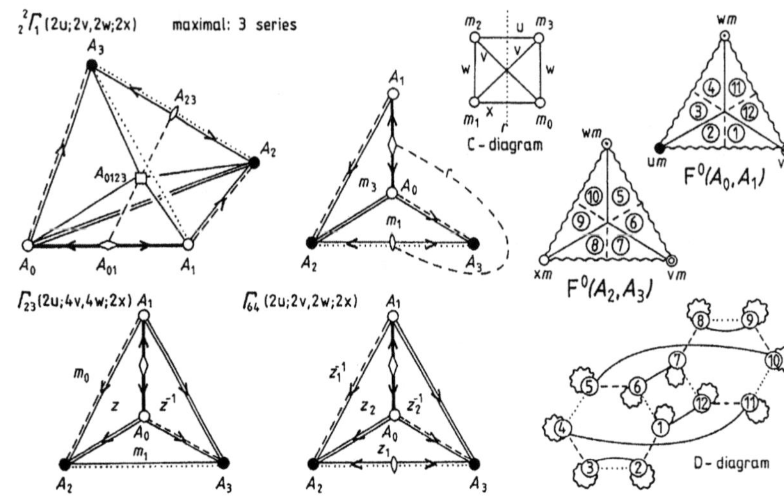

Figure 11. Figures for Family 9

non-geometric splitting orbifold: $\mathbf{E}^3 \# \mathbf{H}^2 \times \mathbf{R}$ — $\mathbf{H}^3$: $(2;2;1)$ id.v. — else out.v.

- $\Gamma_{49}(2u;4v;6w)$ $2 \leq u$; $1 \leq v$; $1 \leq w$. $(u;v;w)$ — $\mathbf{S}^2 \times \mathbf{R}$: $(2;1;1)$ — $\mathbf{E}^3$: $(\mathbf{3};\mathbf{1};\mathbf{1})$ 167. $\mathbf{R\bar{3}c}$ — $\mathbf{H}^2 \times \mathbf{R}$: $(u;1;1)$ $4 \leq u$ — $(2;1;2)$ id.v. **non-geometric splitting orbifold:** $\mathbf{E}^3 \# \mathbf{H}^2 \times \mathbf{R}$ — $\mathbf{H}^3$: $(2;2;1)$ id.v. — else out.v.
- $\Gamma_{61}(2u;4v;6w)$ $2 \leq u$; $1 \leq v$; $1 \leq w$. $(u;v;w)$ — $\mathbf{S}^2 \times \mathbf{R}$: $(2;1;1)$ — $\mathbf{E}^3$: $(\mathbf{3};\mathbf{1};\mathbf{1})$ 148. $\mathbf{R\bar{3}}$ — $\mathbf{H}^2 \times \mathbf{R}$: $(u;1;1)$ $4 \leq u$ — $(2;1;2)$ id.v. **non-geometric splitting orbifold:** $\mathbf{E}^3 \# \mathbf{H}^2 \times \mathbf{R}$ — $\mathbf{H}^3$: $(2;2;1)$ id.v. — else out.v.

Family 12. ${}^{2}_{2}\Gamma_6(3u, 3v)$ $1 \leq u \leq v$. Γ is maximal iff $u \neq v$, else ${}^{\bar{4}3m}_{24}\Gamma(\bar{u} = 3u = 3v)$ is its supergroup (Family 1). $\mathbf{r}_1 : A_0A_2A_3 \to A_2A_0A_3$, $\mathbf{r}_3 : A_0A_1A_2 \to A_0A_2A_1$, $\mathbf{r} : A_0A_1A_2A_3 \to A_1A_0A_3A_2$ — $\mathbf{r}_1^2$, $\mathbf{r}_3^2$, $\mathbf{r}^2$, ▬ $(\mathbf{r}\mathbf{r}_3\mathbf{r}_1\mathbf{r}_3)^u$, ═ $(\mathbf{r}\mathbf{r}_1\mathbf{r}\mathbf{r}_3\mathbf{r}\mathbf{r}_1)^v$. $\Gamma^0 = \mathbf{22uv}$. (u,v) — $\mathbf{S}^3$: $(\mathbf{1},\mathbf{1})$ — $(1,v)$ $2 \leq v$, $\mathbf{S}^3$-orbifold, **degenerate simplex tiling = lens tiling** — $\mathbf{H}^3$: $(\mathbf{2},\mathbf{2})$ all id.v. — else out.v.

- $\Gamma_{19}(6u, 6v)$ $1 \leq u \leq v$. (u,v) — $\mathbf{H}^3$: $(\mathbf{1},\mathbf{1})$ all id.v. — else out.v.
- $\Gamma_{46}(3u; 6v)$ $1 \leq u$; $1 \leq v$. $(u;v)$ — $(1;v)$ $1 \leq v$, $\mathbf{S}^3$-orbifold, **lens tiling** — $\mathbf{H}^3$: $(\mathbf{2};\mathbf{1})$ all id.v. — else out.v.
- $\Gamma_{59}(3u, 3v)$ $1 \leq u \leq v$. (u,v) — $\mathbf{S}^3$: $(\mathbf{1},\mathbf{1})$ **manifold** — $(1,v)$ $2 \leq v$, $\mathbf{S}^3$-orbifold, **lens tiling** — $\mathbf{H}^3$: $(\mathbf{2},\mathbf{2})$ all id.v. — else out.v.

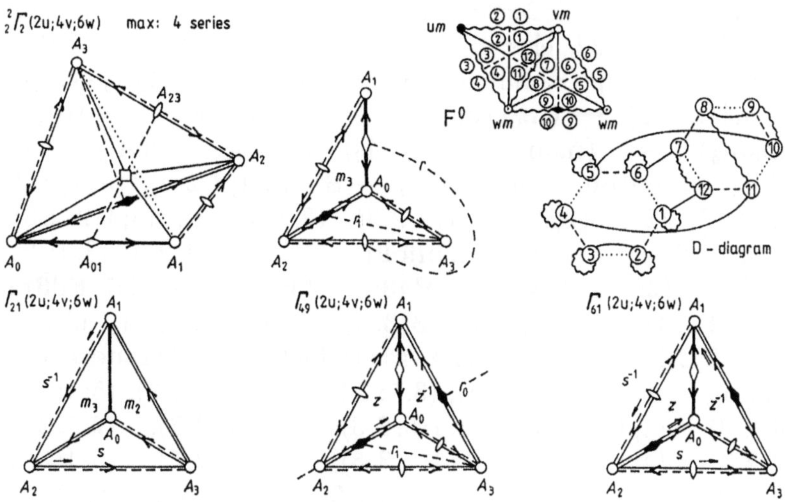

Figure 12. Figures for Family 10

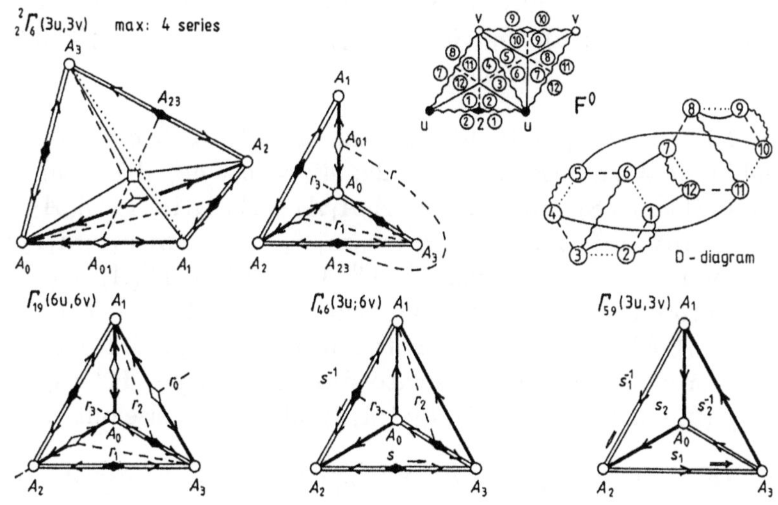

Figure 13. Figures for Family 12

Family 13 (Table 2). $\Gamma_1(2a, 2b, 2c; 2d, 2e, 2f)$ $2 \leq a,\ b,\ c,\ d,\ e,\ f$. $\Gamma_1 = \mathrm{Aut}\,\mathcal{T}$ is maximal iff there is no self-symmetry of type $\mathbf{m}$ and $\mathbf{2}$, else ${}^{m}_{2}\Gamma_1$ (Family 6) or ${}^{2}_{2}\Gamma_1$ (Family 9) is a supergroup. $\Gamma_1 = \big(\mathbf{m}_0, \mathbf{m}_1, \mathbf{m}_2, \mathbf{m}_3 - \mathbf{m}_0^2, \mathbf{m}_1^2, \mathbf{m}_2^2, \mathbf{m}_3^2; (\mathbf{m}_0\mathbf{m}_1)^a, (\mathbf{m}_1\mathbf{m}_2)^b, (\mathbf{m}_0\mathbf{m}_2)^c, (\mathbf{m}_2\mathbf{m}_3)^d, (\mathbf{m}_0\mathbf{m}_3)^e, (\mathbf{m}_1\mathbf{m}_3)^f\big)$ as presentation of a Coxeter–Lanner reflection group. E.g. $\Gamma^0(A_3) = *\mathbf{abc}$. Realizations in $\mathbf{S}^3$, $\mathbf{E}^3$, $\mathbf{H}^3$ see Table 2 — else outer vertex occurs.

Family 14. $\Gamma_2(4a;4b;2c;2d,2e)$ $1 \le a$; $1 \le b$; $2 \le c$; $2 \le d \le e$. Γ_2 is maximal iff $2e \neq 2d$, else ${}^{m}_{2}\Gamma_1(2\bar{u} = 4a;\ 2\bar{v} = 2e = 2d,\ 2\bar{w} = 4b;\ \bar{x} = 2c)$ is its supergroup (Family 6). $\mathbf{m}_0$, $\mathbf{m}_2$, $\mathbf{m}_3$, $\mathbf{r}_1 : A_0A_2A_3 \to A_0A_3A_2$ — $\mathbf{m}_0^2$, $\mathbf{m}_2^2$, $\mathbf{m}_3^2$, $\mathbf{r}_1^2$, ▬ $(\mathbf{m}_0\mathbf{r}_1\mathbf{m}_0\mathbf{r}_1)^a$, ═ $(\mathbf{m}_2\mathbf{r}_1\mathbf{m}_3\mathbf{r}_1)^b$, ╍ $(\mathbf{m}_2\mathbf{m}_3)^c$, ┈ $(\mathbf{m}_0\mathbf{m}_3)^d$, –o– $(\mathbf{m}_0\mathbf{m}_2)^e$. $\Gamma^0(A_0) = \mathbf{2{*}bc}$, $\Gamma^0(A_1) = \mathbf{{*}cde}$, $\Gamma^0(A_2, A_3) = \mathbf{{*}adbe}$. $(a;b;c;d,e)$ — $\mathbf{S}^3$: $(2;1;2;2,e)$, $\mathbf{(a;1;c;2,2)}$, $\mathbf{(1;1;2;3,3)}$, $(2;1;3;2,3)$, $(3;1;2;2,3)$, $(2;1;4;2,3)$, $(4;1;2;2,3)$ — $(1;1;c;d,e)$ $2 \le d < e$ **bad orbifold** — $\mathbf{E}^3$: $(2;1;3;2,4)$ $(3;1;2;2,4)$ 229. $\mathbf{Im\bar{3}m}$, $\mathbf{(2;1;2;3,3)}$ 215. $\mathbf{P\bar{4}3m}$, $(3;1;3;2,3)$ 227. $\mathbf{Fd\bar{3}m}$ / $\mathbf{(1;2;2;2,2)}$ A_0 id.v. / $\mathbf{(1;1;3;3,3)}$, $\mathbf{(1;1;2;4,4)}$ A_1 id.v. — $\mathbf{H}^2{\times}\mathbf{R}$: $(1;2;2;2,e)$ $3 \le e$ A_0 id.v. as orbifold exists, **degenerate simplex tiling** — $\mathbf{H}^3$: $(2;1;3;2,5)$, $(3;1;2;2,5)$, $(2;1;5;2,3)$, $(5;1;2;2,3)$, $(3;1;4;2,3)$, $(4;1;3;2,3)$, $(3;1;5;2,3)$, $(5;1;3;2,3)$, $(2;1;2;3,4)$, $(3;1;3;2,4)$, $(4;1;4;2,3)$, $(2;1;2;3,5)$, $(3;1;3;2,5)$, $(5;1;5;2,3)$, $(4;1;5;2,3)$, $(5;1;4;2,3)$ / $\mathbf{(1;2;2;3,3)}$, $(1;2;2;3,4)$, $(1;2;2;3,5)$ A_0 id.v. / $\mathbf{(2;1;3;3,3)}$, $(2;1;4;2,4)$, $(2;1;3;2,6)$, $(2;1;6;2,3)$ A_1 id.v. / $\mathbf{(3;1;2;3,3)}$, $(4;1;2;2,4)$, $(3;1;2;2,6)$, $(6;1;2;2,3)$ A_2, A_3 id.v. / $\mathbf{(2;2;2;2,2)}$ A_0, A_2, A_3 id.v. / $\mathbf{(3;1;3;3,3)}$, $\mathbf{(2;1;2;4,4)}$, $(4;1;4;2,4)$, $(2;1;2;3,6)$, $(3;1;3;2,6)$, $(6;1;6;2,3)$ A_1, A_2, A_3 id.v. / $\mathbf{(1;2;2;4,4)}$, $(1;2;2;3,6)$ all id.v. — else out.v.

Family 15. $\Gamma_3(2a;6b;4c,4d)$ $2 \le a$; $1 \le b$; $1 \le c \le d$. Γ_3 is maximal iff $4c \neq 4d$, else ${}^{m}_{2}\Gamma_6(2\bar{u} = 2a;\ 4\bar{v} = 4c = 4d;\ 3\bar{w} = 6b)$ is its supergroup (Family 8). $\mathbf{m}_0$, $\mathbf{m}_1$, $\mathbf{r}_2 : A_0A_1A_3 \to A_0A_3A_1$, $\mathbf{r}_3 : A_0A_1A_2 \to A_0A_2A_1$ — $\mathbf{m}_0^2$, $\mathbf{m}_1^2$, $\mathbf{r}_2^2$, $\mathbf{r}_3^2$, ▬ $(\mathbf{m}_0\mathbf{m}_1)^a$, ═ $(\mathbf{m}_1\mathbf{r}_2\mathbf{r}_3\mathbf{m}_1\mathbf{r}_3\mathbf{r}_2)^b$, ╍ $(\mathbf{m}_0\mathbf{r}_2\mathbf{m}_0\mathbf{r}_2)^c$, ┈ $(\mathbf{m}_0\mathbf{r}_3\mathbf{m}_0\mathbf{r}_3)^d$. $\Gamma^0(A_0) = \mathbf{22{*}b}$, $\Gamma^0(A_1, A_2, A_3) = \mathbf{{*}acdab}$. $(a;b;c,d)$ — $\mathbf{E}^3$: $\mathbf{(2;1;1,1)}$ A_0 id.v. — $\mathbf{H}^2{\times}\mathbf{R}$: $\mathbf{(a;1;1,1)}$ $3 \le a$ A_0 id.v. as orbifold exists, **degenerate simplex tiling**; $(2;1;1,d)$ $2 \le d$ A_0 id.v. as orbifold exists, **degenerate simplex tiling** — $\mathbf{H}^3$: $(3;1;1,2)$ A_0 id.v. / $\mathbf{(2;1;2,2)}$, $(3;1;1,3)$, $(4;1;1,2)$ all id.v. — else out.v.

Family 16. $\Gamma_4(2a;6b;4c,4d)$ $2 \le a$; $1 \le b$; $1 \le c \le d$. Γ_4 is maximal iff $4c \neq 4d$ and $2a \neq 6b$, else its supergroup is ${}^{2}_{2}\Gamma_2(2\bar{u} = 2a;\ 4\bar{v} = 4c = 4d;\ 6\bar{w} = 6b)$ (Family 10) or ${}^{mm2}_{4}\Gamma_1(\bar{u} = 4c,\ 2\bar{v} = 2a = 6b,\ \bar{w} = 4d)$ (Family 4), respectively. $\mathbf{m}_0$, $\mathbf{m}_1$, $\mathbf{r}_2$, $\mathbf{r}_3$ — $\mathbf{m}_0^2$, $\mathbf{m}_1^2$, $\mathbf{r}_2^2$, $\mathbf{r}_3^2$, ▬ $(\mathbf{m}_0\mathbf{m}_1)^a$, ═ $(\mathbf{m}_1\mathbf{r}_2\mathbf{r}_3\mathbf{m}_0\mathbf{r}_3\mathbf{r}_2)^b$, ╍ $(\mathbf{m}_1\mathbf{r}_3\mathbf{m}_1\mathbf{r}_3)^c$, ┈ $(\mathbf{m}_0\mathbf{r}_2\mathbf{m}_0\mathbf{r}_2)^d$. $\Gamma^0(A_0, A_2) = \mathbf{2{*}abc}$, $\Gamma^0(A_1, A_3) = \mathbf{2{*}abd}$. $(a;b;c,d)$ — $\mathbf{S}^2{\times}\mathbf{R}$: $\mathbf{(2;1;1,1)}$ — $\mathbf{E}^3$: $\mathbf{(3;1;1,1)}$ 166. $\mathbf{R\bar{3}m}$ — $\mathbf{H}^2{\times}\mathbf{R}$: $\mathbf{(a;1;1,1)}$ $4 \le a$ — $(2;1;1,2)$ A_1, A_3 id.v. **non-geometric splitting orbifold:** $\mathbf{E}^3 \# \mathbf{H}^2{\times}\mathbf{R}$ — $\mathbf{(2;2;1,1)}$ all id.v. **non-geometric splitting orbifold:** $\mathbf{H}^2{\times}\mathbf{R} \# \mathbf{H}^2{\times}\mathbf{R}$ — $\mathbf{H}^3$: $\mathbf{(2;1;2,2)}$ all id.v. — else out.v.

Family 17. $\Gamma_5(2a;8b;4c;4d)$ $2 \le a$; $1 \le b$; $1 \le c$; $1 \le d$. Γ_5 is maximal iff $2a \ne 4d$ or $8b \ne 4c$, else ${}^{3m}_{6}\Gamma(2\bar{u} = 8b = 4c; \bar{v} = 2a = 4d)$ (Family 3) is its supergroup. $\mathbf{m}_0$, $\mathbf{m}_1$, $\mathbf{r}_2 : A_0A_1A_3 \to A_0A_3A_1$, $\mathbf{r}_3 : A_0A_1A_2 \to A_1A_0A_2$ — $\mathbf{m}_0^2$, $\mathbf{m}_1^2$, $\mathbf{r}_2^2$, $\mathbf{r}_3^2$, ▬ $(\mathbf{m}_0\mathbf{m}_1)^a$, ═ $(\mathbf{m}_1\mathbf{r}_2\mathbf{r}_3\mathbf{r}_2\mathbf{m}_1\mathbf{r}_2\mathbf{r}_3\mathbf{r}_2)^b$, ┅ $(\mathbf{m}_0\mathbf{r}_2\mathbf{m}_0\mathbf{r}_2)^c$, ┈ $(\mathbf{m}_0\mathbf{r}_3\mathbf{m}_1\mathbf{r}_3)^d$. $\Gamma^0(A_2)$ = **2∗ad**, $\Gamma^0(A_0, A_1, A_3)$ = **2∗abdc**. $(a;b;c;d)$ — $\mathbf{S}^2{\times}\mathbf{R}$: $(2;1;1;1)$ — $\mathbf{E}^3$: $(3;1;1;1)$ 166. **R$\bar{3}$m** — $\mathbf{H}^2{\times}\mathbf{R}$: $(a;1;1;1)$ $4 \le a$ — $\mathbf{H}^3$: $(\mathbf{2};\mathbf{1};\mathbf{2};\mathbf{1})$ A_0, A_1, A_3 id.v. — $(2;2;1;1)$ A_0, A_1, A_3 id.v. **non-geometric splitting orbifold: $\mathbf{E}^3 \# \mathbf{H}^2{\times}\mathbf{R}$** — $(2;1;1;2)$ all id.v. **non-geometric splitting orbifold: $\mathbf{H}^2{\times}\mathbf{R} \# \mathbf{H}^2{\times}\mathbf{R}$** — else out.v.

Family 18. $\Gamma_7(4a;16b;4c)$ $2 \le a$; $1 \le b$; $1 \le c$. Γ_7 is maximal iff $4a = 16b = 4c$ does not hold, else ${}^{\bar{4}3m}_{24}\Gamma(\bar{u} = 4a = 16b = 4c)$ is its supergroup (Family 1). Γ^0 = **222∗abc** $\mathbf{H}^2$-group. Out.v.

Family 19. $\Gamma_8(4a;12b;8c)$ $1 \le a$; $1 \le b$; $1 \le c$. Γ_8 is maximal iff $4a = 12b = 8c$ does not hold, else ${}^{\bar{4}3m}_{24}\Gamma(\bar{u} = 4a = 12b = 4c)$ is its supergroup (Family 1). Γ^0 = **222∗abc** $\mathbf{H}^2$-group. Out.v.

Family 20. $\Gamma_{11}(4a, 4b, 4c; 3d)$ $1 \le a \le b \le c$; $1 \le d$. Γ_{11} is maximal iff $4a$, $4b$, $4c$ are different, else, e.g. $b = c$, ${}^{m}_{2}\Gamma_6(2\bar{u} = 4a,\ 4\bar{v} = 4b = 4c,\ 3\bar{w} = 3d)$ is its supergroup (Family 8). $\mathbf{m}_0$, $\mathbf{r}_1 : A_0A_2A_3 \to A_0A_3A_2$, $\mathbf{r}_2 : A_0A_1A_3 \to A_0A_3A_1$, $\mathbf{r}_3 : A_0A_1A_2 \to A_0A_2A_1$ — $\mathbf{m}_0^2$, $\mathbf{r}_1^2$, $\mathbf{r}_2^2$, $\mathbf{r}_3^2$, ▬ $(\mathbf{m}_0\mathbf{r}_1\mathbf{m}_0\mathbf{r}_1)^a$, ═ $(\mathbf{m}_0\mathbf{r}_3\mathbf{m}_0\mathbf{r}_3)^b$, ┅ $(\mathbf{m}_0\mathbf{r}_2\mathbf{m}_0\mathbf{r}_2)^c$, ┈ $(\mathbf{r}_1\mathbf{r}_2\mathbf{r}_3)^d$. $\Gamma^0(A_0)$ = **222d**, $\Gamma^0(A_1, A_2, A_3)$ = **d∗abc**. $(a, b, c; d)$ — $\mathbf{S}^3$: $(\mathbf{1}, \mathbf{b}, \mathbf{b}; \mathbf{1})$ $1 \le b$, $(\mathbf{2}, \mathbf{2}, \mathbf{c}; \mathbf{1})$, $2 \le c$ — $(\mathbf{1}, \mathbf{1}, \mathbf{c}; \mathbf{1})$ $2 \le c$ **bad orbifold** — $\mathbf{E}^3$: $(\mathbf{2}, \mathbf{3}, \mathbf{3}; \mathbf{1})$ 224. **Pn$\bar{3}$m** / $(\mathbf{1}, \mathbf{1}, \mathbf{1}; \mathbf{2})$ A_0 id.v. — $\mathbf{H}^2{\times}\mathbf{R}$: $(\mathbf{1}, \mathbf{1}, \mathbf{c}; \mathbf{2})$ $2 \le c$ A_0 id.v. **degenerate simplex tiling** — $\mathbf{H}^3$: $(2, 3, 4; 1)$, $(2, 3, 5; 1)$ / $(\mathbf{3}, \mathbf{3}, \mathbf{3}; \mathbf{1})$, $(\mathbf{2}, \mathbf{4}, \mathbf{4}; \mathbf{1})$, $(2, 3, 6; 1)$ A_1, A_2, A_3 id.v. / $(\mathbf{1}, \mathbf{2}, \mathbf{2}; \mathbf{2})$ all id.v. — else out.v.

Family 21. $\Gamma_{12}(2a;6b;8c)$ $2 \le a$; $1 \le b$; $1 \le c$. Γ_{12} is maximal iff $2a \ne 8c$, else ${}^{3m}_{6}\Gamma(2\bar{u} = 2a = 8c;\ \bar{v} = 6b)$ is its supergroup (Family 3). $\mathbf{m}_0$, $\mathbf{r}_1 : A_0A_2A_3 \to A_3A_2A_0$, $\mathbf{r}_2 : A_0A_1A_3 \to A_1A_0A_3$, $\mathbf{r}_3 : A_0A_1A_2 \to A_1A_0A_2$ — $\mathbf{m}_0^2$, $\mathbf{r}_1^2$, $\mathbf{r}_2^2$, $\mathbf{r}_3^2$, ▬ $(\mathbf{r}_2\mathbf{r}_3)^a$, ═ $(\mathbf{m}_0\mathbf{r}_3\mathbf{r}_1\mathbf{m}_0\mathbf{r}_1\mathbf{r}_3)^b$, ┅ $(\mathbf{m}_0\mathbf{r}_2\mathbf{r}_1\mathbf{r}_2\mathbf{m}_0\mathbf{r}_2\mathbf{r}_1\mathbf{r}_2)^c$, $\Gamma^0(A_2)$ = **22∗b**, $\Gamma^0(A_0, A_1, A_3)$ = **2a∗bc**. $(a;b;c)$ — $\mathbf{H}^3$: $(2;1;1)$ all id.v. — else out.v.

Family 22. $\Gamma_{13}(8a, 8b, 8c)$ $1 \le a \le b \le c$. Γ_{13} is maximal iff $8a = 8b = 8c$ does not hold, else ${}^{\bar{4}3m}_{24}\Gamma(\bar{u} = 8a = 8b = 8c)$ is its supergroup (Family 1). Γ^0 = **222∗abc** $\mathbf{H}^2$-group. Out.v.

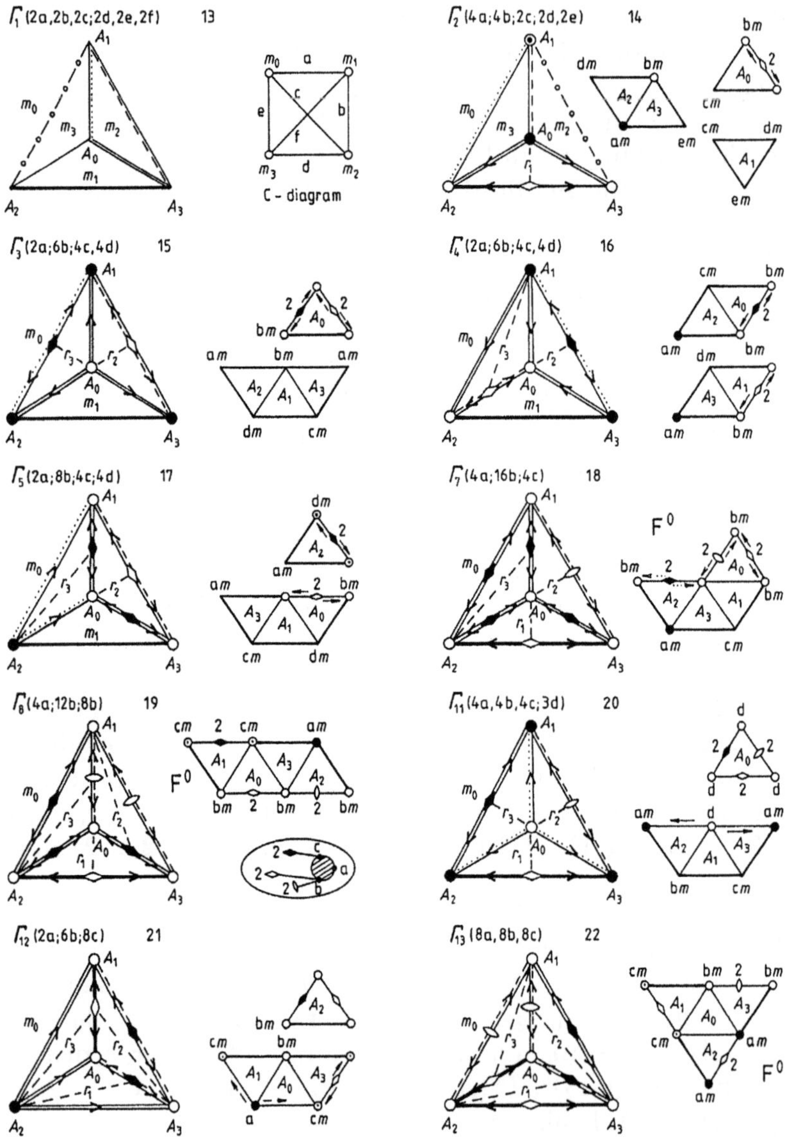

Figure 14. Figures for the maximal Families 13–22

Family 23. $\Gamma_{15}(2a;10b)$ $2 \leq a;$ $1 \leq b.$ Γ_{15} is maximal iff $2a \neq 10b$, else ${}^{\bar{4}3m}_{24}\Gamma(\bar{u} = 2a = 10b)$ is its supergroup (Family 1). $\Gamma^0 = \mathbf{2222ab}$ $\mathbf{H}^2$-group. Out.v.

Family 24. $\Gamma_{18}(4a;8b)$ $1 \leq a;$ $1 \leq b.$ Γ_{18} is maximal iff $4a \neq 8b$, else ${}^{\bar{4}3m}_{24}\Gamma(2\bar{u} = 4a = 8b)$ is its supergroup (Family 1). $\mathbf{r}_0 : A_1A_2A_3 \to A_1A_3A_2$, $\mathbf{r}_1 : A_0A_2A_3 \to A_3A_2A_0$, $\mathbf{r}_2 : A_0A_1A_3 \to A_0A_3A_1$, $\mathbf{r}_3 : A_0A_1A_2 \to A_2A_1A_0$ — $\mathbf{r}_0^2$, $\mathbf{r}_1^2$, $\mathbf{r}_2^2$, $\mathbf{r}_3^2$, ▬ $(\mathbf{r}_0\mathbf{r}_1\mathbf{r}_3\mathbf{r}_1)^a$, ═ $(\mathbf{r}_1\mathbf{r}_2\mathbf{r}_3\mathbf{r}_0\mathbf{r}_2\mathbf{r}_0\mathbf{r}_3\mathbf{r}_2)^b$, $\Gamma^0 = \mathbf{2222ab}$. $(a;b)$ — $(1;1)$ all id.v. **non-geometric splitting orbifold:** $\mathbf{E}^3 \# \widetilde{\mathbf{SL}_2\mathbf{R}}$ **and** $\mathbf{E}^3 \# \mathbf{H}^2 \times \mathbf{R}$ — else out.v.

Family 25. $\Gamma_{25}(4a;10b)$ $1 \leq a;$ $1 \leq b.$ Γ_{25} is maximal iff $4a \neq 10b$, else ${}^{\bar{4}3m}_{24}\Gamma(\bar{u} = 4a = 10b)$ is its supergroup (Family 1). $\Gamma^0 = \mathbf{2{*}ab{*}b}$ $\mathbf{H}^2$-group. Out.v.

Family 26. $\Gamma_{29}(a;4b;12c)$ $3 \leq a;$ $1 \leq b;$ $1 \leq c.$ Γ_{29} is maximal iff $a \neq 4b$, else ${}^{3m}_{6}\Gamma(2\bar{u} = 12c;\ \bar{v} = a = 4b)$ is its supergroup (Family 3). $\Gamma^0(A_1) = \mathbf{a{*}b}$, $\Gamma^0(A_0, A_2, A_3) = \mathbf{2a{*}bc}$ $\mathbf{H}^2$-group. Out.v.

Family 27. $\Gamma_{31}(6a;12b)$ $1 \leq a;$ $1 \leq b.$ Γ_{31} is maximal iff $6a \neq 12b$, else ${}^{\bar{4}3m}_{24}\Gamma(\bar{u} = 6a = 12b)$ is its supergroup (Family 1). $\Gamma^0 = \mathbf{2{*}a{*}ab}$ $\mathbf{H}^2$-group. Out.v.

Family 28. $\Gamma_{32}(4a;16b)$ $1 \leq a;$ $1 \leq b.$ Γ_{32} is maximal iff $4a \neq 16b$, else ${}^{\bar{4}3m}_{24}\Gamma(\bar{u} = 4a = 16b)$ is its supergroup (Family 1). $\Gamma^0 = \mathbf{2{*}aab}\otimes$ $\mathbf{H}^2$-group. Out.v.

Family 29. $\Gamma_{34}(2a;6b;8c)$ $1 \leq a;$ $1 \leq b;$ $1 \leq c.$ Γ_{34} is maximal iff $2a \neq 6b$, else ${}^{\bar{4}2m}_{8}\Gamma(\bar{u} = 8c,\ 2\bar{v} = 2a = 6b)$ is a supergroup (Family 2). $\Gamma^0 = \mathbf{2a{*}bbc}$ $\mathbf{H}^2$-group. Out.v.

Family 31. $\Gamma_{43}(4a;8b)$ $1 \leq a;$ $1 \leq b.$ Γ_{43} is maximal iff $4a \neq 8b$, else ${}^{\bar{4}3m}_{24}\Gamma(\bar{u} = 4a = 8b)$ is its supergroup (Family 1). $\mathbf{r}_0 : A_1A_2A_3 \to A_1A_3A_2$, $\mathbf{r}_1 : A_0A_2A_3 \to A_3A_2A_0$, $\mathbf{z} : A_0A_3A_1 \to A_0A_1A_2$, $\mathbf{r}_0^2$, $\mathbf{r}_1^2$, ▬ $(\mathbf{z}\mathbf{r}_0\mathbf{z}\mathbf{r}_0)^a$, ═ $(\mathbf{r}_0\mathbf{r}_1\mathbf{z}^{-2}\mathbf{r}_1\mathbf{z}^2\mathbf{r}_1)^b$. $\Gamma^0 = \mathbf{22ab}\otimes$. $(a;b)$ — $(1;1)$ all id.v. **non-geometric splitting orbifold:** $\mathbf{E}^3 \# \widetilde{\mathbf{SL}_2\mathbf{R}}$ **and** $\mathbf{E}^3 \# \mathbf{H}^2 \times \mathbf{R}$ — else out.v.

Family 32. $\Gamma_{44}(2a;10b)$ $2 \leq a;$ $1 \leq b.$ Γ_{44} is maximal iff $2a \neq 10b$, else ${}^{\bar{4}3m}_{24}\Gamma(\bar{u} = 2a = 10b)$ is its supergroup (Family 1). $\Gamma^0 = \mathbf{22ab}\otimes$ $\mathbf{H}^2$-group. Out.v.

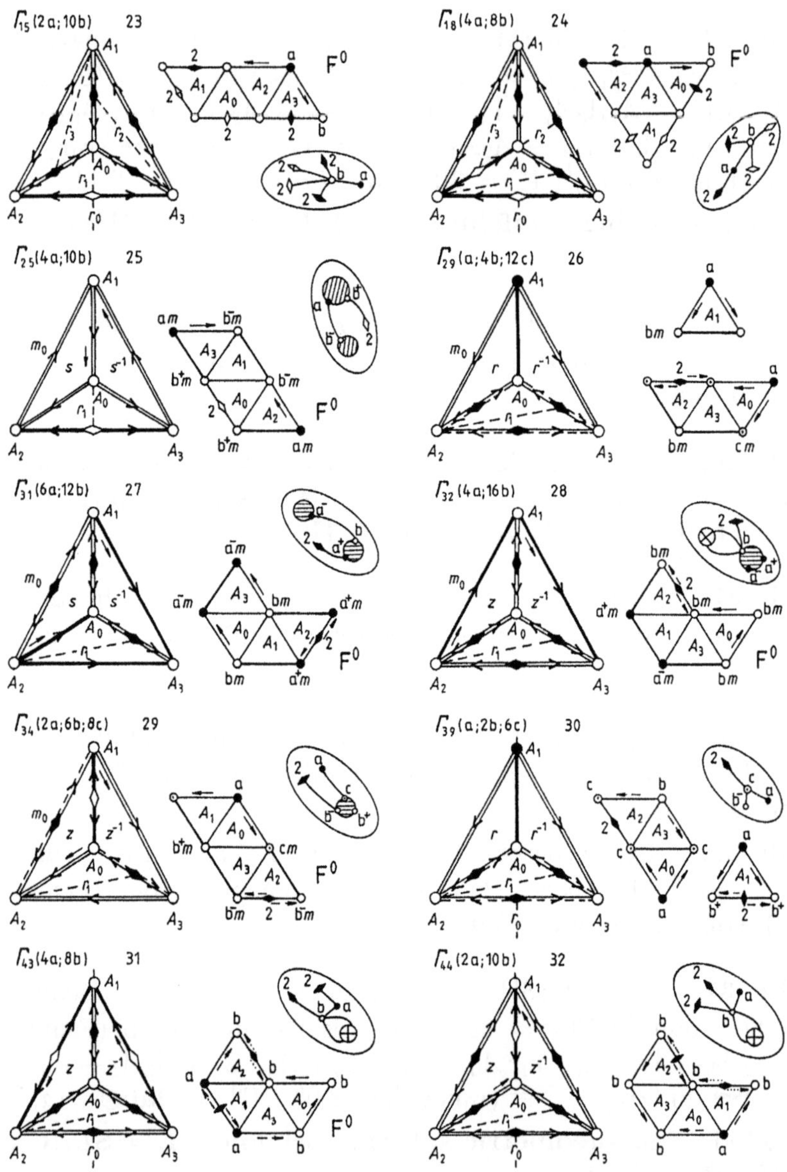

Figure 15. Figures for the maximal Families 23–32

Bibliography

[CM80] Coxeter, H. S. M. — Moser, W. O. J. (1980) *Generators and relations for discrete groups.* 4th ed. Springer, Berlin-Heidelberg-New York.

[D87] Dress, A. W. M. (1987) Presentation of discrete groups acting on simply connected manifolds in terms of parametrized systems of Coxeter matrices. *Advances in Math.* **63** 196–212.

[DHM93] Dress, A. W. M. — Huson, D. H. — Molnár, E. (1993) The classification of face-transitive periodic three-dimensional tilings. *Acta Crystallographica,* **A 49** 806–817.

[H93] Huson, D. H. (1993) The generation and classification of tile-k-transitive tilings of the Euclidean plane, the sphere and the hyperbolic plane. *Geometriae Dedicata,* **47** 269–296.

[LM90] Lučić, Z. — Molnár, E. (1990) Combinatorial classification of fundamental domains of finite area for planar discontinuous isometry groups. *Arch. Math.* **54** 511–520.

[M92] Molnár, E. (1992) Polyhedron complexes with simply transitive group actions and their realizations. *Acta Math. Hungarica,* **59** 175–216.

[M96] Molnár, E. (1996) Discontinuous groups in homogeneous Riemannian spaces by classification of D-symbols. *Publ. Math. Debrecen,* **49/3-4** 265–294.

[M97] Molnár, E. (1997) The projective interpretation of the eight 3-dimensional homogeneous geometries. *Beiträge zur Algebra und Geometrie* (*Contributions to Algebra and Geometry*), **38** No. 2, 261–288.

[MP88] Molnár, E. — Prok, I. (1988) A polyhedron algorithm for finding space groups. *Proc. of Third Int. Conf. on Engineering Graphics and Descriptive Geometry,* Vienna Vol.**2**, pp. 37–44.

[MP91] Molnár, E. — Prok, I. (1994) Classification of solid transitive simplex tilings in simply connected 3-spaces, Part I. The combinatorial description by figures and tables, results in spaces of constant curvature. *Colloquia Math. Soc. János Bolyai* **63.** *Intuitive Geometry,* Szeged (Hungary), 1991. North-Holland Publ. Comp. Amsterdam–Oxford–New York, 311–362.

[MPSz97] Molnár, E. — Prok, I. — Szirmai, J. (1997) Classification of solid transitive simplex tilings in simply connected 3-spaces, Part II. Metric realizations of the maximal simplex tilings. *Periodica Math. Hungar.* **35/1-2** 47–94.

[MPSz98] Molnár, E. — Prok, I. — Szirmai, J. (1998) Two families of fundamental tilings and their realizations in various 3-spaces. *Proc. of International Scientific Conferences of Mathematics, Žilina, Slovakia,* Vol.**2** 43–64.

[MSzW] Molnár, E. — Szirmai, J. — Weeks, J. R. 3-simplex tilings, splitting orbifolds and manifolds. *Annales Univ. Sci. Budapest, Sect. Math.* (submitted)

[MSz] Molnár, E. — Szirmai, J. Minimally presented orientable splitting 3-manifold with one cusp. *Annales Univ. Sci. Budapest, Sect. Math.* (submitted)

[P99] PROK, I. (1999) Fundamental tilings in spaces of constant curvature with regular polyhedra (in Hungarian). *PhD thesis, Budapest University of Technology and Economics, Faculty of Natural Sciences.*

[S83] Scott, P. (1983) The geometry of 3-manifolds. *Bull. London Math. Soc.*, **15** 401–487. Russian translation: Moscow 'Mir' (1986).

[S93] STOJANOVIĆ, M. (1993) Some series of hyperbolic space groups. *Annales Univ. Sci. Budapest, Sect. Math.* **36** 85–102.

[Sz96] SZIRMAI, J. (1996) Metrische Realisierungen von zwei Familien der dreidimensionalen körpertransitiven Simplexpflasterungen, *Annales Univ. Sci. Budapest, Sect. Math.* **39** 145–162.

[Th97] THURSTON, W. P. (1997) *Three-dimensional geometry and topology I.* Ed. by SILVIO LEVY, Princeton Univ. Press, Princeton, New Jersey.

[Zh83] ZHUK, I. K. (1983) Fundamental tetrahedra in Euclidean and Lobachevsky spaces. *Soviet Math. Dokl.* **27** 540–543.

[ZVC80] ZIESCHANG, H. — VOGT, E. — COLDEWEY, H. D. (1980) *Surfaces and planar discontinuous groups.* Lecture Notes in Math. **835,** Springer, Berlin–Heidelberg–New York.

V

DIFFERENTIAL GEOMETRY

NON-EUCLIDEAN ANALYSIS

Sigurdur Helgason
Department of Mathematics,
Massachusetts Institute of Technology
Cambridge, MA 02139
helgason@mit.edu

1. The Non-Euclidean Plane

In case the work of Bolyai [Bo] and Lobatschevsky [Lo] left any doubts about the existence of non-Euclidean geometry these doubts were removed by the work [Be] of Beltrami. With a modification made possible by hindsight one can state the following result.

Theorem 1.1. *Given a simply connected region $D \subset \mathbf{R}^2\,(D \neq \mathbf{R}^2)$ there exists a Riemannian metric g on D which is invariant under all conformal transformations of D. Also g is unique up to a constant factor.*

Because of the Riemann mapping theorem we can assume D to be the unit disk. Given $a \in D$ the mapping $\varphi : z \to \dfrac{a-z}{1-\bar{a}z}$ is conformal and $\varphi(a) = 0$. The invariance of g requires

$$g_a(u,u) = g_0(d\varphi(u),\, d\varphi(u)) \tag{1.1}$$

for each $u \in D_a$ (the tangent space to D at a). Since g_0 is invariant under rotations around 0,

$$g_0(z,z) = c|z|^2\,, \tag{1.2}$$

where c is a constant. Here D_0 is identified with $\mathbf{C}$. Let $t \to z(t)$ be a curve with $z(0) = a$, $z'(0) = u \in \mathbf{C}$. Then $d\varphi(u)$ is the tangent vector

$$\left\{\frac{d}{dt}\varphi(z(t))\right\}_{t=0} = \left(\frac{d\varphi}{dz}\right)_a \left(\frac{dz}{dt}\right)_0 = \left\{\frac{1-|a|^2}{(1-\bar{a}z)^2}\right\}_{z=a} u$$

so by (1.1), (1.2)

$$g_a(u,u) = c\frac{1}{(1-|a|^2)^2}|u|^2\,.$$

Thus g is the Riemannian structure

$$ds^2 = c\frac{dx^2 + dy^2}{(1 - x^2 - y^2)^2} \tag{1.3}$$

and the proof shows that it is indeed invariant.

We shall now take D as the unit disk $|z| < 1$ with $g = ds^2$ given by (1.3) with $c = 1$. In our analysis on D we are mainly interested in the *geodesics* in D (the arcs orthogonal to the boundary $B = \{z \in \mathbf{C} : |z| = 1\}$) and the *horocycles* in D which are the circles inside D tangential to B. Note that a horocycle tangential to B at b is orthogonal to all the geodesics in D which end at b.

2. The Non-Euclidean Fourier Transform

We first recall some of the principal results from Fourier analysis on $\mathbf{R}^n$. The Fourier transform $f \to \tilde{f}$ or $\mathbf{R}^n$ is defined by

$$\tilde{f}(u) = \int_{\mathbf{R}^n} f(x)e^{-i(x,u)}\,dx \tag{2.1}$$

where $(\ ,\ )$ denotes the scalar product and dx the Lebesgue measure. In polar coordinates $u = \lambda w$ $\lambda \in \mathbf{R}$, $w \in \mathbf{S}^{n-1}$ we can write

$$\tilde{f}(\lambda w) = \int_{\mathbf{R}^n} f(x)e^{-i\lambda(x,w)}\,dx\,. \tag{2.2}$$

It is then inverted by

$$f(x) = (2\pi)^{-n}\int_{\mathbf{R}^+\times\mathbf{S}^{n-1}} \tilde{f}(\lambda w)e^{i\lambda(x,w)}\lambda^{n-1}\,d\lambda\,dw \tag{2.3}$$

say for $f \in \mathcal{D}(\mathbf{R}^n) = C_c^\infty(\mathbf{R}^n)$, dw denoting the surface element on the sphere $\mathbf{S}^{n-1}$. The Plancherel formula

$$\int_{\mathbf{R}^n} |f(x)|^2\,dx = (2\pi)^{-n}\int_{\mathbf{R}^+\times\mathbf{S}^{n-1}} |\tilde{f}(\lambda, w)|^2\lambda^{n-1}\,d\lambda\,dw \tag{2.4}$$

expresses that $f \to \tilde{f}$ is an isometry of $L^2(\mathbf{R}^2)$ onto $L^2(\mathbf{R}^+ \times \mathbf{S}^{n-1}, (2\pi)^{-n}\lambda^{n-1}\,d\lambda\,dw)$.

The range of the mapping $f(x) \to \tilde{f}(\lambda w)$ as f runs through $\mathcal{D}(\mathbf{R}^n)$ is expressed in the following theorem [He7]. A vector $a = (a_1, \ldots, a_n) \in \mathbf{C}^n$ is said to be *isotropic* if $(a, a) = a_1^2 + \cdots + a_n^2 = 0$.

Theorem 2.1. *The Fourier transform $f(x) \to \tilde{f}(\lambda w)$ maps $\mathcal{D}(\mathbf{R}^n)$ onto the set of functions $\tilde{f}(\lambda w) = \varphi(\lambda, w) \in C^\infty(\mathbf{R} \times \mathbf{S}^{n-1})$ satisfying:*

(i) *There exists a constant $A > 0$ such that for each $w \in \mathbf{S}^{n-1}$ the function $\lambda \to \varphi(\lambda, w)$ extends to a holomorphic function on* $\mathbf{C}$ *with the property*

$$\sup_{\lambda \in \mathbf{C},\, w \in \mathbf{S}^{n-1}} |\varphi(\lambda, w)|(1 + |\lambda|)^N e^{-A|\operatorname{Im} \lambda|} < \infty \tag{2.5}$$

for each $N \in \mathbf{Z}$. ($\operatorname{Im} \lambda =$ imaginary part of λ).

(ii) *For each $k \in \mathbf{Z}^+$ and each isotropic vector $a \in \mathbf{C}^n$ the function*

$$\lambda \to \lambda^{-k} \int_{\mathbf{S}^{n-1}} \varphi(\lambda, w)(a, w)^k \, dw \tag{2.6}$$

is even and holomorphic in $\mathbf{C}$.

Condition (2.5) expresses that the function $\lambda \to \varphi(\lambda, w)$ is of *uniform exponential type*: The classical Paley–Wiener theorem states that $\mathcal{D}(\mathbf{R}^n)^\sim$ consists of entire functions of exponential type in n variables whereas in the description above only λ enters.

Formula (2.2) motivates a Fourier transform definition on D. The inner product (x, ω) equals the (signed) distance from 0 to the hyperplane through x with normal ω. A horocycle in D through b is perpendicular to the (parallel) family of geodesics ending at b so is an analog of a hyperplane in $\mathbf{R}^n$. Thus if $z \in D$, $b \in B$ we define $\langle z, b\rangle$ as the (signed) distance from 0 to the horocycle through z and b. The Fourier transform $f \to \tilde{f}$ on D is thus defined by

$$\tilde{f}(\lambda, b) = \int_D f(z) e^{(-i\lambda+1)\langle z, b\rangle} \, dz \tag{2.7}$$

for all $b \in B$ and $\lambda \in \mathbf{C}$ for which the integral converges. Here dz is the invariant surface element on D,

$$dz = (1 - x^2 - y^2)^{-2} \, dx \, dy \,. \tag{2.8}$$

The $+1$ term in (2.7) is included for later technical convenience.

The Fourier transform (2.7) is a special case of the Fourier transform on a symmetric space $X = G/K$ of the non-compact type, introduced in [He3]. Here G is a semisimple connected Lie group with finite center and K is a maximal compact subgroup. In discussing the properties of $f \to \tilde{f}$ below we stick to the case $X = D$ for notational simplicity but shall indicate (with references) the appropriate generalizations to arbitrary X. Some of the results require a rank restriction on X.

Theorem 2.2. *The transform $f \to \tilde{f}$ in (2.7) is inverted by*

$$f(z) = \frac{1}{4\pi} \int_{\mathbf{R}} \int_B \tilde{f}(\lambda, b) e^{(i\lambda+1)\langle z,b\rangle} \lambda \, \mathrm{th}\left(\frac{\pi\lambda}{2}\right) d\lambda \, db, \quad f \in \mathcal{D}(D). \tag{2.9}$$

Also the map $f \to \tilde{f}$ extends to an isometry of $L^2(D)$ onto $L^2(\mathbf{R}^+ \times B, \mu)$ where μ is the measure

$$\mu = \frac{\lambda}{2\pi} \, \mathrm{th}\left(\frac{\pi\lambda}{2}\right) d\lambda \, db \tag{2.10}$$

and db is normalized by $\int db = 1$.

This result is valid for arbitrary $X = G/K$ ([He3, He4]), suitably formulated in terms of the fine structure of G. While this result resembles (2.3)–(2.4) closely the range theorem for D takes a rather different form.

Theorem 2.3. *The range $\mathcal{D}(D)^\sim$ consists of the functions $\varphi(\lambda, b)$ which (in λ) are holomorphic of uniform exponential type and satisfy the functional equation*

$$\int_B \varphi(\lambda, b) e^{(i\lambda+1\langle z,b\rangle)} \, db = \int_B \varphi(-\lambda, b) e^{(-i\lambda+1)\langle z,b\rangle} \, db. \tag{2.11}$$

One can prove that condition (2.11) is equivalent to the following conditions (2.12) for the Fourier coefficients $\varphi_k(\lambda)$ of φ,

$$\begin{aligned} \varphi_k(\lambda) &= \frac{1}{2\pi} \int_0^{2\pi} \varphi(\lambda, e^{i\theta}) e^{-ik\theta} \, d\theta, \\ \varphi_k(-\lambda) p_k(-i\lambda) &= \varphi_k(\lambda) p_k(i\lambda), \quad k \in \mathbf{Z}, \end{aligned} \tag{2.12}$$

where $p_k(x)$ is the polynomial

$$p_k(x) = \frac{\Gamma(\frac{1}{2}(x+1) + |k|)}{\Gamma(\frac{1}{2}(x+1))}.$$

Again these results are valid for arbitrary $X = G/K$ ([He5, He7]).

The Paley–Wiener type theorems can be extended to the Schwartz spaces $\mathcal{S}^p(D)$ $(0 < p \leq 2)$. Roughly speaking, f belongs to $\mathcal{S}^p(D)$ if each invariant derivative Df belongs to $L^p(D)$, more precisely, it is rapidly decreasing in the distance from 0 even after multiplication by the pth root of the volume element. Let S_p denote the strip $|\operatorname{Im} \lambda| < \frac{2}{p} - 1$ in $\mathbf{C}$ and $\mathcal{S}(S_p \times B)$ the space of smooth functions on $\mathcal{S}(S_p \times B)$ holomorphic (in λ) in S_p and rapidly decreasing (uniformly for $b \in B$) on each line $\lambda = \xi + i\eta$ $(|\eta| < \frac{2}{p} - 1)$. Then we have

Theorem 2.4. *The Fourier transform $f \to \tilde{f}$ on D is a bijection of $\mathcal{S}^p(D)$ onto the set of $\varphi \in \mathcal{S}(S_p \times B)$ satisfying (2.11).*

The theorem holds for all $X = G/K$ (Eguchi [Eg]). The proof is complicated. For the case of K-invariant functions (done for $p = 2$ by Harish-Chandra [H] and Trombi – Varadarajan [TV] for general p) a substantial simplification was done by Anker [A]. A further range theorem for the space of functions for which each invariant derivative has arbitrary exponential decay was proved by Oshima, Saburi and Wakayama [OSW]. See also Barker [Bar, p. 27] for the operator Fourier transform of the intersection of all the Schwartz spaces on $\mathbf{SL}(2, \mathbf{R})$.

In classical Fourier analysis on $\mathbf{R}^n$ the Riemann–Lebesgue lemma states that for $f \in L^1(\mathbf{R})$, $\tilde{f}$ tends to 0 at ∞. For D the situation is a bit different.

Theorem 2.5. *Let $f \in L^1(D)$. Then there exists a null set N in B such that if $b \in B - N$, $\lambda \to \tilde{f}(\lambda, b)$ is holomorphic in the strip $|\operatorname{Im} \lambda| < 1$ and*

$$\lim_{\xi \to \infty} \tilde{f}(\xi + i\eta, b) = 0 \tag{2.13}$$

uniformly for $|\eta| \leq 1$.

The proof [HRSS] is valid even for symmetric spaces $X = G/K$ of arbitrary rank. Moreover

$$\|\tilde{f}(\lambda, \cdot)\|_1 \to 0 \text{ as } \lambda \to \infty, \tag{2.14}$$

uniformly in the strip $|\operatorname{Im} \lambda| \leq 1$, and this extends to $f \in L^p$ $(1 \leq p < 2)$ this time in the strip $|\operatorname{Im} \lambda| < \frac{2}{p} - 1$ ([SS]). In particular, if $f \in L^p(D)$ then there is a null set N in B such that $\tilde{f}(\lambda, b)$ exists for $b \notin N$ and all λ in the strip $|\operatorname{Im} \lambda| < \frac{2}{p} - 1$.

The classical inversion formula for the Fourier transform on $\mathbf{R}^n$ now extends to $f \in L^p(D)$ $(1 \leq p < 2)$ as follows.

Theorem 2.6. *Let $f \in L^p(D)$ and assume $\tilde{f} \in L^1(\mathbf{R} \times B, \mu)$ (with μ as in (2.10)). Then the inversion formula (2.9) holds for almost all $z \in D$ (the Lebesgue set for f).*

Again this holds for all $X = G/K$. A result of this type was proved by Stanton and Thomas [ST] without invoking $\tilde{f}$ explicitly (since the existence had not been established). The version in Theorem 2.6 is from [SS].

In Schwartz's theory of mean-periodic functions [Sc] it is proved that any closed translation-invariant subspace of $\mathcal{C}^\infty(\mathbf{R})$ contains an exponential $e^{\mu x}$. The analogous question here would be:

Does an arbitrary closed invariant subspace of $\mathcal{C}^\infty(D)$ contain an exponential

$$e_{\mu,b}(z) = e^{\mu\langle z,b\rangle} \tag{2.15}$$

for some $\mu \in \mathbf{C}$ and some $b \in B$?

Here the topology of $\mathcal{C}^\infty(D)$ is the usual Frèchet space topology and "invariant" refers to the action of the group $G = \mathbf{SU}(1,1)$ on D. The answer is yes.

Theorem 2.7. *Each closed invariant subspace E of $\mathcal{C}^\infty(D)$ contains an exponential $e_{\mu,b}$.*

This was proved in [HS] for all symmetric $X = G/K$ of rank 1. Here is a sketch of the proof. By a result of Bagchi and Sitaram [BS], E contains a spherical function

$$\varphi_\lambda(z) = \int_B e^{(i\lambda+1)\langle z,b\rangle}\, db\,, \quad \varphi_\lambda = \varphi_{-\lambda}\,. \tag{2.16}$$

For either λ or $-\lambda$ it is true ([He9, Lemma 2.3, Ch. III]) that the Poisson transform $P_\lambda : F \to f$ where

$$f(z) = \int_B e^{(i\lambda+1)\langle z,b\rangle} F(b)\, db\,, \tag{2.17}$$

maps $L^2(B)$ into the closed invariant subspace of E generated by φ_λ. On the other hand it is proved in [He9, Ex. B1 in Ch. III] that $e_{i\lambda+1,b}$ is a series of terms $P_\lambda(F_n)$ where $F_n \in L^2(B)$ and the series converges in the topology of $\mathcal{C}^\infty(D)$. Thus $e_{i\lambda+1,b} \in E$ as desired.

The following result for the Fourier transform on $\mathbf{R}^n$ is closely related to the Wiener Tauberian theorem.

Let $f \in L^1(\mathbf{R}^n)$ be such that $\tilde{f}(u) \neq 0$ for all $u \in \mathbf{R}^n$. Then the translates of f span a dense subspace of $L^1(\mathbf{R}^n)$.

There has been considerable activity in establishing analogs of this theorem for semisimple Lie groups and symmetric spaces. See e.g. [EM, Sa, Si1, Si2]. The neatest version for D seems to me to be the following result from [SS, MRSS] which remains valid for $X = G/K$ of rank 1.

Let $d(z,w)$ denote the distance in D and if $\epsilon > 0$, let $\mathcal{L}_\epsilon(D)$ denote the space of measurable functions f on D such that $\int_D |f(z)| e^{\epsilon d(0,z)}\, dz < \infty$. Let T_ϵ denote the strip $|\operatorname{Im}\lambda| \leq 1+\epsilon$.

Theorem 2.8. *Let $f \in \mathcal{L}_\epsilon(X)$ and assume f is not almost everywhere equal to any real analytic function. Let*

$$Z = \{\lambda \in T_\epsilon : \tilde{f}(\lambda,\cdot) \equiv 0\}\,.$$

If $Z = \emptyset$ then the translates of f span a dense subspace of $L^1(D)$.

A theorem of Hardy's on Fourier transforms on $\mathbf{R}^n$ asserts in a precise fashion that f and its Fourier transform cannot both vanish too fast at infinity. More precisely ([Ha]):

Assume

$$|f(x)| \leq Ae^{-\alpha|x|^2}\,,\; |\tilde{f}(u)| \leq Be^{-\beta|u|^2}\,,$$

where A, B, α and β are positive constants and $\alpha\beta > \frac{1}{4}$. Then $f \equiv 0$.

Variations of this theorem for L^p spaces have been proved by Morgan [M] and Cowling–Price [CP].

For the Fourier transform on D the following result holds.

Theorem 2.9. *Let f be a measurable function on D satisfying*

$$|f(x)| \leq Ce^{-\alpha d(0,x)^2} \quad |\tilde{f}(\lambda, b)| \leq Ce^{-\beta|\lambda|^2}$$

where C, α, β are positive constants. If $\alpha\beta > 16$ then $f \equiv 0$.

This is contained in Sitaram and Sundari [SiSu, § 5] where an extension to certain symmetric spaces $X = G/K$ is also proved. The theorem for all such X was obtained by Sengupta [Se], together with refinements in terms of $L^p(X)$.

Many such completions of Hardy's theorem have been given, see [RS, CSS, NR, Shi].

3. Eigenfunctions of the Laplacian

Consider first the plane $\mathbf{R}^2$ and the Laplacian

$$L^0 = \frac{\partial^2}{\partial x_1^2} + \frac{\partial^2}{\partial x_2^2}\,.$$

Given a unit vector $\omega \in \mathbf{R}^2$ and $\lambda \in \mathbf{C}$ the function $x \to e^{i\lambda(x,\omega)}$ is an eigenfunction

$$L_x^0 e^{i\lambda(x,\omega)} = -\lambda^2 e^{i\lambda(x,\omega)}\,. \tag{3.1}$$

Because of (2.3) one might expect all eigenfunctions of L to be a "decomposition" into such eigenfunctions with fixed λ but variable ω.

Note that the function $\omega \to e^{i\lambda(x,\omega)}$ is the restriction to $\mathbf{S}^1$ of the holomorphic function

$$z \to \exp\left[\tfrac{1}{2}(i\lambda)x_1(z + z^{-1}) + \tfrac{1}{2}\lambda x_2(z - z^{-1})\right] \quad z \in \mathbf{C} - (0)\,,$$

which satisfies a condition

$$\sup_z \left(|f(z)|e^{-a|z|-b|z|^{-1}}\right) < \infty\,, \tag{3.2}$$

with $a, b \geq 0$. Let $E_{a,b}$ denote the Banach space of holomorphic functions satisfying (3.2), the norm being the expression in (3.2). We let E denote the union of the spaces $E_{a,b}$ and give it the induced topology. We identify the elements of E with their restrictions to $\mathbf{S}^1$ and call the members of the dual space E' *entire functionals.*

Theorem 3.1. *([He6]) The eigenfunctions of L^0 on $\mathbf{R}^2$ are precisely the harmonic functions and the functions*

$$f(x) = \int_{\mathbf{S}^1} e^{i\lambda(x,\omega)}\, dT(\omega) \tag{3.3}$$

where $\lambda \in \mathbf{C} - (0)$ and T is an entire functional on $\mathbf{S}^1$.

For the non-Euclidean metric (1.3) (with $c = 1$) the Laplacian is given by

$$L = (1 - x^2 - y^2)^2 \left(\frac{\partial^2}{\partial x^2} + \frac{\partial^2}{\partial y^2} \right) \tag{3.4}$$

and the exponential function $e_{\mu,b}(z) = e^{\mu\langle z,b\rangle}$ is an eigenfunction:

$$L_z(e^{(i\lambda+1)\langle z,b\rangle}) = -(\lambda^2 + 1)e^{(i\lambda+1)\langle z,b\rangle}\,. \tag{3.5}$$

In particular, the function $z \to e^{2\langle z,b\rangle}$ is a harmonic function and in fact coincides with the classical Poisson kernel from potential theory:

$$e^{2\langle z,b\rangle} = \frac{1 - |z|^2}{|z - b|^2}\,. \tag{3.6}$$

Again the eigenfunctions of L are obtained from the functions $e_{\mu,b}$ by superposition. To describe this precisely consider the space $\mathcal{A}(B)$ of analytic functions on B. Each $F \in \mathcal{A}(B)$ extends to a holomorphic function on a belt $B_\epsilon : 1 - \epsilon < |z| < 1 + \epsilon$ around B. The space $\mathcal{H}(B_\epsilon)$ of holomorphic functions on B_ϵ is topologized by uniform convergence on compact subsets. We can view $\mathcal{A}(B)$ as the union $\cup_n^\infty \mathcal{H}(B_{1/n})$ and give it the inductive limit topology. The dual space $\mathcal{A}'(B)$ then consists of the *analytic functionals* on B (or hyperfunctions in B).

Theorem 3.2. *([He4, IV, §1]) The eigenfunctions of L are precisely the functions*

$$u(z) = \int_B e^{\mu\langle z,b\rangle}\, dT(b)\,, \tag{3.7}$$

where $\mu \in \mathbf{C}$ and $T \in \mathcal{A}'(B)$.

Lewis in [L] has proved (under minor restriction on μ) that T in (3.7) is a distribution if and only if u has at most an exponential growth (in

$d(0,z)$). On the other hand, Ban and Schlichtkrull proved in [BaS] that $T \in \mathcal{C}^\infty(B)$ if and only if all the invariant derivatives of u have the same exponential growth.

We consider now the natural group representations on the eigenspaces. The group $\mathbf{M}(2)$ of isometries of $\mathbf{R}^2$ acts transitively on $\mathbf{R}^2$ and leaves the Laplacian L^0 invariant: $L^0(f \circ \tau) = (L^0 f) \circ \tau$ for each $\tau \in \mathbf{M}(2)$. If $\lambda \in \mathbf{C}$ the eigenspace

$$\mathcal{E}_\lambda = \{f \in \mathcal{C}^\infty(\mathbf{R}^2) : L^0 f = -\lambda^2 f\}$$

is invariant under the action $f \to f \circ \tau^{-1}$ so we have a representation T_λ of $\mathbf{M}(2)$ on $\mathcal{E}_\lambda$ given by $T_\lambda(\tau)(f) = f \circ \tau^{-1}$, the *eigenspace representation.*

Theorem 3.3. *([He6]) The representation T_λ is irreducible if and only if $\lambda \neq 0$.*

Similarly the group $G = \mathbf{SU}(1,1)$ of conformal transformations

$$z \to \frac{az+b}{\bar{b}z+\bar{a}} \quad (|a|^2 - |b|^2 = 1)$$

leaves (1.3) and the operator L in (3.4) invariant. Thus we get again an eigenspace representation τ_λ of G on each eigenspace

$$\mathcal{E}_\lambda = \{f \in \mathcal{C}^\infty(D) : Lf = -(\lambda^2+1)f\}\,.$$

Theorem 3.4. *([He4]) The representation τ_λ is irreducible if and only if $i\lambda + 1 \notin 2\mathbf{Z}$.*

Again all these results extend to Euclidean spaces of higher dimensions and suitably formulated, to all symmetric spaces G/K of the noncompact type.

4. The Radon Transform

A. The Euclidean Case

Let d be a fixed integer, $0 < d < n$ and let $\mathbf{G}(d,n)$ denote the space of d-dimensional planes in $\mathbf{R}^n$. To a function f on $\mathbf{R}^n$ we associate a function $\hat{f}$ on $\mathbf{G}(d,n)$ by

$$\hat{f}(\xi) = \int_\xi f(x)\,dm(x)\,, \quad \xi \in \mathbf{G}(d,n)\,, \tag{4.1}$$

dm being the Euclidean measure on ξ. The transform $f \to \hat{f}$ is called the *d-plane transform.* For $d = 1$, $n = 2$ it is the classical Radon transform. The parity of d turns out to be important.

The inversion of the transform $f \to \hat{f}$ is well known (case $d = n-1$ in [R, J, GS], general d in [F, He1, He2]). We shall give another group-theoretic method here, resulting in alternative inversion formulas.

The group $G = \mathbf{M}(n)$ acts transitively both on $\mathbf{R}^n$ and on $\mathbf{G}(d,n)$. In particular, $\mathbf{R}^n = G/K$ where $K = \mathbf{O}(n)$. Let $p > 0$. Consider a pair $x \in \mathbf{R}^n$, $\xi \in \mathbf{G}(d,n)$ at distance $p = d(x,\xi)$. Let $g \in G$ be such that $g \cdot 0 = x$. Then the family $kg^{-1} \cdot \xi$ constitutes the set of elements in $\mathbf{G}(d,n)$ at distance p from 0. Along with the transform $f \to \hat{f}$ we consider the *dual transform* $\varphi \to \overset{\vee}{\varphi}$ given by

$$\overset{\vee}{\varphi}(x) = \int_{\xi \ni x} \varphi(\xi)\, d\mu(\xi)\,, \tag{4.2}$$

the average of φ over the set of d-planes passing through x. More generally we put

$$\overset{\vee}{\varphi}_p(x) = \int_{d(\xi,x)=p} \varphi(\xi)\, d\mu(\xi) \tag{4.3}$$

the average of φ over the set of d-planes at distance p from x. Since K acts transitively on the set of d-planes through 0 we see by the above that

$$\overset{\vee}{\varphi}_p(g \cdot 0) = \int_K \varphi(gkg^{-1} \cdot \xi)\, dk\,, \tag{4.4}$$

dk being the normalized Haar measure on K. Let $(M^r f)(x)$ denote the mean-value of f over the sphere $S_r(x)$ of radius r with center x. If $z \in \mathbf{R}^n$ has distance r from 0 we then have

$$(M^r f)(g \cdot 0) = \int_K f(gk \cdot z)\, dk\,. \tag{4.5}$$

We thus see that since $d(0, g^{-1} \cdot y) = d(x,y)$,

$$\begin{aligned} (\hat{f})^\vee_p(x) &= \int_K \hat{f}(gkg^{-1} \cdot \xi)\, dk = \int_\xi dk \int_\xi f(gkg^{-1} \cdot y)\, dm(y) \\ &= \int_\xi dm(y) \int_K f(gkg^{-1} \cdot y)\, dk = \int_\xi (M^{d(x,y)} f)(x)\, dm(y)\,. \end{aligned}$$

Let x_0 be the point in ξ at minimum distance p from x. The integrand $(M^{d(x,y)} f(x))$ is constant in y on each sphere in ξ with center x_0. It follows that

$$(\hat{f})^\vee_p(x) = \Omega_d \int_0^\infty (M^q f)(x) r^{d-1}\, dr\,, \tag{4.6}$$

where $r = d(x_0, y)$, $q = d(x,y)$, Ω_d denoting the area of the unit sphere in $\mathbf{R}^d$. We have $q^2 = p^2 + r^2$ so putting $F(q) = (M^q f)(x)$, $\hat{F}(p) = (\hat{f})^\vee_p(x)$

we have

$$\hat{F}(p) = \Omega_d \int_p^\infty F(q)(q^2 - p^2)^{d/2-1} q\, dq\,. \tag{4.7}$$

This Abel-type integral equation has an inversion

$$F(r) = c(d)\left(\frac{d}{d(r^2)}\right)^d \int_r^\infty p(p^2 - r^2)^{d/2-1}\hat{F}(p)\, dp\,, \tag{4.8}$$

where $c(d)$ is a constant, depending only on d. Putting $r = 0$ we obtain the inversion formula

$$f(x) = c(d)\left[\left(\frac{d}{d(r^2)}\right)^d \int_r^\infty p(p^2 - r^2)^{d/2-1}(\hat{f})_p^\vee(x)\, dp\right]_{r=0}. \tag{4.9}$$

Note that in (4.8)

$$p(p^2 - r^2)^{d/2-1} = \frac{d}{dp}(p^2 - r^2)^{(d/2)} \cdot \frac{1}{d}$$

so in (4.8) we can use integration by parts and the integral becomes

$$C\int_r^\infty (p^2 - r^2)^{d/2}\hat{F}'(p)\, dp\,.$$

Applying $\dfrac{d}{d(r^2)} = \dfrac{1}{2r}\dfrac{d}{dr}$ to this integral reduces the exponent $d/2$ by 1. For d odd we continue the differentiation $\frac{d+1}{2}$ times until the exponent is $-\frac{1}{2}$. For d even we continue until the exponent is 0 and then replace $\int_r^\infty \hat{F}'(p)\, dp$ by $-\hat{F}(r)$. This $\hat{F}(r)$ is an even function so taking $(d/d(r^2))^{d/2}$ at $r = 0$ amounts to taking a constant multiple of $(d/dr)^d$ at $r = 0$. We thus get the following refinement of (4.9) where we recall that $(\hat{f})_r^\vee(x)$ is the average of the integrals of f over the d-planes tangent to $S_r(x)$.

Theorem 4.1. *The d-plane transform is inverted as follows:*

(i) If d is even then

$$f(x) = C_1\left[\left(\frac{d}{dr}\right)^d (\hat{f})_r^\vee(x)\right]_{r=0}. \tag{4.10}$$

(ii) If d is odd then

$$f(x) = C_2\left[\left(\frac{d}{d(r^2)}\right)^{(d-1)/2} \int_r^\infty (p^2 - r^2)^{-1/2}\frac{d}{dp}(\hat{f})_p^\vee(x)\, dp\right]_{r=0}. \tag{4.11}$$

(iii) If $d = 1$ then

$$f(x) = -\frac{1}{\pi}\int_0^\infty \frac{1}{p}\,\frac{d}{dp}(\hat{f})_p^\vee(x)\,dp\,. \tag{4.12}$$

For $n = 2$ formula (4.12) is proved in Radon's original paper [R]. Note that the constant $-1/\pi$ is the same for all n. In the case $d = n-1$ the formula in coincides with formula (21) in Rouvière [Ro].

Another inversion formula ([He1, He2]) valid for all d and n is

$$f \;=\; c(-L)^{d/2}((\hat{f})^\vee) \tag{4.13}$$

where

$$c \;=\; \frac{\Gamma\left(\frac{n-d}{2}\right)}{(4\pi)^{d/2}\Gamma\left(\frac{n}{2}\right)}\,.$$

Here the fractional power of L is defined in the usual way by the Fourier transform. The parity of d shows up in the same way as in Theorem 4.1.

For range questions for the transform $f \to \hat{f}$ see an account in [He10] and references there.

B. The Hyperbolic Case

The hyperbolic space $\mathbf{H}^n$ is the higher-dimensional version of (1.3) and its Riemannian structure is given by

$$ds^2 = 4\frac{dx_1^2 + \cdots + dx_n^2}{(1 - x_1^2 - \cdots - x_n^2)^2} \tag{4.14}$$

in the unit ball $|x| < 1$. The constant 4 is chosen such that the curvature is now -1. The d-dimensional totally geodesic submanifolds are spherical caps perpendicular to the boundary $B : |x| = 1$. They are natural analogs of the d-planes in $\mathbf{R}^n$. We have accordingly a Radon transform $f \to \hat{f}$, where

$$\hat{f}(\xi) = \int_\xi f(x)\,dm(x) \quad \xi \in \Xi\,, \tag{4.15}$$

where Ξ is the space of d-dimensional totally geodesic submanifolds of $\mathbf{H}^n$.

The group G of isometries of $\mathbf{H}^n$ acts transitively on Ξ as well. As in (4.2)–(4.3) we consider the dual transform

$$\check{\varphi}(x) = \int_{\xi \ni x} \varphi(\xi)\,d\mu(\xi) \tag{4.16}$$

and more generally for $p \geq 0$,

$$\overset{\vee}{\varphi}_p(x) = \int_{d(\xi,x)=p} \varphi(\xi)\, d\mu(\xi)\,, \tag{4.17}$$

the mean value of φ over the set of $\xi \in \Xi$ at distance p from x. The formula

$$(\hat{f})^\vee_p(x) = \int_\xi (M^{d(x,y)} f)(x)\, dm(y) \tag{4.18}$$

is proved just as before. Let x_0 be the point in ξ at minimum distance p from x and put $r = d(x_0, y)$, $q = d(x, y)$. Since the geodesic triangle (xx_0y) is right angled at x_0 we have by the cosine rule

$$\cosh q = \cosh p \cosh r\,. \tag{4.19}$$

Also note that since ξ is totally geodesic, distances between two points in ξ are the same as in $\mathbf{H}^n$. In particular $(M^{d(x,y)} f)(x)$ is constant as y varies on a sphere in ξ with center x_0. Therefore (4.18) implies

$$(\hat{f})^\vee_p(x) = \Omega_d \int_0^\infty (M^q f)(x) \sinh^{d-1} r\, dr\,. \tag{4.20}$$

For x fixed we put

$$F(\cosh q) = (M^q f)(x)\,, \quad \hat{F}(\cosh p) = (\hat{f})^\vee_p(x)\,,$$

substitute in (4.20) and use (4.19). Writing $t = \cosh p$, $s = \cosh r$ we obtain the integral equation

$$\hat{F}(t) = \Omega_d \int_1^\infty F(ts)(s^2-1)^{d/2-1}\, ds\,. \tag{4.21}$$

Putting here $u = ts$, $ds = t^{-1}\, du$ we get the Abel–type integral equation

$$t^{d-1}\hat{F}(t) = \Omega_d \int_t^\infty u^{-1} F(u)(u^2-t^2)^{d/2-1} u\, du\,,$$

which by (4.8) is inverted by

$$r^{-1}F(r) = c(d)\left(\frac{d}{d(r^2)}\right)^d \int_r^\infty t(t^2-r^2)^{d/2-1} t^{d-1}\hat{F}(t)\, dt\,. \tag{4.22}$$

Here we put $r = 1$ and $s(p) = \cosh^{-1} p$. We then obtain the following variation of Theorem 3.12, Ch. I in [He9]:

Theorem 4.2. *The transform $f \to \tilde{f}$ is inverted by*

$$f(x) = C\left[\left(\frac{d}{d(r^2)}\right)^d \int_r^\infty (t^2 - r^2)^{d/2-1} t^d (\hat{f})^\vee_{s(t)}(x)\, dt\right]_{r=1}. \tag{4.23}$$

As in the proof of Theorem 4.1 we can obtain the following improvement.

Theorem 4.3.

(i) If d is even the inversion can be written

$$f(x) = C\left[\left(\frac{d}{d(r^2)}\right)^{d/2} (r^{d-1}(\hat{f})^\vee_{s(r)}(x))\right]_{r=1}. \tag{4.24}$$

(ii) If $d = 1$ then

$$f(x) = -\tfrac{1}{\pi}\int_0^\infty \frac{1}{\sinh p}\frac{d}{dp}\left((\hat{f})^\vee_p(x)\right)\, dp. \tag{4.25}$$

Proof. Part (i) is proved as (4.10) except that we no longer can equate $(d/d(r^2))^{d/2}$ with $(d/dr)^d$ at $r = 1$.

For (ii) we deduce from (4.22) since $t(t^2 - r^2)^{-1/2} = \frac{d}{dt}(t^2 - r^2)^{1/2}$ that

$$\begin{aligned} F(1) &= -\frac{c(1)}{2}\left[\frac{d}{dr}\int_r^\infty (t^2 - r^2)^{1/2}\frac{d}{dt}\hat{F}(t)\, dt\right]_{r=1} \\ &= \frac{c(1)}{2}\int_1^\infty (t^2 - 1)^{-1/2}\frac{d}{dt}\hat{F}(t)\, dt. \end{aligned}$$

Putting again $t = \cosh p$, $dt = \sinh p\, dp$ our expression becomes

$$\frac{c(1)}{2}\int_0^\infty \frac{1}{\sinh p}\frac{d}{dp}((\hat{f})^\vee_p)(x)\, dp.$$

□

Remark. For $n = 2$, $d = 1$ formula (4.25) is stated in Radon [R, Part C]. The proof (which is only indicated) is very elegant but would not work for $n > 2$.

For d even (4.24) can be written in a simpler form ([He1]) namely

$$f = c\, Q_d(L)((\hat{f})^\vee), \tag{4.26}$$

where $c = \frac{\Gamma\left(\frac{n-d}{2}\right)}{(-4\pi)^{d/2}\Gamma\left(\frac{n}{2}\right)}$ and Q_d is the polynomial

$$Q_d(x) = (x + (d-1)(n-d))(x + (d-3)(n-d+2)) \cdots (x + 1 \cdot (n-2)).$$

The case $d = 1$, $n = 2$ is that of the X-ray transform on the non-Euclidean disk ((4.15) for $n = 2$). Here are two further alternatives to the inversion formula (4.25). Let S denote the integral operator

$$(Sf)(x) = \int_D (\coth d(x,y) - 1) f(y)\, dy. \tag{4.27}$$

Then

$$LS(\hat{f})^\vee = -4\pi^2 f, \quad f \in \mathcal{D}(X). \tag{4.28}$$

This is proved by Berenstein–Casadio [BC]; see [He10] for a minor simplification. By invariance it suffices to prove (4.28) for f radial and then it is verified by taking the spherical transform on both sides. Less explicit versions of (4.28) are obtained in [BC] for any dimension n and d.

One more inversion formula was obtained by Lissianoi and Ponomarev [LP] using (4.23) for $d = 1$, $n = 2$ as a starting point. By parameterizing the geodesics γ by the two points of intersection of γ with B they prove a hyperbolic analog of the Euclidean formula:

$$f(x) = \int_{\mathbf{S}^1} \left\{ \mathcal{H}_p \frac{d}{dp} \hat{f}(\omega, p) \right\}_{p=(\omega,x)} d\omega, \tag{4.29}$$

which is an alternative to (4.12). Here $\mathcal{H}_p$ is a normalized Hilbert transform in the variable p and $\hat{f}(\omega, p)$ is $\hat{f}(\xi)$ for the line $(x, \omega) = p$, where $|\omega| = 1$.

In the theorems in this section we have not discussed smoothness and decay at infinity of the functions. Here we refer to [Je, Ru1, Ru2, BeR1, BeR2] as examples.

Additional inversion formulas for the transform $f \to \hat{f}$ can be found in [Sem, Ru3, K]. The range problem for the transform $f \to \hat{f}$ is treated in [BCK, I].

Bibliography

[A] J. Anker, The spherical Fourier transform of rapidly decreasing functions– a simple proof of a characterization due to Harish–Chandra, Helgason, Trombi and Varadarajan. J. Funct. Anal. 96 (1991), 331–349.

[BS] S.C. Bagchi and A. Sitaram, Spherical mean–periodic functions on semisimple Lie groups. Pac. J. Math. 84 (1979), 241–250.

[BaS] Ban, van den, E.P. and H. Schlichtkrull, Asymptotic expansions and boundary values of eigenfunctions on a Riemannian symmetric space. J. Reine Angew. Math. 380 (1987), 108–165.

[Bar] W.H. Barker, L^p harmonic analysis on $SL(2,\mathbf{R})$. Memoir of Amer. Math. Soc. 393 Providence, R.I. 1988.

[Be] E. Beltrami, Saggio di interpretazione della geometria non euclidea. Giornale di Matematica, 1868.

[BeR1] C.A. Berenstein and B. Rubin, Radon transform of L^p-functions on the Lobachevsky space and hyperbolic wavelet transforms. Forum. Math. 11 (1999), 567–590.

[BeR2] C.A. Berenstein and B. Rubin, Totally geodesic Radon transform of L^p-functions on real hyperbolic space. (Preprint).

[BC] C.A. Berenstein and E. Casadio Tarabusi, Inversion formulas for the k-dimensional Radon transform in real hyperbolic spaces. Duke Math. J. 62 (1991), 613–631.

[BCK] C.A. Berenstein, A. Kurusa and E. Casadio Tarabusi, Radon transform on spaces of constant curvature. Proc. Amer. Math. Soc. 125 (1997), 455–461.

[Bo] J. Bolyai, The Science Absolute of Space, 1831.

[CP] M. Cowling and J. Price, Generalizations of Heisenberg's inequality. Lecture Notes No. 992, Springer–Verlag, 1983.

[CSS] M. Cowling, A. Sitaram and M. Sundari, Hardy's uncertainty principle on semisimple Lie groups. Pac. J. Math. 192 (2000), 293–296.

[Eg] M. Eguchi, Asymptotic Expansions of Eisenstein Integrals and Fourier Transform on Symmetric Spaces. J. Funct. Anal. 34 (1979), 167–216.

[EM] L. Ehrenpreis and F. Mautner, Some properties of the Fourier transform on semisimple Lie groups I. Ann. of Math. 61 (1955), 406–439.

[F] B. Fuglede, An integral formula. Math. Scand. 6 (1958), 207–212.

[GS] I.M. Gelfand and G.E. Shilov, *Generalized Functions*, I, Academic Press, New York, 1964.

[H] Harish-Chandra, Spherical functions on a semisimple Lie group II. Amer. J. Math. 80 (1958), 553–613.

[Ha] G.H. Hardy, A theorem concerning Fourier transforms. J. London Math. Soc. 8 (1933), 227–231.

[He1] S. Helgason, Differential operators on homogeneous spaces. Acta Math. 102 (1959), 239–299.

[He2] S. Helgason, The Radon transform on Euclidean spaces, two–point homogeneous spaces and Grassmann manifolds. Acta Math. 113 (1965), 153–180.

[He3] S. Helgason, Radon–Fourier transforms on symmetric spaces and related group representations. Bull. Amer. Math. Soc, 71 (1965), 757–763.

[He4] S. Helgason, A duality for symmetric spaces with applications to group representations. Advan. Math. 5 (1970), 1–154.

[He5] S. Helgason, The surjectivity of invariant differential operators on symmetric spaces. Ann. of Math. 98 (1973), 451–480.

[He6] S. Helgason, Eigenspaces of the Laplacian; integral representations and irreducibility. J. Functional Anal. 17 (1974), 328–353.

[He7] S. Helgason, A duality for symmetric spaces with applications to group representations II. Differential equations and eigenspace representations. Advan. Math. 22 (1976), 187–219.

[He8] S. Helgason, *Groups and Geometric Analysis.* Acad. Press, 1984, Amer. Math. Soc., 2000.

[He9] S. Helgason, *Geometric Analysis on Symmetric Spaces.* Math Surveys and Monographs No. 39, AMS, Providence, R.I. 1994.

[He10] S. Helgason, *The Radon Transform.* Birkhäuser, Boston, 1999.

[HRSS] S. Helgason, R. Rawat, J. Sengupta and A. Sitaram, Some remarks on the Fourier transform on a symmetric space. Tech. Report, Ind. Stat. Inst. Bangalore 1998.

[HS] S. Helgason and J. Sengupta, Preprint 1997.

[I] S. Ishikawa, The range characterization of the totally geodesic Radon transform on the real hyperbolic space. Duke Math. J. 90 (1997), 149–203.

[J] F. John, *Plane Waves and Spherical Means.* Wiley, New York, 1955.

[Je] S.R. Jensen, Sufficient conditions for the inversion formula for the k-plane transform in $\mathbf{R}^n$. (Preprint).

[K] A. Kurusa, The Radon transform on hyperbolic space. Geom. Dedicata 40 (1991), 325–339.

[L] J. Lewis, Eigenfunctions on symmetric spaces with distribution-valued boundary forms. J. Funct. Anal. 29 (1978), 287–307.

[LP] S. Lissianoi and I. Ponomarev, On the inversion of the geodesic Radon transform on the hyperbolic plane. Inverse Problems 13 (1997), 1053–1062.

[Lo] N. Lobatchevski, Geometrical Researches on the Theory of Parallels, Kasan, 1826.

[M] G.W. Morgan, A note on Fourier transforms. J. London Math. Soc. 9 (1934), 187–192.

[MRSS] P. Mohanty, S.K. Ray, R.P. Sarkar and A. Sitaram, Helgason Fourier Transform for Symmetric Spaces II, J. Lie Theory 14 (2004), 227–242.

[NR] E.K. Narayanan and S.K. Ray, L^p version of Hardy's theorem on semisimple Lie groups. Proc. Amer. Math. Soc. 130 (2002), 1859–1866.

[OSW] T. Oshima, Y. Saburi and M. Wakayama, Paley–Wiener theorems on a symmetric space and their application. Diff. Geom. and Appl. (1991), 247–278.

[R] J. Radon, Über die Bestimmung von Funktionen durch ihre Integralwerte längs gewisser Mannigfaltigkeiten. Ber. Verth. Sächs. Akad. Wiss. Leipzig. Math. Nat. kl. 69 (1917), 262–277.

[RS] S.K. Ray and R.P. Sarkar, Cowling–Price theorem and characterization of heat kernel of symmetric spaces. (Preprint).

[Ro] F. Rouvière, Inverting Radon transforms; the group-theoretic approach. Enseign. Math. 47 (2001), 205–252.

[Ru1] B. Rubin, Reconstruction of functions from their integrals over k-planes. (Preprint).

[Ru2] B. Rubin, Helgason–Marchand inversion formulas for Radon transforms. Proc. Amer. Math. Soc. 130 (2002), 3017–3023.

[Ru3] B. Rubin, Radon, Cosine and Sine transforms on real hyperbolic spaces. Advan. Math. 170 (2002), 206–233.

[SS] R. Sarkar and A. Sitaram, The Helgason Fourier Transform for Symmetric Spaces. A tribute to C.S. Seshadri. A collection of articles on geometry and representation theory. Basel: Birkhuser. Trends in Mathematics, (2003), 467-473.

[Sa] R.P. Sarkar, Wiener Tauberian theorem for rank one symmetric spaces. Pacific J. of Math. 186 (1998), 349–358.

[Sc] L. Schwartz, Théorie generale des fonctions moyenne–periodiques. Ann. of Math. 48 (1947), 857–929.

[Se] J. Sengupta, The Uncertainty Principle on Riemannian symmetric spaces of the noncompact type. Proc. Amer. Math. Soc. 128 (2000), 2493–2499.

[Sem] V.I. Semyanisty, Homogeneous functions and some problems of integral geometry in spaces of constant curvature. Soviet Math. Dokl. 2 (1961), 59–62.

[Shi] N. Shimeno, An analog of Hardy's theorem for the Harish–Chandra transform. Hiroshima Math. J. 31 (2001), 383–390.

[Si1] A. Sitaram, An analog of the Wiener Tauberian theorem for spherical transforms on semisimple Lie groups. Pac. J. Math. 89 (1980), 439–445.

[Si2] A. Sitaram, On an analog of the Wiener Tauberian theorem for symmetric spaces of the noncompact type. Pac. J. Math. 133 (1988), 197–208.

[SiSu] A. Sitaram and M. Sundari, An analog of Hardy's theorem for very rapidly decreasing functions on semisimple Lie groups. Pac. J. Math. 177 (1997), 187–200.

[ST] R.J. Stanton and P.A. Thomas, Pointwise inversion of the spherical transform on $L^p(G/K)$ $(1 \leq p < 2)$. Proc. Amer. Math. Soc. 73 (1979), 398–404.

[TV] P. Trombi and V.S. Varadarajan, Spherical transforms on a semisimple Lie group. Ann. of Math. 94 (1971), 246–303.

HOLONOMY, GEOMETRY AND TOPOLOGY OF MANIFOLDS WITH GRASSMANN STRUCTURE

Neda Bokan*
Faculty of Mathematics, University of Belgrade,
Studentski trg 16, PP 550, 11001 Belgrade, Serbia,
neda@matf.bg.ac.yu

Paola Matzeu
Dipartimento di Matematica,
Viale Merello 92, 09123 Cagliari, Italy
matzeu@vaxca1.unica.it

Zoran Rakić *
Faculty of Mathematics, University of Belgrade,
Studentski trg 16, PP 550, 11001 Belgrade, Serbia,
zrakic@matf.bg.ac.yu

Abstract The main goal is to present how geometry and algebra intertwine in studying some topics. Let $\mathcal{R}(G)$ be the vector bundle of all curvature tensors corresponding to torsion-free connections on a smooth manifold M whose holonomy group is G. Here we present an overview on $\mathcal{R}(G)$ from the representation theory point of view. We pay attention especially to manifolds endowed with Grassmann structure. Some examples of this type of manifolds are given as well as some topological obstructions for the existence of this structure. It is confirmed that many geometrical properties of manifolds with Grassmann structure are related to some symmetry properties of a curvature, which belongs to some simple G-modules of $\mathcal{R}(G)$. A manifold may be endowed with various types of connections. The most interesting are those whose curvatures belong either to some simple submodules or to some of their direct sums as they define some special geometry of man-

*Research partially supported by the Ministry of Science of Serbia, project MM1646.

ifolds with Grassmann structure. Among these types of connections we point out: half-flat Grassmann connections, connections corresponding to some normalization, projectively equivalent connections, connections with symmetric or skew-symmetric Ricci tensor, etc. Having in mind a complete decomposition of $\mathcal{R}(G)$ into simple submodules we can also reveal some new curvature invariants corresponding to some transformations of manifolds with Grassmann structure. Some relations between projective structures and corresponding reductions of a structure group are presented. We also give an overview of relations between projective geometry of manifolds with a special type of Grassmann structure and Riccati type equations. Various examples of previously mentioned facts are also provided.

Keywords: holonomy group, Grassmann manifold, Grassmann connection, normalization, torsion-free connection, action of a group, simple G-modules, Riccati equation.

1. Some relations between algebra and geometry

Let (M, D) be a smooth n-dimensional manifold endowed with a torsion-free connection (not necessary a Levi-Civita one, i.e. compatible with some metric). Let $V = T_m M$ be the tangent space at a point $m \in M$, e a basis at $T_m M$, γ a curve passing through a point m. We denote by e' the image of e obtained by parallel transport along γ with respect to D (see Fig. 1).

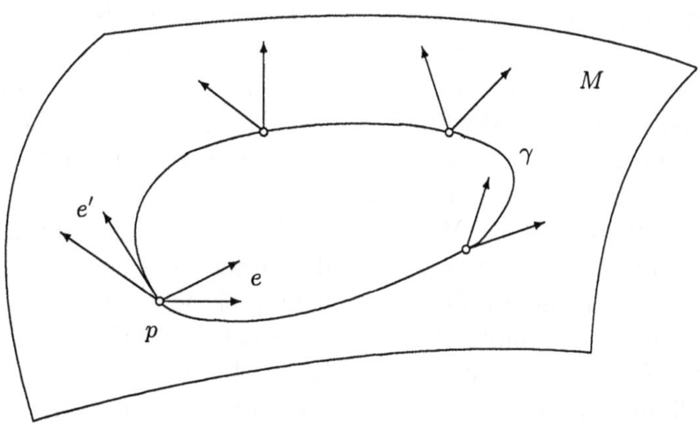

Figure 1. Parallel transport of a basis

Consequently we may write $e' = ge$, where $g \in GL(n, \mathbb{R})$. If we apply the same procedure along all closed curves γ, then all such $g \in GL(n, \mathbb{R})$

form a subgroup $G \subseteq GL(n, \mathbb{R})$. This subgroup is independent of the original point m and it is called the holonomy group of the connection D. We have that $G \subseteq SO(n, \mathbb{R})$ if the connection preserves a metric (and M is orientable). We refer to [2, 28, 40] for more details. As Hano and Ozeki showed in [27], every connected linear Lie group can be realized as a holonomy group of an affinely connected space. The list of irreducible subgroups of Lie groups that can occur as holonomy groups of affine torsion-free connections of non-locally symmetric spaces was proposed by Berger [8]. In the same paper Berger also gave the complete sublist of holonomies of connections preserving Riemannian or pseudo-Riemannian metrics. Bryant [14], and later Chi and Schwachhöfer [15] and others discovered many exceptional holonomy groups of non-metric torsion-free connections. The full list of holonomy groups of non-metric torsion-free connections was given by Merkulov and Schwachhöfer [36, 37].

So one can reveal that a connection D, as a geometrical quantity is closely related to its holonomy group, and hence the geometry based on D heavily depends on algebraic properties of the group G. Besides, we see that holonomy groups act on the tangent space (a geometrical quantity) whose algebraic structure is a vector space. The Ambrose-Singer theorem says that the curvature corresponding to a connection D is closely related to the holonomy algebra $\mathfrak{g}$ — the Lie algebra of the holonomy group G. Thus we may build a new algebraic structure $\mathcal{R}(T_m M)$, of all curvature tensors in a point $m \in M$, corresponding to torsion-free connections whose holonomy group is G.

The decomposition of the vector space $\mathcal{R}(G)$ under the action of $\mathfrak{g}$ into simple $\mathfrak{g}$-submodules is important, because it also carries some information about the manifold. The decomposition of the space $\mathcal{R}(G)$ for all holonomy groups $G \subseteq SO(V)$ of the Levi-Civita connection was described by Alekseevsky [2]. For example, in the case of the Levi-Civita connection $D = \nabla^g$ of a metric g with holonomy group $SO(V)$ the simple modules of $\mathcal{R}(SO(V))$ correspond to the scalar curvature, the traceless Ricci curvature and the Weyl tensor of $D = \nabla^g$. Vanishing of some of those simple modules gives all possible differential equations of the first order on the connection D. Consequently, these components give rise to the first-order invariants of D. For $G = GL(n, \mathbb{R})$ the complete decomposition of $\mathcal{R}(G)$ was given by Strichartz [42], for $G = Sp(1) \otimes GL(n, \mathbb{H})$ (the full holonomy group of a quaternionic connection) it was given by Alekseevsky and Marchiafava [5], while the action of the group $SL(p, \mathbb{C}) \otimes SL(q, \mathbb{C})$ has been used to introduce a para-conformal structure on complex manifolds. We have described in [13] the decomposition of $\mathcal{R}(G)$ under the action of the group $G = GL(p, \mathbb{R}) \otimes GL(q, \mathbb{R})$ (see [7]). So we can see very easily that geometry and algebra intertwine in studying various topics in this framework. Generally speaking, by analyzing some transformations in geometry one can reveal a group structure. But, in this framework, by studying some algebraic problems, one can reveal a corresponding geometrical interpretation of their solution. We illustrate it in the framework of geometry of a manifold endowed with a torsion-free connection D, its holonomy group G and the corresponding vector space $\mathcal{R}(G)$. In order to analyze these problems we see that we also need to involve finite groups to solve some problems related to some continuous group (see for example [13]). Finite groups are also an important tool if we are interested in global and local geometry of some manifolds and their covering space and many other areas. Of course, global geometry heavily depends not only on algebraic properties of curvature and its traces but also on its topology.

2. Manifolds with Grassmann structure

A Grassmann structure of type (p, q) on a manifold M is, by definition, an isomorphism from the tangent bundle TM of M to the tensor product $E \otimes H$ of two vector bundles E and H with rank p and q over M respectively ($TM \backsimeq E \otimes H$). If $\dim M = n$, then $n = pq$. These manifolds play an important role in high energy physics and geometry (see [3, 32, 33, 35]). The group which naturally acts on a manifold with a Grassmann structure is $G = GL(p, \mathbb{R}) \otimes GL(q, \mathbb{R})$. This group is a subgroup of $GL(p\,q)$ and it acts on $V^p \otimes V^q$ such that $GL(p)$ acts on V^p in the usual way and $GL(q)$ acts on V^q by inverse from the right. There is the natural projection

$$\pi : GL(p) \times GL(q) \longrightarrow GL(p) \otimes GL(q)\,,$$

that defines a fibre bundle with fibre $\mathbb{R}^*$ by the scalar multiplication and the dimension of the group G is $p^2 + q^2 - 1$ (see [31]).

The group G is also a significant example of a group of twistor type (see Alekseevsky and Graev [4]), very important in modern geometry, control theory and mathematical physics. We also refer to [6, 17, 38, 49], etc. for more details.

To clarify a Grassmann structure as well as an action of the group G we present here some examples of manifolds endowed with this structure.

2.1 Examples of Grassmann structures

1. As it is known a real Grassmann manifold $\mathcal{G}r(V^{p\otimes q})$ ($V^{p\otimes q} = V^p \otimes V^q$) is the set of all p-dimensional real vector subspaces in $\mathbb{R}^{p+q}$ endowed with the structure of real manifold. There exist various coordinate charts that one can define on a Grassmann manifold (see [28, 49] for some of them).

2. Let M be an n-dimensional smooth manifold and let TM be the tangent bundle of M with the natural projection π of TM onto M. Then for a linear connection D on M one can obtain the decomposition of the tangent space T_vTM at each point v of TM into the n-dimensional horizontal space $H_v = T^H_{\pi(v)}M \equiv T_{\pi(v)}M$ and the n-dimensional vertical space $V_v = T^V_{\pi(v)}M \equiv T_{\pi(v)}M$. Then the tangent bundle TTM of TM is presented as follows:

$$\begin{aligned} TTM &= H \oplus V \\ &\cong \pi^*TM \oplus \pi^*TM \\ &\cong \pi^*TM \otimes 2_{TM}, \end{aligned}$$

where π^*TM is the induced vector bundle of TM by $\pi : TM \longrightarrow M$, and 2_{TM} is the trivial bundle of rank 2 over TM. Consequently the $2n$-dimensional manifold TM has a Grassmann structure of type $(n, 2)$.

3. We generalize this idea to obtain a Grassmann structure on an r-frame bundle of M. An r-frame is a set of linearly independent r tangent vector fields at a point of M. In the case $r = 1$, F^1M is the tangent bundle TM of M. In the case $r = n$, F^nM is nothing but the linear frame bundle FM of M.

4. Let h be a Hermitian inner product of type $(m+1, m)$ on the complex $(2m+1)$-dimensional vector space $\mathbb{C}^{m+1,m}$. Put $n = 2m$. The quadric hypersurface N defined by $h(z, z) = 1$ is the (real) $(2n+1)$-dimensional pseudo-hyperbolic space $H^{n+1,n}$ of type $(n+1, n)$ with negative constant curvature. The group $U(m+1, m)$ acts transitively on $H^{n+1,n}$ and $S^1 = \{e^{i\theta}\}$ acts freely on the same space such that $z \mapsto e^{i\theta}z$. The base space M of the principal bundle $H^{n+1,n}$ with structure group S^1 is the complex pseudo-hyperbolic space $H^{m,m}(\mathbb{C})$ of type (m, m):

$$\begin{array}{ccccc} \mathbb{C}^{m+1,m} & \supset & H^{n+1,n} & \longleftarrow & S^1 \\ & & \pi \big\downarrow & & \\ & & H^{m,m}(\mathbb{C}) & & \end{array} \tag{1}$$

The group $U(m+1, m)$ acts transitively on $H^{m,m}(\mathbb{C})$ and the space $H^{m,m}(\mathbb{C})$ has the form

$$H^{m,m}(\mathbb{C}) \cong U(m+1, m)/U(1) \times U(m, m)$$

as a symmetric space. But $H^{m,m}(\mathbb{C})$ has also a Grassmann structure of type $(n, 2)$ (see [31] for more details).

Other examples can be found in [3].

The notion of Grassmann structures of type $(2, 2)$ is equivalent to the notion of conformal structures of type $(2, 2)$ (see [1]).

2.2 Topological obstructions

Anyhow, it is important to underline now that, anyhow, topological facts obstruct the existence of some types of Grassmann structures. For example, because of some topological obstructions, there does not exist any Grassmann structure of type $(n, 2)$ on any $2n$-dimensional manifold. In fact, if M as a Grassmann structure of type $(n, 2)$, then it yields

$$TM \cong V \otimes W,$$

where V and W are vector bundles of rank $n\,(\geq 2)$ and 2 over M respectively.

Assuming that $H^2(M; \mathbb{Z}) = 0$ we obtain that any vector bundle of rank 2 over M is trivial. Consequently we have

$$TM \equiv V \oplus V.$$

Let M be the $2n$-dimensional sphere S^{2n}. Since the homotopy set $[S^{2n}; B_{SO(n)}]$ ($\cong \pi_{2n-1}(SO(n))$) from S^{2n} to the classifying space $B_{SO(n)}$ of $SO(n)$ is 0 mod torsion, the vector bundle V of rank n over S^{2n} is trivial. So $V \oplus V$ is trivial. The Euler number of the tangent bundle TS^{2n} of S^{2n} is equal to 2. This is a contradiction. Therefore S^{2n} admits no Grassmann structures of type $(n, 2)$.

By using the total Pontryagin classes $\{p_i\}$ one can prove that the quaternionic projective space $P^m(\mathbb{H})$ $(n = 2m)$ admits no Grassmann structure of type $(n, 2)$. In a similar way one can check that the Cayley projective space $P^2(Ca)$ admits no Grassmann structures of type $(n, 2)$.

The geometry and topology of these manifolds have been studied from several points of view by many authors (for more details see for example [1, 11, 16, 20, 21, 22, 23, 24, 25, 26, 30, 34, 39, 43, 44, 49], etc.).

2.3 Grassmann connections

One can introduce a linear connection on a manifold with Grassmann structure as usual. But the most interesting for us are so-called Grassmann connections, which preserve the Grassmann structure. In other words for any vector field X on M and local sections $e \in \Gamma(E)$ and $h \in \Gamma(H)$ we have

$$D_X\,(e \otimes h) = D^E_X\, e \otimes h + e \otimes D^H_X\, h,$$

where D^E, D^H are connections in the bundles E, H respectively. In the general case D is with torsion, but if D is torsion-free, its holonomy group is $G = GL(p) \otimes GL(q)$.

Throughout this paper we consider only smooth manifolds, nevertheless many results can be stated in the setting of complex manifolds and holomorphic connections.

Let M be a real manifold with Grassmann structure of type (p, q) $(p\,q = n)$, and D a torsion-free connection on M whose curvature tensor R is given by

$$R(X, Y) = [D_X, D_Y] - D_{[X,Y]},$$

where $X, Y \in \mathcal{X}$, the algebra of C^∞ vector fields on M. Then R fulfills the symmetry conditions, typical for curvatures corresponding to torsion-free connections:

$$R(X, Y) = -R(Y, X) \tag{2}$$

$$R(X, Y)Z + R(Z, X)Y + R(Y, Z)X = 0\,. \tag{3}$$

These relations confirm that it makes sense to study the curvatures from an algebraic point of view as elements of a vector space.

3. Algebraic approach to a curvature tensor

If we use an algebraic approach to a notion of curvature then we have the following facts. Let $R \in \mathcal{T}_3^1 = V \otimes V \otimes V \otimes V^*$. Then R is an algebraic curvature tensor and we can rewrite symmetry conditions (2) and (3) as:

$$\sigma_{12}.\, R(x_1, x_2, x_3, \omega) = R(x_2, x_1, x_3, \omega) = -R(x_1, x_2, x_3, \omega)\,, \tag{4}$$

$$\sum_{\sigma \in \mathfrak{A}_3} R(x_{\sigma(1)}, x_{\sigma(2)})\, x_{\sigma(3)} = \mathfrak{A}_3^+ \,.\, R(x_1, x_2)\, x_3 = 0\,, \tag{5}$$

where σ_{12} is the transposition (12) and $\mathfrak{A}_3^+ = \sum_{\sigma \in \mathfrak{A}_3} \sigma$ are considered as the elements of the group ring $\mathfrak{G}\,(\mathfrak{S}_3)$, where we put

$$R(x_1, x_2, x_3, \omega) = \omega(R(x_1, x_2)\, x_3) = \langle R(x_1, x_2)\, x_3, \omega \rangle\,. \tag{6}$$

It is clear that the algebraic curvature tensors build a vector subspace $\mathcal{R}(V^n)$ in $\mathcal{T}_3^1$. The Ricci traces, $\rho^{ij} : \mathcal{T}_k^r \longrightarrow \mathcal{T}_{k-1}^{r-1}$, are defined as

$$\rho^{ij} = con(i, j)\, R\,, \tag{7}$$

where $con(i, j)$ means contraction in the $i \leftrightarrow j$ places. In the case when $k = 1$ or $r = 1$, we drop its index. For example, in the case of the Ricci contraction, $con(i,4)$, on $\mathcal{T}_3^1$ we write simple ρ^i. It is easy to see the Ricci traces ρ^2 and ρ^3 depend on ρ^1, for $R \in \mathcal{R}(V^n)$.

The group $GL(n)$ acts naturally on $\mathcal{T}_3^1$ by

$$\pi(g)\, R(x_1, x_2, x_3, \omega) = R(g^{-1} x_1, g^{-1} x_2, g^{-1} x_3, g^{\tau} \omega)\,, \tag{8}$$

and this action defines on $\mathcal{T}_3^1$ the structure of a $GL(n)$-module. As it is known, the simple $GL(n)$-modules are parameterized by the lattice of highest weights. We denote by $\pi(m)$ the simple $GL(n)$-module of the highest weight $m = (m_1, m_2, \ldots, m_n)$ $(m_1 \geq m_2 \geq \cdots \geq m_n$, m_i are integers). For simplicity of notation we delete the string of zeros, so that $\pi(2,-1)$ means $\pi(2,0,\ldots,0,-1)$. It is clear that $\pi(1)$ is the fundamental representation of $GL(n)$ in V^n (V^n is a simple $GL(n)$-module) and $\pi(-1)$ is its contragradient representation in $(V^n)^*$. In [42] the following theorem has been shown.

Theorem 3.1. *The semi-simple $GL(n)$-module $\mathcal{R}(GL(n))$ is the direct sum of simple $GL(n)$-modules,*

$$\mathcal{R}(GL(n)) = \pi(2) \oplus \pi(1,1) \oplus \pi(2,1,-1)\,,$$

(when $n = 2$ the third component is deleted). The $\pi(2,1,-1)$ component is the kernel of the Ricci contraction ρ^1.

It makes sense now to consider the action not only of $GL(n,\mathbb{R})$, but also of various subgroups (we refer to [12] for more details).

Here we consider only the case when $G = GL(p) \otimes GL(q)$, thus $V = V^p \otimes V^q$. Then, for a base $v_{\alpha i} = x_\alpha \otimes y_i$ $(\alpha = 1 \ldots p,\ i = p+1, \ldots p+q)$ of V we have for R symmetry properties analogous to these given by (4) and (5) and also an action of group $G = GL(p) \otimes GL(q)$ on $\mathcal{T}_3^1$ analogous to (8). These symmetry properties of a curvature tensor R arise naturally from 'the combined action' of both subgroups $GL(p)$ and $GL(q)$ at the same time. Since the group G is a subgroup of the group $GL(p\,q)$ one can expect a much more complicated structure of this space than that one of a similar space of curvature tensors corresponding to torsion-free connections whose holonomy group is $GL(n)$. Consequently the vector space $\mathcal{R}(G)$ of all curvature tensors satisfying these symmetry properties is an interesting object to study from the representation theory point of view. We have studied this problem in [13]. To obtain the complete decomposition of $\mathcal{R}(G)$ we reduce this problem to the representation of finite symmetry groups, and we use Schur functors to overcome difficulties in the proof of simplicity of submodules as well as for a better understanding of their joint actions (see also [18, 19] for more details). In [13] we introduced the Ricci traces of the tensors of rank $(1,3)$, that are given by

$$\rho^{i\alpha} = con\,((i,\alpha)(4,4))\; R = \overline{con}\,(i,4) \circ \overline{con}\,(\alpha,4)\; R\,, \tag{9}$$

where $\overline{con}\,(i,4) = con\,(i,4) \otimes id$ and $\overline{con}\,(\alpha,4) = id \otimes con\,(\alpha,4)$, and we proved that the vector space of Ricci traces is spanned by ρ^{11}, ρ^{12} and ρ^{13}.

Denoting by $\Pi_1(m)$ and $\Pi_2(m)$ the simple $GL(p)$ and $GL(q)$-modules respectively, for the highest weight m, we obtained the complete decomposition of $\mathcal{R}(G)$. More precisely we proved,

Theorem 3.2. *Let $G = GL(p) \otimes GL(q)$ and $V = V^{p\otimes q} = V^p \otimes V^q$; then the space of curvature tensors $\mathcal{R}(G)$ decomposes as follows ($p, q > 3$):*

$$\begin{aligned}
\mathcal{R}(G) = {} & 3\,(\Pi_1(2)\otimes\Pi_2(2)\oplus\Pi_1(1,1)\otimes\Pi_2(1,1)\oplus\Pi_1(2)\otimes\Pi_2(1,1)\\
&\oplus\Pi_1(1,1)\otimes\Pi_2(2))\oplus 2\,(\Pi_1(2,1,-1)\otimes\Pi_2(2)\oplus\Pi_1(2)\otimes\Pi_2(2,1,-1)\\
&\oplus\Pi_1(2,1,-1)\otimes\Pi_2(1,1)\oplus\Pi_1(1,1)\otimes\Pi_2(2,1,-1))\oplus\Pi_1(2)\otimes\Pi_2(3,-1)\\
&\oplus\Pi_1(1,1)\otimes\Pi_2(3,-1)\oplus\Pi_1(2,1,-1)\otimes\Pi_2(3,-1)\oplus\Pi_1(3,-1)\otimes\Pi_2(2)\\
&\oplus\Pi_1(3,-1)\otimes\Pi_2(1,1)\oplus\Pi_1(3,-1)\otimes\Pi_2(2,1,-1)\oplus\Pi_1(1,1,1,-1)\otimes\Pi_2(2)\\
&\oplus\Pi_1(1,1,1,-1)\otimes\Pi_2(1,1)\oplus\Pi_1(1,1,1,-1)\otimes\Pi_2(2,1,-1)\\
&\oplus\Pi_1(2)\otimes\Pi_2(1,1,1,-1)\oplus\Pi_1(1,1)\otimes\Pi_2(1,1,1,-1)\\
&\oplus\Pi_1(2,1,-1)\otimes\Pi_2(1,1,1,-1)\oplus\Pi_1(2,1,-1)\otimes\Pi_2(2,1,-1)\,.
\end{aligned}$$

Remark 3.3. Some components in the decomposition from Theorem 3.2 vanish when $p, q \leq 3$ since their dimensions are equal to zero. Furthermore in [13] we found the highest weights of all 33 simple G-modules which appear in the previous theorem.

Let us emphasize that the highest weights of components in the complete decompositions of $\mathcal{R}(G)$ are unique, but the projections on simple modules of the same highest weight are not unique. So we may choose them to obtain the projective invariants, but also to have some invariants of another type. This is a consequence of the non-uniqueness of the direct complement in a decomposition, under the action of group G, while it is unique under the action of an orthogonal group. Moreover we also have various isomorphisms among simple G-modules of highest weights $\Pi_1(2)\otimes\Pi_2(2), \Pi_1(1,1)\otimes\Pi_2(1,1), \Pi_1(2)\otimes\Pi_2(1,1), \Pi_1(1,1)\otimes\Pi_2(2)$ and corresponding simple G-modules of the same highest weights of $\mathcal{R}(G)$, respectively. This is a consequence of algebraic independency of Ricci traces $\rho^{11}, \rho^{12}, \rho^{13}$.
The uniqueness of the traceless part of $\mathcal{R}(G)$ has been considered by Krupka [29].

4. Curvatures of Grassmann connections

Studying the geometry of manifolds with Grassmann structure the authors have been interested in the existence and properties of a connection defined in this type of manifolds. Hence, to study the space $\mathcal{R}(G)$ from a geometrical point of view, we need undoubtedly to involve $\mathcal{R}(M)$, the vector bundle with fiber $\mathcal{R}(T_mM)$. Theorem 3.2 applied on $\mathcal{R}(T_mM)$ gives rise to a decomposition of $\mathcal{R}(M)$ in subbundles. We still denote the simple components as in the previous sections. This space is not empty, i.e. there exist linear connections such that their corresponding curvature tensors belong to $\mathcal{R}(M)$. Among various connections the most interesting are these ones whose curvatures belong to some either simple submodules or to some of their direct sums as they define some special geometry of these manifolds with Grassmann structure. Among these type of connections we point out: half-flat Grassmann connections, connections corresponding to some normalization, projectively equivalent connections, connections with symmetric or skew-symmetric Ricci tensor, etc. Having in mind a complete decomposition of $\mathcal{R}(M)$ into simple submodules we can reveal also some new curvature invariants corresponding to some transformations of manifolds with Grassmann structure. The complete solution of this problem is beyond the scope of this paper but anyhow it is important at least to announce it.

In subsections which follow we review some type of special connections defined on manifolds with Grassmann structure, the corresponding geometry of its manifolds and subbundles of $\mathcal{R}(M)$, whose elements are curvatures of these connections.

4.1 Half-flat Grassmann connections

Let M be a manifold with a Grassmann structure $TM \simeq E \otimes H$ of type (p, q) and with a Grassmann connection $D = D^E \otimes Id + Id \otimes D^H$. Then D is called *positive half-flat* if the connection D^H in the vector bundle $H \longrightarrow M$ is flat and *negative half-flat* if D^E is flat in the vector bundle $E \longrightarrow M$. The corresponding structures $TM \simeq E \otimes H$ are called *a positive half-flat Grassmann structure* and *a negative half-flat Grassmann structure* respectively. Let us recall that in the general case D is not torsion-free, but we are interested only in those which are torsion-free. We call them *half-flat torsion-free connections* and the corresponding Grassmann structures *half-flat torsion-free structures.* The Levi-Civita connection on a hyperKähler manifold is an example of a positive half-flat torsion-free connection [3].

Mashida and Sato [31] developed the twistor theory of manifolds with Grassmann structures of type (p, q) by considering the complete integrability of the associated null distributions. The case where $q = 2$ is more interesting, since they could define the notion of half flatness meaningfully and the geometric structure of the twistor partner is a different projective structure from a Grassmann structure. By the normal Cartan connection they define the notion of half flatness for the Grassmann structures of type (p, q).

Considering the set of all the null p-planes with forms $\{V_m \otimes w \,|\omega \in W_m\}$ in $T_m M$ at each point $m \in M$, we have a null p-plane bundle F_L with fibre $P^{m-1}(\mathbb{R})$ and the projection $\bar{\omega}_L : F_L \longrightarrow M$ over M. Similarly, considering the set of all the null q-planes with forms $\{v \otimes W_m \mid v \in V_m\}$ in $T_m M$, we have a null q-plane bundle F_R with fibre $P^{n-1}(\mathbb{R})$ and the projection $\bar{\omega}_R : F_R \longrightarrow M$ over M. By the normal Cartan connection, the tautological distribution D_L of null p-planes on F_L over M and D_R of null q-planes on F_R over M are defined respectively. Mashida and Sato have considered the complete integrability of D_L and D_R in terms of positive half-flat and negative half-flat Grassmann connections.
By using a different technic this problem has been considered also by Akivis-Goldberg [1].

The null p-plane bundle F_L and the null q-plane bundle F_R over M have also geometric structures, which we call co-Grassmann structures. A co-Grassmann structure of type (k, l) on a manifold R is defined by a pair (E, F) consisting of transversal, completely integrable distributions of dimensions k and l on the tangent bundle TR of R such that (i) $TR = D + [D, D]$ and (ii) rank TR/D = rank E · rank $F (= kl)$ for $D = E \oplus F$.

Let F be a manifold endowed with a co-Grassmann structure of type (k, l) by a pair (D_2, D_1) and equipped with the normal Cartan connection (Q, ω). Put $M_1 = F/D_1$, $M_2 = F/D_2$. Mashida and Sato have studied how the Grassmann structures on M_1 and M_2 depend on each other. More precisely, they proved the following theorem.

Theorem 4.1.

(1) *Let M_1 be a manifold with a negative torsion-free Grassmann structure of type (p, q). Then if the structure on M_1 induces a Grassmann structure of type $(p+1, q-1)$ on $M_2 = F_L/D_L$, the Grassmann structure of type (p, q) on M_1 is flat.*

(2) *Let M_2 be a manifold with a positive half torsion-free Grassmann structure of type $(p+1, q-1)$. Then, if the structure on M_2 induces a Grassmann structure of type (p, q) on $M_1 = F_R/D_R$, the Grassmann structure of type $(p+1, q-1)$ on M_2 is flat.*

An interesting problem is which connections given on a manifold with Grassmann structure satisfy the Yang-Mills equation. Alekseevsky, Cortes and Devchand have studied this problem. To present some of their results we introduce some notions.

Let $\nu : W \longrightarrow M$ be a real vector bundle over a pseudo-Riemannian manifold (M, g). A Yang-Mills connection D in ν is one which satisfies the Yang-Mills equation

$$d^D \star R = 0\,.$$

Proposition 4.2. *A connection D in a vector bundle $W \longrightarrow M$ over a quaternionic Kähler manifold is half-flat if and only if it is λ_1-self-dual. Hence any such connection is a Yang-Mills connection.*

The same authors have also constructed half-flat connections by using the harmonic space method.

We have revealed in [13] an algebraic structure of curvatures corresponding to half-flat torsion-free connections. Really, the space of 2-forms $\Lambda^2(TM)$ or $\Lambda^2(T^*M)$ is decomposed as

$$\Lambda^2(TM) = S^2(E) \otimes \Lambda^2(H) \oplus \Lambda^2(E) \otimes S^2(H)\,. \tag{10}$$

The decomposition is invariant under the group G. Consequently the vector bundle $\mathcal{R}(M)$ is

$$\mathcal{R}(M) = \left\{R \in \Lambda^2 \otimes \mathfrak{g} \mid \sigma_{cy} R(x, y) z = 0\right\}, \tag{11}$$

whose subbundles

$$\mathcal{R}^+(M) = \left\{R \in S^2(E) \otimes \Lambda^2(H) \otimes \mathfrak{g} \mid \sigma_{cy} R(x, y)\, z = 0\right\}, \tag{12}$$

$$\mathcal{R}^-(M) = \left\{R \in \Lambda^2(E) \otimes S^2(H) \otimes \mathfrak{g} \mid \sigma_{cy} R(x, y)\, z = 0\right\}, \tag{13}$$

consist of curvatures of positive and negative torsion-free half-flat Grassmann connections. Hence, as a consequence of Theorem 3.2 we get

Theorem 4.3. *Let $G = GL(p) \otimes GL(q)$ and let $\mathcal{R}^+(M)$ be the subbundle that consists of curvatures of a manifold with a positive half-flat Grassmann structure. Then $\mathcal{R}^+(M)(G)$ decomposes $(p, q > 3)$ as follows:*

$$\begin{aligned}\mathcal{R}^+(M)(G) &= \big(\Pi_1(2) \oplus \Pi_1(3, -1)\big) \otimes \big(\Pi_2(2) \oplus \Pi_2(1, 1) \oplus \Pi_2(2, 1, -1)\big)\\ &\oplus \big(\Pi_1(2) \oplus \Pi_1(1, 1) \oplus \Pi_1(2, 1, -1)\big) \otimes \big(\Pi_2(1, 1)\\ &\oplus \Pi_2(1, 1, 1, -1)\big)\,.\end{aligned}$$

A similar statement holds for $\mathcal{R}^-(M)(G)$. We also obtain some algebraic obstructions for a torsion-free connection defined on M $(TM \cong E \otimes H)$ being positive half-flat torsion-free.

4.2 Projective invariants

We recall that a G-structure on M^{pq} is flat if M^{pq} may be covered by coordinate neighborhoods such that in their intersections the Jacobian matrices of coordinate transformations belong to the linear subgroup G. Hangan [21] has shown that for a flat Grassmann structure on M^{pq} one may find local coordinates

$$x^i_\alpha, \alpha = 1, \dots, p;\ i = p+1, \dots, p+q\,, \tag{14}$$

such that in the intersection of two neighborhoods the following system is satisfied:

$$\frac{\partial x'^i_\alpha}{\partial x^j_\beta} \frac{\partial x'^k_\gamma}{\partial x^l_\delta} = \frac{\partial x'^i_\gamma}{\partial x^j_\delta} \frac{\partial x'^k_\alpha}{\partial x^l_\beta}\,. \tag{15}$$

If $p \neq 1$, $q \neq 1$ the system

$$\frac{\partial^2 x'^i_\alpha}{\partial x^j_\beta \partial x^k_\gamma} = \frac{1}{n}\frac{\partial x'^i_\alpha}{\partial x^j_\gamma}\frac{\partial \log \Delta}{\partial x^k_\beta} + \frac{1}{n}\frac{\partial x'^i_\alpha}{\partial x^k_\beta}\frac{\partial \log \Delta}{\partial x^j_\gamma}\,, \qquad \Delta = \det\left[\frac{\partial x'^i_\alpha}{\partial x^j_\beta}\right] \tag{16}$$

is a differential consequence of (15). It generalizes the system of the projective differential geometry. The general solution of the system (15), written in matrix form, is

$$x' = (ax+b)(cx+d)^{-1} \tag{17}$$

where x, x' are $p \times q$ matrices and a, b, c, d are respectively $p \times p$, $p \times q$, $q \times p$, $q \times q$ matrices. Hangan [21] has proved for $p \neq 1$, $q \neq 1$ the flat Grassmann structures are defined by the system (15) whose differential consequence is the system (16). The general solution of (15) is given by the formula (17). The Grassmann manifolds, the classical domains of the first type, and all the space forms of these manifolds carry flat Grassmann structure. For their complete classification we refer to [48] and [21].

Marchiafava [34] has discussed the same problem of integrability for quaternionic Grassmann manifold. Zelikin [49] has discussed the formula (17) in the framework of cross ratio and the solution of generalized Riccati equations and its applications in control theory.

Since some of projective properties of a manifold might be expressed in terms of the Weyl projective curvature tensor we present here some algebraic properties of the corresponding subbundle $\mathcal{P}(M)$ of these curvature tensors. R. Strichartz has proved that this subbundle $\mathcal{P}(M)$ is simple if $G = GL(n, \mathbb{R})$ is the corresponding holonomy group which acts in $\mathcal{R}(M)$. We proved in [13] that this subbundle $\mathcal{P}(M)$ is a direct sum of 29 simple subbundles if $G = GL(p, \mathbb{R}) \otimes GL(q, \mathbb{R})$ acts in $\mathcal{R}(M)$ as a holonomy group. This is a consequence of the facts that the Ricci trace ρ^{11} of P vanishes, the existence of three independent traces ρ^{1i}, $i = 1, 2, 3$ of $R \in \mathcal{R}(M)$ and that G is a subgroup of $GL(pq, \mathbb{R})$. Therefore geometry of manifolds whose holonomy group is G involves more invariants than in the case of $GL(n, \mathbb{R})$ being the holonomy group.

4.3 Reduction of the structure group

It is possible to reduce the structure group of the Grassmann manifold to the tensor product of two orthogonal groups, but in this case the structure is no longer flat. The integrability is thus related to the projective properties of the Grassmann manifold.

For an Einstein metric given on the Grassmann manifold $\mathcal{G}r(V^{p\otimes q})$, $pq > 1$ (see [30]) and its Levi-Civita connection, the corresponding Weyl projective curvature tensor $W = P_{\pi(2,1,-1)}R$ fulfills the relations

(i1) $P_{\pi(2,1,-1)}R = P_{\Pi_1(2)\otimes\Pi_2(2)}R \; + \; P_{\Pi_1(2)\otimes\Pi_2(2)}R$ for $p \neq q$ and

(i2) $P_{\pi(2,1,-1)}R = P_{\Pi_1(2)\otimes\Pi_2(2)}R$ for $p = q$.

Moreover at the origin point O, we have

$$\begin{aligned} P_{\pi(2,1,-1)}\, R_{(i\alpha)(j\beta)(k\gamma)}{}^{(l\nu)} &= W_{(i\alpha)(j\beta)(k\gamma)}{}^{(l\nu)} \\ &= (\,\delta_{ik}\,\delta^l_j - \delta_{jk}\,\delta^l_i\,)\,\delta_{\alpha\beta}\,\delta^\nu_\gamma + \delta_{ij}\,\delta^l_k\,(\,\delta_{\alpha\gamma}\,\delta^\nu_\beta - \delta_{\beta\gamma}\,\delta^\nu_\alpha\,) \\ &\quad + \frac{p+q-2}{1-pq}\,(\,\delta_{jk}\,\delta^l_i\,\delta_{\beta\gamma}\,\delta^\nu_\alpha - \delta_{ik}\,\delta^l_j\,\delta_{\alpha\gamma}\,\delta^\nu_\beta\,)\,. \end{aligned} \tag{18}$$

We recall that $P_{\pi(m)}$ denotes the projection operator of $\mathcal{R}(T_mM)$ on the subbundle $\pi(m)$. Consequently, there exists a connection whose Weyl projective curvature belongs to a simple module $\Pi_1(2) \otimes \Pi_2(2)$. Grassmann manifolds $\mathcal{G}r(V^{p\otimes q})$, for $p = q$ are also interesting in the study of solutions of generalized Riccati equations which are very important in control theory. We give a short review of these topics in the next subsection.

By using other reductions of the structure group we may obtain also new geometries. For example, if there exist a Kähler structure and a commutative Kähler product structure then the structure group is reduced to $Sp(n, \mathbb{R}) \otimes Sp(1, \mathbb{R})$. S. Vukmirović [47] has studied the dimensional reduction of a manifold with the structure group $Sp(n, \mathbb{R}) \otimes Sp(1, \mathbb{R})$ in order to obtain new examples of manifolds with smaller dimension and the same type structure group.

It is possible to reduce the structure group of the pseudo-hyperbolic space $H^{m,m}(\mathbb{C})$ to the tensor product of two pseudo-unitary groups, as this space permits a metric with negative constant curvature. This means its projective structure is flat and its Weyl projective curvature vanishes.

4.4 Riccati type equation and projective geometry

Let us recall that the Riccati equation is the differential equation with respect to a real function $x = x(t)$ of the following type:

$$\dot{x} + A(t)\,x^2 + B(t)\,x + C(t) = 0\,.$$

This equation is very closely related to the group of fractional linear transformations of a real line. These connections can be recognized analyzing the following two theorems.

Theorem 4.4. *The general solution of a Riccati equation is a fractional linear function of an integration constant, i.e.*

$$x = \frac{Cf(t) + g(t)}{C\phi(t) + \psi(t)}\,.$$

Conversely, every differential equation of the first order, with this property, is a Riccati equation.

Theorem 4.5. *Let $x_1,\ x_2,\ x_3,\ x_4$ be four particular solutions of a Riccati equation corresponding to values $C_1,\ C_2,\ C_3,\ C_4$ of constant C. Then*

$$\frac{x_3 - x_1}{x_3 - x_2} : \frac{x_4 - x_1}{x_4 - x_2} = \frac{C_3 - C_1}{C_3 - C_2} : \frac{C_4 - C_1}{C_4 - C_2}\,,$$

for every $t \in \mathbb{R}$.

One can generalize a Riccati equation to a system of differential equations with the quadratic right part. Then projective lines are substituted by their multidimensional analogs: Grassmann manifolds and Lagrange-Grassmann manifolds. A multidimensional analog of a cross ratio is a matrix cross ratio (17). To explain their connections we use a chart U on $\mathcal{G}r(V^{p\otimes p})$ such that we project every point (h,v) $(h = (h_1,\ldots,h_p),\ v = (v_1,\ldots,v_p))$ on two p-dimensional planes, $H_0 = \{(h,0)\}$ and $H_\infty = \{(0,v)\}$, so called horizontal and vertical planes respectively, such that $\pi_0(h,v) = (h,0),\ \pi_\infty(h,v) = (0,v)$. The domain of a chart U is a set of p-planes W in $\mathbb{R}^{2p}$, transversal to H_∞. Elements of the matrix Z of a map $\pi \circ (\pi_0)^{-1}$ are the coordinates of a point in a plane W.

If we consider that a plane W moves under the action of a 1-parameter group, described by a linear canonical system of differential equations with Hamiltonian $H(\tau, h, v)$, then the Riccati equation appears as an equation of local coordinates W on a Grassmann manifold.

Theorem 4.6. *Let*

$$\Gamma(t_0, t_1) = \begin{pmatrix} a & b \\ c & d \end{pmatrix}$$

be the fundamental matrix of solutions of a system

$$\dot{h} = -A_1 C^T h + A^{-1} p, \tag{19}$$

$$\dot{v} = (B - CA^{-1}C^T)h + CA^{-1}p, \tag{20}$$

then

$$W = (c + dW_0)(a + bW_0)^{-1}$$

is a solution of the Riccati equation

$$\dot{W} = (C + W)A^{-1}(C^T + W) - B.$$

We refer to [49] for more details in studying the solution of generalized Riccati equations and its applications in control theory.

5. Another type of tensor product structure

Starting from $V = T_m M$ and tensoring V by itself we obtain a new vector space $W = V \otimes V$ whose elements are $w = w^{ij} e_i \otimes e_j$. Elements of vector spaces W and $\bar{W}$ in two neighboring points $m, \bar{m}$ of M may be connected by some non-linear connection Γ, that in a special case can be reduced to a linear connection D defined on TM, and its tensor product by itself and $(TM)^*$. In this way we obtain also some type of a tensor product structure on M with an absolute differential operator, induced by Γ, satisfying certain properties. The geometry of manifolds with this tensor-product structure has been studied by L. Tamássy and others [43, 44, 45, 46], etc. (see the references therein).

6. Normalization of a Grassmann manifold

As it is well known a manifold M admits a torsion-free connection D with symmetric Ricci tensor ρ^D if and only if M admits a D-parallel volume form $\omega = \omega(D)$, i.e. $D\omega = 0$. The form ω is unique modulo a constant non-zero factor. But in the framework of the Grassmann manifolds $\mathcal{G}r(V^{p \otimes q})$ whose holonomy group of torsion-free connections is $G = GL(p) \otimes GL(q)$ there exists a distinguished class of Ricci symmetric connections. They are obtained by means of certain constructions called normalizations, which we are now going to describe. It confirms that a smaller group involves more invariants. The Grassmann manifold $\mathcal{G}r(V^{p\otimes q})$ is said to be *normalized* if to each of its p-dimensional subspaces U there corresponds a chosen subspace U^* of dimension q in the space $\mathbb{R}^{p+q}$, such that U^* is a complement of U in $\mathbb{R}^{p+q}$. The U^* is called the *normalizing* subspace for U, while the Grassmann manifold $\mathcal{G}r(V^{p\otimes q})$ is called *normalized* and denoted by $\mathcal{G}r^{\nu}(V^{p\otimes q})$. We refer to [1] for more details. Let us mention also that the idea of a normalization has appeared in affine differential geometry, founded by W. Blaschke and his school (see [41] and references therein).

A normalization of the Grassmann manifold $\mathcal{G}r(V^{p\otimes q})$ is defined by a differential mapping $\nu : \mathcal{G}r(V^{p\otimes q}) \longrightarrow \mathcal{G}r(V^{q\otimes p})$ whose differential equations are

$$\omega_{(i\alpha)} = \lambda_{(i\alpha)(j\beta)}\,\omega_{(j\beta)}\,, \tag{21}$$

where the 1-forms $\omega^{i\alpha}$ are basis forms of the frame bundle associated with $\mathcal{G}r(V^{p\otimes q})$ and the coefficients $\lambda_{(i\alpha)(j\beta)}$ form a tensor called *the fundamental tensor* of $\mathcal{G}r^{\nu}(V^{p\otimes q})$. The normalization ν is *harmonic* if the coefficients in equation (21) are symmetric with respect to the pair of indices $(j\beta)$, $(i\alpha)$, i.e., $\lambda_{(i\alpha)(j\beta)} = \lambda_{(j\beta)(i\alpha)}$. When the fundamental tensor $\lambda_{(i\alpha)(j\beta)}$ satisfies the relation

$$\lambda_{(i\alpha)(j\beta)} = -\,g_{ij}\;g_{\alpha\beta}\,, \tag{22}$$

with g_{ij}, $g_{\alpha\beta}$ symmetric, then the corresponding normalization ν is said to be *polar* and the normalized Grassmann manifold is homogeneous. It is clear that a polar normalization is also harmonic. In [1] Akivis and Goldberg proved the following theorem.

Theorem 6.1. *The normalization ν of a normalized domain $U^{\nu} \subseteq \mathcal{G}r(V^{p\otimes q})$ uniquely determines a torsion-free affine connection D^{ν} with the connection forms*

$$\omega_{(i\alpha)(j\beta)} = \delta_{\alpha\beta}\,\omega_{ij} - \delta_{ij}\,\omega_{\alpha\beta}\ ,$$

where $\omega_{\alpha\beta}$ and ω_{ij} are fiber forms of the frame bundle associated with a normalized Grassmann manifold $\mathcal{G}r^{\nu}(V^{p\otimes q})$ (or a domain U^{ν} of this manifold).

One can use Theorem 6.1 to see which projections of the curvature corresponding to a connection D^{ν} vanish. The curvature tensor of this connection is expressed in terms of the fundamental tensor of the normalization ν according to the formulas

$$\begin{aligned} R_{(k\gamma)(l\mu)(j\beta)}{}^{(i\alpha)} = \frac{1}{2}\,\big(&\delta^{\alpha}_{\beta}\,\delta^{i}_{k}\,\lambda_{(j\,\gamma)(l\mu)} + \delta^{\alpha}_{\gamma}\,\delta^{i}_{j}\,\lambda_{(k\,\beta)(l\mu)} \\ &- \delta^{\alpha}_{\beta}\,\delta^{i}_{l}\,\lambda_{(j\,\mu)(k\gamma)} - \delta^{\alpha}_{\mu}\,\delta^{i}_{j}\,\lambda_{(l\,\beta)(k\gamma)}\big)\,. \end{aligned} \tag{23}$$

One can check that the Ricci tensor ρ^{11} of the connection D^ν is symmetric if and only if the normalization ν of the Grassmann manifold $\mathcal{G}r(V^{p\otimes q})$ is harmonic.

Having in mind the results mentioned above, we have reached geometrical interpretations of some components in our complete decomposition of $\mathcal{R}(\mathcal{G}r(V^{p\otimes q}))$. We have shown that we can characterize the curvature tensor in any point of $\mathcal{G}r(V^{p\otimes q})$ corresponding to connections D^ν for an arbitrary normalization, as well as some subspaces related to the connections D^ν for special types of normalizations. To illustrate the study of normalizations from this point of view we give here the following proposition.

Proposition 6.2. *Let* $R(D^\nu) \in \mathcal{R}(\mathcal{G}r(V^{p\otimes q}))$ *be a curvature tensor of torsion-free connection* D^ν *corresponding to the normalization* ν *on* $\mathcal{G}r(V^{p\otimes q})$. *Then it satisfies*

$$(i1)\quad R(D^\nu) \in \Pi_1(2)\otimes\Pi_2(2)\oplus\Pi_1(1,1)\otimes\Pi_2(1,1) \oplus \Pi_1(2)\otimes\Pi_2(1,1)\oplus\Pi_1(1,1)\otimes\Pi_2(2)\,.$$

with corresponding projections as follows:

$$(i2)\quad P_{\Pi(2)\otimes\Pi(2)}\, R_{(k\gamma)(l\mu)(j\beta)}{}^{(i\alpha)} = \frac{1}{2}\Big[\delta^\alpha_\beta(\delta^i_k S^s_{(j\gamma)(l\mu)} - \delta^i_l S^s_{(j\gamma)(k\mu)}) + \delta^i_j(\delta^\alpha_\gamma S^s_{(k\beta)(l\mu)} - \delta^\alpha_\mu S^s_{(l\gamma)(k\beta)})\Big],$$

$$(i3)\quad P_{\Pi(1,1)\otimes\Pi(1,1)}\, R_{(k\gamma)(l\mu)(j\beta)}{}^{(i\alpha)} = \frac{1}{2}\Big[\delta^\alpha_\beta(\delta^i_k S^a_{(j\gamma)(l\mu)} + \delta^i_l S^a_{(j\gamma)(l\mu)}) + \delta^i_j(\delta^\alpha_\gamma S^a_{(k\beta)(l\mu)} + \delta^\alpha_\mu S^a_{(l\gamma)(k\beta)})\Big],$$

$$(i4)\quad P_{\Pi(2)\otimes\Pi(1,1)}\, R_{(k\gamma)(l\mu)(j\beta)}{}^{(i\alpha)} = \frac{1}{2}\Big[\delta^\alpha_\beta(\delta^i_k A^s_{(j\gamma)(l\mu)} + \delta^i_l A^s_{(j\gamma)(l\mu)}) + \delta^i_j(\delta^\alpha_\gamma A^s_{(k\beta)(l\mu)} + \delta^\alpha_\mu S^s_{(l\gamma)(k\beta)})\Big],$$

$$(i5)\quad P_{\Pi(1,1)\otimes\Pi(2)}\, R_{(k\gamma)(l\mu)(j\beta}{}^{(i\alpha)} = \frac{1}{2}\Big[\delta^\alpha_\beta(\delta^i_k A^a_{(j\gamma)(l\mu)} - \delta^i_l A^a_{(j\gamma)(l\mu)}) + \delta^i_j(\delta^\alpha_\gamma A^a_{(k\beta)(l\mu)} - \delta^\alpha_\mu A^a_{(l\gamma)(k\beta)})\Big],$$

where

$$S^s_{(l\mu)(j\beta)} = \frac{1}{4}\left[\lambda_{(j\beta)(l\mu)} + \lambda_{(l\mu)(j\beta)} + \lambda_{(l\beta)(j\mu)} + \lambda_{(j\mu)(l\beta)}\right] \tag{24}$$

$$S^a_{(l\mu)(j\beta)} = \frac{1}{4}\left[\lambda_{(j\beta)(l\mu)} + \lambda_{(l\mu)(j\beta)} - \lambda_{(l\beta)(j\mu)} - \lambda_{(j\mu)(l\beta)}\right] \tag{25}$$

$$A^s_{(l\mu)(j\beta)} = \frac{1}{4}\left[\lambda_{(j\beta)(l\mu)} - \lambda_{(l\mu)(j\beta)} + \lambda_{(l\beta)(j\mu)} - \lambda_{(j\mu)(l\beta)}\right] \tag{26}$$

$$A^a_{(l\mu)(j\beta)} = \frac{1}{4}\left[\lambda_{(j\beta)(l\mu)} - \lambda_{(l\mu)(j\beta)} - \lambda_{(l\beta)(j\mu)} + \lambda_{(j\mu)(l\beta)}\right] \tag{27}$$

Our decomposition of $\mathcal{R}(\mathcal{G}r(V^{p\otimes q}))$ also says which are the obstructions for their elements being a curvature tensor corresponding to some torsion-free connection that admits a normalization. While these obstructions are of both global and local nature, to have an affirmative answer to the first question it is not enough to use only our complete decomposition but also some topological tools.

7. Grassmann manifolds of smaller dimensions

A Grassmann manifold $\mathcal{G}r(V^{2\otimes 2})$ is homeomorphic to $\mathbb{S}_1^2 \times \mathbb{S}_2^2$ where $\mathbb{S}_i^2$ $(i = 1, 2)$ are standard spheres. Then we use local coordinate systems on $\mathbb{S}_1^2$ and adopt computations from [9, 10] to obtain Grassmann connections on $\mathcal{G}r(V^{2\otimes 2})$ whose Ricci tensors are either symmetric or skew-symmetric [13]. The first are obtained globally and the second one only locally because of topological structure of $\mathcal{G}r(V^{2\otimes 2})$.

Acknowledgments

The authors express their gratitude to the Programme Committee of János Bolyai Conference of Hyperbolic Geometry, who invited us to give a lecture. Many thanks to prof. E. Molnár and prof. L. Tamássy for encouragement in writing the paper.

Bibliography

[1] Akivis M. and Goldberg V., *Conformal Differential Geometry and its generalizations,* (*Pure and Applied Mathematics*), John Wiley & Sons, New York, (1996).

[2] Alekseevsky D. V., *Riemannian spaces with unusual holonomy groups,*(*Func. Anal and Appl.*), Vol 2 97-105, (1968).

[3] Alekseevsky D. V., Cortes V., and Devchand C., *Yang-Mills connections over manifolds with Grassmann structure,* (*Max-Planck-Institut für Mathematik*), Preprint Series 2002 (124).

[4] Alekseevsky D. V. and Graev M. M., *Twistors and G-structures,* (*Izv. Ross. Akad. Nauk, Ser. Matem.*) Tom 56, No 1, (1992), English translation: Russian Acad. Sci. Izv. Math. Vol 40 No 1, (1993).

[5] Alekseevsky D. V. and Marchiafava S., *Quaternionic structures on a manifold and subordinated structures,* (*Ann. Mat. pura ed applicata*), Vol CLXXI, No. IV, 205-273, (1996).

[6] Atiyah M. F., Hitchin N. J., and Singer I. M., *Self-duality in four-dimensional Riemannian geometry,* (*Proc. Roy. Soc. London,*) Ser. A, Vol 362, 425-467, (1978).

[7] Bailey T.N. and Eastwood M.G., *Complex paraconformal manifolds - their differential geometry and twistor theory,* (*Forum Math.*), Vol 3, 61-103, (1991).

[8] Berger M., *Sur les groupes d'holonomie homogène des variétés a connexion affine et des variétés riemanniennes,* (*Bull. Soc. Math. France*), Vol 83, 279-330, (1955).

[9] Blažić N. and Bokan N., *Compact Riemann surfaces with skew-symmetric Ricci tensor,* (*Izv. VUZ*), Vol 9 (376), 8-12, (1993).

[10] Blažić N. and Bokan N., *Projective and affine transformations of a complex symmetric connection,* (*Bull. Austral. Math. Soc.*), Vol 50, 337-347, (1994).

[11] Blair D. and Ledger A., *A characterization of oriented Grassmann manifolds,* (*Rocky Mt. J. Math.*), Vol 15, 573-584, (1984).

[12] Bokan N., *The decomposition theory and its applications,* (The math. heritage of *C.F. Gauss (edited by G.M. Rassias)*), World Sci. Publ. Co. Singapore, pp. 66-99, (1991).

[13] Bokan N., Matzeu P., and Rakić Z., *Geometric quantities of manifolds with Grassmann structure,* (submitted), (2003).

[14] Bryant R., *Two exotic holonomies in dimension four,* (Path geometries, and Twistor Theory, *Proc. Symp. in Pure Math.*), Vol 53, 33-88, (1991).

[15] Chi Q.S. and Schwachhöfer L., *Exotic holonomy on moduli spaces of rational curves,* (*Differ. Geom. Appl.*), Vol 8, No 2, 105-134 (1998).

[16] Dimitrić I., *A note on equivariant embeddings of Grassmannians,* (*Publ. Inst. Math.*), Vol 59 (73), 131-137 (1996).

[17] Eells J. and Salamon S., *Constructions twistorielles des applications harmoniques,* (*C.R. Acad. Sci. Paris* Sér. I Math. Vol 296, 685-687, (1983).

[18] Fulton W. and J. Harris, *Representation Theory,* Springer-Verlag, GTM 129, New York, Berlin, Heidelberg, (1991).

[19] Hamermesh M., *Group theory and its application to physical processes,* Addison-Wesley Publishing Company, Reading, Massachusetts, (1962).

[20] Hangan Th., *Le pseudo-groupe projectif grassmannien (Romanian),* (*An. Sti. Univ. Al. I. Cuza Iasi*), N. Ser., Sect. Ia 11B, 349-355 (1965).

[21] Hangan Th., *Tensor-product tangent bundles,* (Arch. der Math.), Vol 19 (4), 436-448, (1968).

[22] Hangan Th., *Géometrie différentielle grassmannienne,* (Revue Roum. Math. *Pures et Appl.*), Vol 11, No 5, 519-531, (1966).

[23] Hangan Th., *Equations tensorielles de certaines métriques riemannienes d'espaces symétriques,* (*C.R. Acad. Sc. Paris*), Série A, Vol 277, 849–851, (1973).

[24] Hangan Th., *Variétés et transformations infinitésimales homographiques,* (*C.R. Acad. Sc. Paris*), Série A, Vol 281, 859-861, (1975).

[25] Hangan Th., *Sur l'intégrabilité des structures tangentes produits tensoriels réels,* (*Ann. Mat. Pura et Appl.*), Vol 126, 149-185, (1980).

[26] Hangan Th., *On the Riemannian structure of the real and complex Grassmann manifolds,* (*Tensor*), Vol 18, 26-31, (1967).

[27] Hano J. and Ozeki H., *On the holonomy group of linear connections,* (*Nagoya Math. J.*), Vol 10, 97-100, (1956).

[28] Kobayashi S. and Nomizu K., *Foundations of differential geometry Vol. 1 and 2,* (Interscience Publishers, New York, London), (1963).

[29] Krupka D., *The trace decomposition of tensors of type (1,2) and (1,3),* (New Developments in *Differential Geomatry, Proc. Colloq. on Diff. Geom., July 1994, Debrecen, J. Szenthe and L. Tamássy, Eds.*), Kluwer Academic Publishers, Dordrecht, 243-253, (1996).

[30] Leichtweiss K., *Zur Riemannschen Geometrie in Grassmannschen Mannigfaltigkeiten*, (*Math. Zeitschr.*), Vol 76, 334-366, (1961).

[31] Machida Y. and Sato H., *Twistor theory of manifolds with Grassmannian structures,* (*Nagoya Math. J.*), Vol 160, 17-102, (2000).

[32] Manin Yu. I., *Gauge Field Theory and Complex Geometry,* 2nd Edition, Springer-Verlag, New York, 1997.

[33] Manin Yu. I., *New dimensions in geometry,* (*Lect. Notes in Math.,*) 1111, Springer-Verlag, New York, 59-101, (1985).

[34] Marchiafava S., *Varietà localmente grassmanniane quaternionali,* (*Atti Accad. naz. dei Lincei*), Vol 57, 80-89, (1974).

[35] Merkulov S. A., *Paraconformal supermanifolds and nonstandard N-extended supergravity models,* (*Classical Quantum Grav.*), Vol 8, 557-569, (1991).

[36] Merkulov S. and Schwachhöfer L., *Classification of irreducible holonomies of torsion-free affine connections*, (*Ann. Math. (2)*), Vol 150, No 1, 77-149, (1999).

[37] Merkulov S. and Schwachhöfer L., *Addendum to: Classification of irreducible holonomies of torsion-free affine connections.*, (*Ann. Math. (2)*), Vol 150, No 3, 1177-1179, (1999).

[38] Penrose R., *The twistor programme,* (*Rep. Math. Phys.,*) Vol 12, 65-76, (1977).

[39] Pontryagin L.S., *Characteristic cycles on differentiable manifolds,* (*Matematicheskiĭ Sbornik (NS)*), Vol 21 (63), 233-284 (1947), AMS (1950), Translation n. 32.

[40] Salamon S., *Riemannian geometry and holonomy groups,* (*Pitman Research Notes in Mathematics Series*), **201**, John Wiley & Sons., New York, (1989).

[41] Simon U., Schwenk-Schellschmidt A. and Viesel H., *Introduction to the affine differential geometry of hypersurfaces,* (*Lecture Notes*), Science University of Tokyo, (1991).

[42] Strichartz R., *Linear algebra of curvature tensors and their covariant derivative,* (*Can. J. Math.*), Vol XL, No 5, 1105-1143, (1988).

[43] Tamássy L., *Über den Affinzusammenhang von, zu tangentialräumen gehörenden Produkträumen,* (*Acta Math. Acad. Sci. Hungar.*), Vol 11, No 1-2, 65-82, (1960).

[44] Tamássy L., *Über tensorielle Übertragungen spezieller Art,* (*Publicationes Mathematicae*), Vol 11, No 1-4, 273-277, (1964).

[45] Tamássy L., *Über autoparallele Flächen tensorial zusammenhängender Räume,* (*Acta Math. Acad. Sci. Hungar.*), Vol 16, No 1-2, 75-87, (1965).

[46] Tamássy L., *Aus dekomponierbaren Elementen bestehende Gebilde eines Produktraumes,* (*Matem. vesnik*), Vol 2 (17), 121-126, (1965).

[47] Vukmirović S., *Para-quaternionic Kähler reduction,* preprint, (2003), arXiv: `math. DG/0304424`.

[48] Wolf J.A., *Space forms of Grassmann manifolds,* (*Can. J. Math.*), Vol 15, 193-205, (1963).

[49] Zelikin M.I., *Homogeneous spaces and the Riccati equation in variational calulus (russian ed.),* (*Factorial*), Moscow, (1998).

HYPERSURFACES OF TYPE NUMBER 2 IN THE HYPERBOLIC FOUR-SPACE AND THEIR EXTENSIONS TO RIEMANNIAN GEOMETRY

Oldřich Kowalski*
Faculty of Mathematics and Physics
Charles University in Prague,
Sokolovská 83, 186 75 *Praha* 8,
Czech Republic
kowalski@karlin.mff.cuni.cz

Masami Sekizawa†
Tokyo Gakugei University
Koganei-shi Nukuikita-machi 4-1-1,
Tokyo 184-8501, *Japan*
sekizawa@u-gakugei.ac.jp

Abstract Pseudo-symmetric spaces of constant type in dimension 3 are Riemannian three-manifolds whose Ricci tensor has, at all points, one double eigenvalue and one simple constant eigenvalue. In this paper we give a survey of our published results for the case when the constant Ricci eigenvalue is negative. In particular, we show that three-dimensional hypersurfaces of the hyperbolic space $\mathbb{H}^4$ whose second fundamental form has rank 2 belong to this class. An explicit classification is presented in the case when the space admits so-called asymptotic foliation. Based on this, we show some existence theorems about local isometric embeddings of such spaces into $\mathbb{H}^4$.

MSC: 53C42, 53C21, 53C12, 53B25

Keywords: Riemannian manifold, hyperbolic space, principal Ricci curvature, hypersurface, type number

*This research was supported by the grant GA ČR 201/02/0616 and was partly supported by the project MSM 113200007.
†This research was supported by the Grant-in-Aid for Scientific Research (C) 14540066.

1. Introduction

According to [6], a Riemannian manifold (M,g) is said to be *pseudo-symmetric* if the following formula holds for arbitrary vector fields X and Y on M:

$$R(X,Y)\cdot R = F((X\wedge Y)\cdot R), \tag{1.1}$$

where

a) R denotes the Riemannian curvature tensor of type (1,3) on (M,g) and

$$R(X,Y) = [\nabla_X, \nabla_Y] - \nabla_{[X,Y]}$$

denote the corresponding curvature transformations,

b) $X\wedge Y$ denotes the endomorphism of the tangent bundle TM defined by

$$(X\wedge Y)Z = g(Y,Z)X - g(X,Z)Y, \tag{1.2}$$

c) F is a smooth function on M,

d) the dot in each side of the formula (1.1) denotes the derivation on the tensor algebra of TM induced by an endomorphism of this tangent bundle.

We call a pseudo-symmetric space (M,g) *of constant type* if $F = \tilde{c} =$ constant.

We obtain easily the following characterization in dimension 3:

Proposition 1.1. *A three-dimensional Riemannian manifold (M,g) is pseudo-symmetric of constant type $F = \tilde{c}$ if and only if its principal Ricci curvatures ρ_1, ρ_2 and ρ_3 locally satisfy the following conditions (up to a numeration):*

(1) $\rho_3 = 2\tilde{c}$,

(2) $\rho_1 = \rho_2$ *everywhere.*

We are not interested in the case when (M,g) is a space of constant curvature and therefore we assume always $\rho_1 = \rho_2 \neq \rho_3$.

If $\tilde{c} = 0$, and hence $F = 0$, we obtain a definition of *semi-symmetric space.* The theory of semi-symmetric spaces has been developed in [23, 25, 24, 11, 7, 1, 2] and especially in the book [3]. For the three-dimensional case, see the explicit classification in [11, 7] and [3, Chapter 6]. We exclude this case from our considerations.

For $\tilde{c} \neq 0$, the present authors made an explicit classification in [15] for so-called "asymptotically foliated" (or "non-elliptic") spaces in dimension 3. The aim of this paper is to present this explicit classification for the case $\tilde{c} < 0$ ("hyberbolic case") together with some local embedding theorems.

Now, we shall turn to the theory of hypersurfaces. Recall that the rank of the second fundamental form of a hypersurface is also called *type number* (see [10]). A classical result (known already by W. Killing) says that a hypersurface $M \subset \mathbb{R}^{n+1}$ $(n \geq 3)$ with type number $t(p) \geq 3$ everywhere is locally rigid (*cf.*, [10, Chapter VII]). The same result holds if one replaces $\mathbb{R}^{n+1}$ by any space of constant curvature $\tilde{c}$. Now, the complete connected hypersurfaces of $\mathbb{R}^{n+1}$ with type number 0 are hyperplanes and those with type number 1 are generalized cylinders. They are always isometrically deformable (in a continuous way). On the other hand, the local metric properties of hypersurfaces with type number 2 were studied thoroughly by É. Cartan [4]. He proved (by purely extrinsic methods) that just the following three possibilities can occur:

a) M is locally isometrically deformable (in a continuous way),

b) M is locally rigid,

c) M is locally deformable in a unique way: to every sufficiently small domain $\Sigma \subset M$ there is exactly one hypersurface $\bar{\Sigma} \subset \mathbb{R}^{n+1}$ which is isometric to Σ and which can not be obtained from Σ by rigid motions or reflections.

In the doctoral Thesis by V. Hájková [7] (see also [3, Chapter 10]) a different approach was used to construct new classes of examples satisfying a), b) or c). The method is based on the explicit description of three-dimensional Riemannian manifolds with conullity 2 as studied in [11]. Later in the same year, M. Dajczer, L. Florit and R. Tojeiro [5] have given a revised version of Cartan's paper (with the generalization to arbitrary space forms) and they also constructed examples of hypersurfaces of space forms with the property c) using purely extrinsic methods.

Now, the natural extension of [7] is the intrinsic study of hypersurfaces with type number 2 in general space forms. The induced metrics of such hypersurfaces make them Riemannian manifolds of $\tilde{c}$-conullity 2 (see [3, 15]). If we consider a hypersurface with type number 2 in the four-dimensonal space form with constant curvature $\tilde{c}$, then the induced Riemannian metric satisfies the properties from Proposition 1.1 (see [8]).

We shall now describe shortly the method by V. Hájková which works well also for arbitrary four-dimensional space forms. On a given Riemannian three-manifold satisfying the conditions of Proposition 1.1, we are looking for a linear operator $\boldsymbol{S}$ of rank 2 satisfying *formally* the

Codazzi equation and the Gauss equation. This leads to a system of partial differential equations to be solved by the Cauchy-Kowalewski theorem.

2. The basic system of partial differential equations for the problem

This section was extracted from [14, 15, 16, 17, 18] where more details can be found.

Let (M, g) be a three-dimensional Riemannian manifold whose Ricci tensor $\hat{R}$ has eigenvalues $\rho_1 = \rho_2 \neq \rho_3$ with nonzero constant ρ_3. Choose a neighborhood $\tilde{U}$ of a fixed point $m \in M$ and a smooth vector field E_3 of unit eigenvectors corresponding to the Ricci eigenvalue ρ_3 in $\tilde{U}$. Let $S : D^2 \to \tilde{U}$ be a surface through m which is transversal with respect to all trajectories generated by E_3 at all cross-points and not orthogonal to such a trajectory at m. (The vector field E_3 determines an orientation of S). Then there is a normal neighborhood U of m, $U \subset \tilde{U}$, with the property that each point $p \in U$ is projected to exactly one point $\pi(p) \in S$ via some trajectory. We fix any local coordinate system (w, x) on S and then a local coordinate system (w, x, y) on U such that the values $w(p)$ and $x(p)$ are defined as $w(\pi(p))$ and $x(\pi(p))$, respectively, for each point $p \in U$, $y(p)$ is the oriented length $d^+(\pi(p), p)$ of the trajectory joining p with $\pi(p)$. Then $E_3 = \partial/\partial y$ can be extended in U to an orthonormal moving frame $\{E_1, E_2, E_3\}$. Let $\{\omega^1, \omega^2, \omega^3\}$ be the corresponding dual coframe. Then ω^i's are of the form

$$\left.\begin{aligned} \omega^i &= a^i \mathrm{d}w + b^i \mathrm{d}x, \quad i = 1, 2, \\ \omega^3 &= dy + H\mathrm{d}w + G\mathrm{d}x. \end{aligned}\right\} \tag{2.1}$$

The Ricci tensor $\hat{R}$ expressed with respect to $\{E_1, E_2, E_3\}$ has the form $\hat{R}_{ij} = \rho_i \delta_{ij}$. Because each ρ_i is expressed through the sectional curvature K_{ij} by the formula $\rho_i = \hat{R}_{ii} = \sum_{j \neq i} K_{ij}$, there exist a function $k = k(w, x, y)$ of the variables w, x and y, and a constant $\tilde{c}$ such that

$$\left.\begin{aligned} &K_{12} = k, \quad K_{13} = K_{23} = \tilde{c}, \\ &\rho_1 = \rho_2 = k + \tilde{c}, \quad \rho_3 = 2\tilde{c}. \end{aligned}\right\} \tag{2.2}$$

Define now the components ω^i_j of the connection form by the standard formulas

$$\left.\begin{aligned} &\mathrm{d}\omega^i - \sum \omega^j \wedge \omega^i_j = 0, \\ &\omega^i_j + \omega^j_i = 0, \quad i, j = 1, 2, 3. \end{aligned}\right\} \tag{2.3}$$

Because the Riemannian curvature tensor satisfies $R_{ijkl} = 0$ whenever at least three of the indices i, j, k and l are distinct, the formulas (2.2) are equivalent to

$$\left.\begin{aligned} \mathrm{d}\omega_2^1 + \omega_3^1 \wedge \omega_2^3 &= k\,\omega^1 \wedge \omega^2, \\ \mathrm{d}\omega_3^1 + \omega_2^1 \wedge \omega_3^2 &= \tilde{c}\,\omega^1 \wedge \omega^3, \\ \mathrm{d}\omega_3^2 + \omega_1^2 \wedge \omega_3^1 &= \tilde{c}\,\omega^2 \wedge \omega^3. \end{aligned}\right\} \tag{2.4}$$

Next, differentiate the equations (2.4) and substitute from (2.4). We obtain easily

$$\omega_3^1 \wedge \omega^1 \wedge \omega^2 = 0, \quad \omega_3^2 \wedge \omega^1 \wedge \omega^2 = 0 \tag{2.5}$$

and

$$\mathrm{d}((k - \tilde{c})\,\omega^1 \wedge \omega^2) = 0. \tag{2.6}$$

The relations (2.5) mean that ω_3^1 and ω_3^2 are linear combinations of ω^1 and ω^2 only, and from the third equation of (2.3) it follows that $\mathrm{d}\omega^3$ is a multiple of $\omega^1 \wedge \omega^2$ *i.e.*, a multiple of $\mathrm{d}w \wedge \mathrm{d}x$. Then (2.1) implies that the functions G and H are independent of y.

Defining a new variable $\bar{w}$ as a potential function of the form $H\mathrm{d}w + G\mathrm{d}x$ and using a convenient rotation of our moving frame at each point, we obtain easily

Proposition 2.1. *In a normal neighborhood of any point $m \in M$ there exist an orthonormal coframe $\{\omega^1, \omega^2, \omega^3\}$ and a local coordinate system (w, x, y) such that*

$$\left.\begin{aligned} \omega^1 &= f\mathrm{d}w, \\ \omega^2 &= A\mathrm{d}x + C\mathrm{d}w, \\ \omega^3 &= \mathrm{d}y + H\mathrm{d}w. \end{aligned}\right\} \tag{2.7}$$

Here f, A and C are smooth functions of the variables w, x and y, $fA \neq 0$, and H is a smooth function of the variables w and x.

The formula (2.6) can be now written in the form

$$((k - \tilde{c})fA)'_y = 0, \quad i.e., \quad k - \tilde{c} = \frac{\sigma}{fA} \tag{2.8}$$

for some nonzero function $\sigma = \sigma(w, x)$.

Now, define the function $\chi = \chi(w, x, y)$ of the variables w, x and y by

$$\chi = \frac{1}{fA} = \frac{k - \tilde{c}}{\sigma}. \tag{2.9}$$

Then, using (2.7) and (2.9), we obtain easily the following expression for the components of the connection form:

$$\left.\begin{aligned} \omega_2^1 &= -A\alpha \mathrm{d}x + R\mathrm{d}w + \beta \mathrm{d}y, \\ \omega_3^1 &= A\beta \mathrm{d}x + S\mathrm{d}w, \\ \omega_3^2 &= A'_y \mathrm{d}x + T\mathrm{d}w, \end{aligned}\right\} \tag{2.10}$$

where

$$\left.\begin{aligned} \alpha &= \chi(A'_w - C'_x - HA'_y), \\ \beta &= \frac{\chi}{2}(H'_x + AC'_y - CA'_y) \end{aligned}\right\} \tag{2.11}$$

and

$$\left.\begin{aligned} R &= \chi f f'_x - C\alpha + H\beta, \\ S &= f'_y + C\beta, \\ T &= C'_y - f\beta. \end{aligned}\right\} \tag{2.12}$$

The curvature conditions (2.4) then give a system of nine partial differential equations for our problem:

(A1) $(A\alpha)'_y + \beta'_x = 0,$

(A2) $R'_y - \beta'_w = 0,$

(A3) $(A\alpha)'_w + R'_x + SA'_y - A\beta T = -fAk,$

(B1) $A''_{yy} - A\beta^2 = -\tilde{c}A,$

(B2) $-A''_{yw} + T'_x + A(\beta R + \alpha S) = \tilde{c}AH,$

(B3) $T'_y - S\beta = -\tilde{c}C,$

(C1) $(A\beta)'_y + A'_y\beta = 0,$

(C2) $S'_x - (A\beta)'_w - (A\alpha T + A'_y R) = 0,$

(C3) $S'_y + T\beta = -\tilde{c}f.$

Now we define the functions a, b, c and e by the following formula:

$$a = \frac{f'_y}{f}, \quad b = \beta, \quad c = \beta - h\chi, \quad e = \frac{A'_y}{A}, \tag{2.13}$$

where $h = H'_x$. Then, using the standard formula from [9], we obtain

$$\left.\begin{array}{ll} \nabla_{E_1} E_1 = -\chi f'_x E_2 - aE_3, & \nabla_{E_1} E_2 = \chi f'_x E_1 - cE_3, \\ \nabla_{E_2} E_1 = \alpha E_2 - bE_3, & \nabla_{E_2} E_2 = -\alpha E_1 - eE_3, \\ \nabla_{E_1} E_3 = aE_1 + cE_2, & \nabla_{E_2} E_3 = bE_1 + eE_2, \\ \nabla_{E_3} E_1 = -bE_2, & \nabla_{E_3} E_2 = bE_1, \\ \nabla_{E_3} E_3 = 0. & \end{array}\right\} \tag{2.14}$$

The last formula of (2.14) shows that the trajectories of the unit vector field E_3 (consisting of the eigenvectors of the Ricci tensor $\hat{R}$ corresponding to $\rho_3 = 2\tilde{c}$) are geodesics. They are called *principal geodesics.*

3. The asymptotic foliations and four types of spaces

This section was extracted from [14, 15, 16, 17, 18] where more details can be found.

We introduce two basic definitions.

Definition 3.1. A smooth surface $N \subset (M, g)$ is called an *asymptotic leaf* if it is generated by the principal geodesics and its tangent planes are parallel along these principal geodesics with respect to the Levi-Civita connection ∇ of (M, g).

Definition 3.2. An *asymptotic distribution* on M is a two-dimensional distribution which is integrable and whose integral manifolds are asymptotic leaves. The integral manifolds of an asymptotic distribution determine a foliation of M, which is called an *asymptotic foliation.*

For the asymptotic distribution, one can derive the formula

$$c(\omega^1)^2 + (e - a)\omega^1\omega^2 - b(\omega^2)^2 = 0, \tag{3.1}$$

where a, b, c and e are given by (2.13). For the details of the proof see [11, Section 6] or [3, Chapter 5].

The following proposition is almost obvious.

Proposition 3.3. *Let* $\Delta = (e - a)^2 + 4bc$ *be the discriminant of the quadratic equation* (3.1). *Then we have:*

(E) *If* $\Delta < 0$ *on* (M, g), *then there is no real asymptotic distribution on*M.

(H) *If* $\Delta > 0$ *on* (M, g), *then there are exactly two different asymptotic distributions on* M.

(P) *If* $\Delta = 0$ *on* (M, g) *and some of the functions* $e - a$, b *and* c *are nonzero at each point, then there is a unique asymptotic distribution on* M.

(Pℓ) *If* $e - a = b = c = 0$ *on* M, *then any* π*-projectable smooth two-dimensional distribution on* M *is asymptotic, where* π *is the projection* $\pi : (w, x, y) \longmapsto (w, x)$.

Definition 3.4. A space (M, g) is said to be of *type* (E), (H), (P) or (Pℓ), respectively, (called also *elliptic*, *hyperbolic*, *parabolic* and *planar* type respectively) if the corresponding case of Proposition 3.3 holds on the whole of M.

Corollary 3.5. *The space* (M, g) *is of type* (Pℓ) *if and only if* $f = \xi A$, $C = \zeta A$ *and* $\beta = 0$, *where* $\xi = \xi(w, x) \neq 0$ *and* $\zeta = \zeta(w, x)$ *are arbitrary functions of the variables* w *and* x. *Assuming* $\beta = 0$, (M, g) *is of type* (P) *if and only if* $f = \xi A$ *and* $h \neq 0$.

In the next section we are going to describe explicitly all spaces of types (H), (P) and (Pℓ). Because we are interested in the local classification, we investigate only the "pure" cases and not the combined ones in the sequel. (For a global treatment of *some* of our geometric types see [25]).

Now, the following theorem will be crucial for the explicit geometric classification of the manifolds of types (H), (P) and (Pℓ) in Section 4.

Theorem 3.6. *For each manifold of types* (H) *and* (P), *there exists a transformation of local coordinates preserving the form* (2.7) *of the metric and annihilating the function* $\beta = b$.

Outline of the proof: we choose one asymptotic distribution $\mathfrak{F}$ and introduce new local variables $\bar{w}$ and $\bar{x}$ such that the distribution $\mathfrak{F}$ will be determined by the equation $d\bar{w} = 0$. After that we also use a rotation of the moving frame to adapt ourselves to (2.7).

Remark. As concerns the type (Pℓ), we have $\beta = 0$ (in a neighborhood of p) by definition. Thus for every space (M, g) of type (H), (P) or (Pℓ) we can assume $\beta = 0$. Conversely, from (3.1) we see that $\beta = 0$ always implies that (M, g) is one of the types (H), (P) and (Pℓ).

Propositions 6.10 and 6.11 from [11] still hold without change. We have

Proposition 3.7. *If $h = 0$, then (M, g) is of type* (H), (P) *or* (Pℓ). *On a space of type* (H), *$h = 0$ means that the asymptotic foliations $\mathcal{F}_1$ and $\mathcal{F}_2$ are mutually orthogonal.*

Proposition 3.8. *Let the metric g be of one of the types* (H), (P) *and* (Pℓ) *expressed in such a coordinate system that $\beta = 0$. If $\alpha = 0$, then at least one of the asymptotic foliations is totally geodesic.*

4. The explicit classification of asymptotically foliated spaces

This section was extracted from [17, 18] and [8] where more details can be found.

In this section we shall explicitly classify all spaces of types (H), (P) and (Pℓ). Moreover, we shall answer the question *how the distinct locally isometry classes can be parametrized. In this section we always assume* $\beta = 0$, which is allowed by Theorem 3.6.

In the sequel, we always suppose that $\tilde{c}$ is of the form $-\lambda^2$, where $\lambda > 0$. As a rule, we omit the full proofs of the corresponding theorems.

In most cases of the classification, we also formulate the corresponding *embedding theorem*, which says how many spaces from the corresponding class can be locally isometrically embedded into $\mathbb{H}^4$ and, moreover, which are the deformation properties of the corresponding hypersurfaces (see [8] for more details).

We shall start with some general results.

Proposition 4.1. *For types* (H), (P) *and* (Pℓ), *the coefficients A, C and f from* (2.7) *can be expressed by*

$$\left.\begin{aligned} A &= p\cosh(\lambda y) + q\sinh(\lambda y),\\ C &= r\cosh(\lambda y) + s\sinh(\lambda y),\\ f &= t\cosh(\lambda y) + u\sinh(\lambda y), \end{aligned}\right\} \tag{4.1}$$

where p, q, r, s, t and u are functions of the variables w and x such that

$$\lambda(qr - ps) = h. \tag{4.2}$$

Moreover, if $h \neq 0$, we may assume $h = 1$ and $H = x$, and if $h = 0$ on an open subset, we may assume $H = 0$ on this subset.

In the sequel we put

$$\left.\begin{aligned} \mathcal{D} &= p'_w - r'_x - \lambda qH,\\ \mathcal{E} &= q'_w - s'_x - \lambda pH. \end{aligned}\right\} \tag{4.3}$$

Proposition 4.2. *The differential equation* (A1) *is satisfied if and only if the following equation holds:*

$$u\,\mathcal{D} - t\,\mathcal{E} = 0. \tag{4.4}$$

Proposition 4.3. *Assume that* (A1) *is satisfied and*

$$A\alpha = \nu, \quad \nu = \nu(w, x). \tag{4.5}$$

Then (A2) *is satisfied if and only if*

$$\lambda(qt'_x - pu'_x) = h\nu. \tag{4.6}$$

Now we can state the "converse" of Proposition 4.1.

Proposition 4.4. *Let p, q, r, s, t and u be arbitrary functions of the variables w and x. Define the functions A, C and f by* (4.1), *and let $H = H(w, x)$ be any function satisfying*

$$H'_x = h = \lambda(qr - ps). \tag{4.7}$$

If the equations (A1) *and* (A2) *are satisfied, then* (2.7) *defines a foliated metric of type* (H), (P) *or* (Pℓ).

Because $b = \beta = 0$, the equation (3.1) for an asymptotic distribution reads

$$\omega^1(c\omega^1 + (e - a)\omega^2) = 0 \tag{4.8}$$

or, equivalently, by (2.13) and $(2.11)_2$ together with $\beta = 0$,

$$\omega^1\left(\left(\frac{C}{A}\right)'_y \omega^1 - \left(\frac{f}{A}\right)'_y \omega^2\right) = 0. \tag{4.9}$$

We see that the equation $\omega^1 = 0$ defines an asymptotic distribution $\text{span}\{E_2, E_3\}$, whose integral manifolds (the asymptotic leaves) are given by the equation $w =$ constant. From (2.14) it follows at once that this foliation is totally geodesic if and only if the function α from (2.11) vanishes identically. We distinguish two geometric situations on the spaces of types (H) and (P). We say a space of type (H) or (P) is *singular* or *generic* according to whether the asymptotic distribution given by the equation $\omega^1 = 0$ is totally geodesic or not; or, equivalently, according to whether $\alpha = 0$ or $\alpha \neq 0$.

4.1 The non-orthogonally foliated spaces of type (H)

Suppose now that the space (M, g) is of type (H) and the asymptotic foliations $\mathcal{F}_1$ and $\mathcal{F}_2$ are nowhere mutually orthogonal. From (4.9) we see that then necessarily $(f/A)'_y \neq 0$ and $(C/A)'_y \neq 0$. Using (4.1) we see that this is equivalent with the inequalities $pu - qt \neq 0$ and $ps - qr \neq 0$.

Now it is easy to prove the following

Theorem 4.5. *The metric of a three-dimensional non-orthogonally foliated generic space of type* (H) *is locally determined by an orthonormal coframe*

$$\left.\begin{aligned} \omega^1 &= [t\cosh(\lambda y) + u\sinh(\lambda y)]\mathrm{d}w, \\ \omega^2 &= [p\cosh(\lambda y) + q\sinh(\lambda y)]\mathrm{d}x \\ &\qquad + [r\cosh(\lambda y) + s\sinh(\lambda y)]\mathrm{d}w, \\ \omega^3 &= \mathrm{d}y + x\,\mathrm{d}w, \end{aligned}\right\} \quad (4.10)$$

where p, q, r and s are arbitrary functions of the variables w and x such that $\lambda(ps - qr) = 1$, and t and u are calculated from p, q, r and s as follows:
Put (as a special case of (4.3)*)*

$$\mathcal{D} = p'_w - r'_x - \lambda q x, \quad \mathcal{E} = q'_w - s'_x + \epsilon\lambda p x. \quad (4.11)$$

If $\mathcal{E} \neq 0$, then

$$\left.\begin{aligned} t &= \frac{u\,\mathcal{D}}{\mathcal{E}}, \\ u &= \exp\Big(\frac{1}{2}\int P\,\mathrm{d}x\Big)\Big[\int Q\exp\Big(-\int P\,dx\Big)\mathrm{d}x\Big]^{1/2}, \end{aligned}\right\} \quad (4.12)$$

where

$$P = \frac{2q(\mathcal{D}'_x\mathcal{E} - \mathcal{D}\mathcal{E}'_x)}{\mathcal{E}(q\,\mathcal{D} - p\,\mathcal{E})}, \quad Q = \frac{2\mathcal{E}^2}{\lambda(q\,\mathcal{D} - p\,\mathcal{E})}. \quad (4.13)$$

If $\mathcal{D} \neq 0$, then

$$\left.\begin{aligned} t &= \exp\Big(\frac{1}{2}\int \bar{P}\,\mathrm{d}x\Big)\Big[\int \bar{Q}\exp\Big(-\int \bar{P}\,dx\Big)\mathrm{d}x\Big]^{1/2}, \\ u &= \frac{t\,\mathcal{E}}{\mathcal{D}}, \end{aligned}\right\} \quad (4.14)$$

where

$$\bar{P} = \frac{2p(\mathcal{D}\mathcal{E}'_x - \mathcal{D}'_x\mathcal{E})}{\mathcal{D}(q\,\mathcal{D} - p\,\mathcal{E})}, \quad \bar{Q} = \frac{2\mathcal{D}^2}{\lambda(q\,\mathcal{D} - p\,\mathcal{E})}. \quad (4.15)$$

The local isometry classes of the metric (4.10) *are parametrized by three arbitrary functions of two variables modulo two arbitrary functions of one variable.*

Remarks.

a) The isometry part of Theorem 4.5 (and of some other theorems which follow) can be stated more precisely using the concept of germs (*cf.* [19]).

b) In Theorem 11.38 of [3] the formulas (11.114) contain a misprint. The coefficients 2 should be in the numerators (not in the denominators) of the formulas for the both functions P and Q. Our formulas (4.13) give the correct version.

Theorem 4.6. *The metric of a three-dimensional non-orthogonally foliated singular space of type* (H) *is locally determined by an orthonormal coframe*

$$\left.\begin{aligned} \omega^1 &= [t\cosh(\lambda y) + u\sinh(\lambda y)]\mathrm{d}w, \\ \omega^2 &= [p\cosh(\lambda y) + (\varphi - p)\sinh(\lambda y)]\mathrm{d}x \\ &\qquad + r[\cosh(\lambda y) - \sinh(\lambda y)]\mathrm{d}w, \\ \omega^3 &= \mathrm{d}y + H\mathrm{d}w, \end{aligned}\right\} \tag{4.16}$$

where $\varphi = \varphi(w, x)$ *is an arbitrary nonzero function of the variables* w *and* x *such that* $(\ln|\varphi|)''_{wx} \neq 0$ *on an open set, and*

$$p = \frac{1}{2}\varphi + \frac{(\ln|\varphi|)''_{xx}}{\lambda^2\varphi} - \frac{[(\ln|\varphi|)'_x]^2}{2\lambda^2\varphi} + \frac{\psi}{\varphi},$$

$$r = \frac{(\ln|\varphi|)''_{wx}}{\lambda^2\varphi}, \qquad H = \frac{(\ln|\varphi|)'_w}{\lambda},$$

where $\psi = \psi(x)$ *is an arbitrary function of the variable* x. *Further,* t *can be chosen as an arbitrary function of the variables* w *and* x, *and* $u = \displaystyle\int (\varphi/p - 1)t'_x\,\mathrm{d}x$.

4.2 The orthogonally foliated spaces of type (H)

For the metrics of type (H) it may also happen that $h = 0$ everywhere (and thus $H = 0$ due to Proposition 4.1), which means, according to Proposition 3.7, that the asymptotic foliations are mutually orthogonal. This case was excluded in the previous section. As we see from (4.8), one of the asymptotic distributions is given by the equation $\omega^1 = 0$. Consequently, the second must be given by the equation $\omega^2 = 0$. Thus, the equation (4.9) implies that $(C/A)'_y = 0$. Hence, there exists a function

$\mu = \mu(w, x)$ of the variables w and x such that $C = \mu A$, and hence

$$\left.\begin{aligned} \omega^1 &= [t\cosh(\lambda y) + u\sinh(\lambda y)]\mathrm{d}w, \\ \omega^2 &= [p\cosh(\lambda y) + q\sinh(\lambda y)](\mathrm{d}x + \mu\mathrm{d}w), \\ \omega^3 &= \mathrm{d}y. \end{aligned}\right\}$$

Introducing the new variable $\bar{x}$ instead of x, where $\bar{x} = \bar{x}(w, x)$ is a potential function of the Pfaffian equation $\mathrm{d}x + \mu\,\mathrm{d}w = 0$, we get $C = 0$.

The conditions (4.4) and (4.6) now read (in the standard notation)

$$\left.\begin{aligned} up'_w - tq'_w &= 0, \\ pu'_x - qt'_x &= 0. \end{aligned}\right\} \tag{4.17}$$

Moreover, we must have $pu - qt \neq 0$, otherwise the metrics would be not of type (H). We put $\varphi = pu - qt$ and rewrite (4.17) in the equivalent form

$$\left.\begin{aligned} up'_w - tq'_w &= 0, \\ up'_x - tq'_x &= \varphi'_x, \\ pu - qt &= \varphi. \end{aligned}\right\} \tag{4.18}$$

Again we distinguish two geometric situations: the generic case and the singular case.

We first treat the *generic* case where the asymptotic distribution given by the equation $\omega^1 = 0$ is not totally geodesic. We remember that this requirement is equivalent to $\alpha \neq 0$.

Now it is easy to prove the following

Theorem 4.7. *The metric of a three-dimensional orthogonally foliated generic space of type* (H) *is locally determined by an orthonormal coframe*

$$\left.\begin{aligned} \omega^1 &= [t\cosh(\lambda y) + u\sinh(\lambda y)]\mathrm{d}w, \\ \omega^2 &= [p\cosh(\lambda y) + q\sinh(\lambda y)]\mathrm{d}x, \\ \omega^3 &= \mathrm{d}y, \end{aligned}\right\} \tag{4.19}$$

where p and q are arbitrary functions of the variables w and x satisfying $pq'_w - qp'_w \neq 0$, and t and u are calculated from p and q by

$$t = \frac{\varphi p'_w}{pq'_w - qp'_w}, \qquad u = \frac{\varphi q'_w}{pq'_w - qp'_w}, \tag{4.20}$$

where

$$|\varphi| = \exp\Big(\int \frac{p'_x q'_w - p'_w q'_x}{p q'_w - p'_w q}\,\mathrm{d}x\Big). \tag{4.21}$$

The local isometry classes are parametrized by two arbitrary functions of two variables modulo two arbitrary functions of one variable and a real parameter.

We shall now formulate the corresponding embedding theorem from [8]:

Theorem 4.8. *The three-dimensional orthogonally foliated generic spaces of type* (H) *which can be locally realized as hypersurfaces in* $\mathbb{H}^4(-\lambda^2)$ *depend on one arbitrary function of two variables, six arbitrary functions of one variable and five real parameters. For the proper choice of the involved functions (which still remain "arbitrary"), these spaces can be realized as hypersurfaces in* $\mathbb{H}^4(-\lambda^2)$ *in exactly two qualitatively different ways (up to an isometry of* $\mathbb{H}^4(-\lambda^2)$*).*

There remains the *singular* case where the asymptotic distribution given by the equation $\omega^1 = 0$ is totally geodesic.

Here it is easy to prove the following

Theorem 4.9. *The metric of three-dimensional orthogonally foliated singular spaces of type* (H) *belong to the following two classes:*
Class I. *The orthonormal coframe is given by*

$$\left.\begin{aligned} \omega^1 &= [t\cosh(\lambda y) + u\sinh(\lambda y)]\mathrm{d}w,\\ \omega^2 &= [\cosh(\lambda y) + q\sinh(\lambda y)]\mathrm{d}x,\\ \omega^3 &= \mathrm{d}y, \end{aligned}\right\} \tag{4.22}$$

where q *is an arbitrary function of the variable* x*,* t *is an arbitrary function of the variables* w *and* x*, and the function* u *has the form* $u = \int q t'_x\,\mathrm{d}x$.
Class II. *The orthonormal coframe is given by*

$$\left.\begin{aligned} \omega^1 &= [\cosh(\lambda y) + u\sinh(\lambda y)]\mathrm{d}w,\\ \omega^2 &= \sinh(\lambda y)\mathrm{d}x,\\ \omega^3 &= \mathrm{d}y, \end{aligned}\right\} \tag{4.23}$$

where the function u *is an arbitrary function of the variables* w *and* x*.*

Remarks.

a) In all cases we shall ensure, by choosing proper initial conditions, that $\varphi = pu - qt \neq 0$.

b) The asymptotic distribution given by the equation $\omega^2 = 0$ is totally geodesic if and only if $t'_x = u'_x = 0$ in the above expression.

We have again the corresponding embedding theorem from [8]:

Theorem 4.10.

(1) The three-dimensional orthogonally foliated singular spaces in the class I *of type* (H) *which can be locally realized as hypersurfaces in* $\mathbb{H}^4(-\lambda^2)$ *depend on six arbitrary functions of one variable and five real parameters. For every such realization, the corresponding hypersurface is locally rigid.*

(2) The three-dimensional orthogonally foliated singular spaces in the class II *of type* (H) *which can be locally realized as hypersurfaces in* $\mathbb{H}^4(-\lambda^2)$ *depend on four arbitrary functions of one variable and five real parameters. For every such realization, the corresponding hypersurface is locally rigid.*

4.3 The spaces of type (P)

The only asymptotic distribution on any space of type (P) is defined by the equation $\omega^1 = 0$.

According to Corollary 3.5 and Proposition 4.1, all we have to assume is $H = x$, $h = 1$ and the parabolicity condition $f = \xi A$ with some nonzero function $\xi = \xi(w, x)$ of the variables w and x. Thus the only algebraic relations for the basic functions are

$$\lambda(qr - ps) = 1 \tag{4.24}$$

and

$$t = \xi p, \quad u = \xi q. \tag{4.25}$$

Now, we put

$$\left.\begin{aligned} D &= \lambda(p'_x q - p q'_x), \\ E &= \lambda[p q'_w - p'_w q + \lambda(-p^2 + q^2)x]. \end{aligned}\right\} \tag{4.26}$$

Taking use of (4.6) and (4.25), we can rewrite (4.5) as

$$\xi D = \nu. \tag{4.27}$$

We can also rewrite (4.4) as

$$q\mathcal{D} - p\mathcal{E} = 0, \tag{4.28}$$

which means

$$\lambda(ps'_x - qr'_x) = E$$

or, due to (4.24),

$$\lambda(q'_x r - p'_x s) = E. \tag{4.29}$$

Treating the generic case we have

Theorem 4.11. *The metric of a three-dimensional foliated generic space of type* (P) *is given by*

$$\left.\begin{aligned} \omega^1 &= \xi[p\cosh(\lambda y) + q\sinh(\lambda y)]\mathrm{d}w, \\ \omega^2 &= [p\cosh(\lambda y) + q\sinh(\lambda y)]\mathrm{d}x \\ &\qquad + [r\cosh(\lambda y) + s\sinh(\lambda y)]\mathrm{d}w, \\ \omega^3 &= \mathrm{d}y + x\,\mathrm{d}w, \end{aligned}\right\} \tag{4.30}$$

where p and q are arbitrary functions of the variables w and x, $qp'_x - pq'_x \neq 0$, and

$$\left.\begin{aligned} & r = \frac{p'_x - pE}{D}, \qquad s = \frac{q'_x - qE}{D}, \\ & \xi = \left[\frac{p'_w - r'_x - \lambda qx}{pD}\right]^{1/2} = \left[\frac{q'_w - s'_x + \lambda px}{qD}\right]^{1/2}. \end{aligned}\right\} \tag{4.31}$$

The local isometry classes are parametrized by two arbitrary functions of two variables modulo two arbitrary functions of one variable.

The corresponding embedding theorem from [8] is

Theorem 4.12. *The three-dimensional foliated generic spaces of type* (P) *which can be locally realized as hypersurfaces in* $\mathbb{H}^4(-\lambda^2)$ *depend on one arbitrary function of two variables and six arbitrary functions of one variable. For each such realization, the corresponding hypersurface is locally rigid, i.e., it cannot be (locally) isometrically deformed.*

Next treating the singular case we have

Theorem 4.13. *The metric of a three-dimensional foliated singular space of type* (P) *is given by*

$$\left.\begin{aligned} \omega^1 &= \xi p\cosh(\lambda y)\mathrm{d}w, \\ \omega^2 &= p\cosh(\lambda y)\mathrm{d}x + [r\cosh(\lambda y) + s\sinh(\lambda y)]\mathrm{d}w, \\ \omega^3 &= \mathrm{d}y + x\,\mathrm{d}w \end{aligned}\right\} \tag{4.32}$$

with

$$\left.\begin{aligned} p &= \frac{1}{\sqrt{\psi + \lambda^2 x^2}}, \\ r &= -\frac{\psi' x}{2\psi\sqrt{\psi + \lambda^2 x^2}} + \varphi, \quad s = -\lambda\sqrt{\psi + \lambda^2 x^2} \end{aligned}\right\} \tag{4.33}$$

or

$$\left.\begin{aligned} \omega^1 &= \xi q \sinh(\lambda y) \mathrm{d}w, \\ \omega^2 &= q \sinh(\lambda y) \mathrm{d}x + [r \cosh(\lambda y) + s \sinh(\lambda y)] \mathrm{d}w, \\ \omega^3 &= \mathrm{d}y + x\, \mathrm{d}w \end{aligned}\right\} \tag{4.34}$$

with

$$\left.\begin{aligned} q &= \frac{1}{\sqrt{\psi - \lambda^2 x^2}}, \\ r &= \lambda\sqrt{\psi - \lambda^2 x^2}, \quad s = -\frac{\psi' x}{2\psi\sqrt{\psi - \lambda^2 x^2}} + \varphi, \end{aligned}\right\} \tag{4.35}$$

where $\xi = \xi(w, x)$ *is an arbitrary nonzero function of the variables* w *and* x*, and* $\varphi = \varphi(w)$ *and* $\psi = \psi(w)$ *are arbitrary functions of the variable* w*.*

The corresponding embedding theorem from [8] is

Theorem 4.14. *A three-dimensional foliated singular space of type* (P) *and of the form* (4.32) *or* (4.34) *can be realized as a hypersurface in* $\mathbb{H}^4(-\lambda^2)$ *if and only if the nonzero function* $\xi = \xi(w, x)$ *of the variables* w *and* x *is given by the formula*

$$\xi^2 = \frac{1}{p^2}\left(\mu_1 x^2 - \mu_2 \frac{1}{p^2} + 2\mu_3 \frac{x}{p}\right), \tag{4.36}$$

where $\mu_i = \mu_i(w)$*,* $i = 1, 2, 3$*, are arbitrary functions of the variable* w *satisfying the inequalities* $\mu_1 > 0$ *and* $\mu_1\mu_2 + {\mu_3}^2 < 0$*. These metrics depend on five arbitrary functions of one variable. The corresponding hypersurface can be locally isometrically deformed in a continuous way.*

4.4 The spaces of type (Pℓ)

We are now left with the type (Pℓ) in which there are no singular solutions.

Theorem 4.15. *The metric of a three-dimensional foliated space of type* $(\mathrm{P}\ell)$ *is locally determined by an orthonormal coframe*

$$\left.\begin{aligned} \omega^1 &= \xi \sinh(\lambda y)\mathrm{d}w, \\ \omega^2 &= \sinh(\lambda y)\mathrm{d}x, \\ \omega^3 &= \mathrm{d}y \end{aligned}\right\} \tag{4.37}$$

or

$$\left.\begin{aligned} \omega^1 &= \xi \cosh(\lambda y)\mathrm{d}w, \\ \omega^2 &= \cosh(\lambda y)\mathrm{d}x, \\ \omega^3 &= \mathrm{d}y, \end{aligned}\right\} \tag{4.38}$$

where $\xi = \xi(w, x)$ *is a nonzero function of the variables* w *and* x*. The local isometry classes are parametrized by the function* ξ *modulo two arbitrary functions of one variable.*

The corresponding embedding theorem from [8] is

Theorem 4.16. *Every three-dimensional foliated space of type* $(\mathrm{P}\ell)$ *can be (locally) isometrically embedded as a hypersurface in* $\mathbb{H}^4(-\lambda^2)$*. The corresponding hypersurface is (locally) isometrically deformable in a continuous way.*

Remark. A similar classification is valid for the cases $\tilde{c} > 0$ and $\tilde{c} = 0$. The general rule is that the "basis" $\{\cosh(\lambda y),\ \sinh(\lambda y)\}$ in Proposition 4.1 is replaced by the "basis" $\{\cos(\lambda y),\ \sin(\lambda y)\}$ or $\{1,\ y\}$, respectively. The specific formulas expressing some of the basic coefficients p, q, r, s, t and u through arbitrary functions are very similar to the case $\tilde{c} < 0$ but not the same. Also, the formulas valid for $\tilde{c} = 0$ cannot be obtained from the case $\tilde{c} \neq 0$ by a limit procedure $\tilde{c} \to 0$. (See [15, 16, 17, 18]).

5. Spaces with constant Ricci eigenvalues in the hyperbolic case

Three-dimensional spaces with constant Ricci eigenvalues $\rho_1 = \rho_2$ have been studied thoroughly in [12]. There are only a few known explicit examples of metrics and it remains an open question if other explicit examples exist.

Example 5.1. Let $p = p(w)$, $s = s(w)$, $\varphi = \varphi(w)$ and $\psi = \psi(w)$ be arbitrary functions of the variable w. Put $H = 2\lambda p(w)s(w)x + \varphi(w)$ and

$$r = -\lambda^2 p^2(w)s(w)x^2 + [p'(w) - \lambda p(w)\varphi(w)]x + \psi(w).$$

Then the corresponding Ricci eigenvalues are $\rho_1 = \rho_2 = 0$ and $\rho_3 = -2\lambda^2$.

Example 5.2. Put $p = \sqrt{1+x^2}$, $s = \dfrac{x}{\sqrt{1+x^2}}$, $H = \lambda(x^2 + \frac{1}{4})$ and $r = -\frac{1}{4}\lambda^2 x(\sqrt{1+x^2})^3$. The corresponding Ricci eigenvalues are $\rho_1 = \rho_2 = \frac{1}{4}\lambda^2$ and $\rho_3 = -2\lambda^2$.

Example 5.3. Put $p = \sqrt{1+x^4}$, $s = \dfrac{x^2}{\sqrt{1+x^4}}$, $H = \frac{2}{3}\lambda x^3$ and $r = -\frac{1}{9}\lambda^2(\sqrt{1+x^4})^3$. The corresponding Ricci eigenvalues are $\rho_1 = \rho_2 = \frac{2}{9}\lambda^2$ and $\rho_3 = -2\lambda^2$.

It is proved that all these examples are not locally homogeneous. Moreover, what is remarkable, the Examples 5.2 and 5.3 are examples of Riemannian three-manifolds with constant Ricci eigenvalues such that there are no locally homogeneous spaces with the same Ricci eigenvalues. The last fact can be proved as follows: According to K. Sekigawa [22], each three-dimensional homogeneous Riemannian space is locally isometric either to a Riemannian symmetric space or a Lie group with a left invariant metric. According to J. Milnor [20, p.310], *no* three-dimensional Lie group admits a left invariant Riemannian metric whose principal Ricci curvature has the signs $(+,+,-)$. We conclude hence easily that a three-dimensional homogeneous Riemannian space *cannot* admit (constant) Ricci eigenvalues with the signs $(+,+,-)$. But the corresponding triplets of the Ricci eigenvalues for the Examples 5.2 and 5.3 are $(\lambda^2/4,\ \lambda^2/4,\ -2\lambda^2)$ and $(2\lambda^2/9,\ 2\lambda^2/9,\ -2\lambda^2)$, respectively.

Bibliography

[1] E.Boeckx, Foliated semi-symmetric spaces, Doctoral Thesis, Leuven, 1995.

[2] E.Boeckx, O.Kowalski, and L.Vanhecke, Non-homogeneous relatives of symmetric spaces, *Diff. Geom. Appl.* **4**(1994), 45–69.

[3] E.Boeckx, O.Kowalski and L.Vanhecke, *Riemannian Manifolds of Conullity Two*, World Scientific, Singapore, 1996.

[4] É.Cartan, La déformation des hypersurfaces dans l'espace euclidien réel a n dimensions, *Bull. Soc. Math. France* **44**(1916), 65–99.

[5] M.Dajczer, L.Florit, and R.Tojeiro, On deformable hypersurfaces in space forms, *Ann. Mat. Pura Appl.* **(4)174**(1998), 361–390.

[6] R.Deszcz, On psudo-symmetric spaces, *Bull. Soc. Math. Belgium,* Série A. **44**(1992), 1–34.

[7] V.Hájková, Foliated semi-symmetric spaces in dimension 3 (Czech), Doctoral Thesis, Prague, 1995.

[8] V.Hájková, O.Kowalski, and M.Sekiawa, On three-dimensional hypersurfaces with type number two in $\mathbb{H}^4$ and $\mathbb{S}^4$ treated in intrinsic way, to appear in *Rend. Circ. Mat. Palermo*, Serie II, Suppl.72(2004), 107–126.

[9] S.Kobayashi and K.Nomizu, *Foundations of Differential Geometry* I, Interscience Publishers, New York, 1963.

[10] S.Kobayashi and K.Nomizu, *Foundations of Differential Geometry* II, Interscience Publishers, New York, 1969.

[11] O.Kowalski, An explicit classification of 3-dimensional Riemannian spaces satisfying $R(X,Y)\cdot R = 0$, *Czechoslovak Math. J.* **46(121)**(1996), 427–474. (Preprint 1991).

[12] O.Kowalski, A classification of Riemannian 3-manifolds with constant principal Ricci curvatures $\rho_1 = \rho_2 \neq \rho_3$, *Nagoya Math. J.* **132**(1993), 1–36.

[13] O.Kowalski and M.Sekizawa, Local isometry classes of Riemannian 3-manifolds with constant Ricci eigenvalues $\rho_1 = \rho_3 \neq \rho_3$, *Archivum Math.* **32**(1996), 137–145.

[14] O.Kowalski and M.Sekizawa, Three-dimensional Riemannian manifolds of c-conullity two, Chapter 11 of "Riemannian Manifolds of Conullity Two", World Scientific, Singapore, 1996.

[15] O.Kowalski and M.Sekizawa, Riemannian 3-manifolds with c-conullity two, *Bolletino U.M.I.* (7)**11**-B(1997), Suppl. fasc. 2, 161–184.

[16] O.Kowalski and M.Sekizawa, Pseudo-symmetric spaces of constant type in dimension three—elliptic spaces, *Rendiconti di Matematica,* Serie VII, Volume **17**, Roma(1997), 477–512.

[17] O.Kowalski and M.Sekizawa, Pseudo-symmetric spaces of constant type in dimension three—non-elliptic spaces, *Bull. Tokyo Gakugei University. Sect.* IV **50**(1998), 1–28.

[18] O.Kowalski and M.Sekizawa, Pseudo-symmetric spaces of constant type in dimension three, *Personal Note*, Prague-Tokyo, 1998.

[19] O.Kowalski, F.Tricerri, and L.Vanhecke, Exemples nouveaux de variétés riemanniennes non homogènes dont le tenseur de courbure est celui d'un espace symétrique riemannien, *C. R. Acad. Sci. Paris Ser.* I **311**(1990), 355–360.

[20] J.Milnor, Curvature of left invariant metrics on Lie groups, *Advances in Math.,* **21**(1976). 293–329.

[21] B.O'Neill, *Semi-Riemannian Geometry with Applications to Relativity,* Academic Press, New York, London, 1983.

[22] K.Sekigawa, On some 3-dimensional curvature hmogeneous spaces, *Tensor, N.S.* **31**(1977), 343–347

[23] Z.I.Szabó, Structure theorems on Riemannian manifolds satisfying $R(X,Y)\cdot R = 0$, I, Local version, *J. Diff. Geom.* **17**(1982), 531–582.

[24] Z.I.Szabó, Classification and construction of complete hypersurfaces satisfying $R(X,Y)\cdot R = 0$, *Acta. Sci. Math.*(Hung.) **47**(1984), 321–348.

[25] Z.I.Szabó, Structure theorems on Riemannian manifolds satisfying $R(X,Y)\cdot R = 0$, II, Global version, *Geom. Dedicata* **19**(1985), 65–108.

HOW FAR DOES HYPERBOLIC GEOMETRY GENERALIZE?

János Szenthe
Department of Geometry
Eötvös Loránd University
Budapest, Hungary
szenthe@ludens.elte.hu

Introduction

Following the initial discovery of hyperbolic geometry by J. Bolyai (1823) and N. I. Lobachevski (1829), the general concept of Riemannian manifold was created by B. Riemann (1854) as a general framework for geometric researches. Then it was observed that hyperbolic spaces can be considered as Riemannian manifolds of constant negative curvature and accordingly, the concept of spaces of constant curvature, or of space forms, was studied, compising euclidean, elliptic and hyperbolic spaces, i. e. all the spaces known in geometry at that time. The theory of spaces of constant curvature with its counterpart, the non-euclidean geometry, was the great achivement of the 19th century. It turned out however that these spaces can be considered as the simplest special cases of a more general space concept, that of the Riemannian symmetric spaces, which was introduced by É. Cartan (1926). In spite of the wide variety of these symmetric spaces, they have rich and beautiful geometry and their theory was an outstanding accomplishment in differential geometry of the 20th century [11]. The question naturally arises whether there is a more general space concept with a likewise rich geometric structure which comprises the Riemannian symmetric spaces as special cases. This question seems to have remained open up to now, and a simple positive solution is not very probable as the results presented below will show.

1. Some basic concepts and facts concerning homogeneous Riemannian manifolds, spaces of constant curvature and symmetric spaces

As a general framework for studying the question stated above, the theory of homogeneous Riemannian manifolds will be adopted. Accordingly some basic concepts and fundamental facts concerning these manifolds are summarized in what follows.

A Riemannian manifold $(M, <,>)$ is said to be *homogeneous* or called a *Klein space* if there is a transitive isometric action

$$\Phi : G \times M \to M$$

of a connected Lie group G which is assumed to be a closed subgroup of the full isometry group $\mathcal{I}(M, <,>)$ of $(M, <,>)$. Let $o \in M$ be fixed and $H = G_o < G$ the corresponding isotropy subgroup which is a compact group. Then there is a smooth canonical bijection $\chi : M \to G/H$ such that the diagram

$$\begin{array}{ccc} G \times M & \xrightarrow{\Phi} & M \\ {\scriptstyle Id\times\chi}\downarrow & & \downarrow{\scriptstyle\chi} \\ G \times (G/H) & \xrightarrow{\Lambda} & G/H \end{array}$$

is commutative, where Λ is the canonical left action of G on the smooth manifold G/H of left-cosets. Since there is a unique Riemannian metric on G/H which renders χ isometric, and the canonical action Λ is then transitive and isometric, all the homogeneous Riemannian manifolds can be obtained as left-coset spaces of compact subgroups of Lie groups.

Let G be a Lie group, $H < G$ a compact subgroup and $\mathfrak{h} < \mathfrak{g}$ their Lie algebras. Then there is a *reductive complement* of $\mathfrak{h}$, i. e. a subspace $\mathfrak{m} \subset \mathfrak{g}$ complementary to $\mathfrak{h}$ such that

$$Ad(h)\mathfrak{m} = \mathfrak{m}, \;\; h \in H$$

is valid. Let $\pi : G \to G/H$ be the canonical projection, if $\mathfrak{g}$ is identified with T_eG then $T_e\pi : \mathfrak{g} \to T_oG/H$ restricted to $\mathfrak{m}$ is an isomorphism. Therefore any $Ad(H)$-invariant inner product

$$(\,,\,) : \mathfrak{m} \times \mathfrak{m} \to \mathbb{R}$$

can be lifted by $T_e\pi$ to an inner product $<,>_o$: $T_oG/H \times T_oG/H \to \mathbb{R}$ which is invariant under the linear isotropy representation

$$T_o\Lambda_h : T_oG/H \to T_oG/H, \;\; h \in H,$$

and therefore $<,>_o$ extends to a Riemannian metric on G/H which is invariant under the transitive action Λ. Accordingly, all homogeneous Riemannian manifolds on a left-coset space G/H of a compact subgroup H of a Lie group G are obtainable from $Ad(H)$-invariant inner products defined on reductive complements $\mathfrak{m}$ of $\mathfrak{h} < \mathfrak{g}$.

A Riemannian manifold $(M, <,>)$ is said to be of *constant curvature* if its sectional curvature is constant. If $(M, <,>)$ is an m-dimensional homogeneous Riemannian manifold, and $\mathcal{I}(M, <,>)$ its full isometry group, then

$$dim\ \mathcal{I}(M, <,>) \leq \binom{m+1}{2},$$

holds, and the manifold is of constant curvature if and only if in the above inequality the equality sign is true.

Let $(M, <,>)$ be a Riemannian manifold, $z \in M$ and $U \subset M$ a symmetric normal neighbourhood U of z. If $x \in U - \{z\}$ then there is a unique geodesic segment

$$\sigma : [-1, 1] \to U$$

such that $\sigma(-1) = x$ and $\sigma(0) = z$. Then $\sigma(1) = x' \in U$ and the map $\Sigma_z : U \to U$ defined by

$$\Sigma_z(x) = x',\ \Sigma_z(z) = z$$

is an involutive diffeomorphism of U which is called a *local geodesic symmetry.* A Riemannian manifold is said to be *symmetric* if to each point there is a local geodesic symmetry which extends uniquely to an involutive isometry of the entire manifold; these isometries are called *symmetries* of the manifold. The symmetries of a symmetric Riemannian manifold generate a group G of isometries which has a canonical Lie group structre and its action is smooth and transitive on the manifold M. If $o \in M$ is fixed and $H < G$ is the corresponding isotropy subgroup, then the equivariant bijection $\chi : M \to G/H$ defines a Riemannian metric on G/H which is endowed consequently with a symmetric Riemannian manifold structure. Moreover, the reductive complement $\mathfrak{m}$ of $\mathfrak{h} < \mathfrak{g}$ is now unique. Furthermore, beyond the infinitesimal version $[\mathfrak{h}, \mathfrak{m}] \subset \mathfrak{m}$ of the equality $Ad(h)\mathfrak{m} = \mathfrak{m},\ h \in H$ defining the reductive complement, now even the inclusion

$$[\mathfrak{m}, \mathfrak{m}] \subset \mathfrak{h}$$

holds. Conversely, if G is a semi-simple Lie group and $H < G$ is a compact subgroup such that a complement $\mathfrak{m} \subset \mathfrak{g}$ of $\mathfrak{h}$ with the properties

$$[\mathfrak{h}, \mathfrak{m}] \subset \mathfrak{m},\ [\mathfrak{m}, \mathfrak{m}] \subset \mathfrak{h}$$

exists, then G/H has a unique symmetric Riemannian manifold structure. This fact renders the above given synthetic definition of symmetric spaces practicable, namely not only implies the existence of symmetric Riemannian manifolds, but also yields the possibility to produce a complete list of these spaces as well. In particular, each Riemannian manifold of constant curvature is a symmetric Riemannian manifold.

The circumstance that symmetric Riemannian manifolds can be defined in purely geometric terms and also alternatively in terms of Lie algebras is the reason for their exquisite theory.

2. Homogeneous geodesics and spaces with homogeneous geodesics

In what follows a basic geometric property of symmetric Riemannian spaces is considered which will serve later on as a criterion in looking for reasonable generalizations of the symmetric spaces among the homogeneous Riemannian manifolds.

Let $(M, <,>)$ be a Riemannian manifold and $\Phi : G \times M \to M$ isometric action of a Lie group. A geodesic $\gamma : \mathbb{R} \to M$ is said to be *homogeneous* if there is an $X \in \mathfrak{g}$ such that

$$\gamma(\tau) = \Phi(Exp(\tau X), \gamma(0)), \ \tau \in \mathbb{R}$$

is valid. A homogeneous Riemannian manifold is called a *space with only homogeneous geodesics* if all its geodesics are homogeneous. These homogeneous Riemannian manifolds were also called geodesic orbit spaces or shortly g. o. spaces. Spaces of constant curvature are obviously spaces with homogeneous geodesics and it is also a basic property of symmetric Riemannian spaces that all their geodesics are homogeneous. Therefore it seems reasonable to consider the class of those homogeneous Riemannian manifolds which are spaces with only homogeneous geodesics as candidates for a generalization of the class of Riemannian symmetric spaces.

In fact, there are homogeneous Riemannian manifolds where not all the geodesics are homogeneous. Namely, it can be shown under fairly general assumptions that in a homogeneous Riemannian manifold there are homogeneous geodesics [18, 25]; yet, there are homogeneous Riemannian manifolds where through any point exactly one homogeneous geodesic exists [20]. In general, the set of homogeneous geodesics in a homogeneous Riemannian manifold is rather complicated even in the simplest cases [4, 6, 9, 12, 17, 22, 26].

There is a well-known class of homogeneous Riemannian manifolds which consists of spaces with only homogeneous geodesics, this class of the so-called naturally reductive homogeneous Riemannian manifolds is described in what follows.

Let $(G/H, <,>)$ be a homogeneous Riemannian manifold, where the metric is obtained from an inner product $(\,,\,)$ defined on a reductive complement $\mathfrak{m} \subset \mathfrak{g}$ of $\mathfrak{h} < \mathfrak{g}$. Consider also the Levi-Cività covariant derivation ∇ of this Riemannian manifold. Then ∇ induces a bilinear map

$$\alpha : \mathfrak{m} \times \mathfrak{m} \to \mathfrak{m}$$

which is based on the following construction [23]: There is a neighbourhood $U \subset \mathfrak{m}$ of $0 \in \mathfrak{m}$ such that

$$\pi \circ Exp : \mathfrak{g} \to G/H$$

restricted to U is a diffeomorphism $\delta : U \to U'$ where U' is a neighbourhood of $o \in G/H$. A vector field $X' : U' \to TU'$ is called *canonical* if it is obtainable as $X' = T\delta \circ X \circ \delta^{-1}$ with $X \in \mathfrak{m}$ where Lie algebra elements are considered as left-invariant vector fields on G. Now

$$\alpha : \mathfrak{m} \times \mathfrak{m} \ni (X, Y) \mapsto T\delta^{-1} \nabla_{X'} Y' \in \mathfrak{m}.$$

A homogeneous Riemannian manifold $(G/H, <,>)$ with a fixed reductive decomposition $\mathfrak{g} = \mathfrak{m} \oplus \mathfrak{h}$ is said to be *naturally reductive* if

$$\alpha(X, Y) = \frac{1}{2}[X, Y]_{\mathfrak{m}}, \;\; X, Y \in \mathfrak{m}$$

holds, where the component $[X, Y]_{\mathfrak{m}}$ of $[X, Y]$ is defined by the above reductive decomposition ([23], [21, pp 60-61], [14, vol. II, pp 200-204]).

All the naturally reductive homogeneous Riemannian manifolds are spaces with only homogeneous geodesics, and it was a long standing conjecture that the converse statement holds, namely, that any space with only homogeneous geodesics is also obtainable as a naturally reductive homogeneous Riemannian manifold [2]. However, this conjecture could be proved only under a differentiability assumption concerning a map, the so-called geodesic map, which will be defined below [24]. At last an example was given of a homogeneous Riemannian manifold which is a space with only homogeneous geodesics but cannot be obtained as a naturally reductive one [13]. Later on using results of [24], namely the concept of a geodesic map, several examples of non-naturally reductive spaces with only homogeneous geodesics were given [6, 12, 19].

As the above mentioned results show, the class of naturally reductive homogeneous Riemannian manifolds, which was defined in terms of Lie algebras, lacks a purely geometric definition, namely it is not equal to the class of spaces with only homogeneous geodesics.

3. A classification of trajectories, the potential homogeneous geodesics

Let G be a connected Lie group, $H < G$ a compact subgroup, $\pi : G \to G/H$ the canonical projection on the smooth manifold formed by left-cosets of H and

$$\Lambda : G \times G/H \to G/H$$

the canonical left action. Then for $X \in \mathfrak{g}$ the equality

$$\Lambda(Exp(\tau X), o) = \pi \circ Exp(\tau X), \tau \in \mathbb{R}$$

holds, where $o = H \in G/H$ is the origin of the homogeneous manifold. The map

$$\varphi(X) : \mathbb{R} \ni \tau \mapsto \pi \circ Exp(\tau X) \in G/H$$

will be called the *trajectory associated with the Lie algebra element* X; furthermore, its image, the set

$$O(X) = \{ \pi \circ Exp(\tau X) \in G/H \mid \tau \in \mathbb{R} \},$$

is called the *orbit associated with the Lie algebra element* X.

As the already mentioned results concerning spaces with only homogeneous geodesics show, the answer to the fundamental question discussed here seems to be rather tangled. The results obtained up to now and also those presented here are based on the concept of a geodesic map which was introduced in [24]. Accordingly, the construction of the geodesic map is given subsequently. The starting point in this construction is to single out those trajectories which can be homogeneous geodesics; in other words, to present necessary conditions for the trajectories to be geodesics of some invariant Riemannian metrics. Those trajectories which satisfy these necessary conditions will be called then potential homogeneous geodesics.

Let $X \in \mathfrak{g} - \mathfrak{h}$ be fixed, an element $g \in G$ is called an *automorphism of the orbit* $O(X)$ if

$$\Lambda(g, O(X)) = O(X)$$

holds. It can be shown that there is a unique maximal Lie subgroup $P_X < G$ consisting of automorphisms of $O(X)$ which is called the *automorphism group of the orbit* $O(X)$ [24]. The following earlier result [24] will be used here, where $\mathfrak{p}_X$, the Lie algebra of P_X, is called the automorphism algebra of the orbit $O(X)$.

Proposition 3.1. *Let G be a Lie group $H < G$ a compact subroup, G/H the corresponding smooth manifold of left-cosets, and $X \in \mathfrak{g}$, $Z \in \mathfrak{h}$. Then*

$$\varphi(X) = \varphi(X + Z)$$

holds if and only if $Z \in \mathfrak{q}_X = \mathfrak{p}_X \cap \mathfrak{h}$.

Consider the linear isotropy representation $T_o\Lambda_h : T_oG/H \to T_oG/H$, $h \in H$ induced by the action Λ at o. Fix a $v \in T_oG/H - \{0_o\}$, then the set of those $h \in H$ for which

$$T_o\Lambda_h v = \lambda(h)v$$

holds with some $\lambda(h) \in \mathbb{R}$ form a closed subgroup $H_{[v]} < H$, which is called the *isotropy subgroup in the direction* of v. It can be shown that if $X \in \mathfrak{g} - \mathfrak{h}$ and $v = T_e\pi X \in T_oG/H - \{0_o\}$, then

$$Q_X = H \cap P_X < H_{[v]}$$

holds [24]. The orbit $O(X)$ is said to be *isotropy rigid* if the equality

$$Q_X = H \cap P_X = H_{[v]}$$

holds. Let $<,>$ be a Λ-invariant Riemannian metric on G/H and ∇ its Levi-Civitá covariant derivation. If for $X \in \mathfrak{g} - \mathfrak{h}$ the trajectory $\varphi(X)$ associated with X is a geodesic of ∇, in other words, if it is a homogeneous geodesic, then the orbit $O(X)$ is isotropy rigid [24].

Isotropy rigid trajectories of a coset manifold G/H are obtainable by the following construction as well. Fix an $X \in \mathfrak{g} - \mathfrak{h}$, then a monotone sequence

$$\mathfrak{h} = \mathfrak{q}_X^0 > \cdots > \mathfrak{q}_X^i > \cdots$$

is given by the following successive definition:

$$\begin{aligned}
\mathfrak{q}_X^0 &= \mathfrak{h}, \\
\mathfrak{q}_X^1 &= \{ Z \in \mathfrak{q}_X^0 \mid [Z, X] = \lambda X + Z', \ \lambda \in \mathbb{R}, \ Z' \in \mathfrak{q}_X^0 \}, \\
&\vdots \\
\mathfrak{q}_X^i &= \{ Z \in \mathfrak{q}_X^{i-1} \mid [X, X] = \lambda X + Z', \ \lambda \in \mathbb{R}, \ Z' \in \mathfrak{q}_X^{i-1} \}, \\
&\vdots \, .
\end{aligned}$$

Let $k \in \mathbb{N}$ be the first integer i such that $\mathfrak{q}_X^i = \mathfrak{q}_X^{i+1}$ holds, then k is called the *order of the orbit* $O(X)$.

It can be shown that an orbit $O(X)$ is isotropy rigid if and only if it has order 1 [24]. Furthermore, if $\mathfrak{m} \subset \mathfrak{g}$ is a reductive complement of the subalgebra $\mathfrak{h} < \mathfrak{g}$, and $X \in \mathfrak{m} - \{0\}$, then the orbit $O(X)$ is isotropy rigid [24].

Let $X \in \mathfrak{g} - \mathfrak{h}$ be fixed and consider an automorphism $g \in P_X$ of the corresponding orbit $O(X) \subset G/H$. It can be shown that there is an analytic bijection $\kappa_g : \mathbb{R} \to \mathbb{R}$ such that

$$\Lambda_g \circ \pi \circ Exp(\tau X) = \pi \circ Exp(\kappa(\tau)X), \ \tau \in \mathbb{R}$$

holds [24]. Accordingly, it is said that the trajectory $\varphi(X)$ is *parametrically rigid*, if

$$\kappa_g(\tau) = \tau, \ \tau \in \mathbb{R}$$

holds for each $g \in Q_X = H \cap P_X$.

Let $<,>$ be a Λ-invariant Riemannian metric on G/H and ∇ its Levi-Cività covariant derivation. If $X \in \mathfrak{g} - \mathfrak{h}$ and $\varphi(X)$ is a geodesic of ∇, then it is parametrically rigid [24].

Let G be a connected Lie group, $H < G$ a compact subgroup and $X \in \mathfrak{g} - \mathfrak{h}$. If the orbit $O(X)$ is isotropy rigid and the trajectory $\varphi(X)$ is parametrically rigid, then the trajectory $\varphi(X)$ is called a *potential homogeneous geodesic.*

The following result has a fundamental role in the construction of the geodesic map [24]:

Proposition 3.2. *Let G be a connected Lie group $H < G$ a compact subgroup, $\mathfrak{m} \subset \mathfrak{g}$ a reductive complement of $\mathfrak{h}$. If $X \in \mathfrak{m} - \{0\}$, then the trajectory $\varphi(X)$ is a potential homogeneous geodesic. Moreover, if $Y = X + Z$ with $Z \in \mathfrak{h}$ then the trajectory $\varphi(Y)$ is a potential homogeneous geodesic if and only if Z is an element of the normalizer of the subalgebra*

$$\mathfrak{q}_X = \mathfrak{h} \cap \mathfrak{p}_X$$

in the isotropy subalgebra $\mathfrak{h} < \mathfrak{g}$, where $\mathfrak{p}_X$ is the Lie algebra of the automorphism group P_X of the orbit $O(X)$ associated with X.

4. The construction of the geodesic map and some of its basic properties

Let $(G/H, <,>)$ be a homogeneous Riemannian manifold which is a space with only homogeneous geodesics, and $\mathfrak{m} \subset \mathfrak{g}$ a reductive complement of $\mathfrak{h}$. Since the subgroup $H < G$ is compact, the restricted adjoint action $Ad : H \times \mathfrak{h} \to \mathfrak{h}$ admits an invariant inner product

$$(\,,\,) : \mathfrak{h} \times \mathfrak{h} \to \mathbb{R}$$

on the Lie algebra $\mathfrak{h}$. In the spatial case when $\mathfrak{h}$ is also semi-simple, then $(\,,\,)$ can be uniquely defined by $(\,,\,) = -K$, where K is the Killing form of $\mathfrak{h}$. If $X \in \mathfrak{m} - \{0\}$, then $\varphi(X)$ is a potential homogeneous geodesic as already mentioned. Consider the automorphism group $P_X < G$ of the orbit $O(X)$, also its Lie algebra $\mathfrak{p}_X < \mathfrak{g}$, and the corresponding isotropy subalgebra

$$\mathfrak{q}_X = \mathfrak{h} \cap \mathfrak{p}_X.$$

Since $\varphi(X)$ is a potential homogeneous geodesic, $\mathfrak{q}_X < \mathfrak{h}$ is also the isotropy subalgebra in the direction of $v = T_e\pi X \in T_oG/H$, therefore $\mathfrak{q}_X$

is also a compact subalgebra. Let $\mathfrak{a}_X \subset \mathfrak{h}$ be the orthogonal complement of $\mathfrak{q}_X$ with respect to $(\,,\,)$, then a reductive decomposition

$$\mathfrak{h} = \mathfrak{a}_X \overset{\perp}{\oplus} \mathfrak{q}_X$$

is obtained. The set of those $Z \in \mathfrak{h}$ for which the trajectory $\varphi(X+Z)$ is a potential homogeneous geodesic is the normalizer $\mathfrak{n}_X$ of $\mathfrak{q}_X$in $\mathfrak{h}$ according to Proposition 3.2. Let now $\mathfrak{c}_X$ be the centralizer of $\mathfrak{q}_X$ in $\mathfrak{h}$, and consider the orthogonal direct sum decompositions

$$\mathfrak{h} = \mathfrak{a}_X \overset{\perp}{\oplus} \mathfrak{q}_X = (\mathfrak{c}_X \cap \mathfrak{a}_X) \overset{\perp}{\oplus} \mathfrak{b}_X \overset{\perp}{\oplus} \mathfrak{q}_X$$

with respect to the above fixed inner product $(\,,\,)$. Then the normalizer of the subalgebra $\mathfrak{q}_X$ in $\mathfrak{h}$ is given by

$$\mathfrak{n}_X = (\mathfrak{c}_X \cap \mathfrak{a}_X) \overset{\perp}{\oplus} \mathfrak{q}_X$$

([21, pp 66-70]). Assume now that a potential homogeneous geodesic can be obtained as

$$\varphi(X+Z) = \varphi(X+Z'),\ \ Z, Z' \in \mathfrak{n}_X.$$

Then $Z = C_X + Q_X$, $Z' = C'_X + Q'_X$ with C_X, $C'_X \in \mathfrak{c}_X$, Q_X, $Q'_X \in \mathfrak{q}_X$, and therefore the equalities

$$< X + C_X > \oplus \mathfrak{q}_X = \mathfrak{p}_{X+Z} = \mathfrak{p}_{X+Z'} =< X + C'_X > \oplus \mathfrak{q}_X$$

by Proposition 3.1 imply that $C'_X \in< C_X > \oplus \mathfrak{q}_X$ and therefore $C'_X = \alpha C_X + Z''$ with $\alpha \in \mathbb{R}$, $Z'' \in \mathfrak{q}_X$. But then $Z'' = C'_X - \alpha C_X \in \mathfrak{c}_X$ yields that $Z'' = 0$, and therefore $C_X = C'_X$ holds. Assume now that the trajectory $\varphi(X+Z)$ is the homogeneous geodesic of the homogeneous Riemannian manifold $(G/H, <,>)$ emanating from $o \in G/H$ in the direction $v = T_e\pi X$. Then there is a unique decomposition

$$Z = C_X + Q_X$$

with $C_X \in \mathfrak{c}_X \cap \mathfrak{a}_X$, $Q_X \in \mathfrak{q}_X$, and C_X is uniquely defined by the trajectory. Put now $\xi(X) = C_X$, then a map $\xi : \mathfrak{m} \to \mathfrak{h}$ is defined by

$$\xi : \mathfrak{m} \ni X \mapsto C_X \in \mathfrak{c}_X \cap \mathfrak{a}_X \subset \mathfrak{h},$$

which is called the *geodesic map* associated with $(G/H, <,>)$, the given space with only homogeneous geodesics with respect to the reductive complement $\mathfrak{m}$. In case of a fixed reductive complement $\mathfrak{m} \subset \mathfrak{g}$, the map ξ is obviously unique if $\mathfrak{h}$ is semi-simple.

If a homogeneous Riemannian manifold $(G/H, <,>)$ with homogeneous geodesics admits a geodesic map $\xi : \mathfrak{m} \to \mathfrak{h}$ which is a vector space homomorphism, then it is a naturally reductive homogeneous Riemannian manifold [24].

The geodesic map ξ as given here was introduced in [24] and called a geodesic graph later [19]; the same term was also used for a modification of this concept [15, 16].

The geodesic map ξ is equivariant with respect to the restricted adjoint actions

$$Ad : H \times \mathfrak{m} \to \mathfrak{m}, \quad Ad : H \times \mathfrak{h} \to \mathfrak{h}$$

by obvious simple arguments [24].

Proposition 4.1. *Let $(G/H, <,>)$ be a space with only homogeneous geodesics and $\mathfrak{g} = \mathfrak{m} \oplus \mathfrak{h}$ a reductive decomposition. Then the corresponding geodesic map $\xi : \mathfrak{m} \to \mathfrak{h}$ is smooth on the open set $\mathfrak{m} - \{0\}$.*

Proof. A suitable coordinate expression of the map ξ will be obtained in what follows which will directly yield the smoothness of the map ξ on the set $\mathfrak{m} - \{0\}$.

In fact, a reformulation of the definition of ξ can be obtained as follows: Let the trajectory $\varphi(X + Z)$ with $X \in \mathfrak{m}$, $Z \in \mathfrak{h}$ be a geodesic, then the corresponding automorphism algebra $\mathfrak{p}_{X+Z}$ is given by

$$\mathfrak{p}_{X+Z} = < X + Z > \oplus \mathfrak{q}_X$$

and therefore the set of those $Z' \in \mathfrak{h}$ which satisfy $\varphi(X+Z) = \varphi(X+Z')$ is the affine subspace $Z+\mathfrak{q}_X$. Then by the above definition $\xi(X) \in Z+\mathfrak{q}_X$ is that element which satisfies $\xi(X) \perp \mathfrak{q}_X$. Therefore

$$\|\xi(X)\| = min\{ \|Z'\| \mid Z' \in Z + \mathfrak{q}_X \}$$

holds and this minimum property also defines $\xi(X)$ uniquely. Now to apply the Lagrangian theory of conditional extremal values, an analytic description of the affine subspace $Z + \mathfrak{q}_X$ will be given.

First a basic fact concerning the analytic expression of the trajectory $\varphi(Y)$ with $Y = X + Z$, $X \in \mathfrak{m}$, $Z \in \mathfrak{h}$ is established. There is a neighbourhood $\hat{W} \subset \mathfrak{g}$ of $0 \in \mathfrak{g}$ such that Exp is injective on $\hat{W}$ and to each $g = Exp(Y)$, $Y \in \hat{W}$ there are unique $X \in \mathfrak{m} \cap \hat{W}$ and $Z \in \mathfrak{h} \cap \hat{W}$ with

$$g = Exp(Y) = Exp(X) \cdot Exp(Z),$$

and the maps $\hat{W} \ni Y \mapsto X \in \mathfrak{m}$, $\hat{W} \ni Y \mapsto Z \in \mathfrak{h}$ are smooth ([11, pp 123-126]). Fix such a neighbourhood $\hat{W}$. □

Fix also $X \in \mathfrak{m} \cap \hat{W}$, $Z \in \mathfrak{h} \cap \hat{W}$ and a symmetric neighbourhood $I \subset \mathbb{R}$ of 0 such that $\tau(X+Z) \in \hat{W}$, $\tau \in I$ holds. Then there are unique smooth functions $U : I \to \mathfrak{m} \cap \hat{W}$, $V : I \to \mathfrak{h} \cap \hat{W}$ such that

$$Exp(\tau Y) = Exp(\tau(X+Z)) = Exp(U(\tau)) \cdot Exp(V(\tau)), \ \tau \in I$$

holds. Then by Taylor's formula

$$\begin{aligned} U(\tau) &= D^1U(0)\tau + \frac{1}{2}D^2(0)\tau^2 + \tilde{U}(\tau) \\ &= (D^1U(0) + \frac{1}{2}D^2U(0)\tau + \frac{\tilde{U}(\tau)}{\tau})\tau = \hat{U}(\tau)\tau, \ \tau \in I, \\ V(\tau) &= D^1V(0)\tau + \frac{1}{2}D^2(0)\tau^2 + \tilde{V}(\tau) \\ &= (D^1V(0) + \frac{1}{2}D^2V(0)\tau + \frac{\tilde{V}(\tau)}{\tau})\tau = \hat{V}(\tau)\tau, \ \tau \in I, \end{aligned}$$

where $\tilde{U}(\tau)$, $\tilde{V}(\tau) = o(\tau^2)$. But then Taylor's formula for the exponential map ([11, pp 102-107]) yields the following:

$$\begin{aligned} Exp(\tau(X+Z)) &= Exp(\tau\hat{U}(\tau)) \cdot Exp(\tau\hat{V}(\tau)) \\ &= Exp\Big(\tau(\hat{U}(\tau) + \hat{V}(\tau)) + \frac{\tau^2}{2}[\hat{U}(\tau), \hat{V}(\tau)] + O(\tau^3)\Big). \end{aligned}$$

Then, since the restriction of Exp to $\hat{W}$ is injective, the following is obtained:

$$\tau(X+Z) = \tau\Big(\hat{U}(\tau) + \hat{V}(\tau) + \frac{\tau^2}{2}[\hat{U}(\tau), \hat{V}(\tau)] + O(\tau^3)\Big).$$

Now comparison of the terms of equal order on the two sides of the above equality yields the following:

$$\begin{aligned} &X + Z = D^1U(0) + D^1V(0), \\ &0 = \frac{1}{2}(D^2U(0) + D^2V(0) + [D^1U(0), D^1V(0)]), \\ &0 = \frac{\tilde{U}(\tau)}{\tau} + \frac{\tilde{V}(\tau)}{\tau} \\ &\quad + \frac{\tau}{2}[D^1U(0) + \frac{1}{2}D^2U(0)\tau + \frac{\tilde{U}(\tau)}{\tau}, D^1V(0) + \frac{1}{2}D^2V(0)\tau + \frac{\tilde{v}(\tau)}{\tau}] \\ &\quad + \frac{O(\tau^3)}{\tau}. \end{aligned}$$

Consequently, by the reductive decomposition $\mathfrak{g} = \mathfrak{m} \oplus \mathfrak{h}$ the following is obtained:

$$
\begin{aligned}
&X = D^1U(0), \quad Z = D^1V(0), \\
&\frac{1}{2}D^2U(0) + [D^1U(0), D^1V(0)] = \frac{1}{2}D^2U(0) + [X, Z] = 0, \\
&\frac{1}{2}D^2V(0) = 0.
\end{aligned}
$$

Now consider the exponential map defined by the invariant Riemannian metric $<,>$ of the space with only homogeneous geodesics:

$$exp_o : T_oG/H \to G/H,$$

which is defined on the entire tangent space T_oG/H since $(G/H, <,>)$ is complete as a homogeneous Riemannian manifold. Consider also the vector space isomorphism

$$T_e\pi\lceil\mathfrak{m} : \mathfrak{m} \to T_oG/H.$$

Then there is a neighbourhood $W^0 \subset \mathfrak{m}$ of $0 \in \mathfrak{m}$ such that

(1) $exp_o\lceil\Big(T_e\pi(W^0)\Big)$ is a diffeomorphism.

(2) The restricted exponential map $Exp\lceil\mathfrak{m}$ of the Lie algebra $\mathfrak{g}$ has an inverse on the neighbourhood $W = exp_o \circ T_e\pi(W^0)$of $o \in G/H$; this inverse will be denoted by Exp^{-1}.

Consequently, the following local expression of the exponential map exp_o is obtainable:

$$\Xi : Exp^{-1} \circ exp_o \circ T_e\pi : W^0 \to \mathfrak{m},$$

which renders the following diagramme commutative:

$$
\begin{array}{ccc}
W^0 & \xrightarrow{\Xi} & W \\
{\scriptstyle T_e\pi}\Big\downarrow & & \Big\uparrow{\scriptstyle Exp^{-1}} \\
W' & \underset{exp_o}{\longrightarrow} & \tilde{W}
\end{array}
$$

Consider now the homogeneous geodesic emanating from $o \in G/H$ with the initial tangent vector $v = T_e\pi X$, $X \in \mathfrak{m}$ given by the trajectory $\varphi(X + \xi(X))$. Then the above definition of the map Ξ yields that

$$U(\tau) = Exp^{-1} \circ exp_o(T_e\pi(\tau X))) = \Xi(\tau X), \ \ \tau \in I$$

holds. Then under the canonical identification $T_0\mathfrak{m} = \mathfrak{m}$ we obtain

$$D^1U(0) = D^1\Xi(0)X,\ \ D^2U(0) = D^2\Xi(0)(X,X),$$

where $D^2\Xi : \mathfrak{m}\times\mathfrak{m} \to \mathfrak{m}$ is a symmetric bilinear map which is $Ad(H)$-equivariant. Put now

$$A = D^2\Xi : \mathfrak{m}\times\mathfrak{m} \to \mathfrak{m}.$$

Since $\mathfrak{q}_X = \{\ Z' \in \mathfrak{h} \mid [Z', X] = 0\ \}$, the affine subspace consisting of those $Z' \in \mathfrak{h}$ which yield a $\varphi(X + Z')$ equal to the gedodesic, is given by

$$\xi(X) + \mathfrak{q}_X = \{\ Z \in \mathfrak{h} \mid A(X,X) + [X, Z'] = 0\ \},$$

in consequence of the equality $D^2U(0) + 2[X, Z] = 0$ obtained earlier above.

In order to apply the Lagrangian theory of conditional extremal values, fix a base $(E_1, \dots, E_n)$ of the Lie algebra $\mathfrak{g}$ which is compatible with the reductive decomposition $\mathfrak{g} = \mathfrak{m} \oplus \mathfrak{h}$ such that $(E_1, \dots, E_m)$ is a base of $\mathfrak{m}$ and $(E_{m+1}, \dots, E_n)$ is a base of $\mathfrak{h}$ and orthonormal with respect to the inner product $(\ ,\ )$. Then

$$\begin{aligned}[X,Z] &= \Big(\sum_{i=1}^{m}\sum_{k=m+1}^{n} c_{ik}^1 X^i Z^k, \dots, \sum_{i=1}^{m}\sum_{k=m+1}^{n} c_{ik}^m X^i Z^k, 0, \dots, 0\Big),\\ A(X,X) &= \Big(\sum_{i,j=1}^{m} a_{ij}^1 X^i X^j, \dots, \sum_{i,j=1}^{m} a_{ij}^m X^i X^j, 0, \dots, 0\Big).\end{aligned}$$

The observations made above can be reformulated as follows in terms of coordinates: The function

$$Z \mapsto (Z, Z) = \sum_{k=m+1}^{n} (Z^k)^2$$

in case of a fixed $(X^1, \dots, X^m) \neq 0$ under the conditions

$$F_p(Z^{m+1}, \dots, Z^n) = \sum_{i,j=1}^{m} a_{ij}^p X^i X^j + \sum_{i=1}^{m}\sum_{k=m+1}^{n} c_{ik}^p X^i Z^k = 0,\ \ p = 1, \dots m$$

attains its minimal value at a single point which is given by

$$Z^k = \xi^k(X^1, \dots, X^m),\ \ k = m+1, \dots, n.$$

Consider now the function

$$\Phi(Z^{m+1}, \dots, Z^n) = \sum_{k=m+1}^{n} (Z^k)^2 + \sum_{p=1}^{m} \lambda_p F_p(Z^{m+1}, \dots, Z^n),$$

$$(Z^{m+1}, \dots, Z^n) \in \mathbb{R}^{n-m},$$

where $\lambda_1, \ldots, \lambda_m$ are the Lagrange multipliers. Then by the theory of conditional extremal values, the system of equations

$$\frac{\partial \Phi}{\partial Z^k} = 2Z^k + \sum_{p=1}^{m} \lambda_p \frac{\partial F_p}{\partial Z^k} = 2Z^k + \sum_{p=1}^{m} \lambda_p \sum_{i=1}^{m} c_{ik}^p X^i = 0,$$
$$F_j(Z^{m+1}, \ldots, Z^n) = 0, \;\; k = m+1, \ldots, n, \;\; j = 1, \ldots, m$$

has a single solution given by $Z^k = \xi^k(X^1, \ldots, X^m), \; k = m+1, \ldots, n$.

A substitution yields now the system of equations

$$\sum_{i,j=1}^{m} a_{ij}^p X^i X^j + \sum_{i=1}^{m} \sum_{k=m+1}^{n} c_{ik}^p X^i \Big(-\frac{1}{2} \sum_{r=1}^{m} \lambda_r \sum_{j=1}^{m} c_{jk}^r X^j\Big)$$
$$= \sum_{i,j=1}^{m} a_{ij}^p X^i X^j - \frac{1}{2} \sum_{i,j,r=1}^{m} \sum_{k=m+1}^{n} c_{ik}^p X^i c_{jk}^r X^j \lambda_r = 0, \;\; p = 1, \ldots m.$$

Then λ_r, $r = 1, \ldots, m$ are uniquely defined and as functions of $(X^1, \ldots, X^m)$, are analytic in $\mathfrak{m} - \{0\}$. Consequently, the functions $\xi_k(X^1, \ldots, X^m)$, $k = m+1, \ldots, n$ are homogeneous of degree 1 and also analytic in $\mathfrak{m} - \{0\}$. In fact,

$$\xi_k(X^1, \ldots, X^m) = -\frac{1}{2} \sum_{i,l=1}^{m} c_{ik}^l X^i \lambda_l, \;\; k = m+1, \ldots, n.$$

Thus the assertion of the Proposition follows. ♠

Corollary. *The Lagrange multipliers $\lambda^l : \mathfrak{m} \to \mathbb{R}, l = 1, \ldots, m$ are positive homogenous of degree 0 and the geodesic map ξ is positive homogenous of degree 1.*

The preceding proposition yields also the following result [24]:

Theorem 4.2. *Let $(G/H, <,>)$ be a homogeneous Riemannian manifold which is a space with homogeneous geodesics, and $\xi : \mathfrak{m} \to \mathfrak{h}$ a geodesic map. If ξ is differentiable at $0 \in \mathfrak{m}$ too, then ξ is a vector space homomorphism and $(G/H, <,>)$ is naturally reductive.*

5. The compact semi-simple case

The geodesic map is applied now to study spaces with only homogeneous geodesics in that case when G is a compact semi-simple Lie group.

Theorem 5.1. *Let $(G/H, <,>)$ be a homogeneous Riemannian manifold which is a space with only homogeneous geodesics where the Lie group G is compact and semi-simple. Let the isotropy representation*

$$T_o\Lambda_h : T_oG/H \to T_oG/H,\ h \in H$$

of the canonical action Λ at the origin $o = H \in G/H$ be such that the principal isotropy subgroups $H_v < H, v \in T_oG/H$ of this compact group action do not leave any tangent vector of the principal orbits invariant, then the homogeneous Riemannian manifold $(G/H, <,>)$ is naturally reductive.

Proof. Let $K : \mathfrak{g} \times \mathfrak{g} \to \mathbb{R}$ be the Killing form and $\prec, \succ = -K$ the corresponding $Ad(G)$-invarint euclidean inner product on $\mathfrak{g}$. Then the inner product $(\,,\,)$ fixed on $\mathfrak{h}$ in the preceding arguments can be taken now as the restriction of $\prec, \succ$ to $\mathfrak{h}$. Put $\mathfrak{m} = \mathfrak{h}^{\perp}$, and let $(E_1, \ldots, E_n)$ be an orthonormal base of $\mathfrak{g}$ such that $(E_1, \ldots, E_m)$is a base of $\mathfrak{m}$ and $(E_{m+1}, \ldots, E_n)$ that of $\mathfrak{h}$, where the orthogonal complement and the base is taken with respect to $\prec, \succ$. Let now c_{ij}^k, $i, j, k = 1, \ldots, n$ be the structural constants of $\mathfrak{g}$ with respect to the above base. □

The earlier obtained coordinate expression of the geodesic map ξ can be given now a more concise coordinate-free form. If $X = \sum_{i=1}^m X^i E_i \in \mathfrak{m}$, then

$$\begin{aligned}
\xi(X) &= \sum_{k=m+1}^{n} \xi^k(X)E_k = -\frac{1}{2}\sum_{k=m+1}^{n}\left(\sum_{i,l=1}^{m} c_{ik}^{l} X^i \lambda^l(X)\right)E_k \\
&= \frac{1}{2}\sum_{k=m+1}^{n}\left(\sum_{i,l=1}^{m} c_{il}^{k} X^i \lambda^l(X)\right)E_k \\
&= \frac{1}{2}\sum_{k=m+1}^{n}\left(\sum_{i,l=1}^{m} \prec [E_i, E_l], E_k \succ X^i \lambda^l(X)\right)E_k \\
&= \frac{1}{2}\sum_{k=m+1}^{n} \prec [\sum_{i=1}^{m} X^i E_i, \sum_{l=1}^{m} \lambda^l(X) E_l], E_k \succ E_k \\
&= \frac{1}{2}[X, \sum_{l=1}^{m} \lambda^l(X)E_l]_{\mathfrak{h}} = \frac{1}{2}[X, L(X)]_{\mathfrak{h}},
\end{aligned}$$

where $L(X) = \sum_{l=1}^m \lambda^l(X)E_l$.

Since ξ is equivariant with respect to the restricted adjoint actions $Ad : H \times \mathfrak{m} \to \mathfrak{m}$ and $Ad : H \times \mathfrak{h} \to \mathfrak{h}$, the following holds:

$$Ad(h)\xi(X) = \xi(Ad(h)X),\ X \in \mathfrak{m},\ h \in H.$$

Therefore by the above expression of the geodesic map the following is obtained:

$$Ad(h)\xi(X) = \frac{1}{2}Ad(h)\Big([X, L(X)]_{\mathfrak{h}}\Big) = \frac{1}{2}\Big(Ad(h)[X, L(X)]\Big)_{\mathfrak{h}}$$
$$= \frac{1}{2}[Ad(h)X, Ad(h)L(X)]_{\mathfrak{h}}$$
$$= \xi(Ad(h)X) = \frac{1}{2}[Ad(h)X, L(Ad(h)X)]_{\mathfrak{h}}.$$

The preceding equalities yield now

$$[Ad(h)X, Ad(h)L(X) - L(Ad(h)X)]_{\mathfrak{h}} = 0, \ X \in \mathfrak{m},$$
$$Ad(h^{-1})[Ad(h)X, Ad(h)L(X) - L(Ad(h)X)]_{\mathfrak{h}}$$
$$= [X, L(X) - Ad(h^{-1})L(Ad(h)X)]_{\mathfrak{h}} = 0, X \in \mathfrak{m}, \ h \in H.$$

But the above equality is equivalent to

$$\prec [X, L(X) - Ad(h^{-1})L(Ad(h)X)], \mathfrak{h} \succ = 0,$$
$$\prec L(X) - Ad(h^{-1})L(Ad(h)X), [X, \mathfrak{h}] \succ = 0.$$

But $[X, \mathfrak{h}] = T_X H(X)$ holds which is the tangent space of the orbit $H(X)$ at the point X under the restricted adjoint action $Ad : H \times \mathfrak{m} \to \mathfrak{m}$. Consequently, the equality obtained above means that the vector $L(X) - Ad(h^{-1})L(Ad(h)X)$ is in the normal space to the orbit $H(X)$ at its point $Ad(h)X$.

Consider now the case of such an $X \in \mathfrak{m}$ that its orbit $H(X)$ is a principal orbit of the restricted adjoint action $Ad : H \times \mathfrak{m} \to \mathfrak{m}$ of the compact group H. Then the linear isotropy representation

$$T_X Ad(h') : T_X\mathfrak{m} \to T_X\mathfrak{m}, \ h' \in H_X$$

of this restricted adjoint action at $X \in H(X)$ acts trivially on the normal space of the orbit (see e.g. [3, pp 179-186, 303-312]). Therefore, in the special case when $h \in H_X$,

$$L(X) - Ad(h^{-1})(X) = L(X) - Ad(h^{-1})L\Big(Ad(h)X\Big)$$
$$= Ad(h)L(X) - L\Big(Ad(h)X\Big) = Ad(h)L(X) - L(X),$$
$$2L(X) = Ad(h)L(X) + Ad(h^{-1})L(X)$$

holds. But then, since the adjoint action leaves the inner product $\prec, \succ$ invariant, the equality

$$L(X) = Ad(h)L(X), \ h \in H_X$$

follows. Consequently, the inclusion $H_X < H_{L(X)}$ is obtained which implies that $\mathfrak{h}_X < \mathfrak{h}_{L(X)}$ holds. But then the identity

$$\Big[[X, L(X)], \mathfrak{h}_X\Big] + \Big[[\mathfrak{h}_X, X], L(X)\Big] + \Big[[L(X), \mathfrak{h}_X], X\Big] = 0$$

implies that

$$\Big[[X, L(X)], \mathfrak{h}_X\Big] = 0$$

holds. Consider now the orthogonal direct sum $\mathfrak{h} = \mathfrak{d}_X \overset{\perp}{\oplus} \mathfrak{h}_X$, then this is a reductive decomposition and by assumption the equality

$$[\mathfrak{h}_X, \mathfrak{d}_X] = \mathfrak{d}_X$$

is valid. Now

$$\begin{aligned} \prec [X, L(X)], \mathfrak{h}_X \succ &= - \prec L(X), [X, \mathfrak{h}_X \succ = 0, \\ \prec [X, L(X)], \mathfrak{d}_X \succ &= \prec [X, L(X)], [\mathfrak{h}_X, \mathfrak{d}_X] \succ \\ &= - \prec \Big[[X, L(X)], \mathfrak{h}_X\Big], \mathfrak{d}_X \succ = 0 \end{aligned}$$

yield that

$$[X, L(X)]_{\mathfrak{h}} = 0$$

holds. Consequently, $\xi(X) = 0$ is obtained. Since the set of those $X \in \mathfrak{m}$ for which $H(X)$ is a principal orbit, is dense in $\mathfrak{m}$, the map ξ is identically zero. ♠

Bibliography

[1] Alekseevsky, D. V., Arvanitoyeorgos, A., Metrics with homogeneous geodesics on flagmanifolds, *Comment. Math. Univ. Carolin.* **43**(2002), pp 189-199

[2] Ambrose W., Singer, I. M., On homogeneous Riemannian manifolds, *Duke Math. J.* **25**(1958), pp 647-669

[3] Bredon, G.E., *Introduction to Compact Transformation Groups*, Academic Press, New York, 1972

[4] Calvaruso, G., Kowalski, O., Marinosci, R. A., Homogeneous geodesics in solvable Lie groups, *Acta Math. Hung.* **101** (2003), pp 313-322

[5] Calvaruso, G., Marinosci, R.A., Homogeneous geodesics in five-dimensional generalized symmetric spaces, *Balkan J. Geom. Appl.* **8** (2003), pp 1-19

[6] Dusek, Z., Structure of geodesics in a 12-dimensional group of Heisenberg type, *Steps in Differential Geometry, (Debrecen, 2002)*, 95-103

[7] Dusek, Z., Explicit geodesic graphs on some H-type groups, *Rend. Circ. Math. Palermo* **69**(2002), pp 77-88

[8] Dusek, Z., Kowalski, O., Geodesic graphs on 13-dimensional group of Heisenberg type, *Math. Nachr.* **254-255**(2003), pp 87-96

[9] Dusek, Z., Kowalski, O., Nikcevic, S. Z., New examples of g.o. spaces in dimension 7, *to appear*

[10] Gordon, C., Homogeneous Riemannian manifolds whose geodesics are orbits, *Progr. Nonlinear Differ. Equ. Appl.* **20**(1996), 155-174

[11] Helgason, H., *Differential Geometry, Lie Groups, and Symmetric Spaces*, Academic Press, New York, 1978

[12] Homolya, S., Geodesic vectors of the six-dimensional spaces, *Steps in Differential Geometry, (Debrecen, 2002)*, pp 139-146

[13] Kaplan, A., On the geometry of groups of Heisenberg type, *Bull. London Math. Soc.* **15**(1983), pp35-42

[14] Kobayashi, S., Nomizu, K., *Foundations of Differential Geometry I, II*, Wiley, New York, 1963, 1969

[15] Kowalski, O., Nikcevic, S. Z., On geodesic graphs of Riemannian g. o. spaces, *Arch. Math.* **73**(1999), pp 223-234

[16] Kowalski, O., Nikcevic, S. Z., On geodesic graphs of Riemannian g. o. spaces (*Arch. Math.* **73**(1999), pp 223-234) appendix, *Arch. Math.* **79**(2002), pp 158-160

[17] Kowalski, O., Nikcevic, S. Z., Vlasek, Z., Homogeneous geodesics in homogeneous Riemannian manifolds (examples), *Geometry and Topology of Submanifolds, (Beijing/Berlin, 1999)*, 104-112

[18] Kowalski, O., Szenthe, J., On the existence of homogeneous geodesics in homogeneous Riemannian manifolds, *Geom. Dedicata* **81**(2000), pp 209-214; Erratum *ibid.* **84**(2001), pp 331-332

[19] Kowalski, O., Vanhecke, L., Riemannian manifolds with homogeneous geodesics, *Boll. Un. Mat. Ital.* **5** (1991), pp 189-246

[20] Kowalski, O., Vlasek, Z., Homogeneous Riemannian manifolds with only one homogeneous geodesic, *Publ. Math.* **62** (2003), 437-446

[21] Lichnerowicz, A., *Géométrie des Groupes de Transformations*, Paris: Dunod. 1958

[22] Marinosci, R. A., Homogeneous geodesics in a three-dimensional Lie group, *Comm. Math. Univ. Carolinae* **43**(2002), 261-270

[23] Nomizu, K., Invariant affine connections on homogeneous spaces, *Amer. J. Math.* **76**(1954), pp 33-65

[24] Szenthe, J., Sur la connexion naturelle à torsion nulle, *Acta Sci.Math.* **38**(1976), pp 383-398

[25] Szenthe, J., Homogeneous geodesics of left-invariant metrics, *Univ. Iagel. Acta Math.* **38**(2000), pp 99-103

[26] Szenthe, J., On the set of homogeneous geodesics of a left-invariant metric, *Univ. Iagel. Acta Math.* **40** (2002), pp 171-181

GEOMETRY OF THE POINT FINSLER SPACES

Lajos Tamássy*
Institute of Mathematics
Debrecen University
H-4010 Debrecen, P.f. 12
Hungary
tamassy@math.klte.hu

Keywords: Finsler spaces, metrical linear connection, locally Minkowskian spaces, affine deformation of the metric

In the beginning, in ancient times, mathematics was a collection of practical rules. Proofs appeared first in the Vth century B.C., when Greek philosophy intruded into mathematics and initiated a very steep, maybe the ever steepest development of its history. Only 150 years later Euclid was able to write his famous Elements. The geometry based on Euclid's principles is now called euclidean geometry. For more than two thousand years this was the only conceivable geometry, until in 1820–1840 János Bolyai and N. I. Lobachevsky created hyperbolic geometry, the first geometry different from that of Euclid. In the history of geometry this step had the same crucial importance as the replacement of the geocentric world concept by the heliocentric system in astronomy. After this, in a short time a number of new geometries were created, among them Finsler geometry, several problems of which will be discussed in this paper commemorating the 200th anniversary of the birth of János Bolyai.

Introduction

Every differential geometry has two bases: metric and parallelism.

*Lecture presented on July 10, 2002 in Budapest at the János Bolyai Conference to commemorate the 200th anniversary of his birth.
Supported by OTKA 32058.
Mathematical Subject Classification: 53C60, 53B40.

A Finsler manifold or a Finsler space $F^n = (M, \mathcal{L})$ is an n-dimensional manifold M endowed with a Finsler metric function

$$\begin{aligned} &\mathcal{L} : TM \to R^+, \\ &(p, v) \mapsto \mathcal{L}(p, v); \qquad p \in M, \ v \in T_pM. \end{aligned} \tag{1}$$

$\mathcal{L}$ is supposed to be first-order homogeneous in v:

$$\mathcal{L}(p, \lambda v) = |\lambda| \mathcal{L}(p, v), \quad \lambda \in R, \tag{2}$$

continuous on TM, and differentiable on $T^\circ M := TM \setminus 0$ (TM without the null section, where $v = 0$). Any other functions, maps and geometric objects are supposed to be smooth, at least on $T^\circ M$ (unless otherwise stated). Moreover we assume M to be connected, orientable and paracompact.

$$\mathcal{L}(p, v) =: \|v\|_F \tag{3}$$

is the Finsler norm (the Finsler length) of v. Using a local coordinate system (x) on M,

$$ds = \mathcal{L}(x, dx) \tag{4}$$

is the Finsler distance of a two neighbouring points x and $x + dx$, and

$$s = \int_a^b \mathcal{L}(x(t), \dot{x}(t)) dt \tag{5}$$

is the arc length of a curve $C : x = x(t)$, $t \in [a, b] \subset R$. Moreover we assume

$$g_{ij}(x, y)\xi^i \xi^j > 0, \quad \xi \neq 0, \qquad g_{ij}(x, y) := \frac{1}{2} \frac{\partial^2 \mathcal{L}^2}{\partial y^i \partial y^j}(x, y), \tag{6}$$

where (x, y) is a canonical extension of the local coordinate system (x) on M to TM. (6) assures the triangle inequality of the norm (3). (3), (1), (2) and (6) make T_xM into a finite-dimensional Banach space.

(2) assures the independence of the arc length (5) from the parametrization of the curve C, and (6) assures the triangle inequality. Thus both are very natural requirements. Hence a metrical differential geometry on M built on the infinitesimal distance (4) or on the arc length (5), for which the above two natural requirements are fulfilled, is automatically a Finsler geometry. This shows that Finsler geometry is a very natural differential geometry (see S. S. Chern [4]). If $\mathcal{L}^2(x, y)$ is quadratic in y, i.e. if it equals $g_{ij}(x)y^i y^j$, then F^n reduces to a Riemannian manifold $V^n = (M, g)$. A Finsler space F^n, whose carrying manifold M is homeomorphic to an open set O^n of the euclidean space

E^n, and on M there exists an (adapted) coordinate system (x) in which $\mathcal{L}(x, y)$ is independent of y, i.e. $\mathcal{L}(x, y) = \bar{\mathcal{L}}(y)$, is a Minkowski space, denoted by $\mathcal{M}^n$. If adapted coordinate systems exist locally only, then F^n is locally Minkowskian, denoted by $\ell\mathcal{M}^n$. In this case M need not be homeomorphic to O^n.

A metric on M can also be given by a distance function $\varrho(p, q)$, $p, q \in M$ having the usual properties of non-negativity, symmetry and triangle inequality. Given a Finsler metric $\mathcal{L}$,

$$\inf\left\{\int_a^b \mathcal{L}(x(t), \dot{x}(t))dt \mid x(a) = w,\ x(b) = z\right\} =: \varrho(w, z) \tag{7}$$

is clearly non-negative, symmetric and satisfies the triangle inequality, hence it is a distance function. In this way $\mathcal{L}$ determines a metric space (M, ϱ). But also conversely, given a distance function $\hat{\varrho}$, consider in a local coordinate system the function $\hat{\mathcal{L}}(x, y) := \hat{\varrho}(x, x + y)$. $\hat{\mathcal{L}}$ is of class C^0 and smooth for $y \neq 0$. Let us form the Taylor expansion of $\hat{\mathcal{L}}(x, ty)$ along rays (x_0, ty_0):

$$\hat{\mathcal{L}}(x, ty) = \hat{\mathcal{L}}(x, 0) + \left(\lim_{t\to 0} \frac{\partial \hat{\mathcal{L}}}{\partial v^i}(x, ty)\right) y^i + R(x, ty).$$

Here $\hat{\mathcal{L}}(x, 0) = \varrho(x, x) = 0$. Then $(\lim_{t\to 0} \frac{\partial \hat{\mathcal{L}}}{\partial v^i}(x, ty))y^i$ is homogeneous in y, continuous, and smooth if $y \neq 0$. For small y, $\varrho = \hat{\mathcal{L}} \approx L$. But ϱ satisfies the triangle inequality. Hence so does $\mathcal{L}$. Hence ϱ determines an $\mathcal{L}$. However the $\varrho(w, z)$ obtained from $\mathcal{L}$ according to (7) need not be equal the initial distance function $\hat{\varrho}$, as can be shown by an example.

Concerning the Finsler metric a highly useful and visual notion is the indicatrix hypersurface

$$I(x_0) := \{y \in T_{x_0}M \mid \mathcal{L}(x_0, y) = 1\}.$$

It lies in the tangent space $T_{x_0}M$, and plays the role of the unit sphere of euclidean geometry. Because of the homogeneity condition (2), knowledge of the indicatrix field $I(x)$ is equivalent to knowledge of the metric function $\mathcal{L}$. The indicatrices of a Riemann manifold V^n are ellipsoids, those of the euclidean space are unit spheres (in a Descartes coordinate system), and those of a Minkowski space are parallel translated, and hence congruent convex hypersurfaces.

The other basic notion is connection. (For a modern survey see J. Szilasi [18].) It needs no metric. In the simplest case it acts on the tangent bundle τ_M, and it can be interpreted as a mapping $\Gamma : T_pM \to T_qM$ depending on the curve C connecting p and q. The connection is called homogeneous if $\Gamma(\lambda y) = \lambda\Gamma y,\ y \in T_pM,\ \lambda \in R^+$, and it is called linear or affine if Γ is an affine mapping. A connection will be denoted by $\overset{h}{\Gamma}$ and $\overset{a}{\Gamma}$, resp. In case of a Finsler space, Γ is called metrical ($\overset{m}{\Gamma}$) if the parallel translation of a vector y preserves the norm: $\|y\|_F = \|\Gamma y\|_F$. For this it is necessary and sufficient that Γ takes indicatrices into indicatrices:

$$\Gamma I(p) = I(q), \tag{8}$$

independently of the curve connecting p and q. There are plenty of metrical connections in any Riemannian manifold, and a unique one with vanishing torsion. Nevertheless there exist several metrical and linear connections on the vertical bundle VTM (called also the Finsler vector bundle) which is a vector bundle of rank n over the $2n$-dimensional base space TM. These connections use line-elements (x, y) which make the apparatus of the (line-element) Finsler geometry somewhat more complicated than that of Riemannian geometry, and sometimes are inconvenient in physical applications.

In this article we consider several problems of Finsler geometry which do not use line elements. This means that if we use a connection, then it acts in TM, and not in VTM (or TTM). Our approach yields the geometry of the point Finsler spaces as given in the title.

1. Homogeneous metrical connections

A) A Finsler manifold $F^n = (M, \mathcal{L})$ admits in the generic case no metrical linear connection $\overset{ma}{\Gamma}$ in the tangent bundle τ_M, however it allows in it a number of homogeneous metrical connections $\overset{mh}{\Gamma}$ (L. Tamássy [18]). Indeed, let $\overset{h}{\Gamma}$ be a homogeneous connection on M, $y \in T_xM$, $y_e := \frac{y}{\|y\|_F}$ a unit vector, C a curve from x to x', and $\overset{h}{\Gamma}_C y$ the parallel translate of y to x' along C. If

$$\Gamma_C y := \overset{h}{\Gamma}_C\, y \frac{1}{\|\overset{h}{\Gamma}_C y_e\|_F}, \tag{9}$$

then

$$\Gamma_C \lambda y = \overset{h}{\Gamma}_C \lambda y \frac{1}{\|\overset{h}{\Gamma}_C y_e\|_F} = \lambda \Gamma_C y,$$

and

$$\|\Gamma_C y\|_F = \|\overset{h}{\Gamma}_C y\|_F \frac{1}{\|\overset{h}{\Gamma}_C\, y_e\|_F} = \frac{\|y\|_F}{\|y_e\|_F} = \|y\|_F,$$

therefore (9) *yields a metrical and homogeneous connection* $\overset{hm}{\Gamma}$ *for* F^n. In L. Kozma and S. Baran [10] the coefficients of this connection $\overset{hm}{\Gamma}$ and of its curvature tensor are explicitly calculated, and some of their interesting properties described.

If $\overset{h}{\Gamma}$ is also linear, i.e. if it is a $\overset{a}{\Gamma}$, then the constructed $\overset{hm}{\Gamma}$ is linear with respect to the directions. – We remark that even if starting with an affine connection $\overset{a}{\Gamma}(x,y)$ of the vertical bundle VTM (i.e. with a linear connection in the Finsler vector bundle) one can also construct a connection $\overset{hm}{\Gamma}$ in τ_M without any additional data.

B) In section A) we used for the construction of a metrical connection $\overset{hm}{\Gamma}$ a homogeneous connection $\overset{h}{\Gamma}$ which was an exterior object. Now we show that *a Finsler manifold* F^n *intrinsically determines a unique metrical homogeneous connection* $\overset{hm}{\Gamma}$ *also without any exterior data.*

We know the notion of area in a Finsler manifold or in its submanifold of any dimension (H. Busemann [3], H. Rund [14] , Z. Shen [15]) and also the explicit expression for it (L. Tamássy [21]). A Finsler manifold makes its tangent spaces T_xM into Minkowski spaces $\mathcal{M}^n$, and $\mathcal{M}^n$ makes the indicatrix $I(x_0) \subset T_{x_0}M$ as its submanifold into an $n-1$ dimensional Finsler space $F^{n-1}(x_0) := \mathcal{M}^n(I(x_0))$. So we can form the integral

$$\int_{I(x_0)} g_{ij}(x_0, y)da =: G_{ij}(x_0), \qquad g_{ij}(x,y) := \frac{1}{2}\frac{\partial^2 \mathcal{L}^2}{\partial y^i \partial y}(x,y),$$

where da means integration according to the Finsler area of $F^{n-1}(x_0)$. It can be proved that $G_{ij}(x_0)$ is a $(0,2)$ tensor, and $G_{ij}(x_0)\xi^i\xi^j$ is a positive definite quadratic form. This G_{ij} can be formed at any point of M. *This yields a Riemannian manifold* $V^n = (M, G)$.

The Levi-Civita connection of this (M, G) is a linear, and thus homogeneous, connection $\overset{h}{\Gamma}$ on M. Starting with this $\overset{h}{\Gamma}$, and performing the construction of section A), we obtain for $F^n = (M, \mathcal{L})$ a homogeneous metrical connection $\overset{hm}{\Gamma}$ without using any exterior data.

We remark that under the regularity condition $\mathrm{Det}\left(\delta^i_h - C_j{}^i{}_h(x,\xi)\xi^j\right) \neq 0$ one can construct a homogeneous metrical connection $\overset{hm}{\Gamma}$ in TM also from a linear connection $\Gamma(x,y)$ in VTM, where the parallelism is defined by

$$d\xi^i = V_j{}^i{}_h(x,y)\xi^j dx^h + C_j{}^i{}_h(x,y)\xi^j dy^h, \quad \xi(x,y) \in VTM.$$

2. A generalization of the Gauss curvature

As is well-known, the Gauss curvature $\mathcal{K}(\phi, x_0)$ of a hypersurface ϕ of the euclidean space E^n is the limit of the ratio of the surface area of the spherical image of a domain B, $x_0 \in B \subset \phi$ and of the surface area of B. This notion can be extended to Finsler spaces (L. Tamássy [20]).

Let ϕ be a submanifold of M, $\dim \phi = k < n = \dim M$, and $F^n = (M, \mathcal{L})$ a Finsler manifold. F^n makes ϕ into a Finsler space $F^k = F^n(\phi)$ with an induced metric $\hat{\mathcal{L}}$. We consider a geodesic sphere U on ϕ around $x_0 \in \phi$ with respect to $\hat{\mathcal{L}}$ within the range of a geodesic polar coordinate system centered at x_0. ϱ is the radius of U, and let x be an arbitrary point of U. Then there is a unique geodesic C of F^k in U from x to x_0. Let $C(s)$ be a unit speed parametrization of C, and $T(s) := C'(s)$. Then $\|T(s)\|_F = 1$. – Let Γ be any connection in TM, metrical with respect to $\mathcal{L}$, and ∇ the covariant derivation with respect to Γ. Then

$$N(x) := \frac{\nabla_T T}{\|\nabla_T T\|_F}, \quad x \in U, \quad N(x) := 0 \text{ if } \nabla_T T = 0$$

is a vector field on U, and

$$\Gamma_C N(x) =: Y_x(x_0) \in T_{x_0}M$$

is the parallel translate of $N(x)$ to x_0 along C. F^n makes $T_{x_0}M$ into a Minkowski space $\mathcal{M}^n$, and the endpoints of $Y_x(x_0)$ form a surface

$$\Psi := \{Y_x(x_0) \mid x \in U\} \subset I(x_0) \subset T_{x_0}M$$

which is a piece of the indicatrix $I(x_0)$ of F^n (and also of $\mathcal{M}^n$). $\|\Psi\|_M$ means the area of Ψ measured in $\mathcal{M}^n$, and $\|U\|_F$ the Finsler area of U in F^n (which equals the Finsler area of U in F^k). Finally we define

$$K(\phi, x_0) := \lim_{\varrho \to 0} \frac{\|\Psi\|_M}{\|U\|_F} \tag{10}$$

if this limit exists. K will be called the curvature of ϕ in F^n with respect to Γ. $K(\phi, x_0)$ can be expressed in terms of $\mathcal{L}$ and Γ.

If $F^n = R^3$, ∇ is the Levi-Civita connection of R^3, and $\dim \phi = 2$, we obtain a 2-dimensional Riemannian manifold $R^3(\phi) = V^2$, where $N(x) \perp \phi$ and thus $K(\phi, x_0)$ is the Gauss curvature of $\phi \subset R^3$ at x_0. So (10) is *a generalization of the Gauss curvature.* K means no generalization of the sectional curvature $\mathcal{K}(\sigma, x_0)$ of a V^n belonging to a plane position σ at a point x_0, because through x_0 there exist many 2-dimensional ϕ tangent to σ. For them $K(\phi, x_0)$ may be different, while the sectional curvature $\mathcal{K}(\sigma, x_0)$ of V^n is uniquely determined by σ and x_0.

For several known metrics and connections we have $\nabla_T T = 0$ along geodesics. In these cases a totally geodesic submanifold ϕ has vanishing $K(\phi, x)$ at any of its points x. In these cases also $K(\hat{\phi}, x_0) = 0$, provided $\hat{\phi}$ is a geodesic surface at x_0, i.e. has the property that the geodesic curves of the ambient space which tangent $\hat{\phi}$ at x_0 lie completely in $\hat{\phi}$.

Let ϕ_α, $\alpha \in \mathcal{A}$ ($\mathcal{A}$ is an index set) be submanifolds of ϕ through $x_0 \in \phi$. It is an interesting question: what relations hold between $K(\phi, x_0)$ and $K(\phi_\alpha, x_0)$, and between the different $K(\phi_\alpha, x_0)$?

3. Affine deformation of the metric

A generic Finsler space $F^n = (M, \mathcal{L})$ has no linear metrical connection in TM. Nevertheless there are many special Finsler spaces which admit them. Such are the Riemannian spaces V^n, the Minkowski spaces $\mathcal{M}^n$ and the locally Minkowski spaces $\ell\mathcal{M}^n$. Are there still other ones? Yes, there are. They will be denoted by $\tilde{F}^n$. As we shall show, they are exactly the affine deformations $\mathfrak{A}\ell\mathcal{M}^n$ of locally Minkowski spaces $\ell\mathcal{M}^n$. Similar questions, from another point of view, were investigated by Y. Ichijyo [6, 7, 8], Sz. Szakál and J. Szilasi [16, 17] and M. Anastasiei [1].

Let $\mathfrak{a}(x)$ be an affine transformation on T_xM:

$$\mathfrak{a}(x) : T_xM \to T_xM,$$

and $\mathfrak{A}$ a field of such transformations. (In this paper affine transformations always mean orientation preserving centro-affine transformations.) Then $\mathfrak{a}(x)$ takes the indicatrix $I(x)$ of F^n into a new indicatrix

$$\bar{I}(x) := \mathfrak{a}(x)I(x). \tag{11}$$

Since the field of the indicatrices $\bar{I}(x)$ determines a metric function $\bar{L}$, we obtain a new Finsler space $\bar{F}^n = (M, \bar{L}) = \mathfrak{A}F^n$ called the *affine deformation of* F^n. (11) is a kind of gauge transformation. In the case of a Riemannian space V^n the indicatrices are ellipsoids $Q(x)$ ($V^n = (M, g) = (M, Q)$) and an affine deformation yields

$$\mathfrak{a}(x)Q(x) = \bar{Q}(x); \quad \text{i.e.} \quad \mathfrak{A}V^n = \bar{V}^n. \tag{12}$$

For any two given V_1^n and V_2^n there exist many fields $\mathfrak{A}$ which deform V_1^n into V_2^n (i.e. for which $V_2^n = \mathfrak{A}V_1^n$). In particular V_1^n may be the euclidean space E^n with the unit spheres S^{n-1} as indicatrices (in a Descartes coordinate system.) So any V^n can be obtained by an affine deformation of E^n. Nevertheless for two Finsler spaces F_1^n and F_2^n or even for two Minkowski spaces $\mathcal{M}_1^n$ and $\mathcal{M}_2^n$ this is not true, i.e. in general *there exists no* $\mathfrak{A}$ *such that* $F_2^n = \mathfrak{A}F_1^n$ or $\mathcal{M}_2^n = \mathfrak{A}\mathcal{M}_1^n$.

$\mathfrak{A}$ *may consist of special affine transformations, e.g. of rotations* $\mathfrak{r}$ (in certain coordinate systems). Then $\mathfrak{A} = \mathfrak{R} = \{\mathfrak{r}(x)\}$. It is an interesting question to study those Finsler spaces which originate from a locally Minkowski space $\ell\mathcal{M}^n = (M, \mathcal{L})$ by an $\mathfrak{A} = \mathfrak{R}$. In adapted coordinate systems these Finsler spaces $F^n = \mathfrak{R}\ell\mathcal{M}^n$ have congruent, but not parallel translated, indicatrices. Similarly one could study the class $\{\mathfrak{R}V_0^n\}$ of those Riemannian spaces which arise from a Riemannian space V_0^n via deformation by different $\mathfrak{R}$. [V_0^n can also be replaced by a Finsler space F_0^n.]

We want to study the affine deformations of the locally Minkowski spaces $\ell\mathcal{M}^n$ or specially of $\mathcal{M}^n$. Let us consider an $\ell\mathcal{M}^n = (M, \mathcal{L}) = (M, I)$. $I(x)$ is the indicatrix, and let $y \in T_xM$ be a unit vector: $\mathcal{L}(y) = 1$. Let $\mathfrak{A}$ be an affine deformation. Then $\bar{y} = \mathfrak{a}(x)y \in \mathfrak{a}(x)I(x) = \bar{I}(x)$, $\mathfrak{A}\ell\mathcal{M}^n = (M, \bar{I}) = (M, \bar{\mathcal{L}}) = \bar{F}^n$, and hence $\bar{\mathcal{L}}(x, \bar{y}) = \bar{\mathcal{L}}(x, \mathfrak{a}(x)y) = 1 = \mathcal{L}(y)$. Let $\mathfrak{b}(x)$ be the inverse of $\mathfrak{a}(x)$. Then $\mathcal{L}(\mathfrak{b}(x)y) = \bar{\mathcal{L}}(x, y)$. In an adapted coordinate system $\mathcal{L}$ depends only on $y^1, y^2, \ldots, y^n$, and it is a function of n variables, while the metric function $\bar{\mathcal{L}}$ depends on x and y, and it is a function of $2n$ variables. An $\bar{F}^n = (M, \bar{\mathcal{L}})$, which has the property that $\bar{\mathcal{L}}(x, y) = \mathcal{L}(\mathfrak{b}(x)y)$ is called a Finsler space of 1-form metric, since each component $b_h^{i_0}(x)y^h$ of $\mathcal{L}$ is a 1-form. Thus *any* $\mathfrak{A}\ell\mathcal{M}^n$ *is a Finsler space with 1-form metric. The converse is also true.* If $\bar{F}^n$ is a Finsler space with 1-form metric, then $\bar{\mathcal{L}}(x, \mathfrak{a}(x)y) = \mathcal{L}(y)$, which yields $\bar{I}(x) = \mathfrak{a}(x)I(x)$. Hence $\bar{F}^n = \mathfrak{A}\ell\mathcal{M}^n$. Finsler spaces with 1-form metric were introduced and studied first by M. Matsumoto and H. Shimada [12, 13], see also Sz. Szakál and J. Szilasi [16]. We will study affine deformations $\mathfrak{A}\ell\mathcal{M}^n$ of locally Minkowski spaces from a different point of view.

4. Linear metrical connections in TM

The base manifold M of a locally Minkowski space $\ell\mathcal{M}^n = (M, I)$ has a cover $\{U_\alpha, x_\alpha\}$, $\alpha \in \mathcal{A}$ with adapted coordinate systems (x_α) on U_α. Then the coordinate transformations $(a) : (x_{\alpha_1}) \leftrightarrow (x_{\alpha_2})$ are linear. A differentiable structure $\{(U_\alpha, \varphi_\alpha)\}$, $\varphi_\alpha : U_\alpha \to E^n(x_\alpha)$ on M with property (a) will be called an *affine structure.* In the sequel manifolds will be assumed to have such a structure.

We want to show that Finsler spaces (over manifolds with affine structure) admitting linear metrical connections in TM (i.e. the $\tilde{F}^n$) are affine deformations of locally Minkowski spaces, and conversely: $\tilde{F}^n \Longleftrightarrow \mathfrak{A}\ell\mathcal{M}^n$. In section A) we show the implication $\Longleftarrow$, first in the case of $M = O \subset E^n$, and after that without this restriction. In section B) we show the implication $\Longrightarrow$ again in order of the above two cases.

A) Let $\mathcal{M}^n = (M, I)$ and $\mathfrak{A}\mathcal{M}^n = (M, \tilde{I})$, $M = O \subset E^n$, and let (x) be an adapted coordinate system on M. Then the indicatrices of $\mathcal{M}^n$ are parallel translates of each other, and thus independent of x: $I(x) = I(x_0) \equiv I_0$. Let $\mathfrak{t}$ be the parallel translation from a fixed point x_0 to an arbitrary point x. Then

$$\mathfrak{g}(x_0, x) := \mathfrak{a}(x) \circ \mathfrak{t} \circ \mathfrak{a}^{-1}(x_0), \qquad \mathfrak{a} \in \mathfrak{A} \tag{13}$$

is an affine transformation taking $T_{x_0}M$ into T_xM. Let $\mathfrak{v}_i(x_0) = v^s_{(i)}(x_0)\frac{\partial}{\partial x^s} \in \tilde{I}(x_0) \subset T_{x_0}M$ be a basis in $T_{x_0}M$. Then

$$\mathfrak{g}(x_0, x)\mathfrak{v}_i(x_0) =: \mathfrak{v}_i(x)$$

is a basis in T_xM, and $\mathfrak{v}_i(x) \in \tilde{I}(x)$, for

$$\mathfrak{g}(x_0, x)\tilde{I}(x_0) = \mathfrak{a}(x) \circ \mathfrak{t} \circ \mathfrak{a}^{-1}(x_0)\tilde{I}(x_0) = \mathfrak{a}(x) \circ \mathfrak{t} I(x_0) = \mathfrak{a}(x) I(x) = \tilde{I}(x).$$

Thus the mapping induced by $\mathfrak{g}$ is not only linear, but also metrical.

We show that $\mathfrak{g}$ determines on M a linear connection $\hat{\Gamma}$ which is metrical with respect to $\mathfrak{A}\mathcal{M}^n$. The vector system $\mathfrak{v}_i(x) \in \tilde{I}(x)$ forms a frame of T_xM denoted by $v(x)$ given by the matrix $\|v^s_{(i)}\|$. $F(x) = \{v(x), w(x), z(x), \dots\}$ is the manifold of all frames of T_xM consisting of vectors with endpoint on $\tilde{I}(x)$. All $F(x)$ are diffeormorphic to each other and so to an abstract manifold F. M fibered by F yields a bundle $\pi : P \to M$ with $\pi^{-1}(x) = F(x) \approx F$. Let $G(x) = \{g(x), h(x), \dots\}$ be the set of those invertible linear transformations (represented by $n \times n$ matrices), which take elements of $F(x)$ into each other. $G(x)$ with the matrix multiplication is a subgroup of $GL(n)$. G acts on P simply transitively (actually one can see that P is a principal bundle). $\mathfrak{g}(x_0, x)$ $\mathfrak{v}_i(x_0) = \mathfrak{v}_i(x) = v(x)$ form a section σ of P. $g : P \to P$, $v(x) \mapsto gv(x)$ is a fiber preserving tranformation. $dgT_{v(x)}\sigma = H(x, gv)$, $\forall g \in G$ yields a horizontal distribution H_P in TP, and thus a connection Γ_P in P. This determines a linear connection $\hat{\Gamma}$ in TM, which is a vector bundle associated to P (see Bishop and Crittenden [2]). The parallel translations with respect to $\hat{\Gamma}$ are the same as the mappings induced by $\mathfrak{g}$. Hence *the linear connection* $\hat{\Gamma}$ *is also metrical with respect to* $\mathfrak{A}\mathcal{M}^n$.

σ is an integral manifold of Γ_P, for $T_{(x,v)}\sigma = H(x, v)$. Hence H_P (and therefore also Γ_P) is completely integrable, and thus so is the associated connection $\hat{\Gamma}$. Thus each $\mathfrak{v}_i(x)$ is an absolute parallel vector field, and hence *the curvature tensor of* $\hat{\Gamma}$ *vanishes*: $\hat{R} = 0$. This can be seen also from the fact that the construction of the parallel vector fields $\mathfrak{v}_i(x)$ is independent of the route connecting x_0 and x.

We derived H_P and $\hat{\Gamma}$ from $\mathfrak{g}$ which is determined by $\mathfrak{a}(x) \in \mathfrak{A}$. Thus *the connection coefficients* $\hat{\Gamma}_k{}^j{}_r(x)$ *of* $\hat{\Gamma}$ *must be expressible by* $a_s^j(x)$, the coefficients of $\mathfrak{a}(x)$. – Let x_1 be a point in M, and x a point on the r-th parameter line starting from x_1 with components $x^h(t) = x_1^h + \delta_r^h t$. Because of (13)

$$\mathfrak{v}_i(x) = \mathfrak{g}(x_0, x)\mathfrak{v}_i(x_0) = \mathfrak{g}(x_1, x) \circ \mathfrak{g}(x_0, x_1)\mathfrak{v}_i(x_0) = \mathfrak{g}(x_1, x)\mathfrak{v}_i(x).$$

In components

$$v_{(i)}^j(x) = a_s^j(x) b_k^s(x_1) v_{(i)}^k(x_1),$$

where b_k^j are components of $\mathfrak{a}^{-1}$. $\mathfrak{v}_i(x)$ is a parallel vector field with respect to $\hat{\Gamma}$. Then along the r-th parameter line

$$\frac{d}{dt} v_{(i)}^j(x(t))|_{t_0} = \hat{\Gamma}_k{}^j{}_r(x(t_0)) v_{(i)}^k(x_1) = \frac{\partial a_s^j}{\partial x^r}(x_1) b_k^s(x_1) v_{(i)}^k(x_1).$$

Hence

$$\hat{\Gamma}_k{}^j{}_r(x) = \frac{\partial a_s^j}{\partial x^r}(x) b_k^s(x). \tag{14}$$

In the case of an $\mathfrak{A}\ell\mathcal{M}^n = (M, \hat{I})$, M cannot be covered by a single adapted coordinate system. Let C be a curve connecting p_0 with an arbitrary p; $p_0, p \in M$. C can be covered by a finite series of neighbourhoods $U_i \subset M$, $i = 1, 2, \ldots, N$. We can assume that $p_0 \in U_1$, $p \in U_N$, and $U_j \cap U_{j+1} \neq 0$, $j = 1, 2, \ldots, N-1$. Choose points $p_j \in U_j \cap U_{j+1}$. Let $\mathfrak{t}_j$ be a parallel translation in U_j from p_{j-1} to p_j; $\mathfrak{t}_N$ parallel translation in U_N from p_N to p, and

$$\mathfrak{t}(p_0, p) := \mathfrak{t}_N \circ \cdots \circ \mathfrak{t}_2 \circ \mathfrak{t}_1. \tag{15}$$

Then we can construct $\mathfrak{g}$ according to (13), and apply the above procedure leading again to a linear and metrical connection $\hat{\Gamma}$ with local coefficients $\hat{\Gamma}_k{}^j{}_r(x)$. Thus also *any* $\mathfrak{A}\ell\mathcal{M}^n$ *admits a metrical linear connection in* TM. (The unambiguity, continuity and differentiability of $\hat{\Gamma}$ still need further discussion.)

B) We consider a Finsler space $\tilde{F}^n = (M, \tilde{I})$, $M = O \subset E^n$ admitting a connection $\overset{am}{\Gamma}$ in TM. We show that this $\tilde{F}^n$ is an affine deformation $\mathfrak{A}\mathcal{M}^n$ of an $\mathcal{M}^n = (M, I)$. First we give an $\mathcal{M}^n$, then we define an $\mathfrak{A}$, and we show that $\mathfrak{A}\mathcal{M}^n = \tilde{F}^n$. Let (x) be a linear coordinate system on M, x_0 a fixed and x an arbitrary point of M, and let the indicatrix $\tilde{I}(x_0)$ of $\tilde{F}^n$ be the indicatrix of $\mathcal{M}^n$ at $x_0 : I(x_0) := \tilde{I}(x_0)$. Let $\mathfrak{t}$ be a parallel translation from x_0 to x. We define $I(x) := \mathfrak{t}I(x_0)$. Thus we have an $\mathcal{M}^n = (M, I)$. Let C again be a curve from x_0 to x, and denote by P_C^Γ the parallel transport along C according to the linear metrical connection $\overset{am}{\Gamma}$ of $\tilde{F}^n$. Then $P_C^\Gamma \tilde{I}(x_0) = \tilde{I}(x)$, and $P_C^\Gamma \circ \mathfrak{t}^{-1} : T_x M \to T_x M$ is an affine transformation

$$\mathfrak{a}(x) := P_C^\Gamma \circ \mathfrak{t}^{-1}, \qquad \forall x \in M. \tag{16}$$

So we obtain a field $\mathfrak{A} = \{\mathfrak{a}(x)\}$. (The continuity and differentiability of $\mathfrak{A}$ is still to be discussed.) $\mathfrak{a}(x)$ takes $I(x)$ into $\tilde{I}(x)$, for

$$\mathfrak{a}(x)I(x) = P_C^\Gamma \circ \mathfrak{t}^{-1} I(x) = P_C^\Gamma I(x_0) = P_C^\Gamma \tilde{I}(x_0) = \tilde{I}(x). \tag{17}$$

This means that *our $\tilde{F}^n$ is an $\mathfrak{A}\mathcal{M}^n$ with the $\mathcal{M}^n$ and $\mathfrak{A}$ constructed.*

Now we consider an $\tilde{F}^n = (M, \tilde{I})$, where M is not necessarily homeomorphic to an open set $O \subset E^n$, nevertheless it has an affine structure $\{(U_\alpha, \varphi_\alpha)\}$, $\varphi_\alpha : U_\alpha \to E^n(x_\alpha)$, and $\tilde{F}^n$ admits a $\overset{am}{\Gamma}$ in TM. Then we can construct a $\mathfrak{t}(p_0, p)$, $p_0, p \in M$, as we did it in (15). With the help of this $\mathfrak{t}$ we construct an $\ell\mathcal{M}^n = (M, I)$. Let $I(p_0) := \tilde{I}(p_0)$, and $I(p) := \mathfrak{t}I(p_0)$. Then we have a field $I(p)$ on M. They are parallel translates of each other on U_α, and (x_α) are adapted coordinate systems on U_α. Now we define $\mathfrak{a}(p)$ by (16). This $\mathfrak{a}(p)$ takes $I(p)$ into $\tilde{I}(p)$ of $\tilde{F}^n$ just as in (17). This means that our $\tilde{F}^n$ is an $\mathfrak{A}\ell\mathcal{M}^n$ with the constructed $\ell\mathcal{M}^n$ and $\mathfrak{A}$.

A) *and* B) *show that Finsler spaces over manifolds with affine structures admitting linear metrical connections $\overset{am}{\Gamma}$ in TM (i.e. the $\tilde{F}^n$ spaces) are exactly the affine deformations of the locally Minkowski spaces:*

$$\{\tilde{F}^n\} = \{\mathfrak{A}\ell\mathcal{M}^n\}.$$

The questions discussed above lead to some functional or differential equations. – A) Is a given Finsler space $F^n = (M, \mathcal{L})$ an affine deformation of a locally Minkowskian space $\ell\mathcal{M}^n$, i.e. does there exist an $\mathfrak{A}$ such that $F^n = \mathfrak{A}\ell\mathcal{M}^n$? This is the case iff the indicatrices of F^n are affine to each other, i.e. there exists a field of affine transformations $\mathfrak{a}(x)$ such that $\mathcal{L}(x, \mathfrak{a}(x)y) = \mathcal{L}(x_0, y)$ or

$$g_{m\ell}(x_0, y)y^m y^\ell = g_{ij}(x, a_s^r(x)y^s)a_m^i(x)y^m a_\ell^j(x)y^\ell$$

or

$$g_{m\ell}(x_0, y) = g_{ij}(x, a_s^r(x)y^s)a_m^i(x)a_\ell^j(x), \tag{18}$$

where $g_{ij}(x, y)$ is the metric tensor of F^n. (18) is a functional equation for $a_s^r(x)$, $\det(a_s^r(x)) \neq 0$. If (18) has a solution, then $F^n = \mathfrak{A}\ell\mathcal{M}^n$

with $\mathfrak{A} = \{\mathfrak{a}(x)\}$. Differentiating with respect to x^a, (18) turns into a first-order partial differential equation system

$$\frac{\partial g_{ij}}{\partial x^a}(x, a^r_s(x)y^s)a^i_m(x)\delta^n_\ell + \frac{\partial g_{ij}}{\partial y^p}(x, a^r_s(x)y^s)\frac{\partial a^p_s}{\partial x^a}(x)y^s a^i_m(x)\delta^n_\ell$$
$$+ g_{ij}(x, a^r_x(x)y^s)\left[\frac{\partial a^i_m}{\partial x^a}(x)\delta^n_\ell + \delta^n_m\frac{\partial a^i_\ell}{\partial x^a}(x)\right] = 0.$$

B) A Riemannian space $V^n = (R^n, g) = (R^n, Q)$ is an $\mathfrak{A}E^n$ with an $\mathfrak{A}$, where $\mathfrak{a}(x)S = Q(x)$ ($E^n = (R^n, S)$ is a euclidean space with a Descartes coordinate system (x)). According to (13) we obtain by (14) for any V^n a metrical linear connection $\hat{\Gamma}$. Its curvature vanishes: $\hat{R} = 0$, but in the generic case its torsion does not: $\hat{T} \neq 0$. Any metrical linear connection of V^n with vanishing curvature can be represented in this way using an $\overset{*}{\mathfrak{A}}$ in place of $\mathfrak{A}$ with $\overset{*}{\mathfrak{a}}(x) = \mathfrak{a}(x) \circ \mathfrak{r}(x)$, where $\mathfrak{r}(x)$ are rotations. However among the metrical linear connections of V^n there exits exactly one, the Levi-Civita connection, which has vanishing torsion. Does there exist a field $\mathfrak{R} = \{\mathfrak{r}(x)\}$ such that *beside* $\overset{*}{R} = 0$ *we also have a vanishing torsion* $\overset{*}{T} = 0$? The existence of such an $\mathfrak{r}(x)$ *characterizes the euclidean space* among the Riemannian spaces. One can try to answer this question by direct geometrical considerations or by investigation of the relation

$$\overset{*}{\Gamma}{}_k{}^j{}_\ell = \frac{\partial \overset{*}{a}{}^j_s}{\partial x^\ell}\overset{*}{b}{}^s_k = \frac{\partial \overset{*}{a}{}^j_s}{\partial x^k}\overset{*}{b}{}^s_\ell = \overset{*}{\Gamma}{}_\ell{}^j{}_k$$

which is a PDE

$$\frac{\partial a^j_p}{\partial x^{[\ell}}a^{-1p}{}_{k]} + a^j_m\frac{\partial r^m_s}{\partial x^{[\ell}}a^{-1p}{}_{k]}r^{-1s}{}_p = 0$$

($[\cdot,\cdot]$ means alternation) for the unknown components $r^m_s(x)$ of the rotation $\mathfrak{r}(x)$. The answer to this question may contribute to a better understanding of the geometrical role of the torsion, of which we still have only a restricted knowledge.

5. Affine automorphisms of the indicatrix

Consider an $\ell\mathcal{M}^n = (M, I)$ and $\mathfrak{A}\ell\mathcal{M}^n = (M, \tilde{I})$. Then $\tilde{I}(x) = \mathfrak{a}(x)I(x)$. There may exist also another $\overset{*}{\mathfrak{A}}$ ($\neq \mathfrak{A}$) such that $\overset{*}{\mathfrak{A}}\ell\mathcal{M}^n = (M\overset{*}{I})$ equals $\mathfrak{A}\ell\mathcal{M}^n = (M, \tilde{I})$ with the same $\ell\mathcal{M}^n = (M, I)$. In this case $\tilde{I} = \mathfrak{a}I = \overset{*}{\mathfrak{a}}I$. Thus $\mathfrak{a}^{-1} \circ \overset{*}{\mathfrak{a}}I = I$ and $\tilde{I} = \overset{*}{\mathfrak{a}} \circ \mathfrak{a}^{-1}\tilde{I}$, i.e. there exist affine transformations $\mathfrak{k}_1 = \mathfrak{a}^{-1} \circ \overset{*}{\mathfrak{a}}$ and $\mathfrak{k}_2 = \overset{*}{\mathfrak{a}} \circ \mathfrak{a}^{-1}$ such that $\mathfrak{k}_1 I = I$

and $\mathfrak{k}_2\tilde{I} = \tilde{I}$. A simple example is $\ell\mathcal{M}^n = E^n$. The indicatrices of E^n in a Descartes coordinate system are the unit spheres S^{n-1} (briefly S). Then $\mathfrak{A}E^n = (M, Q)$ is a Riemannian space V^n, where the indicatrices $Q(x) = \mathfrak{a}(x)S$ are ellipsoids. However we also have $\mathfrak{a}(x) \circ \mathfrak{r}(x)S = Q(x)$, where $\mathfrak{r}(x)$ are rotations. Thus also $\overset{*}{\mathfrak{A}}E^n = V^n = \mathfrak{A}E^n$ with $\mathfrak{a}\circ\mathfrak{r} = \overset{*}{\mathfrak{a}} \in \overset{*}{\mathfrak{A}}$, $\mathfrak{k}_1 = \mathfrak{a}^{-1} \circ \overset{*}{\mathfrak{a}} = \mathfrak{r}$ and $\mathfrak{k}_2 = \overset{*}{\mathfrak{a}} \circ \mathfrak{a}^{-1} = \mathfrak{a} \circ \mathfrak{r} \circ \mathfrak{a}^{-1}$.

An affine transformation $\mathfrak{k}$ in T_xM is an *affine automorphism* of I if $\mathfrak{k}I = I$ (I, as a whole, e.g. $\mathfrak{r}S = S$). There are indicatrices which admit non-trivial, i.e. $\mathfrak{k} \neq$ id. affine automorphisms, and there are those which do not. Let $\dim M = 2$, and consider in the tangent space T_xM a regular triangle inscribed in a circle around the origin. This triangle admits three affine automorphisms (including the identity). Supplementing the vertices of this equilateral triangle with a further point on the circle, we obtain a tetragon, which admits no affine automorphism (except the identity). (We supposed our affine transformations to be orientation preserving. See Sec. 3.) These figures are neither smooth, nor strictly convex. However a small alternation makes them into smooth and strictly convex indicatrices retaining the above discussed properties.

If an $\tilde{I}(x_0)$ has an affine automorphism $\mathfrak{k} \neq$ id., then we call it *mobile*, otherwise it is called *rigid.* Since all $\tilde{I}(x)$ and $I(x)$ of an $\mathfrak{A}\ell\mathcal{M}^n = (M, \tilde{I})$, $\ell\mathcal{M}^n = (M, I)$ are affine to $\tilde{I}(x_0)$, therefore if an $\tilde{I}(x_0)$ or $I(x_0)$ is rigid (resp. mobile), then all $\tilde{I}(x)$ and $I(x)$ are so. Let Γ be a linear metrical connection for $\tilde{F}^n = \mathfrak{A}\ell\mathcal{M}^n$ and C a closed curve through x_0. If there exist $x_0, C,\ y_0 \in \tilde{I}(x_0) \subset T_{x_0}M$ such that $P^\Gamma_C y_0 = y_1 \neq y_0$ (in this case also $y_1 \in \tilde{I}(x_0)$), then $\tilde{I}(x_0)$ is called Γ-*mobile*, otherwise it is called Γ-*rigid.* Clearly Γ-mobility (resp. Γ-rigidity) implies mobility (resp. rigidity), and $\mathfrak{k} = P^\Gamma_C$ is an affine automorphism of $\tilde{I}(x_0)$. If $P^\Gamma_{C_1} = \mathfrak{k}(1) \neq$ id., and C_1 is contractible to x_0 through a family of curves $C(t)$, $0 \le t \le 1$, $C(1) = C_1$, $C(0) = x_0$, then $P^\Gamma_{C(t)} = \mathfrak{k}(t)$ yield infinitely many different affine automorphisms of $\tilde{I}(x_0)$.

Suppose that the indicatrix $I(x)$ of an $\ell\mathcal{M}^n = (M, I)$ admits a field $\mathfrak{K} = \{\mathfrak{k}(x)\}$ of an affine automorphism, and $\mathfrak{k}(x) \not\equiv$ id. Then with $\overset{*}{\mathfrak{a}}(x) = \mathfrak{a}(x) \circ \mathfrak{k}(x)$ we obtain $\mathfrak{A}\ell\mathcal{M}^n = \overset{*}{\mathfrak{A}}\ell\mathcal{M}^n = \tilde{F}^n$. Using (13) and the procedure of Section 4, from $\mathfrak{a}(x)$ we obtain $\hat{\Gamma}_i{}^j{}_h(x)$ (see (14)), and starting with $\overset{*}{\mathfrak{a}}(x)$ in place of $\mathfrak{a}(x)$ we obtain for the same $\tilde{F}^n$ another linear metrical connection $\overset{*\wedge}{\Gamma}$ with coefficients

$$\overset{*\wedge}{\Gamma}{}_i{}^j{}_k(x) = k^{-1}{}^s_r(x) a^{-1}{}^r_i(x) \frac{\partial}{\partial x^k}\left(a^j_p(x) k^p_s(x)\right)$$

(k_s^p are the components of $\mathfrak{k}$). $\overset{*\wedge}{\Gamma}$ *has again vanishing curvature* $\overset{*\wedge}{R} = 0$. If k_s^p are independent of x, i.e. $\mathfrak{k} = \text{const.}$, then $\overset{*\wedge}{\Gamma} = \hat{\Gamma}$.

May an $\tilde{F}^n = \mathfrak{A}\ell\mathcal{M}^n = (M, \tilde{I})$, $\ell\mathcal{M}^n = (M, I)$ have a linear metrical connection Γ with non-vanishing curvature? Yes, under certain conditions it may. – We assume that $\tilde{I}(x)$ admits affine automorphisms $\mathfrak{k}(t) \neq \text{const.}$, $t \in T = [0,1]$. Then $\mathfrak{k}(t)\tilde{I} = \tilde{I}$, $\mathfrak{k}(t) \circ \mathfrak{a}(x)I(x) = \tilde{I}(x)$, and $\overset{*}{\mathfrak{A}}\ell\mathcal{M}^n = \mathfrak{A}\ell\mathcal{M}^n$ with $\overset{*}{\mathfrak{a}}(x,t) = \mathfrak{k}(t) \circ \mathfrak{a}(x)$. Starting with this $\overset{*}{\mathfrak{a}}$, using (13) and following the considerations of the second and third paragraphs of Section 4A, we obtain for each t a section $\sigma(t) = (x, v(t))$ of the frame bundle $P = \{(x,w)\}$. Each $T_{(x,v(t))}\sigma(t) = H(x, v(t))$ can be extended by the differential mappings d_g of the translations $g : P \to P$, $(x, v(t)) \mapsto (x, w)$, $w = gv(t)$ to a distribution $H(x; w, t)$ in TP, similarly to Section 4A. Then at $(x, w) \in P$ we have an n-dimensional linear subspace $H(x; w, t_0) \subset T_{(x,w)}P$ for each $t_0 \in T$. Let $P \to T$, $(x, w) \mapsto t(x, w)$ be a function on P (i.e. a section η of $P \times T$). Then $H(x; w, t(x, w))$ is a distribution in TP. Each $t(x, w)$ (or η) determines a connection Γ_P in TP, and thus a linear metrical connection Γ in TM. $\sigma(t)$ are integrable sections of P. Hence the corresponding connections $\hat{\Gamma}$ in TM have vanishing curvatures (see Section 4A paragraph 3). However, if a section η of P is not integrable, then neither the corresponding Γ_P in TP nor the corresponding Γ in TM is so. Then *the curvature of this* Γ *is not vanishing*: $R^\Gamma \neq 0$.

6. The holonomy group and the structure of the affine deformation of locally Minkowski spaces

Clearly $\mathfrak{k} = P_C^\Gamma$ with closed C are elements of the holonomy group $\mathfrak{H}^\Gamma$ of Γ. Holonomy groups play an important role also in Finsler geometry. (For some new results see L. Kozma [9].) We know that if there exists a $y_0 \in \tilde{I}(x_0)$ and a closed curve C contractible to $x_0 \in M$ such that $P_C^\Gamma y_0 \neq y_0$ (in this case $\tilde{I}$ is Γ-mobile), then $\mathfrak{H}^\Gamma$ is infinite. However, if C is not contractible, then $\mathfrak{H}^\Gamma$ may be finite, even if the other conditions above are fulfilled, so even if $\tilde{I}$ is Γ-mobile. For $\mathfrak{H}^\Gamma$ to be finite, but not necessarily the identity, it is necessary that the fundamental group of M is not the identity. Any $\mathfrak{A}\ell\mathcal{M}^n$ with $\hat{\Gamma}$ constructed from $\mathfrak{A}$ through (13) and having coefficients (14) is $\hat{\Gamma}$-rigid, for its curvature vanishes: $\hat{R} = 0$. If for a linear metrical connection Γ of an $\tilde{F}^n = \mathfrak{A}\ell\mathcal{M}^n$ $R^\Gamma = 0$ and also the torsion T^Γ vanishes, then there exists a coordinate system (x') around each point of M in which ${\Gamma_k}^j{}_r(x') = 0$. In this case the parallel vector fields have constant components, and the indicatrices $\tilde{I}(x')$ are also parallel displaced, consequently this $\tilde{F}^n$ *is a locally Minkowskian space.* Then $\tilde{F}^n$ originates from a locally Minkowski space $(M, \bar{I})$, where the indicatrices $\bar{I}(x')$ are parallel, and $\mathfrak{a}(x') = \text{const}$. But also conversely, if (x') is an adapted coordinate system on $\ell\mathcal{M}^n$ and $\mathfrak{a}(x') = \text{const.}$, then $\mathfrak{A}\ell\mathcal{M}^n$ is also a locally Minkowski space.

We show that a Γ-mobile 2-dimensional $\tilde{F}^2 = \mathfrak{A}\ell\mathcal{M}^2 = (M, \tilde{I})$ with simply connected M is a Riemannian space. – As we have seen above, in this case we have infinitely many affine automorphisms $\mathfrak{k}(t) = P^{\Gamma}_{C(t)}$, $0 \le t \le 1$ for $\tilde{I}(x_0)$. $\tilde{I}(x_0)$ makes $T_{x_0}M$ into a Minkowski space $\mathcal{M}^2_{x_0} = (T_{x_0}M, \tilde{I}(x_0))$. Any $\mathfrak{k}(t)$ is a linear transformation of $T_{x_0}M$, and at the same time an isometry with respect to the metric of $\mathcal{M}^2_{x_0}$, since $\mathfrak{k}(t)\tilde{I}(x_0) = \tilde{I}(x_0)$. However, by a result of P. Gruber ([5] or Thompson [22, p. 83]), in this case $\tilde{I}(x_0)$ is an ellipse $Q(x_0)$. This is true for every point $x \in M$. Hence $\tilde{I}(x) = Q(x)$, and *thus* $\tilde{F}^2 = \mathfrak{A}\ell\mathcal{M}^2 = (M, Q)$ *is a Riemannian space* $V^2 = (M, Q)$.

This result can be extended to higher dimensions. – Consider the set $Y(y_*)$ of the points $y \in \tilde{I}(x_0)$ to which $y_* \in \tilde{I}(x_0)$ can be parallel translated by P^{Γ}_C, where Γ is any linear metrical connection of $\mathfrak{A}\ell\mathcal{M}^n = (M, \tilde{I})$, M simply connected and C a closed curve containing x_0. Denote by $L^r(y_*)$ a smallest linear subspace of $T_{x_0}M$ containing $Y(y_*)$, where r is the dimension of the subspace. r may depend on y_*. Let k be the maximum of $r(y_*)$:

$$k = max_{y_*}\{min_r[L^r(y_*)] \mid y_* \in \tilde{I}(x_0)\},$$

and y_0 such an element of $\tilde{I}(x_0)$ for which this maximum is achieved. Let us denote this subspace by $L^k(x_0)(\subset T_{x_0}M)$. Then we obtain

$$L^k(x_0) \cap \tilde{I}(x_0) = \overset{*}{Q}{}^{k-1}, \tag{19}$$

where $\overset{*}{Q}{}^{k-1}$ is a $k-1$ dimensional ellipsoid (see L. Kozma and L. Tamássy [11]). Thus $\tilde{I}(x_0)$ *contains a* $k-1$ *dimensional ellipsoid.* It is easy to see that the value of k is independent of $x \in M$.

Performing an appropriate linear coordinate transformation $(x) \to (u)$ in M, $\overset{*}{Q}{}^{k-1}$ becomes a sphere S^{k-1} and $\tilde{I}(x_0)$ *becomes a rotation surface determined by* S^{k-1} *and an indicatrix* $\tilde{I}^{n-k-1}(u_0)$ in the $n-k$ dimensional plane ${}^{\perp}L^{n-k}(u_0)$ perpendicular to $L^k(u_0)$. We express this situation by writting

$$\tilde{I}(u_0) = S^{k-1} \times_{\mathfrak{r}} \tilde{I}^{n-k1}(u_0).$$

The subspaces L^k of (19) determine a distribution $\mathcal{D}^k_n(M)$ by ordering $L^k(x)$ to $x \in M$. If $\mathcal{D}^k_n$ is integrable and N is an integral manifold, then we obtain Riemannian spaces $V^k = (N, \overset{*}{Q})$. This is expressed by

$$\tilde{F}^n = V_n^k \times_{\mathfrak{r}} \overset{\circ}{F}^{n-k},$$

where V_n^k means $\mathcal{D}_n^k$ endowed with a Riemannian metric $\overset{*}{Q}^{k-1}$ and $\overset{\circ}{F}^{n-k}$ is a structure, where M^n is endowed with $n-k-1$ dimensional indicatrices $\tilde{I}(x) \cap {}^{\perp}L^{n-k}(x)$.

$\tilde{I}(x_0)$ is called maximally Γ-mobile if $k = n$ in (19), i.e. $L^k(x_0) = T_{x_0}M$. Then (19) takes the form $T_{x_0}M \cap \tilde{I}(x_0) = \overset{*}{Q}^{n-1}(x_0)$, and thus $\tilde{I}(x) = \overset{*}{Q}^{n-1}(x)$. In this case the affine deformated locally Minkowski space $\mathfrak{A}\ell\mathcal{M}^n$ is a Riemannian space $V^n = (M, \overset{*}{Q})$. – On the other hand, we have seen that if $\tilde{I}(x_0)$ is Γ-rigid for every Γ of $\tilde{F}^n$, and for such a Γ also the torsion vanishes, then $\tilde{F}^n$ is a locally Minkowski space. So the *Riemannian and locally Minkowski spaces are the two extreme cases of the affine deformation of locally Minkowski spaces, appearing in the case of maximal, resp. minimal degree of* Γ*-mobility of the indicatrices.*

Bibliography

[1] Anastasiei M., *Metrizable linear connections in vector bundles.* Publ. Math. Debrecen **62** (2003), 461–472.

[2] Bishop R. L. and Crittenden, *Geometry of Manifolds*, Ac. Press, New York, 1964.

[3] Busemann H., *Intrinsic area.* Ann. of Math. **48** (1947), 234–267.

[4] Chern S. S., *Riemannian geometry as a special case of Finsler geometry.* Contemp. Math. **196** (1996), 51-57.

[5] Gruber P. M., *Minimal ellipsoids and their duals.* Rend. Circ. Math. Palermo **37** (1988), 35–64.

[6] Ichijyo Y., *Finsler manifolds modelled on Minkowski spaces.* J. Math. Kyoto Univ. **16** (1976), 639–652.

[7] Ichijyo Y., *Finsler manifolds with a linear connection.* J. Math. Tokushima Univ. **10** (1976), 1–11.

[8] Ichijyo Y., *On the condition for a $\{V, H\}$-manifold to be locally Minkowskian or conformally flat.* J. Math. Tokushima Univ. **13** (1979), 13–21.

[9] Kozma L., *On Landsberg spaces and holonomy of Finsler manifolds.* Contemp. Math. **196** (1996), 177–186.

[10] Kozma L., Baran S., *On metrical homogeneous connections of a Finsler point space.* Publ. Math. Debrecen **49** (1996), 59–68.

[11] Kozma L., Tamássy L., *Finsler geometry without line elements faced to applications.* Reports on Math. Phys. **51** (2003), 233—250.

[12] Matsumoto M., Shimida H., *On Finsler spaces with 1-form metric.* Tensor N. S. **32** (1978), 161–169.

[13] Matsumoto M., Shimida H., *On Finsler spaces with 1-form metric,* II. Tensor N. S. **32** (1978), 275–278.

[14] Rund H. *The differential geometry of Finsler spaces.* Die Grundlehren der Math. Wiss. Bd. **101**. Springer 1959.

[15] Shen Z., *Lectures on Finsler Geometry,* World Scientific, Singapore, 2001.

[16] Szakál Sz., Szilasi J., *A new approach to generalized Berwald manifolds* I, SUT Journal of Math. **37** (2001), 19–41.

[17] Szakál S., Szilasi J., *A new approach to generalized Berwald manifolds* II, Publ. Math. Debrecen **60** (2002), 429–453.

[18] Szilasi J. *A Setting for Spray and Finsler Geometry,* in: Handbook of Finsler Geometry. P.L. Antonelli ed., Kluwer Ac. Publ. **1, 2** (2003), 1183–1426

[19] Tamássy L., *Area and curvature in Finsler spaces.* Reports on Math. Phys. **33** (1993), 233–239.

[20] Tamássy L. *Curvature of submanifolds in point Finsler spaces.* New Dewelopments in Differential Geometry. Proc. of the Coll. on Diff. Geom., Debrecen, July 1994. Kluwer 1996. pp. 391–397.

[21] Tamássy L. *Area and metrical connection in Finsler spaces.* Finsler Geometries. A Meeting of Minds. Kluwer 2000. pp. 263–280.

[22] Thompson A. C., *Minkowski Geometry.* Cambridge Univ. Press, Cambridge, 1996.

VI

PHYSICS

BLACK HOLE PERTURBATIONS

Zoltán Perjés
KFKI Research Institute for Particle and Nuclear Physics,
H–1525, Budapest 114, P.O.B. 49, Hungary

Abstract The study of black hole perturbations is an indispensable means of understanding the processes that give rise to gravitational and electromagnetic radiation. The observation of sources of gravitational radiation is an ongoing endeavour worldwide.

Perturbations of the most general black hole states have been first investigated by Chrzanowski and Misner [1, 2], Detweiler [3] and Chandrasekhar [5]. Among these pioneering studies, Chandrasekhar's is most detailed. He uses a Newman-Penrose (NP) approach both to the unperturbed Kerr metric and to the perturbed space-time.

In this talk we present the full set of black-hole perturbations in the ingoing radiation gauge. Space-time is treated here as a Kerr-metric background on which NP formalism is featured plus a perturbative part h_{ab} in the metric. In the spirit of the weak-field approximation, we take all perturbations as fields being defined on the background manifold of the Kerr metric.

In treatments of black-hole perturbations, it is a common practice to seek a suitable gauge fixing. A convenient choice, promoted by Chrzanowski, is the incoming (or outgoing) radiation gauge for the normal modes. This choice does not uniquely fix the coordinate gauge, however. In fact, the present work has been launched with the intent to take a second look at the remaining gauge freedom.

In this paper, the following picture emerges for the classical electrovacuum perturbations. There exists a *subset* of the field equations, consisting of two gravitational equations and two Maxwell equations not containing any mode mixing. (This is because the complex conjugate electromagnetic stresses are absent). From these relations alone, using the definition of the Weyl tensor perturbation ψ_0, it is possible to obtain a *pair* of coupled equations for ψ_0 and the electromagnetic perturbation ϕ_0. Thus a normal mode expansion for this doublet of fields *is* available. Mode mixing occurs only in the further gravitational and electromagnetic perturbation components derived from the fundamental doublet. This picture fully agrees with the results obtained in an entirely different gauge [10].

1. Introduction

Bolyai's ingenious geometrical ideas came to full fruition in the theory of general relativity. One of the most intriguing predictions of relativity is the existence of black holes. The basic description of the geometrical and thermodynamical properties and the unicity of black holes was consolidated by the middle of the twentieth century. Perturbed states of black holes are important sources of gravitational waves, the detection of which by recently built interferometric observatories is underway. The purpose of this discussion is to review the status of the theory of black hole perturbations, and to present the core system of their field equations.

The most general stationary black hole state allowed by the unicity theorems is the electrically charged Kerr-Newman black hole. An efficient approach to its perturbations is available within the framework of the Newman-Penrose (NP) null tetrad formalism [4]. One can introduce such a tetrad description of the perturbed space-times in one of two alternative ways. One may view the perturbed system as a distinct space-time, and introduce a perturbed null tetrad in the tangent space. The null tetrad vectors of the latter are linear combinations of those for the black-hole background. The combination coefficients plus the Weyl tensor perturbations are fifty real unknown functions. Many authors, including Chandrasekhar [5], pursue this approach.

However, here we shall *not* consider a null tetrad on the perturbed space-time. In the spirit of the weak-field approximation, we take all perturbations as fields being defined on the background manifold of the Kerr metric $\overset{(0)}{g}_{ab}$. These perturbations are then projected on the background tetrad. For source-free fields, no convenience is lost by the lack of the gauge freedom that would accompany a perturbed tetrad. This is simply because the perturbation equations are a homogeneous system in the first-order quantities. This approach has been taken, for example, by Chrzanowski and Misner [1, 2].

For generality, we shall display the full coupled system of Einstein-Maxwell perturbations. There are twenty-one perturbation equations similar to the NP equations. The Bianchi identities have twenty-four components and four further equations linear in the three complex Maxwell tensor perturbations. Part of these equations is used to eliminate the Weyl curvature terms. Adopting the incoming radiation gauge, we greatly reduce the number of independent field equations: there remain only *nine* components of the gravitational equations and the four Maxwell equations. These govern the five remaining unknown metric plus six electromagnetic perturbation functions. With the availability of algebraic computation packages, it is now feasible to pursue an unsophisticated approach in regard to the notation.

The Maxwell equations for electromagnetic perturbations on a fixed space-time are homogeneous. The circularly polarized normal modes have the structure $\phi_A(r,\vartheta)\, e^{i(L_z\varphi+Et)}$. The interaction with the gravitational field, via the stress-energy tensor $\phi_A\overline{\phi}_B$, introduces a mixing among the electromagnetic modes, even in the linearized theory. Thus, the full electrovacuum perturbation equations cease to be homogeneous in the modes. In quantum field theory, perturbative methods are available for such interacting systems. Strangely, no comparable treatment for the corresponding classical system is known. This regrettable backlog in the classical theory, however, will help illuminate the origin of the long-standing difficulties in finding a factorizable equation for the perturbations.

2. The background variables

In the original coordinates the charged Kerr metric is [6]

$$\begin{aligned} ds^2 = &\left(1-\tfrac{2mr-\mathrm{e}^2}{\zeta\overline{\zeta}}\right)\left(dt-a\sin^2\vartheta d\varphi\right)^2 \\ &+2\left(dt-a\sin^2\vartheta d\varphi\right)\left(dr+a\sin^2\vartheta d\varphi\right) \\ &-\zeta\overline{\zeta}\left(d\vartheta^2+\sin^2\vartheta d\varphi^2\right) \end{aligned} \tag{1}$$

with m the mass, e the electric charge, ma the angular momentum, and

$$\zeta = r - ia\cos\vartheta. \tag{2}$$

An overbar denotes complex conjugation. The corresponding null tetrad is,

$$\begin{aligned} D &\equiv \ell^a\tfrac{\partial}{\partial x^a} = \tfrac{\partial}{\partial r} \\ \Delta &\equiv n^a\tfrac{\partial}{\partial x^a} = \tfrac{1}{2}\left[\tfrac{(\zeta+\overline{\zeta})m-\mathrm{e}^2}{\zeta\overline{\zeta}}-1\right]\tfrac{\partial}{\partial r}+\tfrac{\partial}{\partial t} \\ \delta &\equiv m^a\tfrac{\partial}{\partial x^a} = \tfrac{1}{2^{1/2}\zeta}\left[\tfrac{\partial}{\partial\vartheta}+\tfrac{i}{\sin\vartheta}\tfrac{\partial}{\partial\varphi}-ia\sin\vartheta\left(\tfrac{\partial}{\partial r}-\tfrac{\partial}{\partial t}\right)\right] \\ \overline{\delta} &\equiv \overline{m}^a\tfrac{\partial}{\partial x^a}. \end{aligned} \tag{3}$$

In the Newman-Penrose notation, the nonvanishing components of the Weyl spinor are

$$\begin{aligned}\Psi_2 &= \frac{e^2 - m\overline{\zeta}}{\zeta^3\overline{\zeta}} \\ \Psi_3 &= -3ia\sin\vartheta\frac{m\overline{\zeta} - e^2}{2^{1/2}\zeta^4\overline{\zeta}} \\ \Psi_4 &= 3a^2\sin^2\vartheta\frac{m\overline{\zeta} - e^2}{\zeta^5\overline{\zeta}},\end{aligned} \tag{4}$$

and the nonzero Maxwell spinor components have the form

$$\begin{aligned}\Phi_1 &\equiv \tfrac{1}{2}F_{ab}\left(\ell^a n^b + \overline{m}^a m^b\right) = \frac{e}{2^{1/2}\zeta^2} \\ \Phi_2 &\equiv F_{ab}\overline{m}^a n^b = \frac{iea\sin\vartheta}{\zeta^3}.\end{aligned} \tag{5}$$

The tetrad vector ℓ points in one of the double principal null directions, as seen from the condition $\Psi_0 = 0$. The second double principal null direction $\tilde{n}$ can be constructed by the null rotation

$$\begin{aligned}\tilde{\ell} &= \ell \\ \tilde{m} &= m + A\ell \\ \tilde{n} &= n + \overline{A}m + A\overline{m} + A\overline{A}\ell.\end{aligned} \tag{6}$$

The Weyl spinor components transform as

$$\begin{aligned}\widetilde{\Psi}_2 &= \Psi_2 \\ \widetilde{\Psi}_3 &= \Psi_3 + 3\overline{A}\Psi_2 \\ \widetilde{\Psi}_4 &= \Psi_4 + 4\overline{A}\Psi_3 + 6\overline{A}^2\Psi_2.\end{aligned} \tag{7}$$

Choosing $A = ia\sin\vartheta/2^{1/2}\overline{\zeta}$, we can make both $\widetilde{\Psi}_3$ and $\widetilde{\Psi}_4$ vanish. Thus we find that the second principal null vector is

$$\tilde{n} = \tfrac{1}{\zeta\overline{\zeta}}\left[\left(r^2 + a^2\right)\tfrac{\partial}{\partial t} - \tfrac{r^2 - 2mr + a^2 + e^2}{2}\tfrac{\partial}{\partial r} + a\tfrac{\partial}{\partial\varphi}\right]. \tag{8}$$

3. Two metrics

The perturbative treatment to be adopted in this paper is developed from the standard first-order formalism [5]. The metric of the perturbed space-time is written

$$g_{ab} = g_{ab}^{(0)} + h_{ab} \tag{9}$$

where $g^{(0)}_{ab}$ is the metric of the background (Kerr-Newman) spacetime. All higher powers of the metric perturbation h_{ab} and its derivatives will be neglected. The respective covariant derivatives are defined by

$$\nabla^{(0)}_a g^{(0)}_{bc} = 0, \qquad \nabla_a g_{bc} = 0. \tag{10}$$

Hence

$$\nabla_a g_{bc} = \nabla^{(0)}_a g_{bc} - C_{ab}^{\ \ d} g_{dc} - C_{ac}^{\ \ d} g_{bd} \tag{11}$$

where the first-order small connection perturbations are defined as

$$C_{abc} = \tfrac{1}{2}\left(\nabla^{(0)}_a h_{bc} + \nabla^{(0)}_b h_{ac} - \nabla^{(0)}_c h_{ab}\right). \tag{12}$$

The Ricci identity for the unperturbed space-time is

$$\left(\nabla^{(0)}_a \nabla^{(0)}_b - \nabla^{(0)}_b \nabla^{(0)}_a\right)\omega_c = R^{(0)\ d}_{abc}\,\omega_d. \tag{13}$$

The Riemann tensor of the perturbed space time has the form

$$R_{abc}^{\ \ \ d} = R^{(0)\ d}_{abc} - 2\nabla^{(0)}_{[a} C_{b]c}^{\ \ \ d}. \tag{14}$$

One should proceed with care when using the symmetry properties of the perturbed tensor. The tensors $R^{(0)}_{abcd} = g^{(0)}_{de} R^{(0)\ e}_{abc}$ and $R_{abcd} = g_{de}\, R_{abc}^{\ \ \ e}$ possess the full algebraic symmetries. Here

$$\begin{aligned} R_{abcd} &= \left(g^{(0)}_{de} + h_{de}\right)\left(R^{(0)\ e}_{abc} + R^{(1)\ e}_{abc}\right) \\ &= R^{(0)}_{abcd} + h_{de} R^{(0)\ e}_{abc} + g^{(0)}_{de} R^{(1)\ e}_{abc} \\ &= R^{(0)}_{abcd} + R'_{abcd}. \end{aligned} \tag{15}$$

The trace of the perturbed Riemann tensor is hence expressed in terms of the stress-energy tensor T_{ab} and Einstein's equations $R_{ab} - \frac{1}{2} g_{ab} R = kT_{ab}$, as

$$g^{(0)rs} R_{arbs} = k\left(T^*_{ab} - T^{*(0)}_{ab}\right) + h^{rs} R^{(0)}_{arbs} \tag{16}$$

where $T^*_{ab} = T_{ab} - \frac{1}{2} g_{ab} g^{rs} T_{rs}.$.

The Bianchi identity for the perturbed space-time is

$$\nabla_{[a} R_{bc]d}^{\ \ \ \ e} = 0. \tag{17}$$

We next introduce a null tetrad [4]

$$z_{\mathbf{m}} = \{\ell, n, m, \overline{m}\} \tag{18}$$

with $\mathbf{m} = 0, 1, 2, 3$, on the background black-hole space-time with the standard definition of spin coefficients

$$\gamma_{\mathbf{mnp}} = \nabla_j z_{\mathbf{m}i} z^i_{\mathbf{n}} z^j_{\mathbf{p}}. \tag{19}$$

Boldface indices denote tetrad components; indices in round or square brackets stand for symmetrization or skewing, respectively. A semicolon or comma in the subscript denote, respectively, covariant and partial derivatives. The NP form of the perturbed Riemann tensor (15) is

$$R_{\mathbf{mno}}{}^{\mathbf{p}} = R^{(0)}_{\mathbf{mno}}{}^{\mathbf{p}} - 2\left(C_{\mathbf{o}[\mathbf{n}}{}^{\mathbf{p}}{}_{,\mathbf{m}]} - \gamma_{[\mathbf{n}}{}^{\mathbf{r}}{}_{\mathbf{m}]} C_{\mathbf{ro}}{}^{\mathbf{p}} - \gamma_{\mathbf{o}}{}^{\mathbf{r}}{}_{[\mathbf{m}} C_{\mathbf{n}]\mathbf{r}}{}^{\mathbf{p}} - \gamma^{\mathbf{p}}{}_{\mathbf{r}[\mathbf{m}} C_{\mathbf{n}]\mathbf{o}}{}^{\mathbf{r}}\right). \tag{20}$$

Hence we get the tetrad equation $E_{\mathbf{mknl}} = 0$ with

$$\begin{aligned} E_{\mathbf{mknl}} &= -R_{\mathbf{mknl}} + R^{(0)}_{\mathbf{mknl}} \\ &\quad + 2\eta^{\mathbf{ab}}\left(\partial_{[\mathbf{m}} C_{\mathbf{k}]\mathbf{nl}} + \gamma_{\mathbf{a}[\mathbf{mk}]} C_{\mathbf{bnl}} + \gamma_{\mathbf{na}[\mathbf{m}} C_{\mathbf{k}]\mathbf{bl}} - \gamma_{\mathbf{la}[\mathbf{k}} C_{\mathbf{m}]\mathbf{nb}}\right). \end{aligned} \tag{21}$$

We next decompose the curvature perturbation into the trace-less part $C_{\mathbf{mknl}}$, trace perturbations $R_{\mathbf{kl}}$ and Ricci scalar perturbation R ,

$$\begin{aligned} R_{\mathbf{mknl}} &= C_{\mathbf{mknl}} + \tfrac{1}{2}\left(\eta_{\mathbf{mn}} R_{\mathbf{kl}} - \eta_{\mathbf{ml}} R_{\mathbf{kn}} + \eta_{\mathbf{kl}} R_{\mathbf{mn}} - \eta_{\mathbf{kn}} R_{\mathbf{ml}}\right) \\ &\quad + \tfrac{1}{6}\left(\eta_{\mathbf{ml}} \eta_{\mathbf{kn}} - \eta_{\mathbf{mn}} \eta_{\mathbf{kl}}\right) R. \end{aligned} \tag{22}$$

The tetrad components of $\nabla^{(0)}_{[a} R'_{bc]de}$ are

$$\begin{aligned} &\partial_{[\mathbf{m}} R'_{\mathbf{no}]\mathbf{pq}} - \gamma_{[\mathbf{n}}{}^{\mathbf{r}}{}_{\mathbf{m}]} R'_{|\mathbf{r}|\mathbf{o}]\mathbf{pq}} - \gamma_{[\mathbf{o}}{}^{\mathbf{r}}{}_{\mathbf{m}]} R'_{\mathbf{n}]\mathbf{rpq}} - \gamma_{\mathbf{p}}{}^{\mathbf{r}}{}_{[\mathbf{m}} R'_{\mathbf{no}]\mathbf{rq}} - \gamma_{\mathbf{q}}{}^{\mathbf{r}}{}_{[\mathbf{m}} R'_{\mathbf{no}]\mathbf{pr}} \\ &- C_{\mathbf{p}[\mathbf{m}}{}^{\mathbf{r}} R^{(0)}_{\mathbf{no}]\mathbf{rq}} - C_{\mathbf{q}[\mathbf{m}}{}^{\mathbf{r}} R^{(0)}_{\mathbf{no}]\mathbf{pr}} \\ &= 0. \end{aligned} \tag{23}$$

The components of the trace-free part of the perturbed curvature tensor are given an individual notation, in close analogy with the Newman-

Penrose notation for the Weyl tensor,

$$\begin{aligned}
C_{0101} &= C_{1010} = C_{2323} = C_{3232} = -\psi_2 - \bar{\psi}_2 \\
C_{0112} &= C_{1021} = C_{1223} = C_{2132} = \bar{\psi}_3 \\
C_{0113} &= C_{1031} = -C_{1323} = -C_{3132} = \psi_3 \\
C_{0123} &= C_{1032} = \psi_2 - \bar{\psi}_2 \\
C_{0201} &= C_{2010} = -C_{0223} = -C_{2032} = -\psi_1 \\
C_{0202} &= C_{2020} = -\psi_0 \\
C_{0213} &= C_{2031} = \psi_2 \\
C_{0301} &= C_{3010} = C_{0323} = C_{3032} = -\bar{\psi}_1 \\
C_{0303} &= C_{3030} = -\bar{\psi}_0 \\
C_{0312} &= C_{3021} = \bar{\psi}_2 \\
C_{1212} &= C_{2121} = -\bar{\psi}_4 \\
C_{1313} &= C_{3131} = -\psi_4
\end{aligned} \tag{24}$$

and the shorthand notation for the vacuum equations is introduced

$$\begin{array}{lll}
E_a = E_{0101} & E_h = E_{0203} & E_o = E_{0323} \\
E_b = E_{0102} & E_i = E_{0212} & E_p = E_{1212} \\
E_c = E_{0103} & E_j = E_{0213} & E_q = E_{1213} \\
E_d = E_{0112} & E_k = E_{0223} & E_r = E_{1223} \\
E_e = E_{0113} & E_l = E_{0303} & E_s = E_{1313} \\
E_f = E_{0123} & E_m = E_{0312} & E_t = E_{1323} \\
E_g = E_{0202} & E_n = E_{0313} & E_u = E_{2323}\,.
\end{array} \tag{25}$$

4. Gauge transformations

The allowable infinitesimal coordinate transformations from the local coordinate system $\{x^a\}$ to the new coordinates

$$\tilde{x}^a = x^a - \xi^a \tag{26}$$

have the infinitesimal generators ξ^a. They change the metric perturbations as follows,

$$\tilde{h}_{ab} = h_{ab} + \xi_{a;b} + \xi_{b;a}. \tag{27}$$

We choose the incoming radiation gauge [1] in which the metric perturbation is orthogonal to the incoming null rays,

$$h_{ab}\ell^b = 0\,. \tag{28}$$

The remaining gauge transformations have the form

$$\begin{aligned}
\xi_0 &= \xi_0^0 \\
\xi_1 &= \xi_1^0 - \frac{2mr - \mathrm{e}^2}{2\zeta\overline{\zeta}}\xi_0^0 - r\xi_{0,t}^0 \\
\xi_2 &= \overline{\zeta}\xi_2^0 + 2^{-1/2}\left(ia\sin\vartheta\,\xi_{0,t}^0 + \xi_{0,\vartheta}^0 + \frac{i}{\sin\vartheta}\xi_{0,\varphi}^0\right) \\
\xi_3 &= \zeta\xi_3^0 + 2^{-1/2}\left(-ia\sin\vartheta\,\xi_{0,t}^0 + \xi_{0,\vartheta}^0 - \frac{i}{\sin\vartheta}\xi_{0,\varphi}^0\right).
\end{aligned} \tag{29}$$

Quantities with a zero superscript are integration functions independent of the coordinate r.

The $(\mathbf{0},\mathbf{2},\mathbf{0},\mathbf{3})$ component of Eq. (14), $E_h = 0$, is a radial differential equation for the function $h_{\mathbf{23}}$. Integration yields

$$h_{\mathbf{23}} = \frac{1}{\zeta\overline{\zeta}}\left[c^0 r + h_{\mathbf{23}}^0(r^2 - a^2\cos^2\vartheta)\right] . \tag{30}$$

Hence we can arrange

$$h_{\mathbf{23}} = 0 \tag{31}$$

by the remaining gauge freedom. With this choice, ξ_1^0 is determined algebraically and ξ_0^0 is given up to a t-independent additive term.

We successively eliminate the perturbed Weyl spinor terms in the following order,

$$\begin{aligned}
E_m &= 0 \quad \Rightarrow \quad \bar{\psi}_2 , & E_b &= 0 \quad \Rightarrow \quad \psi_1, \\
E_c &= 0 \quad \Rightarrow \quad \bar{\psi}_1 , & E_d &= 0 \quad \Rightarrow \quad \bar{\psi}_3, \\
E_e &= 0 \quad \Rightarrow \quad \psi_3 , & E_j &= 0 \quad \Rightarrow \quad \psi_2, \\
E_g &= 0 \quad \Rightarrow \quad \psi_0 , & E_s &= 0 \quad \Rightarrow \quad \psi_4, \\
E_p &= 0 \quad \Rightarrow \quad \bar{\psi}_4 , & E_l &= 0 \quad \Rightarrow \quad \bar{\psi}_0.
\end{aligned} \tag{32}$$

We get nine equations $E_k = 0$, $k = 1, ..., 9$ where

$$\begin{aligned}
E_1 &\equiv E_i , & E_2 &\equiv E_a , & E_3 &\equiv E_k \\
E_4 &\equiv E_q , & E_5 &\equiv E_o , & E_6 &\equiv E_r \\
E_7 &\equiv E_t , & E_8 &\equiv E_u , & E_9 &\equiv E_n
\end{aligned} \tag{33}$$

(*cf.* Appendix), the integrability conditions of which are all satisfied identically. Substituting back in the field equations (25) and Bianchi identities (23), each is satisfied identically.

The gravitational and electromagnetic excitations will be characterized by the perturbative components

$$\psi_0 = -C_{abcd}\ell^a m^b \ell^c m^d, \qquad \phi_0 = F_{ab}\ell^a m^b \tag{34}$$

of the Weyl tensor C_{abcd} and the Maxwell tensor F_{ab}, respectively. The quantity ψ_0 is invariant under gauge transformations and infinitesimal tetrad rotations.

The perturbed Weyl spinor can be expressed in terms of the metric perturbations as follows,

$$\psi_0 = \tfrac{1}{2} h_{22,rr} + \tfrac{1}{\zeta} h_{22,r} \tag{35a}$$

$$\psi_1 = \tfrac{1}{2}\left(h_{12,rr} + \tfrac{1}{\zeta} h_{12,r} - \tfrac{1}{\zeta^2} h_{12} - 2^{1/2} \tfrac{e}{\zeta^2}\phi_0\right) \tag{35b}$$

$$\psi_2 = -\frac{1}{2^{3/2}\zeta^2\bar{\zeta}}\left\{2^{1/2}\left[h_{11,r}\zeta\bar{\zeta} + (\zeta - \bar{\zeta})h_{11}\right]\right. \tag{35c}$$

$$\left. - \left[(\zeta h_{13})_{,rr}\zeta - 2(\zeta h_{13})_{,r}\right]\sin\vartheta a i - \zeta(\zeta h_{13})_{,r\vartheta} - \zeta\cot\vartheta(\zeta h_{13})_{,r}\right\}$$

...

For black holes with no electric charge, e $= 0$, the Teukolsky equation arises when expressing $h_{12,rr}$ from $E_3 = 0$ and $h_{12,r\vartheta}$ from $E_1 = 0$. In the integrability condition of $h_{12,rr}$ and $h_{12,r\vartheta}$, only one derivative of h_{12} occurs. This is $h_{12,\vartheta}$, which, when substituted back in E_1, gives the uncoupled Teukolsky equation in terms of h_{22}.

A normal mode with energy E and angular momentum L_z has a φ- and a t-dependence

$$e^{i(Et+L_z\varphi)}.$$

Thus for normal modes, the reality conditions for the perturbations h_{ab} do not hold. The expressions E_k are homogeneous in the five complex unknowns $(h_A) = (h_{11}, h_{12}, h_{13}, h_{22}, h_{33})$, where the subscript A takes the respective values $1, 2, ..., 5$. Each homogeneous expression has the structure

$$\sum_{A=1}^{5}\sum_{B=0}^{5} M^{AB} h_{A,B} \tag{36}$$

where the subscript B labels the derivatives taken as follows:

$$(h_{A,B}) = (h_A, h_{A,r}, h_{A,\vartheta}, h_{A,rr}, h_{A,r\vartheta}, h_{A,\vartheta\vartheta})\,. \tag{37}$$

The structure of the field equations is highlighted in Table 1 below, indicating the presence of a term in the 5×6 matrices $\mathbf{M}$ by a bullet. For an uncharged Kerr space-time, the gravitational and electromagnetic perturbations decouple and they individually satisfy the equations [8]

$$\lim_{e=0} \Box_1\phi_0 = 0, \qquad \lim_{e=0} \Box_2\psi_0 = 0, \tag{38}$$

Table 1. Structure of the field equations

	h	$h_{,r}$	$h_{,\vartheta}$	$h_{,rr}$	$h_{,r\vartheta}$	$h_{,\vartheta\vartheta}$	h	$h_{,r}$	$h_{,\vartheta}$	$h_{,rr}$	$h_{,r\vartheta}$	$h_{,\vartheta\vartheta}$	h	$h_{,r}$	$h_{,\vartheta}$	$h_{,rr}$	$h_{,r\vartheta}$	$h_{,\vartheta\vartheta}$
h_{11}	-	-	-	-	-	-	•	•	-	•	-	-	-	-	-	-	-	-
h_{12}	•	•	•	•	•	-	•	•	•	•	-	-	•	•	-	•	-	-
h_{13}	-	-	-	-	-	-	•	•	•	•	•	-	-	-	-	-	-	-
h_{22}	•	•	-	•	-	-	•	•	-	•	-	-	•	•	•	•	•	-
h_{33}	-	-	-	-	-	-	-	-	-	-	-	-	-	-	-	-	-	-

E_1 E_2 E_3

	h	$h_{,r}$	$h_{,\vartheta}$	$h_{,rr}$	$h_{,r\vartheta}$	$h_{,\vartheta\vartheta}$	h	$h_{,r}$	$h_{,\vartheta}$	$h_{,rr}$	$h_{,r\vartheta}$	$h_{,\vartheta\vartheta}$	h	$h_{,r}$	$h_{,\vartheta}$	$h_{,rr}$	$h_{,r\vartheta}$	$h_{,\vartheta\vartheta}$
h_{11}	•	•	•	•	-	•	-	-	-	-	-	-	•	•	•	•	•	-
h_{12}	•	•	•	-	-	-	-	-	-	-	-	-	•	•	•	•	-	•
h_{13}	•	•	•	-	-	-	•	•	-	•	-	-	•	•	•	•	-	•
h_{22}	•	•	•	-	-	-	-	-	-	-	-	-	•	•	•	•	-	-
h_{33}	•	•	•	-	-	-	•	•	•	•	•	-	-	-	-	-	-	-

E_4 E_5 E_6

	h	$h_{,r}$	$h_{,\vartheta}$	$h_{,rr}$	$h_{,r\vartheta}$	$h_{,\vartheta\vartheta}$	h	$h_{,r}$	$h_{,\vartheta}$	$h_{,rr}$	$h_{,r\vartheta}$	$h_{,\vartheta\vartheta}$	h	$h_{,r}$	$h_{,\vartheta}$	$h_{,rr}$	$h_{,r\vartheta}$	$h_{,\vartheta\vartheta}$
h_{11}	•	•	-	•	•	-	•	-	-	•	-	-	-	-	-	-	-	-
h_{12}	•	•	•	•	-	-	•	•	•	•	-	-	-	-	-	-	-	-
h_{13}	•	•	•	•	-	-	•	•	•	•	-	-	•	•	•	•	•	-
h_{22}	•	•	•	•	-	-	•	•	•	•	-	•	-	-	-	-	-	-
h_{33}	•	•	•	-	-	-	•	•	•	•	-	•	•	•	-	•	-	-

E_7 E_8 E_9

where

$$\begin{aligned} \Box_s = {} & \mathbf{\Delta}^{-s}\frac{\partial}{\partial r}\mathbf{\Delta}^{s+1}\frac{\partial}{\partial r} + \frac{\partial^2}{\partial\vartheta^2} + \frac{\cos\vartheta}{\sin\vartheta}\frac{\partial}{\partial\vartheta} \\ & - 2i\left[aL_z + \left(r^2 + a^2\right)E\right]\frac{\partial}{\partial r} \\ & + s - 2\left(2s+1\right)Eir + 4sEa\cos\vartheta - \left(\tfrac{\mathcal{L}_z + s\cos\vartheta}{\sin\vartheta}\right)^2 \end{aligned} \tag{39}$$

is a linear operator separable in the r and ϑ coordinates and

$$\mathbf{\Delta} = \zeta\bar{\zeta} - (\zeta + \bar{\zeta})m + a^2\sin^2\vartheta + \mathrm{e}^2 \tag{40}$$

is the horizon function, $\mathbf{\Delta} = \mathbf{\Delta}\left(r\right)$. Here the function of the mode numbers is introduced

$$\mathcal{L}_z = L_z + Ea\sin^2\vartheta. \tag{41}$$

5. Equations for ϕ_0 and ψ_0

Initially we obtain a pair of coupled equations for the metric perturbation h_{22} and for the Maxwell perturbation field ϕ_0. The first equation follows from Eqs. $E_1 = 0$ and $E_3 = 0$ by expressing from these the respective derivatives $h_{12,r\vartheta}$ and $h_{12,rr}$. The integrability condition of h_{12} then yields the equation for the first derivative $h_{12,\vartheta}$:

$$
\begin{aligned}
&2^{3/2}\zeta^2\bar\zeta^{-1}\sin\vartheta\left\{\bar\zeta^3(\sin\vartheta\tfrac{\partial}{\partial\vartheta}-\mathcal{L}_z-\cos\vartheta)+ia\sin^2\vartheta\left[(\zeta^2-\bar\zeta^2)\bar\zeta\tfrac{\partial}{\partial r}+\zeta^2\right]\right\}h_{12}\\
&+4\mathrm{e}\zeta^3\bar\zeta^{-1}\sin\vartheta\left[\zeta\bar\zeta(\sin\vartheta\tfrac{\partial}{\partial\vartheta}-\mathcal{L}_z-\cos\vartheta)+(\zeta+2\bar\zeta)ia\sin^2\vartheta\right]\phi_0\\
&=\zeta^2\bar\zeta\sin^2\vartheta\left[\left(4m-\zeta-3\bar\zeta\right)\zeta\bar\zeta-(\zeta-\bar\zeta)\boldsymbol{\Delta}+2i(E\zeta\bar\zeta+a\mathcal{L}_z)\zeta\bar\zeta\right]\tfrac{\partial^2}{\partial r^2}h_{22}\\
&-2\left[(2\zeta+\bar\zeta)(\zeta-\bar\zeta)\mathrm{e}^2+(\zeta-2\bar\zeta)Ei\zeta^3\bar\zeta\right]\sin^2\vartheta h_{22}\\
&+2\sin^2\vartheta\left\{-i\zeta^3a\mathcal{L}_z+\left[(\zeta+\bar\zeta)m+\zeta^2\right](\zeta-\bar\zeta)\bar\zeta\right\}h_{22}\\
&-\zeta^2\cos\vartheta\left\{\zeta\bar\zeta^2\sin\vartheta\left(\tfrac{\partial^2}{\partial r\partial\vartheta}-2ia\sin\vartheta\tfrac{\partial^2}{\partial r^2}\right)-2ia\sin^2\vartheta\left[(\zeta-\bar\zeta)\bar\zeta\tfrac{\partial}{\partial r}-2\zeta\right]\right.\\
&\left.+(\zeta-\bar\zeta)\bar\zeta\sin\vartheta\tfrac{\partial}{\partial\vartheta}-\bar\zeta\left(4\mathcal{L}_z-2Ea\sin^2\vartheta+4\cos\vartheta\right)\left(\zeta\bar\zeta\tfrac{\partial}{\partial r}+\zeta-\bar\zeta\right)\right\}h_{22}\\
&-\left\{2(\zeta-\bar\zeta)\sin^2\vartheta\left[(2\zeta^2+\zeta\bar\zeta+\bar\zeta^2)m-(2\zeta+\bar\zeta)\mathrm{e}^2\right]\right.\\
&-2\mathcal{L}_z\zeta\bar\zeta(\zeta-\bar\zeta)ia\sin^2\vartheta-2\zeta^2\bar\zeta\sin^2\vartheta\left[(\zeta+2\bar\zeta)Ei\bar\zeta+\zeta-\bar\zeta\right]\\
&\left.-2\zeta^2a^2\sin^4\vartheta-\mathcal{L}_z^2\zeta^2\bar\zeta^2\right\}\zeta\tfrac{\partial}{\partial r}h_{22}\\
&-\zeta^3\sin^2\vartheta\left(\boldsymbol{\Delta}\bar\zeta^2\tfrac{\partial^3}{\partial r^3}+\bar\zeta^2\tfrac{\partial^3}{\partial r\partial\vartheta^2}+2ia\sin\vartheta\tfrac{\partial}{\partial\vartheta}\right)h_{22}\\
&-\zeta^2(\zeta-\bar\zeta)\bar\zeta\left(\sin^2\vartheta\tfrac{\partial^2}{\partial\vartheta^2}-\mathcal{L}_z^2\right)h_{22}.
\end{aligned}
\tag{42}
$$

When substituting back $h_{12,\vartheta}$ in Eq. $E_1=0$, we get the lengthy equation

$$
\begin{aligned}
&\zeta^3\bar\zeta^3\sin^2\vartheta\left(\boldsymbol{\Delta}\tfrac{\partial^4}{\partial r^4}+\tfrac{\partial^4}{\partial r^2\partial\vartheta^2}-6m\tfrac{\partial^3}{\partial r^3}+\cot\vartheta\tfrac{\partial^3}{\partial r^2\partial\vartheta}\right)h_{22}\\
&+2\zeta^3\bar\zeta^2\sin^2\vartheta\left(\boldsymbol{\Delta}\tfrac{\partial^3}{\partial r^3}+\tfrac{\partial^3}{\partial r\partial\vartheta^2}\right)h_{22}\\
&+2\left[\zeta+2\bar\zeta-i\zeta\bar\zeta E-ia\mathcal{L}_z-ia\cos\vartheta\right]\zeta^3\bar\zeta^3\sin^2\vartheta\tfrac{\partial^3}{\partial r^3}h_{22}\\
&+\left\{2\left[5\bar\zeta^2\zeta-(2\zeta-\bar\zeta)\mathrm{e}^2+2(\zeta-2\bar\zeta)m\right]\sin^2\vartheta\right.\\
&-\left[4a\left(\mathcal{L}_z+\cos\vartheta\right)+2E\bar\zeta(3\zeta+4\bar\zeta)\right]\zeta\bar\zeta i\sin^2\vartheta\\
&\left.-\zeta\bar\zeta^2\left(\mathcal{L}_z+2\cos\vartheta\right)^2-4\zeta a^2\sin^4\vartheta\right\}\zeta^2\bar\zeta\tfrac{\partial^2}{\partial r^2}h_{22}\\
&+2(\bar\zeta\cos\vartheta+2ia\sin^2\vartheta)\zeta^3\bar\zeta\sin\vartheta\tfrac{\partial^2}{\partial r\partial\vartheta}h_{22}\\
&-2\zeta^3\left\{2\left[(4Ei\bar\zeta+1-E\bar\zeta a\cos\vartheta)\bar\zeta^2+(\zeta-2\bar\zeta)m\right]\sin^2\vartheta\right.\\
&\left.+\left[(\mathcal{L}_z+2\cos\vartheta)^2\bar\zeta-2i(\mathcal{L}_z+2\cos\vartheta)a\sin^2\vartheta\right]\bar\zeta\right\}\tfrac{\partial}{\partial r}h_{22}\\
&+4(\zeta^2-\bar\zeta^2)\mathrm{e}^2\sin^2\vartheta\left(\zeta\tfrac{\partial}{\partial r}-1\right)h_{22}\\
&=-4\mathrm{e}\zeta^3\sin\vartheta\left\{(\mathcal{L}_z+\cos\vartheta)\left(\zeta-2\bar\zeta-\zeta\bar\zeta\tfrac{\partial}{\partial r}\right)\right.\\
&\left.+\sin\vartheta\left[\zeta\bar\zeta\tfrac{\partial^2}{\partial r\partial\vartheta}-(\zeta-2\bar\zeta)\tfrac{\partial}{\partial\vartheta}+(\zeta+2\bar\zeta)ia\sin\vartheta\tfrac{\partial}{\partial r}\right]\right\}\phi_0.
\end{aligned}
\tag{43}
$$

express $\psi_{0,rr\vartheta}$ from (48), substitute in (49) from which we isolate $\psi_{0,rrr}$. Repeating the computation of the integrability condition for these third derivatives, we get a nontrivial new relation, of second order in ψ_0 :

$$\mathcal{D}_2^{(2)}\psi_0 = 0. \tag{51}$$

In Eq. (51), five types of derivatives of the function ψ_0 occur. These are: $\psi_{0,rr}$, $\psi_{0,r\vartheta}$, $\psi_{0,r}$, $\psi_{0,\vartheta}$ and ψ_0. The sixth type of terms is those containing the (derivatives of) the function ϕ_0. From the present point of view, these latter are the inhomogeneous terms. Thus we need altogether six equations of this coefficient structure to be able to algebraically eliminate the function ψ_0 from the system.We can obtain new relations by taking the r and ϑ coordinate derivatives of Eq. (51). The third derivatives can be eliminated from the new equations: by expressing $\psi_{0,r\vartheta\vartheta}$ and $\psi_{0,\vartheta\vartheta\vartheta}$ from Eqs. (48) and (49) and subsequently eliminating $\psi_{0,\vartheta\vartheta}$ using Eq. (50).

If one now introduces the new perturbation function w by letting

$$\phi_0 = \frac{w}{\zeta^3}, \tag{52}$$

then all terms with the factor w cancel both in Eqs. (50) and (51). Thus, in effect, the order in the Maxwell field is lowered by one if we introduce the complex function χ by

$$\chi = \frac{w_{,r}}{\zeta^2 \sin\vartheta}. \tag{53}$$

We get the pair of coupled equations for the perturbation ψ_0 of the Weyl spinor and the perturbation χ of the Maxwell field:

$$\begin{aligned}&\bar{\zeta}^2\left\{\left[(\zeta+\bar{\zeta})\tfrac{\partial}{\partial r}-2\right](\zeta\bar{\zeta}\Box_2+2\mathrm{e}^2)+2(\zeta+\bar{\zeta})^2\Box_2\right\}\psi_0\\ &=-2\mathrm{e}\left\{\bar{\zeta}\left[(\zeta+\bar{\zeta})\zeta\tfrac{\partial}{\partial r}+2\bar{\zeta}\right]\mathcal{L}_0+i(\zeta+\bar{\zeta})^2 a\sin^2\vartheta\tfrac{\partial}{\partial r}\right\}\chi\end{aligned} \tag{54}$$

$$\begin{aligned}&\bar{\zeta}^2\left[\bar{\zeta}(\zeta-2\bar{\zeta})\mathcal{L}_2\Box_2+6\mathrm{e}^2\mathcal{L}_2-6a^2\sin^2\vartheta\cos\vartheta\Box_2\right]\psi_0\\ &=2\mathrm{e}\left\{\bar{\zeta}(\zeta+\bar{\zeta})\sin^2\vartheta\left[\mathbf{\Delta}\tfrac{\partial^2}{\partial r^2}-2i\left(E\zeta\bar{\zeta}+a\mathcal{L}_z\right)\tfrac{\partial}{\partial r}-10iE\bar{\zeta}\right]\right.\\ &\left.-(\zeta+\bar{\zeta})\sin^2\vartheta\left[\mathbf{\Delta}+3\bar{\zeta}(2m-\zeta-\bar{\zeta})\right]\tfrac{\partial}{\partial r}+3\bar{\zeta}^2\mathcal{L}_2\mathcal{L}_0\right\}\chi.\end{aligned} \tag{55}$$

Coupled equations for the gravitational and electromagnetic perturbation functions corresponding to ψ_0 and χ have been obtained by Chitre [9]. A pair of separable equations for the shear σ and electromagnetic field component ϕ_0 , using a perturbed null tetrad, has been derived in [10].

Appendix: A. Electrovacuum perturbations

The independent perturbation equations state the vanishing of the following nine complex expressions of second differential order:

$$
\begin{aligned}
E_1 = {} & 2\zeta^3\bar{\zeta}[-\sin\vartheta\left(\bar{\zeta}h_{\mathbf{12},r\vartheta} + h_{\mathbf{12},\vartheta}\right) + 2^{1/2}\bar{\zeta}\sin\vartheta(\bar{\zeta}h_{\mathbf{22},rt} + h_{\mathbf{22},t}) \\
& + ia\sin^2\vartheta\left(\bar{\zeta}h_{\mathbf{12},rr} - \bar{\zeta}h_{\mathbf{12},rt} - h_{\mathbf{12},t} + h_{\mathbf{12},r}\right) - ih_{\mathbf{12},\varphi} - i\bar{\zeta}h_{\mathbf{12},r\varphi}] \\
& - 8\zeta^3 ia\sin^2\vartheta(2^{1/2}\mathrm{e}\phi_0 + h_{\mathbf{12}}) \\
& + 2^{3/2}\bar{\zeta}(\mathrm{e}^2 - m\bar{\zeta})(\zeta - \bar{\zeta})\sin\vartheta h_{\mathbf{22}} + 2\zeta^3\bar{\zeta}\cos\vartheta(\bar{\zeta}h_{\mathbf{12},r} + h_{\mathbf{12}}) \\
& - 2^{1/2}\zeta^2\bar{\zeta}^2\sin\vartheta\left[\mathrm{e}^2 + \zeta\bar{\zeta} - (\zeta + \bar{\zeta})m\right]h_{\mathbf{22},rr} \\
& - 2^{3/2}\zeta\bar{\zeta}\sin\vartheta\left[(\zeta - \bar{\zeta})\mathrm{e}^2 + \zeta^2\bar{\zeta} - (\zeta^2 + \zeta\bar{\zeta} - \bar{\zeta}^2)m\right]h_{\mathbf{22},r}
\end{aligned}
$$

$$
\begin{aligned}
E_2 = {} & \{[(h_{\mathbf{13},r\vartheta} - h_{\mathbf{13},rr}\sin\vartheta ai + h_{\mathbf{13},rt}\sin\vartheta ai)\sin\vartheta + h_{\mathbf{13},r\varphi}i]\,\zeta\bar{\zeta} \\
& + \bar{\zeta}^2\left[(h_{\mathbf{12},rr} - h_{\mathbf{12},rt})\sin^2\vartheta ai - h_{\mathbf{12},r\varphi}i + \left(h_{\mathbf{12},r\vartheta} - h_{\mathbf{11},rr}\zeta 2^{1/2}\right)\sin\vartheta\right] \\
& - [(h_{\mathbf{13},t}\sin\vartheta ai + h_{\mathbf{13},\vartheta} - h_{\mathbf{13},r}\sin\vartheta ai)\sin\vartheta + h_{\mathbf{13},\varphi}i]\,(\zeta - 2\bar{\zeta})\}\zeta \\
& - [(h_{\mathbf{12},t}\sin\vartheta ai - h_{\mathbf{12},\vartheta} - h_{\mathbf{12},r}\sin\vartheta ai)\sin\vartheta + h_{\mathbf{12},\varphi}i]\,(2\zeta - \bar{\zeta})\bar{\zeta} \\
& + \left[4(\bar{\phi}_1\bar{\zeta}^2 + \phi_1\zeta^2)\mathrm{e} + 2^{1/2}(\zeta - \bar{\zeta})^2 h_{\mathbf{11}} - 2^{1/2}(\zeta + \bar{\zeta})\zeta\bar{\zeta}h_{\mathbf{11},r}\right]\sin\vartheta \\
& + \left\{\left[(h_{\mathbf{12},r}\bar{\zeta} + h_{\mathbf{13},r}\zeta)\bar{\zeta} - (\zeta - 2\bar{\zeta})h_{\mathbf{13}}\right]\zeta + (2\zeta - \bar{\zeta})h_{\mathbf{12}}\bar{\zeta}\right\}\cos\vartheta
\end{aligned}
$$

$$
\begin{aligned}
E_3 = {} & \zeta\bar{\zeta}^2\sin\vartheta\left[h_{\mathbf{22},r\vartheta} + ia\sin\vartheta\left(h_{\mathbf{22},rr} - h_{\mathbf{22},rt}\right) - 2^{3/2}h_{\mathbf{12},r} - 2^{1/2}\zeta h_{\mathbf{12},rr}\right] \\
& - (\zeta - \bar{\zeta})\bar{\zeta}\left[\sin\vartheta(ia\sin\vartheta h_{\mathbf{22},t} - h_{\mathbf{22},\vartheta} - ia\sin\vartheta h_{\mathbf{22},r}) + ih_{\mathbf{22},\varphi}\right] \\
& - i\zeta\bar{\zeta}^2 h_{\mathbf{22},r\varphi} + 2\bar{\zeta}\cos\vartheta\left[(\zeta - \bar{\zeta})h_{\mathbf{22}} + \zeta\bar{\zeta}h_{\mathbf{22},r}\right] \\
& + 2\sin\vartheta\left[2^{1/2}(\zeta^2 - \zeta\bar{\zeta} + \bar{\zeta}^2)h_{\mathbf{12}} + 2\mathrm{e}\zeta^2\phi_0\right]
\end{aligned}
$$

$$
\begin{aligned}
E_4 = {} & \zeta^2\bar{\zeta}^2\{2\zeta^3\bar{\zeta}^3\sin^2\theta\left[-h_{11,\theta\theta} + \sin^2\theta a^2\left(2h_{11,rt} - h_{11,tt}\right) - 2h_{11,t}\bar{\zeta}\right] \\
& + 2^{3/2}\zeta^3\bar{\zeta}^3\sin\theta[\bar{\zeta}\left(h_{12,t\theta}\sin\theta - ih_{12,\varphi t}\right) + \zeta\left(h_{13,t\theta}\sin\theta + ih_{13,\varphi t}\right) \\
& + \bar{\zeta}\sin^2\theta ai\left(h_{12,rt} - h_{12,tt}\right) + \zeta\sin^2\theta ai\left(h_{13,tt} - h_{13,rt}\right)] \\
& + \zeta^3\bar{\zeta}^3\left[4\sin^2\theta a\left(h_{11,\varphi r} - h_{11,\varphi t}\right) - 2h_{11,\varphi\varphi}\right] \\
& + 2(\mathrm{e}^2 - m\bar{\zeta})\bar{\zeta}^3\sin^2\theta a\left[(h_{22,t}\sin\theta a + h_{22,\theta}i - h_{22,r}\sin\theta a)\sin\theta + h_{22,\varphi}\right] \\
& + 2(\mathrm{e}^2 - m\zeta)\zeta^3\sin^2\theta a\left[(h_{33,t}\sin\theta a - h_{33,\theta}i - h_{33,r}\sin\theta a)\sin\theta + h_{33,\varphi}\right] \\
& - \sin^3\theta ai\zeta\bar{\zeta}2^{1/2}\left[(\zeta+\bar{\zeta})m - \mathrm{e}^2\right](\zeta-\bar{\zeta})(h_{12,r}\bar{\zeta} + h_{13,r}\zeta) \\
& + 2\sin^2\theta\zeta^2\bar{\zeta}^2\left[(\zeta+\bar{\zeta})^2 m - (\zeta+\bar{\zeta})\mathrm{e}^2 - 2\zeta^2\bar{\zeta}\right]h_{11,r} \\
& + 2^{1/2}\sin\theta\zeta\bar{\zeta}^2\left[(\zeta^2 - 3\bar{\zeta}^2)m - (\zeta - 3\bar{\zeta})\mathrm{e}^2\right]\left[(h_{12,t}\sin\theta ai - h_{12,\theta})\sin\theta + h_{12,\phi}i\right] \\
& + 2^{1/2}\sin\theta\zeta^2\bar{\zeta}\left[(3\zeta^2 - \bar{\zeta}^2)m - (3\zeta - \bar{\zeta})\mathrm{e}^2\right]\left[(h_{13,t}\sin\theta ai + h_{13,\theta})\sin\theta + h_{13,\phi}i\right]\} \\
& + 4\sin^4\theta a^2\left\{\left[(3\zeta - \bar{\zeta})h_{22}\bar{\zeta}^5 - (\zeta - 3\bar{\zeta})h_{33}\zeta^5\right]\mathrm{e}^2 - 3(h_{22}\bar{\zeta}^5 + h_{33}\zeta^5)m\zeta\bar{\zeta}\right\} \\
& - 2\sin^2\theta\zeta^5\bar{\zeta}^5\left[\mathrm{e}^2 + \zeta\bar{\zeta} - (\zeta+\bar{\zeta})m + \sin^2\theta a^2\right]h_{11,rr} \\
& + 2\sin^3\theta ai\zeta^2\bar{\zeta}^2\left\{4(\bar{\phi}_2\bar{\zeta}^3 - \phi_2\zeta^3)\mathrm{e}\zeta\bar{\zeta} - 2^{1/2}m\left[(\zeta + 4\bar{\zeta})h_{12}\bar{\zeta}^3 - (4\zeta + \bar{\zeta})h_{13}\zeta^3\right]\right\} \\
& + 2^{3/2}\sin^3\theta ai\zeta\bar{\zeta}\left[(2\zeta^2 - 4\zeta\bar{\zeta} - \bar{\zeta}^2)h_{13}\zeta^3 + (\zeta^2 + 4\zeta\bar{\zeta} - 2\bar{\zeta}^2)h_{12}\bar{\zeta}^3\right]\mathrm{e}^2 \\
& + 2\{\left[(\zeta-\bar{\zeta})^2 h_{11} + 2\mathrm{e}\phi_1\zeta^2 2^{1/2}\right]\mathrm{e}^2 - (\zeta+\bar{\zeta})(\zeta-\bar{\zeta})^2 h_{11}m \\
& - 2^{3/2}\left[(\zeta+\bar{\zeta})m - \zeta\bar{\zeta}\right]\mathrm{e}\phi_1\zeta^2 - 2^{3/2}\left[(\zeta+\bar{\zeta})m - \zeta\bar{\zeta} - \mathrm{e}^2\right]\bar{\phi}_1\mathrm{e}\bar{\zeta}^2\}\sin^2\theta\zeta^3\bar{\zeta}^3 \\
& + \{2\left[(h_{12,t}\bar{\zeta} + h_{13,t}\zeta)2^{1/2} - h_{11,\theta} + 2h_{11,t}\sin\theta ai - 2h_{11,r}\sin\theta ai\right]\zeta^3\bar{\zeta}^3 \\
& - 4\left[(h_{22}\bar{\zeta}^4 - h_{33}\zeta^4)m - (h_{22}\bar{\zeta}^3 - h_{33}\zeta^3)\mathrm{e}^2\right]\sin\theta ai \\
& + 2^{1/2}\zeta\bar{\zeta}\left[(3\zeta^2 - \bar{\zeta}^2)h_{13}\zeta - (\zeta^2 - 3\bar{\zeta}^2)h_{12}\bar{\zeta}\right]m \\
& - 2^{1/2}\zeta\bar{\zeta}\left[(3\zeta - \bar{\zeta})h_{13}\zeta - (\zeta - 3\bar{\zeta})h_{12}\bar{\zeta}\right]\mathrm{e}^2\}\cos\theta\sin\theta\zeta^2\bar{\zeta}^2
\end{aligned}
$$

$$
\begin{aligned}
E_5 = {} & \left\{\left[h_{33,r\vartheta} + (h_{33,rt} - h_{33,rr})\sin\vartheta ai\right] - 2^{1/2}\left(2h_{13,r} + h_{13,rr}\bar{\zeta}\right)\right\}\sin\vartheta\zeta^2\bar{\zeta} \\
& - \{[(h_{33,t} - h_{33,r})\sin\vartheta ai + h_{33,\vartheta}]\sin\vartheta + h_{33,\varphi}i\}(\zeta-\bar{\zeta})\zeta + \zeta^2\bar{\zeta}h_{33,r\varphi}i \\
& - 2\left[(\zeta-\bar{\zeta})h_{33} - h_{33,r}\zeta\bar{\zeta}\right]\cos\vartheta\zeta + 2\left[(\zeta^2 - \zeta\bar{\zeta} + \bar{\zeta}^2)h_{13}2^{1/2} + 2\bar{\phi}_0\mathrm{e}\bar{\zeta}^2\right]\sin\vartheta
\end{aligned}
$$

$$
\begin{aligned}
E_6 &= 2^{1/2}\zeta^5\bar{\zeta}\left(-h_{13,\vartheta\vartheta}\sin^2\vartheta + h_{13,\varphi\varphi} - 2h_{13,\vartheta\varphi}\sin\vartheta i\right)\\
&+ 2\{\left[2^{1/2}(\zeta^3 - \zeta\bar{\zeta}^2 + 4\bar{\zeta}^3)h_{12}m - 2\mathrm{e}^3\phi_0\zeta^2\right]\zeta + 2\left[(\zeta+\bar{\zeta})m - \zeta\bar{\zeta}\right]\mathrm{e}\phi_0\zeta^3\\
&- 2^{1/2}(\zeta^3 - \zeta^2\bar{\zeta} + 4\zeta\bar{\zeta}^2 - 2\bar{\zeta}^3)\mathrm{e}^2h_{12} - 4\bar{\phi}_2\mathrm{e}\zeta^2\bar{\zeta}^3\}\sin^2\vartheta\zeta\\
&+ 4\sin^3\vartheta ai\bar{\zeta}\{\left[(2\zeta - \bar{\zeta})h_{11}\zeta^2 + 2(\zeta - 2\bar{\zeta})h_{22}m\bar{\zeta}\right]\zeta\\
&- (2\zeta^2 - 4\zeta\bar{\zeta} + \bar{\zeta}^2)\mathrm{e}^2h_{22} - 2\bar{\phi}_1\mathrm{e}\zeta^3\bar{\zeta}2^{1/2}\}\\
&- \left[4(2\mathrm{e}\phi_0 + h_{12}2^{1/2})\sin^4\vartheta a^2 + (h_{12}\bar{\zeta} - h_{13}\zeta)\bar{\zeta}2^{1/2}\right]\zeta^4\\
&+ \sin^3\vartheta ai\zeta^4\left(2h_{22,tt}\bar{\zeta}^3 - 4h_{22,t}\bar{\zeta}^2\right) - 2\sin^2\vartheta\zeta^4\bar{\zeta}^3h_{22,t\vartheta}\\
&- 4\sin^3\vartheta ai\zeta^2\bar{\zeta}\left[(\zeta + 2\bar{\zeta})(\zeta - \bar{\zeta})m - \zeta^2\bar{\zeta} - (\zeta - 2\bar{\zeta})\mathrm{e}^2\right]h_{22,r}\\
&- 2\sin^3\vartheta ai\zeta^3\bar{\zeta}^2\left[(\zeta + \bar{\zeta})m - \zeta\bar{\zeta} - \mathrm{e}^2\right]h_{22,rr}\\
&+ i\zeta^4\bar{\zeta}^3\left(2h_{22,\varphi t}\sin\vartheta - 6h_{22,rt}\sin^3\vartheta a\right)\\
&- 2^{3/2}\bar{\zeta}a\sin^2\vartheta\left[\sin\vartheta i\left(h_{13,t\vartheta}\zeta^5 + 2h_{13,\vartheta}\zeta^4\right) + 2h_{13,\varphi}\zeta^4 + h_{13,\varphi t}\zeta^5\right]\\
&+ 2^{1/2}\bar{\zeta}\sin^4\vartheta a^2\left(4h_{13,t}\zeta^4 + h_{13,tt}\zeta^5 - 4h_{13,r}\zeta^4 - h_{13,rr}\zeta^5\right)\\
&+ 2^{1/2}\left[h_{12,\vartheta\vartheta}\sin^2\vartheta\zeta^4\bar{\zeta}^2 - 2(\zeta - \bar{\zeta})h_{12,\vartheta}\sin^3\vartheta ai\zeta^3\bar{\zeta}\right]\\
&- 2^{3/2}\left[(3\zeta - \bar{\zeta})\sin^2\vartheta a^2 + (\zeta - 2\bar{\zeta})\zeta\bar{\zeta}\right]h_{12,t}\sin^2\vartheta\zeta^3\bar{\zeta}\\
&+ h_{12,tt}\sin^4\vartheta a^2\zeta^4\bar{\zeta}^2 2^{1/2}\\
&- 2^{3/2}\left[(\zeta + \bar{\zeta})^2m - 2\zeta^2\bar{\zeta} - (\zeta + \bar{\zeta})\mathrm{e}^2 - (3\zeta - \bar{\zeta})\sin^2\vartheta a^2\right]h_{12,r}\sin^2\vartheta\zeta^3\bar{\zeta}\\
&- \left\{2\left[(\zeta + \bar{\zeta})m - \zeta\bar{\zeta} - \mathrm{e}^2\right] - 5\sin^2\vartheta a^2\right\}h_{12,rr}\sin^2\vartheta\zeta^4\bar{\zeta}^2 2^{1/2}\\
&- 2^{3/2}\sin^2\vartheta\left[(3\sin^2\vartheta a^2 + \zeta\bar{\zeta})h_{12,rt}\zeta^4\bar{\zeta}^2 + (3\zeta - \bar{\zeta})h_{12,\varphi}\sin^2\vartheta a\zeta^3\bar{\zeta}\right]\\
&+ h_{12,\varphi\varphi}\zeta^4\bar{\zeta}^2 2^{1/2} - 2h_{12,\varphi}i\zeta\bar{\zeta}2^{1/2}\cos\vartheta\zeta^3\bar{\zeta}\\
&+ 2^{1/2}\zeta^4\bar{\zeta}^2 a\sin^2\vartheta\left(2h_{12,\varphi t} - 6h_{12,r\varphi}\right) + 2(\zeta - \bar{\zeta})h_{11,\vartheta}\sin^2\vartheta\zeta^4\bar{\zeta}\\
&+ 2\zeta^4\bar{\zeta}\sin^3\vartheta ai\left[(\zeta - \bar{\zeta})h_{11,t} + (\zeta + 3\bar{\zeta})h_{11,r} + \left(h_{11,rr} + h_{11,rt}\right)\zeta\bar{\zeta}\right]\\
&+ 2h_{11,r\vartheta}\sin^2\vartheta\zeta^5\bar{\zeta}^2 + 2\left[(\zeta - \bar{\zeta})h_{11,\varphi}\zeta^4\bar{\zeta} + h_{11,r\varphi}\zeta^5\bar{\zeta}^2\right]\sin\vartheta i\\
&- \{2^{3/2}\left[(3\zeta - \bar{\zeta})h_{12} + 2h_{13}\zeta + 2\zeta\bar{\zeta}h_{12,r} + \zeta^2h_{13,t}\right]\sin\vartheta ai\\
&+ 4h_{22,t}\zeta\bar{\zeta}^2 + 2^{1/2}\zeta\left(\zeta h_{13,\vartheta} - \bar{\zeta}h_{12,\vartheta}\right)\}\sin\vartheta\cos\vartheta\zeta^3\bar{\zeta}
\end{aligned}
$$

$$
\begin{aligned}
E_7 = \{&\zeta^2\bar{\zeta}^2[-2\zeta\bar{\zeta}\sin\vartheta\,(h_{11,r\vartheta} + h_{11,rr}\sin^2\vartheta ia) - 2(\zeta - \bar{\zeta})h_{11,\vartheta}\sin^2\vartheta \\
&+ 2^{1/2}(h_{12}\bar{\zeta} - h_{13}\zeta) - \zeta^2\sin^2\vartheta h_{33,t\vartheta} + ai\zeta^2\sin^3\vartheta\,(h_{33,rt} - h_{33,tt}) \\
&- i\sin\vartheta\,(h_{22,\varphi t}\bar{\zeta}^2 + h_{33,\varphi t}\zeta^2) + ai\sin^3\vartheta\bar{\zeta}^2\,(h_{22,rt} - h_{22,tt}) \\
&+ h_{22,t\vartheta}\sin^2\vartheta\bar{\zeta}^2 + 2^{1/2}\sin^3\vartheta ai\,(2h_{12,\vartheta} + h_{12,t\vartheta}\bar{\zeta} + 2h_{13,\vartheta} + h_{13,t\vartheta}\zeta) \\
&+ 2^{1/2}\sin\vartheta\,(\sin\vartheta\zeta h_{13,\vartheta\vartheta} - 2h_{13,\varphi}\sin\vartheta a + h_{13,\vartheta\varphi}i\zeta - h_{13,r\varphi}\sin\vartheta a\zeta) \\
&+ 2^{1/2}\sin\vartheta\,(-\sin\vartheta\bar{\zeta}h_{12,\vartheta\vartheta} + 2h_{12,\varphi}\sin\vartheta a + h_{12,\vartheta\varphi}i\bar{\zeta} + h_{12,r\varphi}\sin\vartheta a\bar{\zeta})] \\
&+ 2\left[(\mathrm{e}^2 - m\bar{\zeta})h_{22,r}\bar{\zeta}^3 + (\mathrm{e}^2 - m\zeta)h_{33,r}\zeta^3 - \zeta^2\bar{\zeta}^2(\zeta + \bar{\zeta})h_{11,r}\right]\sin^3\vartheta ai \\
&+ 2^{1/2}(h_{12,rt}\bar{\zeta} - h_{13,rt}\zeta)(a^2\sin^2\vartheta + \zeta\bar{\zeta})\sin^2\vartheta\zeta^2\bar{\zeta}^2 \\
&+ \left[(\zeta - 2\bar{\zeta})\bar{\zeta} + 2\sin^2\vartheta a^2\right]h_{12,t}\sin^2\vartheta\zeta^2\bar{\zeta}^2 2^{1/2} \\
&+ \left[(2\zeta - \bar{\zeta})\zeta - 2\sin^2\vartheta a^2\right]h_{13,t}\sin^2\vartheta\zeta^2\bar{\zeta}^2 2^{1/2} \\
&- \left[(\zeta + \bar{\zeta})^2 m - (\zeta + \bar{\zeta})\mathrm{e}^2 - 2\zeta\bar{\zeta}^2 - 2\sin^2\vartheta a^2\bar{\zeta}\right]h_{13,r}\sin^2\vartheta\zeta^2\bar{\zeta}2^{1/2} \\
&+ \left[(\zeta + \bar{\zeta})^2 m - (\zeta + \bar{\zeta})\mathrm{e}^2 - 2\zeta^2\bar{\zeta} - 2\sin^2\vartheta a^2\zeta\right]h_{12,r}\sin^2\vartheta\zeta\bar{\zeta}^2 2^{1/2} \\
&- \left[\mathrm{e}^2 + \zeta\bar{\zeta} - (\zeta + \bar{\zeta})m + \sin^2\vartheta a^2\right](h_{12,rr}\bar{\zeta} - h_{13,rr}\zeta)\sin^2\vartheta\zeta^2\bar{\zeta}^2 2^{1/2}\}\zeta^2\bar{\zeta}^2 \\
&- 2\{\left[(\zeta - 3\bar{\zeta})h_{22}m\bar{\zeta} + 2h_{11}\zeta^3\right]\zeta\bar{\zeta}^4 - (3\zeta - \bar{\zeta})h_{33}m\zeta^4\zeta\bar{\zeta} \\
&- (h_{22}\bar{\zeta}^4 + h_{33}\zeta^4)(\zeta^2 - 3\zeta\bar{\zeta} + \bar{\zeta}^2)\mathrm{e}^2\}\sin^3\vartheta ai \\
&+ \sin^2\vartheta\zeta\bar{\zeta}\{\left[2(2\bar{\phi}_2\bar{\zeta}^3 + \mathrm{e}^2\phi_0\zeta)\mathrm{e}\zeta - (\zeta^3 - \zeta\bar{\zeta}^2 + 4\bar{\zeta}^3)h_{12}m2^{1/2}\right]\zeta\bar{\zeta}^2 \\
&+ (4\zeta^3 - \zeta^2\bar{\zeta} + \bar{\zeta}^3)h_{13}m\zeta^2\bar{\zeta}2^{1/2} \\
&+ \left[(2\zeta^3 - 4\zeta^2\bar{\zeta} + \zeta\bar{\zeta}^2 - \bar{\zeta}^3)h_{13}\zeta^2 + (\zeta^3 - \zeta^2\bar{\zeta} + 4\zeta\bar{\zeta}^2 - 2\bar{\zeta}^3)h_{12}\bar{\zeta}^2\right]\mathrm{e}^2 2^{1/2} \\
&+ 2\left[(\phi_0\bar{\zeta} - 2\phi_2\zeta)\zeta - (\zeta + \bar{\zeta})m\phi_0\right]\mathrm{e}\zeta^3\bar{\zeta}^2 + 2\left[(\zeta + \bar{\zeta})m - \zeta\bar{\zeta} - \mathrm{e}^2\right]\bar{\phi}_0\mathrm{e}\zeta^2\bar{\zeta}^3\} \\
&+ 2^{1/2}\cos\vartheta\zeta^4\bar{\zeta}^4\{\left[2^{1/2}(h_{22,t}\bar{\zeta}^2 - h_{33,t}\zeta^2) + \zeta\,(h_{13,\vartheta} + h_{13,t}\sin\vartheta ai)\right]\sin\vartheta \\
&- \left[(\bar{\zeta}h_{12,\varphi} + h_{13,\varphi}\zeta)\,i + h_{12,\vartheta}\sin\vartheta\bar{\zeta}\right] + \left[h_{12,t}\bar{\zeta} + 2(h_{12} + h_{13})\right]\sin^2\vartheta ai\}
\end{aligned}
$$

$$
\begin{aligned}
E_8 = 2^{1/2}\{&[\zeta h_{33,\vartheta\vartheta} + 2ai\sin\vartheta\,(h_{33,\vartheta} + \zeta h_{33,t\vartheta}) + 2\zeta a^2\sin^2\vartheta\,(h_{33,r} - h_{33,t} + h_{33,rr}) \\
&- a^2\sin^2\vartheta\zeta h_{33,tt} - 2ah_{33,\varphi}]\zeta\sin^2\vartheta + 2\zeta^2\sin\vartheta\,(ih_{33,\vartheta\varphi} - a\sin\vartheta h_{33,\varphi t}) \\
&- \zeta^2 h_{33,\varphi\varphi} - 2ai\sin^3\vartheta\bar\zeta h_{22,\vartheta} + \sin^2\vartheta\bar\zeta^2 h_{22,\vartheta\vartheta} - a^2\sin^4\vartheta(2\bar\zeta h_{22,t} + \bar\zeta^2 h_{22,tt}) \\
&- 2ai\sin^3\vartheta\bar\zeta^2 h_{22,t\vartheta} + a^2\sin^4\vartheta\,(2\bar\zeta h_{22,r} + \bar\zeta^2 h_{22,rr}) - 2a\sin^2\vartheta\bar\zeta h_{22,\varphi} \\
&- \bar\zeta^2 h_{22,\varphi\varphi} - 2\bar\zeta^2\sin\vartheta i h_{22,\vartheta\varphi} - 2\bar\zeta^2 a\sin^2\vartheta h_{22,\varphi t}\} \\
&+ 4ai\zeta\bar\zeta\sin^3\vartheta\,(h_{12,r} + \bar\zeta h_{12,rr} - h_{13,r} - \zeta h_{13,rr}) - 2^{3/2}\sin^2\vartheta\zeta^2\bar\zeta^2 h_{11,rr} \\
&- 4(\zeta - 2\bar\zeta)\sin\vartheta\bar\zeta\,[\sin\vartheta(ai\sin\vartheta h_{12,t} - h_{12,\vartheta}) + ih_{12,\varphi}] \\
&- 4(2\zeta - \bar\zeta)\sin\vartheta\zeta\,[\sin\vartheta(ai\sin\vartheta h_{13,t} + h_{13,\vartheta}) + ih_{13,\varphi}] \\
&+ 8ai\sin^3\vartheta\left[2^{1/2}\mathrm{e}(\bar\zeta\bar\phi_0 - \zeta\phi_0) - (\zeta - \bar\zeta)(h_{12} + h_{13})\right] \\
&- 2\sin^2\vartheta\left[2^{1/2}(\bar\zeta^2 h_{22} + \zeta^2 h_{33}) - 8\mathrm{e}(\zeta^2\phi_1 + \bar\zeta^2\bar\phi_1) - 2^{3/2}(\zeta^2 - \zeta\bar\zeta + \bar\zeta^2)h_{11}\right] \\
&+ \{2^{1/2}[\zeta^2\sin\vartheta(4ai\sin\vartheta h_{33,t} + 3h_{33,\vartheta}) + \bar\zeta^2\sin\vartheta\,(3h_{22,\vartheta} - 4\sin\vartheta ai h_{22,t}) \\
&+ 2i\zeta^2 h_{33,\varphi} - 2i\bar\zeta^2 h_{22,\varphi} - 4ai\sin^2\vartheta](\bar\zeta h_{22} - \zeta h_{33}) \\
&- 4\sin\vartheta\,[(2\zeta - \bar\zeta)\zeta h_{13} - (\zeta - 2\bar\zeta)\bar\zeta h_{12}]\}\cos\vartheta
\end{aligned}
$$

$$
\begin{aligned}
E_9 = {} & 2^{3/2}\zeta^2\bar\zeta^3\sin\vartheta(\zeta h_{33,rt} + h_{33,t}) + 2\zeta\bar\zeta^3\sin^2\vartheta ai\,(h_{13,t} - h_{13,r} - \zeta h_{13,rr}) \\
&+ 2\zeta\bar\zeta^3\,[\sin\vartheta\,(ai\zeta\sin\vartheta h_{13,rt} - h_{13,\vartheta} - \zeta h_{13,r\vartheta}) + ih_{13,\varphi} + i\zeta h_{13,r\varphi}] \\
&- 2^{3/2}\zeta(\mathrm{e}^2 - m\zeta)(\zeta - \bar\zeta)\sin\vartheta h_{33} \\
&+ 4\bar\zeta^3\sin^2\vartheta ai(2^{1/2}\mathrm{e}\bar\phi_0 + h_{13}) + 2\zeta\bar\zeta^3\cos\vartheta(\zeta h_{13,r} + h_{13}) \\
&- 2^{1/2}\zeta^2\bar\zeta^2\sin\vartheta\,[\mathrm{e}^2 + \zeta\bar\zeta - (\zeta + \bar\zeta)m]\,h_{33,rr} \\
&+ 2^{3/2}\zeta\bar\zeta\sin\vartheta\,[(\zeta - \bar\zeta)\mathrm{e}^2 - \zeta\bar\zeta^2 - (\zeta^2 - \zeta\bar\zeta - \bar\zeta^2)m]\,h_{33,r}
\end{aligned}
$$

The integrability conditions are obtained by expressing the first term from Eqs. $E_k = 0$, $k = 1, ..., 8$, and taking the mixed derivatives. Eq. $E_9 = 0$ can be used to eliminate $h_{13,r\vartheta}$ from E_2. From the resulting equation one can express the second derivative $h_{11,rr}$. The pairs E_1 and E_9 as well as E_3 and E_5 are complex conjugates.

Appendix: B. Maxwell field

The source-free Maxwell equations are

$$F_{[ab;c]} = 0, \qquad F^{ab}{}_{;b} = 0. \tag{1}$$

The independent components are represented by the self-dual equations

$$M^a \equiv F^{ab}{}_{;b} + \tfrac{1}{2} i\epsilon^{abcd} F_{bc;d} = 0. \tag{2}$$

An infinitesimal transformation (26) has the following effect on the Maxwell field:

$$\overline{F}_{ab} = F_{ab} + F_{ab;r}\xi^r + F_{rb}\xi^r_{;a} + F_{ar}\xi^r_{;b}. \tag{3}$$

The last three terms on the right, to first order, contain only the unperturbed part of the field F_{ab}. It follows that the perturbations of F_{ab} are invariant for an uncharged Kerr metric.

The stress-energy tensor is

$$T_{ab} = g^{rs} F_{ar} F_{bs} - \tfrac{1}{4} g_{ab} F^{rs} F_{rs}.$$

For the Newman-Penrose form we take the null tetrad components of Eqs. (2) and use the notation

$$\phi_0 = F_{\mathbf{02}}, \qquad \phi_1 = \tfrac{1}{2}\left(F_{\mathbf{01}} + F_{\mathbf{32}}\right), \qquad \phi_2 = F_{\mathbf{31}}.$$

The tetrad components of Eqs. (2) are

$$\begin{aligned} M_1 &= \left\{2\left[(\phi_{1,r}\zeta + 2\phi_1)2^{1/2} - \phi_{0,\vartheta}\right]\cos\vartheta + (\phi_{0,r} - \phi_{0,t})(\zeta - \bar{\zeta})\sin\vartheta\right\}\sin\vartheta \\ &\quad - 2(\cos\vartheta\phi_0 - i\phi_{0,\varphi})\cos\vartheta \\ M_3 &= 4\left[(\phi_{1,\varphi}i + \phi_{1,\vartheta}\sin\vartheta - \phi_{0,t}\sin\vartheta\bar{\zeta}2^{1/2})\zeta^2 - h_{12,r}\sin\vartheta e\bar{\zeta}\right]\cos\vartheta\zeta^2 \\ &\quad + \left\{\left[2(\phi_{1,r} - \phi_{1,t})\zeta^3 - h_{22,r}e\bar{\zeta}2^{1/2}\right]\zeta - (\zeta - 3\bar{\zeta})eh_{22}2^{1/2}\right\}(\zeta - \bar{\zeta})\sin^2\vartheta \\ &\quad + 2^{3/2}\left[e^2 + \zeta\bar{\zeta} - (\zeta + \bar{\zeta})m\right]\cos\vartheta\phi_{0,r}\sin\vartheta\zeta^3 \\ &\quad - 2\left[2^{1/2}\zeta(e^2 - \zeta^2)\phi_0 + 2(\zeta - 2\bar{\zeta})eh_{12} + 2^{1/2}\zeta(\zeta - \bar{\zeta})m\phi_0\right]\cos\vartheta\sin\vartheta\zeta. \end{aligned}$$

For normal modes, we get

$$\begin{aligned} M_1 &= \zeta\bar{\zeta}\sin\vartheta\left(2^{1/2}\zeta\phi_{1,r} + 2^{3/2}\phi_1 - \phi_{0,\vartheta}\right) \\ &\quad - ia\sin^2\vartheta\left[\zeta\bar{\zeta}\phi_{0,r} + (\zeta + \bar{\zeta})\phi_0\right] - \left(\zeta\bar{\zeta}\cos\vartheta + \mathcal{L}_z\right)\phi_0 \end{aligned} \tag{4a}$$

$$\begin{aligned} M_3 &= 2\zeta^3\sin\vartheta\left\{\bar{\zeta}\left[e^2 + \zeta\bar{\zeta} - (\zeta + \bar{\zeta})m\right]\phi_{0,r} - ia\sin\vartheta\left[\zeta\bar{\zeta}\phi_{1,r} + (\zeta + \bar{\zeta})\phi_1\right]\right\} \\ &\quad + e\bar{\zeta}\sin\vartheta\left\{2^{1/2}ia\sin\vartheta\left[\zeta\bar{\zeta}h_{22,r} + (\zeta - 3\bar{\zeta})h_{22}\right] - 2\zeta\left(\zeta\bar{\zeta}h_{12,r} + (\zeta - 2\bar{\zeta})h_{12}\right)\right\} \\ &\quad + 2\zeta^3\left(\zeta\bar{\zeta}\sin\vartheta\phi_{1,\vartheta} - \mathcal{L}_z\phi_1\right) \\ &\quad - 2^{1/2}\zeta^2\bar{\zeta}\sin\vartheta\left[(2i\bar{\zeta}E - 1)\zeta^2 + e^2 + (\zeta - \bar{\zeta})m\right]\phi_0. \end{aligned} \tag{4b}$$

There are no metric perturbations in Eq. (4a).

Acknowledgements

This work has been supported by the OTKA grant T031724. I thank G. Gálfi for helpful discussions and for creating algebraic computation software. I am grateful to M. Vasúth for computing the electrovacuum equations of Appendix A and for checks.

Bibliography

[1] P.L.Chrzanowski and C.W.Misner, Phys. Rev. **D10**, 1701 (1974)

[2] P.L.Chrzanowski, Phys. Rev. **D11**, 2042 (1975)

[3] S.Detweiler, Proc. Roy. Soc. Lond. A **352**, 381

[4] E.T.Newman and R.Penrose, J. Math. Phys. **3,** 566 (1962)

[5] S.Chandrasekhar: The mathematical theory of black holes, Clarendon Press, Oxford, 1983

[6] E.T.Newman *et al.*, J. Math. Phys. **6,** 918 (1965)

[7] S.Mano and E.Takasugi, Prog. Theor. Phys. **97**, 213 (1997), gr-qc/9611014

[8] S.A.Teukolsky, Astrophys. J. **185**, 635 (1973)

[9] D.M.Chitre, Phys.Rev. **D13**, 2713 (1976)

[10] Z.Perjés, Fundamental equations for the gravitational and electromagnetic perturbations of a charged black hole, gr-qc/0206088

We sadly inform the Reader that Professor Zoltán Perjés, one of the plenary speakers of the Bolyai Conference, the internationally recognized specialist of the applications of non-Euclidean geometry in physics, passed away on October 27th, 2004. András Prékopa, Emil Molnár

PLACING THE HYPERBOLIC GEOMETRY OF BOLYAI AND LOBACHEVSKY CENTRALLY IN SPECIAL RELATIVITY THEORY: AN IDEA WHOSE TIME HAS RETURNED

Abraham A. Ungar
Department of Mathematics
North Dakota State University
Fargo, ND 58105, USA
abraham_ungar@ndsu.nodak.edu

Abstract We show that Einstein addition of relativistically admissible velocities is regulated by the hyperbolic geometry of Bolyai and Lobachevski just as Newtonian velocity addition is regulated by Euclidean geometry. Hence, the time to study special relativity in terms of its underlying hyperbolic geometry has returned.

1. Introduction

The majestic scientific achievement of the 20th century in mathematical beauty and experimental verifications has been the special theory of relativity, proposed by Einstein in 1905. Einstein's special relativity theory is one of the foundation blocks of modern theoretical physics. It is this theory that we explore in this article beyond the Einstein velocity addition and its accompanying Thomas precession [29]. We will find that the study of the most general Einstein addition is a great way to get a glimpse into the fascinating world of the hyperbolic geometry of Bolyai and Lobachevski, and the role it plays in relativity physics.

In 1908 Herman Minkowski recast special relativity as a new geometry of spacetime. The resulting Minkowski space is based on his four-dimensional matrix formalism designed to take advantage of the postulated covariance of physical laws with respect to the Lorentz group [37]. In Minkowski's spacetime, space and time are inseparable, giving rise to the so called *pseudo – Euclidean space* in which the time coordinate is defined to be imaginary. The emergence of Minkowski's four-dimensional formalism as a standard technique in special relativity theory is explored by S. Walter in [37].

Vladimir Varičak's interest in non-Euclidean geometry dates back to 1908 [34]. In 1910 [35] he discovered that Einsteinian velocities have a natural interpretation in the hyperbolic geometry of Bolyai and Lobachevski, thus introducing the so-called *non-Euclidean style* to the study of special relativity. The non-Euclidean program of Varičak (1865 – 1942) is described by Walter in [37]. The myth of the unity of space and time, initiated by Minkowski in 1908, breaks down in the non-Euclidean style approach to special relativity theory, that contrasts Minkowski's pseudo-Euclidean approach.

The history of the two competing geometrical approaches to the study of relativity physics has been studied quantitatively by Walter [37], calling them

(1) the *space-time formalism*, in which the study of relativity physics is based on the Lorentz transformation law, from which the Einstein addition is derived; and

(2) the *non-Euclidean style*, in which the study of relativity physics is based on the general Einstein velocity addition law and its underlying hyperbolic geometry of Bolyai and Lobachevski, from which the Lorentz transformation is derived.

Contrasting the success of Minkowski's space-time formalism, Walter explains the source of the failure of the non-Euclidean style. He notes in [37] that

> In contrast to the amount of publicity they received, applications of non-Euclidean geometry to physics by leading practitioners produced slim theoretical results, the value of which was outstripped by the technical intricacy of the methods developed to obtain them.

Among the leading practitioners in the non-Euclidean style Walter mentions Alfred A. Robb, Vladimir Varičak, Gilbert N. Lewis, Edwin B. Wilson, and Émile Borel [37]. A modern study of connections between the theory of relativity and the hyperbolic geometry of Bolyai and Lobachevski is presented, for instance, in Černikov [3] where references to earlier related work are given.

In fact, connections between the theory of relativity and the hyperbolic geometry of Bolyai and Lobachevski were encountered even before the introduction of the theory of relativity by Einstein in 1905. Owing mainly to the work of Tibor Toró, cited in [10], it is now known that János Bolyai was the forerunner of geometrizing physics. According to Kiss [10], Lajos Dávid drew attention in a 1924 series of articles in Italian journals to the precursory role which János Bolyai played in the constructions of Einstein's relativity theory.

According to A.I. Miller (p. 266 in [12]), one of the first demonstrations that non-Euclidean geometry could be used to present concisely results of relativity theory was obtained by Sommerfeld in 1909 [16] when he was led to the result that relativistically admissible velocities add according to a spherical geometry. Sommerfeld's 1909 work is described by Rosenfeld in his book [13, pp. 270 – 273]:

> Although Sommerfeld established the connections between the formula for the addition of velocities in the theory of relativity and the trigonometric formulas for hyperbolic functions he was not aware [in 1909; but, see our next quotation] that these formulas are formulas of Lobačevskian geometry. This was shown by the Yugoslav geometer Vladimir Varičak (1865 – 1942) ...

From Varičak's acknowledgment of Sommerfeld's 1909 paper [16] it appears that there was a causal link between the latter paper and Varičak's 1910 discovery in [35] of the role that hyperbolic geometry plays in special relativity theory. Thus, it was Sommerfeld's 1909 paper that sparked Varičak's non-Euclidean program for special relativity; see [13, p. 270]. Ironically, however, not only did Sommerfeld employ an imaginary temporal coordinate, following the space-time formalism of Minkowski, he deplored the non-Euclidean style in print, as Walter noted in [37, p. 114]:

> ...just after Varičak's first exposé of the non-Euclidean style ([35], 1910), Sommerfeld completed his signal work on the four-dimensional vector calculus for the *Annalen der Physik*. In a footnote to his work, Sommerfeld remarked that the geometrical relations he presented in terms of three real and one imaginary coordinate could be reinterpreted in terms of non-Euclidean geometry. The latter approach, Sommerfeld cautioned in [17, p. 752]), could "hardly be recommended".

Furthermore, Walter notes in [37] that following the competition between the two geometrical approaches to relativity physics:

> Minkowski neither mentioned the [Einstein] law of velocity addition, nor expressed it in formal terms.

Instead, however [37],

> Minkowski retained the geometric interpretation of the Lorentz transformations that had accompanied the *now-banished* non-Euclidean interpretation of velocity vectors. [italics added].

The trend initiated by Minkowski continues today, with the full Einstein addition and its associated Thomas precession receiving scant attention, and modern texts on relativity physics reflect this with the only single, outstanding exception being the book of Sexl and Urbantke [14].

Soon after its introduction by Einstein in 1905 [4] special relativity theory, as named by Einstein ten years later, became overshadowed by the appearance of general relativity. Subsequently, the study of special relativity followed the lines laid down by Minkowski, in which the role of Einstein velocity addition is ignored [2]. Following Minkowski, therefore, the general Einstein velocity addition law of relativistically admissible velocities that need not be parallel is unheard of in most texts on special and general relativity theory. Rather, it is only the special case of Einstein addition, corresponding to parallel velocities, which is presented. Among outstanding exceptions we note the relativity physics books by Fock [6] and by Sexl and Urbantke [14].

A full-scale treatment of the special theory of relativity, with particular emphasis on the expression and interpretation of its concepts and results in terms of the Minkowski space-time formalism, is presented in Synge's 1956 textbook [18]. It is this book that effectively created the trend toward four-dimensional approaches, suggesting that the current domination of the space-time formalism in textbooks on special relativity theory is a late-20th century phenomenon. Additionally, the boom of research in general relativity is also a late-20th century phenomenon. Since special relativity is considered to be a preliminary subject to general relativity, the promotion of the space-time formalism in special relativity followed the upsurge of research in general relativity. Connections between the special theory of relativity and the hyperbolic geometry of Bolyai and Lobachevski, like the one explored by Černikov [3], are thus ignored in the mainstream literature. Of course, there are other factors that must be considered in our attempt to understand the current domination of the space-time formalism in textbooks on special relativity.

The task we face in this article is to reverse the process initiated by Minkowski, turning the "now-banished" non-Euclidean style into the cornerstone of relativity physics. Following the emergence of the theory of gyrogroups and gyrovector spaces [29], we now have a better case to demonstrate that the general Einstein addition, which until recently has rested in undeserved obscurity for too long, plays the central role in relativity physics.

2. The Einstein Addition

Let $\mathbb{V}$ be a real inner product space, and let $\mathbb{V}_c$,

$$\mathbb{V}_c = \{\mathbf{v} \in \mathbb{V} : \|\mathbf{v}\| < c\} \tag{1}$$

be the set of all relativistically admissible velocities in $\mathbb{V}$, that is, all vectors $\mathbf{v} \in \mathbb{V}$ with magnitude $< c$, c being any fixed positive number representing the vacuum speed of light. The set $\mathbb{V}_c$ of all relativistically admissible velocities in $\mathbb{V}$ is thus the open ball $\mathbb{V}_c$ of the space $\mathbb{V}$ with radius c, centered at the origin of its space.

We assume that the inner product in $\mathbb{V}$ (also known as a scalar product) is *positive definite* in the sense that $\mathbf{v} \cdot \mathbf{v} = \|\mathbf{v}\|^2 \geq 0$ for all $\mathbf{v} \in \mathbb{V}$ and $\|\mathbf{v}\|^2 > 0$ if $\mathbf{v} \neq \mathbf{0}$. Clearly, a positive definite inner product is *non-*

degenerate, that is, if $\mathbf{u}, \mathbf{v} \in \mathbb{V}$ and $\mathbf{u}\cdot\mathbf{v} = 0$ for all $\mathbf{v} \in \mathbb{V}$ then $\mathbf{u} = \mathbf{0}$. Indeed, if $\mathbf{u}\cdot\mathbf{v} = 0$ for all $\mathbf{v} \in \mathbb{V}$ and any given $\mathbf{u} \in \mathbb{V}$ then, in particular, for $\mathbf{v} = \mathbf{u}$ we have $\|\mathbf{u}\|^2 = 0$ implying $\mathbf{u} = \mathbf{0}$.

Let $\mathbf{u}, \mathbf{v} \in \mathbb{V}_c$ be any two relativistically admissible velocities in $\mathbb{V}$. Their Einstein sum, $\mathbf{u}\oplus\mathbf{v}$, is given by the equation

$$\mathbf{u}\oplus\mathbf{v} = \frac{1}{1+\frac{\mathbf{u}\cdot\mathbf{v}}{c^2}} \left\{ \mathbf{u} + \frac{1}{\gamma_{\mathbf{u}}}\mathbf{v} + \frac{1}{c^2}\frac{\gamma_{\mathbf{u}}}{1+\gamma_{\mathbf{u}}}(\mathbf{u}\cdot\mathbf{v})\mathbf{u} \right\} \tag{2}$$

where $\gamma_{\mathbf{v}}$ is the Lorentz factor of $\mathbf{v}$,

$$\gamma_{\mathbf{v}} = \frac{1}{\sqrt{1 - \frac{\|\mathbf{v}\|^2}{c^2}}} \tag{3}$$

satisfying the *gamma identity* [15]

$$\gamma_{\mathbf{u}\oplus\mathbf{v}} = \gamma_{\mathbf{u}}\gamma_{\mathbf{v}}\left(1 + \frac{\mathbf{u}\cdot\mathbf{v}}{c^2}\right). \tag{4}$$

It follows from (3) that the reality of $\gamma_{\mathbf{v}}$ is equivalent to the condition $\mathbf{v} \in \mathbb{V}_c$. Hence, it follows from the gamma identity (4) that $\mathbf{u}, \mathbf{v} \in \mathbb{V}_c$ implies $\mathbf{u}\oplus\mathbf{v} \in \mathbb{V}_c$ so that Einstein addition is, indeed, a binary operation in $\mathbb{V}_c$, forming the Einstein *groupoid* $(\mathbb{V}_c, \oplus)$, a groupoid being a non-empty set with a binary operation. $\mathbb{V}_c$ is thus the open c-ball of a real inner product space $\mathbb{V}$, equipped with the Einstein velocity addition $\oplus$ and with the positive definite inner product it inherits from its space $\mathbb{V}$.

We note that (2) defines the Einstein sum $\|\mathbf{u}\|\oplus\|\mathbf{v}\|$ as well, where the equality $\|\mathbf{u}\|\cdot\|\mathbf{v}\| = \|\mathbf{u}\|\|\mathbf{v}\|$ is understood. Thus,

$$\|\mathbf{u}\|\oplus\|\mathbf{v}\| = \frac{\|\mathbf{u}\| + \|\mathbf{v}\|}{1 + \frac{\|\mathbf{u}\|\|\mathbf{v}\|}{c^2}}. \tag{5}$$

In the special case when the velocity vectors are parallel, Einstein velocity addition (2) takes the simpler form

$$\mathbf{u}\oplus\mathbf{v} = \frac{\mathbf{u}+\mathbf{v}}{1+\frac{\mathbf{u}\cdot\mathbf{v}}{c^2}}, \qquad \mathbf{u}\|\mathbf{v} \tag{6}$$

for $\mathbf{u}, \mathbf{v} \in \mathbb{R}^3_c$ and $\mathbf{u}\|\mathbf{v}$, where the operation $\oplus$ is both commutative and associative [1]. Indeed, as Poincaré pointed out in 1906/7 [36, fn. 34, p. 56], Einstein addition (6) of relativistically admissible parallel velocities is a group operation, thus admitting a useful algebraic structure. More than 80 years later it was discovered [20] that also the general Einstein velocity addition (2) is not algebraically structureless. Rather, it is a gyrocommutative gyrogroup operation [20, 24, 25, 29], as we will see in the sequel.

Einstein addition is noncommutative. In general $\mathbf{u}\oplus\mathbf{v} \neq \mathbf{v}\oplus\mathbf{u}$ $\mathbf{u}, \mathbf{v} \in \mathbb{V}_c$. Moreover, Einstein addition is also nonassociative. In general $(\mathbf{u}\oplus\mathbf{v})\oplus\mathbf{w} \neq \mathbf{u}\oplus(\mathbf{v}\oplus\mathbf{w})$ $\mathbf{u}, \mathbf{v}, \mathbf{w} \in \mathbb{V}_c$. Hence, the Einstein groupoid $(\mathbb{V}_c, \oplus)$ is not a group. Why, counterintuitively, is Einstein addition neither commutative nor associative? This seemingly notorious problem in special relativity theory transmogrifies itself into extreme profundity.

Until recently, before the advent of gyrogroup theory in 1988 [20], physicists had difficulty explaining why the Einstein velocity addition fell into such despair. Physicists and mathematicians tend to think of symmetry as being virtually synonymous with the theory of groups. It therefore seems that some symmetry and mathematical regularity has been lost in the transition from ordinary vector addition, which is a commutative group operation, into Einstein addition, which is not a group operation. However, we will see that the seemingly lost symmetry and mathematical regularity reappear when we realize that Einstein addition is, in fact, a gyrocommutative gyrogroup operation regulated by the hyperbolic geometry of Bolyai and Lobachevski. More generally, we will see that despite the fact that a gyrogroup is a nongroup generalized group, it retains the group ability to capture symmetry and mathematical regularity. Moreover, we will see that the structure of the abstract gyrogroup is richer than that of the abstract group, since it involves a gyro-opration, which is an automorphism generator that possesses its own rich structure.

Einstein addition $\oplus$, (2), is a binary operation in the ball $\mathbb{V}_c$ of the real inner product space $\mathbb{V}$. The identity element of $\mathbb{V}_c$ for Einstein addition is the zero element of $\mathbb{V}$, $\mathbf{0} \in \mathbb{V}_c$, and the inverse $\ominus\mathbf{v}$ of $\mathbf{v} \in \mathbb{V}_c$ for Einstein addition is its negative, $\ominus\mathbf{v} = -\mathbf{v} \in \mathbb{V}_c$. Furthermore, Einstein addition possesses a *left cancellation law*

$$\mathbf{u}\oplus(\ominus\mathbf{u}\oplus\mathbf{v}) = \mathbf{v}\,. \tag{7}$$

It seems that the right counterpart of (7) does not exist since, in general,

$$(\mathbf{u}\ominus\mathbf{v})\oplus\mathbf{v} \neq \mathbf{u}\,. \tag{8}$$

Fortunately, however, the right cancellation law, in its two mutually dual forms, will naturally emerge in the sequel, (28).

In the limit of large c, $c \to \infty$, the ball $\mathbb{V}_c$ expands into the whole of its space $\mathbb{V}$, and Einstein addition $\oplus$ in $\mathbb{V}_c$ reduces to the ordinary vector addition $+$ in $\mathbb{V}$. The ordinary vector addition $+$ in $\mathbb{V}$ is both commutative and associative, forming the group $(\mathbb{V}, +)$ of vectors. In contrast, Einstein addition (2) is neither commutative nor associative. It seems that following the breakdown of commutativity and associativity in Einstein addition some mathematical regularity has been lost in the transition from vector addition in $\mathbb{V}$ to Einstein vector addition (2) in $\mathbb{V}_c$. This is, however, not the case since, as we will see in Section 3, Thomas gyration comes to the rescue.

3. Thomas Gyration

For any $\mathbf{u}, \mathbf{v} \in \mathbb{V}_c$ let $\text{gyr}[\mathbf{u}, \mathbf{v}] : \mathbb{V}_c \to \mathbb{V}_c$ be the self-map of $\mathbb{V}_c$ given in terms of Einstein addition $\oplus$ by the equation [20]

$$\text{gyr}[\mathbf{u}, \mathbf{v}]\mathbf{w} = \ominus(\mathbf{u}\oplus\mathbf{v})\oplus\{\mathbf{u}\oplus(\mathbf{v}\oplus\mathbf{w})\} \tag{9}$$

where $\ominus\mathbf{v} = -\mathbf{v}$, for all $\mathbf{w} \in \mathbb{V}_c$. The self-map $\text{gyr}[\mathbf{u}, \mathbf{v}]$ of $\mathbb{V}_c$, which takes $\mathbf{w} \in \mathbb{V}_c$ into $\ominus(\mathbf{u}\oplus\mathbf{v})\oplus\{\mathbf{u}\oplus(\mathbf{v}\oplus\mathbf{w}) \in \mathbb{V}_c$, is called the *Thomas gyration* generated by $\mathbf{u}$ and $\mathbf{v}$.

Owing to the nonassociativity of Einstein addition $\oplus$, in general, a Thomas gyration is not the identity map. Interestingly, it keeps the inner product of elements of the ball $\mathbb{V}_c$ invariant, that is,

$$\text{gyr}[\mathbf{u}, \mathbf{v}]\mathbf{a}\cdot\text{gyr}[\mathbf{u}, \mathbf{v}]\mathbf{b} = \mathbf{a}\cdot\mathbf{b} \tag{10}$$

for all $\mathbf{a}, \mathbf{b}, \mathbf{u}, \mathbf{v} \in \mathbb{V}_c$. As such, $\text{gyr}[\mathbf{u}, \mathbf{v}]$ is an *isometry* of $\mathbb{V}_c$, keeping the norm of elements of the ball $\mathbb{V}_c$ invariant,

$$\|\text{gyr}[\mathbf{u}, \mathbf{v}]\mathbf{w}\| = \|\mathbf{w}\| \tag{11}$$

Hence, $\text{gyr}[\mathbf{u}, \mathbf{v}]$ represents a rotation of the ball $\mathbb{V}_c$ about its origin. Interestingly, the inverse of the gyration $\text{gyr}[\mathbf{u}, \mathbf{v}]$ turns out to be the gyration $\text{gyr}[\mathbf{v}, \mathbf{u}]$ for all $\mathbf{u}, \mathbf{v} \in \mathbb{V}_c$. We, accordingly, say in gyrolanguage that $\text{gyr}[\mathbf{u}, \mathbf{v}]$ is a *gyroisometry*. Furthermore, the bijective (one-to-one) self-map $\text{gyr}[\mathbf{u}, \mathbf{v}]$ of $\mathbb{V}_c$ preserves the Einstein addition in $\mathbb{V}_c$,

$$\text{gyr}[\mathbf{u}, \mathbf{v}](\mathbf{a}\oplus\mathbf{b}) = \text{gyr}[\mathbf{u}, \mathbf{v}]\mathbf{a}\oplus\text{gyr}[\mathbf{u}, \mathbf{v}]\mathbf{b} \tag{12}$$

so that it is an automorphism of the Einstein groupoid $(\mathbb{V}_c, \oplus)$. We recall that an automorphism of the groupoid $(\mathbb{V}_c, \oplus)$ is a bijective self-map of the groupoid $\mathbb{V}_c$ that respects its binary operation, that is, it satisfies (12). Under bijection composition the automorphisms of a groupoid $(\mathbb{V}_c, \oplus)$ form a group known as the automorphism group, and denoted $Aut(\mathbb{V}_c, \oplus)$. Being an automorphism, a Thomas gyration $\text{gyr}[\mathbf{u}, \mathbf{v}]$, $\mathbf{u}, \mathbf{v} \in \mathbb{V}_c$, is said to be a *gyroautomorphism*, gyr being the gyroautomorphism generator.

The gyroautomorphism gyr$[\mathbf{u},\mathbf{v}]$ regulates the Einstein addition in the ball $\mathbb{V}_c$, giving rise to the following nonassociative algebraic laws that "repair" the breakdown of commutativity and associativity in Einstein addition:

$$\begin{aligned} \mathbf{u}\oplus\mathbf{v} &= \mathrm{gyr}[\mathbf{u},\mathbf{v}](\mathbf{v}\oplus\mathbf{u}) && \text{Gyrocommutative Law} \\ \mathbf{u}\oplus(\mathbf{v}\oplus\mathbf{w}) &= (\mathbf{u}\oplus\mathbf{v})\oplus\mathrm{gyr}[\mathbf{u},\mathbf{v}]\mathbf{w} && \text{Left Gyroassociative Law} \\ (\mathbf{u}\oplus\mathbf{v})\oplus\mathbf{w} &= \mathbf{u}\oplus(\mathbf{v}\oplus\mathrm{gyr}[\mathbf{v},\mathbf{u}]\mathbf{w}) && \text{Right Gyroassociative Law} \end{aligned} \tag{13}$$

A most important property of the Thomas gyration is the so-called *loop property* (left and right),

$$\begin{aligned} \mathrm{gyr}[\mathbf{u}\oplus\mathbf{v},\mathbf{v}] &= \mathrm{gyr}[\mathbf{u},\mathbf{v}] && \text{Left Loop Property} \\ \mathrm{gyr}[\mathbf{u},\mathbf{v}\oplus\mathbf{u}] &= \mathrm{gyr}[\mathbf{u},\mathbf{v}] && \text{Right Loop Property} \end{aligned} \tag{14}$$

for all $\mathbf{u},\mathbf{v}\in\mathbb{V}_c$.

It follows from the Thomas gyration definition in (9) and from the left cancellation law (7) that gyrations generated by a zero vector vanish, being reduced to the identity map I of $\mathbb{V}_c$. Thus, for all $\mathbf{u},\mathbf{v}\in\mathbb{V}_c$,

$$\mathrm{gyr}[\mathbf{0},\mathbf{v}] = \mathrm{gyr}[\mathbf{u},\mathbf{0}] = I\,. \tag{15}$$

Furthermore, it then follows from the left loop property that

$$\begin{aligned} \mathrm{gyr}[\mathbf{u},\ominus\mathbf{u}] &= \mathrm{gyr}[\mathbf{u}\oplus(\ominus\mathbf{u}),\ominus\mathbf{u}] \\ &= \mathrm{gyr}[\mathbf{u}\ominus\mathbf{u},\ominus\mathbf{u}] \\ &= \mathrm{gyr}[\mathbf{0},\ominus\mathbf{u}] \\ &= I\,. \end{aligned} \tag{16}$$

The Einstein velocity addition, rather than the Lorentz transformation, is a primitive notion in our approach to special relativity. Hence, our novel way of describing the Thomas gyration (precession) gyr$[\mathbf{u},\mathbf{v}]$ in terms of Einstein addition, (9), is crucially important in our attempt to present a better case for the superiority of the abandoned non-Euclidean style in the study of Einstein's own route to the special theory. For several years Einstein was very unhappy about the four-dimensional spacetime picture that Minkowski introduced in 1908 following Einstein's discovery of special relativity in 1905. Contrasting Einsteinian relativity, based on relativistically admissible three-velocities and their Einstein addition, Minkowskian relativity is based on four-velocities and their Lorentz transformation.

We show in [32] that Einstein velocity addition is a primitive notion in the foundation of special relativity, linking the special theory to the hyperbolic geometry of Bolyai and Lobachevski. It provides novel insights into Minkowskian relativity thus increasing our understanding of Minkowski's reformulation of the theory. Furthermore, we show in [32] that, unlike Minkowskian relativity, our approach to Einstein velocity addition meshes extraordinarily well with the notion of the relativistic mass. In particular, we show in [32] that following the relativistic mass correction, (i) the relativistic center of momentum of a system of three uniformly moving particles with equal rest masses has a hyperbolic geometric interpretation as the hyperbolic triangle centroid in the same way that (ii) the classical center of momentum of a system of three uniformly moving particles with equal Newtonian masses has a Euclidean geometric interpretation as the triangle centroid. Unfortunately, we will never know what Einstein's reaction to our way of understanding his velocity addition in terms of the hyperbolic geometry of Bolyai and Lobachevski would have been.

We are now in a position to demonstrate the power and elegance of the nonassociative algebra that regulates the Einstein addition, by solving two related equations.

4. Solving Equations In Einstein Addition

We present the flavor of the nonassociative gyrovector algebra of Einstein addition by solving each of the two similar equations in (17) and (22) below. Unexpectedly, we will find that owing to the presence of a Thomas gyration the two respective solutions of the two similar equations are not similar.

$$\mathbf{a}\oplus\mathbf{x} = \mathbf{b} \tag{17}$$

in $\mathbb{V}_c$ for the unknown $\mathbf{x}$ and any given $\mathbf{a}, \mathbf{b} \in \mathbb{V}_c$. If $\mathbf{x}$ is a solution, then by the right gyroassociative law, and by (16) and (17), we have

$$\begin{aligned}
\mathbf{x} &= \mathbf{0}\oplus\mathbf{x} \\
&= (\ominus\mathbf{a}\oplus\mathbf{a})\oplus\mathbf{x} \\
&= \ominus\mathbf{a}\oplus(\mathbf{a}\oplus\mathrm{gyr}[\mathbf{a}, \ominus\mathbf{a}]\mathbf{x}) \\
&= \ominus\mathbf{a}\oplus(\mathbf{a}\oplus\mathbf{x}) \\
&= \ominus\mathbf{a}\oplus\mathbf{b}\,.
\end{aligned} \tag{18}$$

Thus, if a solution exists, it must be given uniquely by

$$\mathbf{x} = \ominus\mathbf{a}\oplus\mathbf{b}\,. \tag{19}$$

Conversely, if $\mathbf{x} = \ominus\mathbf{a}\oplus\mathbf{b}$, then $\mathbf{x}$ is indeed a solution since by the left gyroassociative law and (16) we have

$$\begin{aligned} \mathbf{a}\oplus\mathbf{x} &= \mathbf{a}\oplus(\ominus\mathbf{a}\oplus\mathbf{b}) \\ &= (\mathbf{a}\oplus(\ominus\mathbf{a}))\oplus\mathrm{gyr}[\mathbf{a},\ominus\mathbf{a}]\mathbf{b} \\ &= \mathbf{0}\oplus I\mathbf{b} \\ &= \mathbf{b}\,. \end{aligned} \tag{20}$$

Substituting the solution (19) in its equation (17) and replacing $\mathbf{a}$ by $-\mathbf{a} = \ominus\mathbf{a}$ we obtain the *left cancellation law* for Einstein addition, (7),

$$\ominus\mathbf{a}\oplus(\mathbf{a}\oplus\mathbf{b}) = \mathbf{b}\,. \tag{21}$$

A most interesting application of the left loop property and the left gyroassociative law appears when we derive the solution of the equation

$$\mathbf{x}\oplus\mathbf{a} = \mathbf{b} \tag{22}$$

in $\mathbb{V}_c$ for given $\mathbf{a}, \mathbf{b} \in \mathbb{V}_c$. By solving this equation, the importance of the left loop property in making the left gyroassociative law effective is uncovered.

If $\mathbf{x} \in \mathbb{V}_c$ is a solution of (22), then by the left gyroassociative law, the left loop property, and (22) we have

$$\begin{aligned} \mathbf{x} &= \mathbf{x}\oplus\mathbf{0} \\ &= \mathbf{x}\oplus(\mathbf{a}\oplus(\ominus\mathbf{a})) \\ &= (\mathbf{x}\oplus\mathbf{a})\oplus\mathrm{gyr}[\mathbf{x},\mathbf{a}](\ominus\mathbf{a}) \\ &= (\mathbf{x}\oplus\mathbf{a})\oplus(\ominus\mathrm{gyr}[\mathbf{x},\mathbf{a}]\mathbf{a}) \\ &= (\mathbf{x}\oplus\mathbf{a})\ominus\mathrm{gyr}[\mathbf{x},\mathbf{a}]\mathbf{a} \\ &= \mathbf{b}\ominus\mathrm{gyr}[\mathbf{x},\mathbf{a}]\mathbf{a} \\ &= \mathbf{b}\ominus\mathrm{gyr}[\mathbf{x}\oplus\mathbf{a},\mathbf{a}]\mathbf{a} \\ &= \mathbf{b}\ominus\mathrm{gyr}[\mathbf{b},\mathbf{a}]\mathbf{a} \end{aligned} \tag{23}$$

where we abbreviate: $\mathbf{a}\ominus\mathbf{b} = \mathbf{a}\oplus(-\mathbf{b}) = \mathbf{a}\oplus(\ominus\mathbf{b})$.

Thus, if a solution of $\mathbf{x}\oplus\mathbf{a} = \mathbf{b}$ exists, it must have the form

$$\mathbf{x} = \mathbf{b}\ominus\mathrm{gyr}[\mathbf{b},\mathbf{a}]\mathbf{a}. \tag{24}$$

Conversely, one must show that $\mathbf{x} = \mathbf{b}\ominus\mathrm{gyr}[\mathbf{b},\mathbf{a}]\mathbf{a}$ is indeed a solution of (22). This is a delicate matter, verified in [29].

The chain of equations in (23) strikingly illustrates the elementary use of the gyrogroup axioms, demonstrating that the gyroassociative law owe its effectiveness to the loop property.

In order to display analogies with groups, we define a dual binary operation $\boxplus$ by the equation

$$\mathbf{a} \boxplus \mathbf{b} = \mathbf{a} \oplus \mathrm{gyr}[\mathbf{a}, \ominus \mathbf{b}]\mathbf{b}, \tag{25}$$

calling it the Einstein *coaddition*, that *coexists* with the Einstein addition $\oplus$. Accordingly, the solution $\mathbf{x}$ in (24) of the equation (22) can be written as

$$\mathbf{x} = \mathbf{b} \ominus \mathrm{gyr}[\mathbf{b}, \mathbf{a}]\mathbf{a} = \mathbf{b} \boxminus \mathbf{a} \tag{26}$$

when we abbreviate

$$\mathbf{a} \boxminus \mathbf{b} = \mathbf{a} \boxplus (-\mathbf{b}) = \mathbf{a} \boxplus (\ominus \mathbf{b}). \tag{27}$$

The Einstein binary cooperation $\boxplus$, called the Einstein coaddition, proves useful in the algebra of Einstein addition. Substituting the solution $\mathbf{x}$, (24) and (26), into its equation (22) and replacing $\mathbf{a}$ by $-\mathbf{a} = \ominus \mathbf{a}$ we obtain, by means of the notation in (27), the *right cancellation law* for Einstein addition

$$(\mathbf{b} \boxplus \mathbf{a}) \ominus \mathbf{a} = \mathbf{b}. \tag{28a}$$

The latter can be dualized into the *dual right cancellation law*

$$(\mathbf{b} \oplus \mathbf{a}) \boxminus \mathbf{a} = \mathbf{b}. \tag{28b}$$

We thus see that while Einstein addition $\oplus$ possesses a left cancellation law, (21), we need its coexisting operation, the Einstein coaddition $\boxplus$, for the right cancellation law.

Einstein addition $\oplus$ and coaddition $\boxplus$ are mutually dual in the sense of the following two identities [27, 29]

$$\begin{aligned} \mathbf{a} \boxplus \mathbf{b} &= \mathbf{a} \oplus \mathrm{gyr}[\mathbf{a}, \ominus \mathbf{b}]\mathbf{b} \\ \mathbf{a} \oplus \mathbf{b} &= \mathbf{a} \boxplus \mathrm{gyr}[\mathbf{a}, \mathbf{b}]\mathbf{b} \end{aligned} \tag{29}$$

in which each of the two mutually dual Einstein additions is expressed in terms of the other one and a gyration. Clearly, in order to capture analogies with ordinary vector addition, the two mutually dual Einstein additions, $\oplus$ and $\boxplus$, are needed. Considering the two identities in (29) mutually dual, the two gyrations $\mathrm{gyr}[\mathbf{a}, \mathbf{b}]$ and $\mathrm{gyr}[\mathbf{a}, \ominus \mathbf{b}]$ turn out to be mutually dual as well.

5. Gyrogroups

Motivated by the definition of the group in algebra, we take key features of the algebra that regulates the Einstein addition as axioms in the definition of the gyrogroup in nonassociative algebra.

Definition (Gyrogroups and Gyrocommutative Gyrogroups). A groupoid $(G, \oplus)$ is a *gyrogroup* if its binary operation satisfies the following axioms $(Gy1)$ – $(Gy3b)$. In G there is at least one element, 0, called a left identity, satisfying

(Gy1) $0 \oplus a = a\,.$ Left Identity

for all $a \in G$. There is an element $0 \in G$ satisfying axiom (Gy1) such that for each a in G there is an element $\ominus a$ in G, called a left inverse of a, satisfying

(Gy2) $\ominus a \oplus a = 0\,.$ Left Inverse

Moreover, for any $a, b, z \in G$ there exists a unique element $\mathrm{gyr}[a,b]z \in G$ such that

(Gy3) $a \oplus (b \oplus z) = (a \oplus b) \oplus \mathrm{gyr}[a,b]z\,.$ Left Gyroassociative Law

If $gyr[a,b]$ denotes the map $gyr[a,b] : G \to G$ given by $z \mapsto gyr[a,b]z$ then

(Gy3a) $\mathrm{gyr}[a,b] \in Aut(G, \oplus)$ Gyroautomorphism

and $gyr[a,b]$ is called the *Thomas gyration*, or the *gyroautomorphism* of G, *generated* by $a, b \in G$. The operation $\mathrm{gyr} : G \times G \to Aut(G, \oplus)$ is called the gyrooperation of G. Finally, the gyroautomorphism $gyr[a,b]$ generated by any $a, b \in G$ satisfies

(Gy3b) $\mathrm{gyr}[a,b] = \mathrm{gyr}[a \oplus b, b]\,.$ Left Loop Property

A gyrogroup $(G, \oplus)$ is *gyrocommutative* if for all $a, b \in G$

(Gy4) $a \oplus b = \mathrm{gyr}[a,b](b \oplus a)\,.$ Gyrocommutative Law

First theorems in gyrogroup theory [29, 25] establish the existence of a unique identity and a unique inverse, each of which is both left and right.

Einstein addition turns out to be a gyrocommutative gyrogroup operation and, accordingly, the groupoid consisting of the set of all relativistically admissible velocities with their Einstein addition forms a gyrocommutative gyrogroup, called the relativity gyrogroup. The gyrogroup notion has thus sprung from the soil of Einstein's special theory of relativity in [20, 21, 22, 23].

6. The Einstein Gyrovector Spaces

Some commutative groups admit scalar multiplication turning them into vector spaces. Vector spaces, in turn, form the setting for the standard model of Euclidean geometry. In full analogy, some gyrocommutative gyrogroups admit scalar multiplication turning them into gyrovector spaces. Gyrovector spaces, in turn, form the setting for hyperbolic geometry. Thus for instance, as we will see in this article, (i) Einstein addition gives rise to the Einstein gyrovector space that forms the setting for the Beltrami (also known as the Klein) Euclidean-ball model of hyperbolic geometry, and (ii) Möbius addition gives rise to the Möbius gyrovector space that forms the setting for the Beltrami ball model of hyperbolic geometry.

Einstein addition admits scalar multiplication that possesses useful properties. It is given by the equation [25, 29]

$$\begin{aligned} r\otimes\mathbf{v} &= c\frac{(1+\|\mathbf{v}\|/c)^r-(1-\|\mathbf{v}\|/c)^r}{(1+\|\mathbf{v}\|/c)^r+(1-\|\mathbf{v}\|/c)^r}\frac{\mathbf{v}}{\|\mathbf{v}\|} \\ &= c\tanh\left(r\tanh^{-1}\frac{\|\mathbf{v}\|}{c}\right)\frac{\mathbf{v}}{\|\mathbf{v}\|} \end{aligned} \tag{30}$$

where $r\in\mathbb{R}$, $\mathbf{v}\in\mathbb{V}_c$, $\mathbf{v}\neq\mathbf{0}$, and $r\otimes\mathbf{0}=\mathbf{0}$, and with which we use the notation $r\otimes\mathbf{v}=\mathbf{v}\otimes r$.

The Einstein addition and scalar multiplication possess the following properties. For any positive integer n and for all $r, r_1, r_2\in\mathbb{R}$ and $\mathbf{a},\mathbf{b},\mathbf{u},\mathbf{v}\in\mathbb{V}_c$,

$$\begin{aligned}
n\otimes\mathbf{v} &= \mathbf{v}\oplus\cdots\oplus\mathbf{v} && n \text{ terms} \\
(r_1+r_2)\otimes\mathbf{v} &= r_1\otimes\mathbf{v}\oplus r_2\otimes\mathbf{v} && \text{Scalar Distributive Law} \\
(r_1r_2)\otimes\mathbf{v} &= r_1\otimes(r_2\otimes\mathbf{v}) && \text{Scalar Associative Law} \\
r\otimes(r_1\otimes\mathbf{v}\oplus r_2\otimes\mathbf{v}) &= r\otimes(r_1\otimes\mathbf{v})\oplus r\otimes(r_2\otimes\mathbf{v}) && \text{Monodistributive Law} \\
\|r\otimes\mathbf{v}\| &= |r|\otimes\|\mathbf{v}\| && \text{Homogeneity Property} \\
\frac{|r|\otimes\mathbf{v}}{\|r\otimes\mathbf{v}\|} &= \frac{\mathbf{v}}{\|\mathbf{v}\|} && \text{Scaling Property} \\
\|\mathbf{u}\oplus\mathbf{v}\| &\leq \|\mathbf{u}\|\oplus\|\mathbf{v}\| && \text{Triangle Inequality} \\
\mathrm{gyr}[\mathbf{a},\mathbf{b}](r\otimes\mathbf{v}) &= r\otimes\mathrm{gyr}[\mathbf{a},\mathbf{b}]\mathbf{v} && \text{Gyroautomorphism} \\
\mathrm{gyr}[r_1\otimes\mathbf{v},r_2\otimes\mathbf{v}] &= I\,. && \text{Identity Automorphism}
\end{aligned} \tag{31}$$

Remarkably, the classical distributive law has no counterpart in a gyrovector space. In fact, it is owing to the lack of a distributive law that gyrovector spaces are more general than vector spaces. Ambiguously, the Einstein scalar multiplication $\otimes$ in the homogeneity property, and the Einstein addition $\oplus$ in the triangle inequality are

(1) the Einstein scalar multiplication $\otimes$ and the Einstein addition $\oplus$, (2), in the c-ball $\mathbb{V}_c$ of $\mathbb{V}$ on the left; and

(2) the Einstein scalar multiplication $\otimes$ and the Einstein addition $\oplus$, (5), in the non-negative part $[0, c)$ of the c-ball $\mathbb{R}_c$ of $\mathbb{R}$ on the right.

The triple $(\mathbb{V}_c, \oplus, \otimes)$ is known as the Einstein gyrovector space, $\mathbb{V}_c$ being the c-ball of a real inner product space $\mathbb{V}$, (1). Having the gyrovector notion in hand, we can now relativize the Newtonian uniform acceleration

$$\begin{aligned} \mathbf{v}_-(t) &= \mathbf{v}_0 + \mathbf{a}t \\ &= \mathbf{a}t + \mathbf{v}_0 \,, \end{aligned} \tag{32}$$

$\mathbf{v}_0, \mathbf{a} \in \mathbb{V}$, $t \in \mathbb{R}$, $\mathbf{v}_0$ and $\mathbf{a}$ being an initial velocity and an acceleration vector. The uniform acceleration (32) is represented geometrically by a line that is a geodesic in the velocity space $\mathbb{V}$ relative to the Euclidean metric $d_-(\mathbf{u}, \mathbf{v})$,

$$d_-(\mathbf{u}, \mathbf{v}) = \|\mathbf{u} - \mathbf{v}\| \,. \tag{33}$$

There are various ways to "relativize" the classical notion of uniform acceleration; see for instance [11] and [8]. However, a widely accepted one states that in general relativity uniformly accelerated velocities are those which traverse along geodesics in velocity spaces [33]. Guided by analogies, we relativize the classical uniform acceleration that (32) represents into the following pair of two relativistic mutually dual counterparts, represented by the equations

$$\mathbf{v}_{\ominus}(t) = \mathbf{v}_0 \oplus \mathbf{a} \otimes t \tag{34a}$$

$$\mathbf{v}_{\boxminus}(t) = \mathbf{a} \otimes t \oplus \mathbf{v}_0 \,, \tag{34b}$$

$\mathbf{v}_0, \mathbf{a} \in \mathbb{V}_c$, $t \in \mathbb{R}^+$. Owing to the noncommutativity of Einstein addition, the two relativistic versions (34) of the Newtonian uniform acceleration (32) are distinct. One of these, (34a), agrees with the uniform acceleration definition in [33].

Owing to the analogies that (34) shares with (32), and being guided by analogies, we view each of the two distinct expressions in (34) as the legitimate representation of the relativistic uniform acceleration. Accordingly, in the transition from prerelativity uniform acceleration to relativity uniform acceleration, the Newtonian uniform acceleration, represented by (32), bifurcates into two mutually dual Einsteinian uniform accelerations, represented by (34). Some duality symmetries that the two expressions in (34) possess will be uncovered in the sequel. Some others are found in [29].

To justify our bifurcation approach [27] let us explore the geometric significance of each of the two distinct, mutually dual Einsteinian uniform accelerations in (34). Calling $\mathbf{v}_{\ominus}(t)$ in (34a) a *gyroline*, and calling $\mathbf{v}_{\boxminus}(t)$ in (34b) a *cogyroline*, we explore these two mutually dual gyrolines that coexist in hyperbolic geometry. They are geodesics relative to their respective mutually dual metrics, the *gyrometric* $d_{\ominus}(\mathbf{u},\mathbf{v})$ and its dual, *cogyrometric* $d_{\boxminus}(\mathbf{u},\mathbf{v})$, given by the equations

$$
\begin{aligned}
d_{\ominus}(\mathbf{u},\mathbf{v}) &= \|\mathbf{u}\ominus\mathbf{v}\| &\quad (35a)\\
d_{\boxminus}(\mathbf{u},\mathbf{v}) &= \|\mathbf{u}\boxminus\mathbf{v}\|\,, &\quad (35b)
\end{aligned}
$$

$\mathbf{u},\mathbf{v}\in\mathbb{V}_c$.

The gyrometric $d_{\ominus}(\mathbf{u},\mathbf{v})$ is called the Einstein metric and, similarly, the cogyrometric $d_{\boxminus}(\mathbf{u},\mathbf{v})$ is called the Einstein cometric. Interestingly, by methods of differential geometry the Einstein metric can be translated into the Riemannian line element of the Beltrami (also known as the Klein) model of hyperbolic geometry [29]. Classically, it is only the gyrometric that is qualified to the title "metric". In contrast, the cogyrometric does not possess a classical triangle inequality. Rather, it possesses the classical triangle inequality that has been "corrected" by a gyration, as shown in the cogyrotriangle inequality (48) below.

There are two mutually dual families of geodesics that result from the two mutually dual gyrometrics (35). The gyrolines, (34a), being geodesics relative to the gyrometric $d_{\ominus}(\mathbf{u},\mathbf{v})$, are called *gyrogeodesics*. Similarly, the cogyrolines, (34b), being geodesics relative to the cogyrometric $d_{\boxminus}(\mathbf{u},\mathbf{v})$, are called *cogyrogeodesics*.

Further analogies and duality symmetries emerge when we write the Euclidean line (32) in the equivalent form

$$
L_{-}(t) = \mathbf{a} + (-\mathbf{a} + \mathbf{b})t\,, \qquad (36)
$$

$t\in\mathbb{R}$, that emphasizes its characterization as the unique Euclidean geodesic passing through the two distinct points $\mathbf{a},\mathbf{b}\in\mathbb{V}$. It passes through the point $\mathbf{a}$ at "time" $t=0$, and through the point $\mathbf{b}$ at "time" $t=1$.

In full analogy with (36), the two mutually dual gyrolines (34) in the Einstein gyrovector space $(\mathbb{V}_c, \oplus, \otimes)$ are written as

$$L_{\ominus}(t) = \mathbf{a}\oplus(\ominus\mathbf{a}\oplus\mathbf{b})\otimes t \qquad (37a)$$

$$L_{\boxminus}(t) = (\mathbf{b} \boxminus \mathbf{a})\otimes t\oplus\mathbf{a}\,, \qquad (37b)$$

$t \in \mathbb{R}$, when we wish to emphasize that the gyroline $L_{\ominus}(t)$ and its dual gyroline, the cogyroline $L_{\boxminus}(t)$, are the unique ones that pass through the two distinct points $\mathbf{a}, \mathbf{b} \in \mathbb{V}_c$:

(1) The gyroline $L_{\ominus}(t)$ in (37a), Fig. 1, passes through the point $\mathbf{a}$ at "time" $t = 0$ and, owing to the left cancellation law, (7), it passes through the point $\mathbf{b}$ at "time" $t = 1$. A point $\mathbf{p}$ is *between* $\mathbf{a}$ and $\mathbf{b}$ if

$$\mathbf{p} = \mathbf{a}\oplus(\ominus\mathbf{a}\oplus\mathbf{b})\otimes t_0 \qquad (38)$$

for some $0 < t_0 < 1$. It satisfies the *gyrotriangle equality*

$$\|\ominus\mathbf{a}\oplus\mathbf{p}\|\oplus\|\ominus\mathbf{p}\oplus\mathbf{b}\| = \|\ominus\mathbf{a}\oplus\mathbf{b}\| \qquad (39)$$

as remarked below in the gyrotriangle inequality (47). Furthermore, the point

$$m_{\mathbf{a},\mathbf{b}} = \mathbf{a}\oplus(\ominus\mathbf{a}\oplus\mathbf{b})\otimes\tfrac{1}{2} \qquad (40)$$

reached at "time" $t = \frac{1}{2}$ is the midpoint between $\mathbf{a}$ and $\mathbf{b}$, Fig. 1, satisfying $m_{\mathbf{a},\mathbf{b}} = m_{\mathbf{b},\mathbf{a}}$ and

$$\|\mathbf{a}\ominus m_{\mathbf{a},\mathbf{b}}\| = \|\mathbf{b}\ominus m_{\mathbf{a},\mathbf{b}}\|\,. \qquad (41)$$

(2) Similarly, the cogyroline $L_{\boxminus}(t)$ in (37b), Fig. 2, passes through the point $\mathbf{a}$ at "time" $t = 0$ and, owing to the right cancellation law, (28a), it passes through the point $\mathbf{b}$ at "time" $t = 1$. It is the need to employ the right cancellation law that dictates the necessity to replace the term $\ominus\mathbf{a}\oplus\mathbf{b}$ in (37a) by the term $\boxminus\mathbf{a}\boxplus\mathbf{b} = \mathbf{b}\boxminus\mathbf{a}$ in (37b). The replacement of the term $\ominus\mathbf{a}\oplus\mathbf{b}$ in (37a) by the term $\mathbf{b} \boxminus \mathbf{a}$ in (37b) is thus a matter of necessity rather than choice.

A point $\mathbf{p}$ is *cobetween* $\mathbf{a}$ and $\mathbf{b}$ if

$$\mathbf{p} = (\mathbf{b} \boxminus \mathbf{a})\otimes t_0\oplus\mathbf{a} \qquad (42)$$

for some $0 < t_0 < 1$. It satisfies the *cogyrotriangle equality*

$$\| \boxminus \mathbf{a} \boxplus \mathbf{p}\| \boxplus \| \boxminus \mathbf{p} \boxplus \mathbf{b}\| = \| \boxminus \mathbf{a} \boxplus \mathbf{b}\| \qquad (43)$$

as remarked below in the cogyrotriangle inequality (48). Furthermore, the point

$$m'_{\mathbf{a},\mathbf{b}} = (\mathbf{b} \boxminus \mathbf{a})\otimes\tfrac{1}{2}\oplus\mathbf{a} \qquad (44)$$

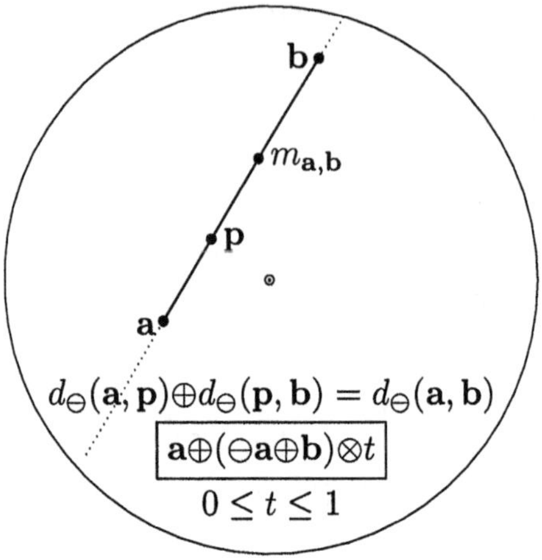

Figure 1. The gyroline segment, (37a), linking the two points $\mathbf{a}$ and $\mathbf{b}$ in an Einstein gyrovector space $(\mathbb{V}_c, \oplus, \otimes)$. $\mathbf{p}$ is a generic point between $\mathbf{a}$ and $\mathbf{b}$, (38), and $m_{\mathbf{a},\mathbf{b}}$ is the midpoint of $\mathbf{a}$ and $\mathbf{b}$, (40).

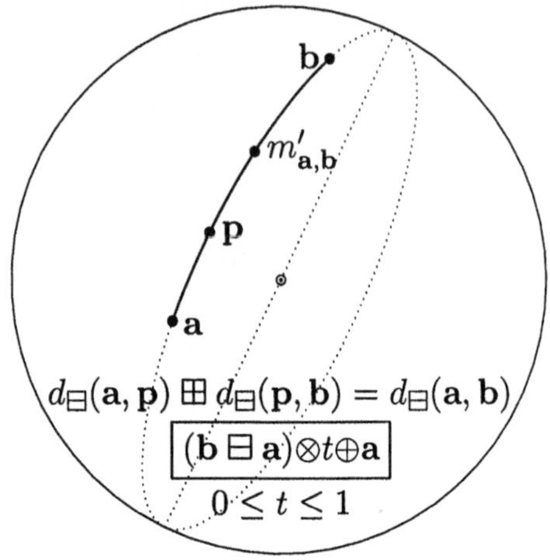

Figure 2. The cogyroline segment, (37b), linking the two points $\mathbf{a}$ and $\mathbf{b}$ in an Einstein gyrovector space $(\mathbb{V}_c, \oplus, \otimes)$. $\mathbf{p}$ is a generic point cobetween $\mathbf{a}$ and $\mathbf{b}$, (42), and $m'_{\mathbf{a},\mathbf{b}}$ is the comidpoint of $\mathbf{a}$ and $\mathbf{b}$, (44).

reached at "time" $t = \frac{1}{2}$ is the *comidpoint* between $\mathbf{a}$ and $\mathbf{b}$, Fig. 2, satisfying $m'_{\mathbf{a},\mathbf{b}} = m'_{\mathbf{b},\mathbf{a}}$ and

$$\|\mathbf{a} \boxminus m'_{\mathbf{a},\mathbf{b}}\| = \|\mathbf{b} \boxminus m'_{\mathbf{a},\mathbf{b}}\| . \tag{45}$$

The Einstein gyrovector space $(\mathbb{V}_c, \oplus, \otimes)$ is bimetric, having the two mutually dual metrics $d_\ominus(\mathbf{a},\mathbf{b})$ and $d_\boxminus(\mathbf{a},\mathbf{b})$ called, respectively, the gyrometric and the cogyrometric of the space. The Euclidean metric $d_-(\mathbf{a},\mathbf{b})$ of a real inner product space $\mathbb{V}$ admits the triangle inequality

$$d_-(\mathbf{a},\mathbf{p}) + d_-(\mathbf{p},\mathbf{b}) \ge d_-(\mathbf{a},\mathbf{b}) \tag{46}$$

for any three points $\mathbf{a},\mathbf{p},\mathbf{b} \in \mathbb{V}$, where equality holds if and only if the point $\mathbf{p}$ lies between $\mathbf{a}$ and $\mathbf{b}$, that is, on the Euclidean geodesic segment that links $\mathbf{a}$ to $\mathbf{b}$.

In full analogy, the gyrometric $d_\ominus(\mathbf{a},\mathbf{b})$ admits the gyrotriangle inequality

$$d_\ominus(\mathbf{a},\mathbf{p})\oplus d_\ominus(\mathbf{p},\mathbf{b}) \ge d_\ominus(\mathbf{a},\mathbf{b}) \tag{47}$$

for any three points $\mathbf{a},\mathbf{p},\mathbf{b} \in \mathbb{V}_c$, where equality holds if and only if the point $\mathbf{p}$ lies on the *gyrogeodesic* segment that links $\mathbf{a}$ to $\mathbf{b}$, Fig. 1.

Moreover, the cogyrometric $d_{\boxminus}(\mathbf{a}, \mathbf{b})$ admits the cogyrotriangle inequality

$$d_{\boxminus}(\mathbf{a}, \mathbf{p}) \boxplus d_{\boxminus}(\mathbf{p}, \mathbf{b}) \geq d_{\boxminus}(\mathbf{a}, \mathrm{gyr}[\mathbf{a} \boxminus \mathbf{p}, \mathbf{p} \boxminus \mathbf{b}]\mathbf{b}) \tag{48}$$

for any three points $\mathbf{a}, \mathbf{p}, \mathbf{b} \in \mathbb{V}_c$, where equality holds if and only if the point $\mathbf{p}$ lies on the *cogyrogeodesic* segment that links $\mathbf{a}$ to $\mathbf{b}$, Fig. 2.

It can be shown that the gyrometric with its gyrotriangle inequality (47) is equivalent to a classical metric that admits a classical triangle inequality [26]. This is, however, not the case with the cogyrometric. Accordingly, gyrogeodesics are geodesics in the classical sense of differential geometry, while cogyrogeodesics are not geodesics in the classical sense. Furthermore, it can be shown that geodesics relative to the cogyrometric in the classical sense are incomplete, that is, they do not exist for all "time" t. In contrast, the cogyrogeodesics are complete, being valid for all "time" $t \in \mathbb{R}$. Thus, the gyro-correction that appears in the cogyrotriangle inequality (48) is one of the manifestations of the way nonassociative gyro-algebra gyro-corrects itself (that is, corrects itself by appropriate insertion of a gyration) to achieve perfection.

The Einstein ball gyrovector spaces $(\mathbb{V}_c, \oplus, \otimes)$ form the setting for the Beltrami (also known as the Klein) ball model of hyperbolic geometry, where geodesics in the ball are Euclidean lines [27]. Furthermore, the Einstein ball gyrovector spaces admit hyperbolic trigonometry fully analogous to Euclidean trigonometry, as shown in [29] and [28].

The gyrolines (37a) are the common hyperbolic lines, also called geodesics, in the Beltrami ball model of hyperbolic geometry. The analogies they share with Euclidean lines, (36), suggest the unification of hyperbolic and Euclidean geometry into a super geometry, called gyrogeometry [30]. Euclidean geometry, then, emerges in gyrogeometry as a special, degenerate case corresponding to vanishing gyrations. Further analogies that manifest the unification are presented in [29].

The coordinate transformation group in Galilean relativity, called the Galilei group, is parametrized by Newtonian velocities and 3-space rotations. Similarly, the coordinate transformation group in Einsteinian relativity, called the Lorentz group, can be parametrized by Einsteinian velocities and 3-space rotations. This parametrization of the Lorentz group, along with the gyrovector space structure of Einsteinian velocities, uncovers novel analogies that relativistic mechanics shares with its Newtonian counterpart [29, 31, 19].

Placing the gyrovector space notion in the context of Lie groups and their Cartan decompositions results in the *Lie gyrovector spaces* studied in [9]. An algebraic object weaker than that of the gyrogroup, and hence more general, is the left gyrogroup, studied, for instance, in [5] and [7].

Bibliography

[1] G. A. Jn. Baker. Einstein numbers. *Amer. Math. Monthly*, 61(1):39–41, 1954.

[2] J. F. Barrett. *Special relativity and hyperbolic geometry.* Univ. Sunderland, Sunderland, UK, 1998. Physical Interpretations of Relativity Theory. Proceedings, London, UK, 11–14 Sept. 1998.

[3] N. A. Černikov. Connection between the theory of relativity and Lobačevskiĭ geometry. In *Gravitation and Theory of Relativity, No. 2 (Russian)*, pages 9–22. Izdat. Kazan. Univ., Kazan, 1965.

[4] Albert Einstein. Zur Elektrodynamik Bewegter Körper [on the electrodynamics of moving bodies]. *Ann. Physik (Leipzig)*, 17:891–921, 1905.

[5] Tomás Feder. Strong near subgroups and left gyrogroups. *J. Algebra*, 259(1):177–190, 2003.

[6] V. Fock. *The theory of space, time and gravitation.* The Macmillan Co., New York, 1964. Second revised edition. Translated from the Russian by N. Kemmer. A Pergamon Press Book.

[7] Tuval Foguel and Abraham A. Ungar. Involutory decomposition of groups into twisted subgroups and subgroups. *J. Group Theory*, 3(1):27–46, 2000.

[8] Garry Helzer. Special relativity with acceleration. *Amer. Math. Monthly*, 107(3):219–237, 2000.

[9] Azniv K. Kasparian and Abraham A. Ungar. Lie gyrovector spaces. *J. Geom. Symm. Phys.* preprint.

[10] Elemér Kiss. *Mathematical gems from the Bolyai chests.* Akadémiai Kiadó, Budapest, 1999. János Bolyai's discoveries in number theory and algebra as recently deciphered from his manuscripts, Translated by Anikó Csirmaz and Gábor Oláh.

[11] L. Marder. On uniform acceleration in special and general relativity. *Proc. Cambridge Philos. Soc.*, 53:194–198, 1957.

[12] Arthur I. Miller. *Albert Einstein's special theory of relativity.* Springer-Verlag, New York, 1998. Emergence (1905) and early interpretation (1905–11), Includes a translation by the author of Einstein's "On the electrodynamics of moving bodies", Reprint of the 1981 edition.

[13] B. A. Rosenfeld. *A history of non-Euclidean geometry.* Springer-Verlag, New York, 1988. Evolution of the concept of a geometric space, Translated from the Russian by Abe Shenitzer.

[14] Roman U. Sexl and Helmuth K. Urbantke. *Relativity, groups, particles.* Springer-Verlag, Vienna, 2001. Special relativity and relativistic symmetry in field and particle physics, Revised and translated from the third German (1992) edition by Urbantke.

[15] L. Silberstein. *The Theory of Relativity.* MacMillan, London, 1914.

[16] A. Sommerfeld. Über die Zusammensetzung der Geschwindigkeiten in der Relativtheorie. *Physikalische Zeitschrift*, 10:826–829, 1909.

[17] A. Sommerfeld. Zur Relativität (i): Vierdimensioale Vektoralgebra. *Annelen der Physik*, 32:749–776, 1910.

[18] J. L. Synge. *Relativity: the special theory.* North-Holland Publishing Co., Amsterdam, 1956.

[19] Abraham Ungar. The relativistic centre of momentum. In *Ivailo M. Mladenov and Gregory L. Naber (eds.): Geometry, integrability and quantization (Varna, 2002)*, pages 316–325. Coral Press Sci. Publ., Sofia, 2003.

[20] Abraham A. Ungar. Thomas rotation and the parametrization of the Lorentz transformation group. *Found. Phys. Lett.*, 1(1):57–89, 1988.

[21] Abraham A. Ungar. Axiomatic approach to the nonassociative group of relativistic velocities. *Found. Phys. Lett.*, 2(2):199–203, 1989.

[22] Abraham A. Ungar. The relativistic noncommutative nonassociative group of velocities and the Thomas rotation. *Resultate Math.*, 16(1-2):168–179, 1989. The term "K-loop" is coined here.

[23] Abraham A. Ungar. Weakly associative groups. *Resultate Math.*, 17(1-2):149–168, 1990.

[24] Abraham A. Ungar. Thomas precession and its associated grouplike structure. *Amer. J. Phys.*, 59(9):824–834, 1991.

[25] Abraham A. Ungar. Thomas precession: its underlying gyrogroup axioms and their use in hyperbolic geometry and relativistic physics. *Found. Phys.*, 27(6):881–951, 1997.

[26] Abraham A. Ungar. The hyperbolic Pythagorean theorem in the Poincaré disc model of hyperbolic geometry. *Amer. Math. Monthly*, 106(8):759–763, 1999.

[27] Abraham A. Ungar. The bifurcation approach to hyperbolic geometry. *Found. Phys.*, 30(8):1253–1281, 2000.

[28] Abraham A. Ungar. Hyperbolic trigonometry in the Einstein relativistic velocity model of hyperbolic geometry. *Comput. Math. Appl.*, 40(2–3):313–332, 2000.

[29] Abraham A. Ungar. *Beyond the Einstein Addition Law and its Gyroscopic Thomas Precession: The Theory of Gyrogroups and Gyrovector Spaces.* Kluwer Acad. Publ., Dordrecht, Boston, London, 2001.

[30] Abraham A. Ungar. On the unification of hyperbolic and Euclidean geometry. 2001. preprint.

[31] Abraham A. Ungar. On the appeal to a pre-established harmony between pure mathematics and relativity physics. *Found. Phys. Lett.*, 16(1):1–23, 2003.

[32] A. A. Ungar, "The Einstein velocity addition, the relativistic mass, and the hyperbolic geometry of Bolyai and Lobachevski," 2003. preprint.

[33] Helmuth K. Urbantke. Physical holonomy, Thomas precession, and Clifford algebra. *Amer. J. Phys.*, 58(8):747–750, 1990.

[34] Vladimir Varičak. Beiträge zur nichteuklidischen geometrie [contributions to non-euclidean geometry]. *Jber. dtsch. Mat. Ver.*, 17:70–83, 1908.

[35] Vladimir Varičak. Anwendung der Lobatschefskjschen Geometrie in der Relativtheorie. *Physikalische Zeitschrift*, 11:93–96, 1910.

[36] Scott Walter. Minkowski, mathematicians, and the mathematical theory of relativity. In *The expanding worlds of general relativity (Berlin, 1995)*, pages 45–86. Birkhäuser Boston, Boston, MA, 1999.

[37] Scott Walter. The non-Euclidean style of Minkowskian relativity. In *The symbolic universe (Editor – Jeremy J. Gray)*, pages 91–127. Oxford Univ. Press, New York, 1999.

MIX
Papier aus verantwortungsvollen Quellen
Paper from responsible sources
FSC® C105338

If you have any concerns about our products,
you can contact us on
ProductSafety@springernature.com

In case Publisher is established outside the EU,
the EU authorized representative is:
Springer Nature Customer Service Center GmbH
Europaplatz 3, 69115 Heidelberg, Germany

Printed by Libri Plureos GmbH
in Hamburg, Germany